Lecture Notes in Computer Science 1776

Edited by G. Goos, J. Hartmanis, and J. van Leeuwen

Berlin
Heidelberg
New York
Barcelona
Hong Kong
London
Milan
Paris
Singapore
Tokyo

Gastón H. Gonnet   Daniel Panario
Alfredo Viola (Eds.)

LATIN 2000:
Theoretical Informatics

4th Latin American Symposium
Punta del Este, Uruguay, April 10-14, 2000
Proceedings

Series Editors

Gerhard Goos, Karlsruhe University, Germany
Juris Hartmanis, Cornell University, NY, USA
Jan van Leeuwen, Utrecht University, The Netherlands

Volume Editors

Gastón H. Gonnet
Informatik, ETH
8092 Zürich, Switzerland
E-mail: gonnet@inf.ethz.ch

Daniel Panario
University of Toronto
Department of Computer Science
10 Kings College Road Toronto - Ontario, Canada
E-mail: daniel@cs.toronto.edu

Alfredo Viola
Universidad de la República, Facultad de Ingenieria
Instituto de Computación, Pedeciba Informática
Casilla de Correo 16120, Distrito 6
Montevideo, Uruguay

Cataloging-in-Publication Data applied for

Die Deutsche Bibliothek - CIP-Einheitsaufnahme

Theoretical informatics : proceedings / LATIN 2000, 4th Latin American
Symposium, Punta del Este, Uruguay, April 10 - 14, 2000 Gastón H.
Gonnet ... (ed.). - Berlin ; Heidelberg ; New York ; Barcelona ; Hong
Kong ; London ; Milan ; Paris ; Singapore ; Tokyo : Springer, 2000
   (Lecture notes in computer science ; Vol. 1776)
   ISBN 3-540-67306-7

CR Subject Classification (1991): F.2, G.2, G.1, F.1, F.3, C.2, E.3

ISSN 0302-9743
ISBN 3-540-67306-7 Springer-Verlag Berlin Heidelberg New York

Springer-Verlag is a company in the BertelsmannSpringer publishing group
© Springer-Verlag Berlin Heidelberg 2000
Printed in Germany

Typesetting: Camera-ready by author, data conversion by PTP-Berlin, Danny Lewis
Printed on acid-free paper   SPIN: 10719839   06/3142   5 4 3 2 1 0

Preface

This volume contains the proceedings of the LATIN 2000 International Conference (Latin American Theoretical INformatics), to be held in Punta del Este, Uruguay, April 10-14, 2000.

This is the fourth event in the series following São Paulo, Brazil (1992), Valparaíso, Chile (1995), and Campinas, Brazil (1998). LATIN has established itself as a fully refereed conference for theoretical computer science research in Latin America. It has also strengthened the ties between local and international scientific communities. We believe that this volume reflects the breadth and depth of this interaction.

We received 87 submissions, from 178 different authors in 26 different countries. Each paper was assigned to three program committee members. The Program Committee selected 42 papers based on approximately 260 referee reports. In addition to these contributed presentations, the conference included six invited talks.

The assistance of many organizations and individuals was essential for the success of this meeting. We would like to thank all of our sponsors and supporting organizations. Ricardo Baeza-Yates, Claudio Lucchesi, Arnaldo Moura, and Imre Simon provided insightful advice and shared with us their experiences as organizers of previous LATIN meetings. Joaquín Goyoaga and Patricia Corbo helped in the earliest stages of the organization in various ways, including finding Uruguayan sources of financial support. SeCIU (Servicio Central de Informática Universitario, Universidad de la República) provided us with the necessary communication infrastructure. The meeting of the program committee was hosted by the Instituto de Matemática e Estatística, Universidade de São Paulo, which also provided us with the Intranet site for discussions among PC members. We thank the researchers of the Institute for their collaboration, and in particular, Arnaldo Mandel for the Intranet setup. Finally, we thank Springer-Verlag for their commitment in publishing this and previous LATIN proceedings in the Lecture Notes in Computer Science series.

We are encouraged by the positive reception and interest that LATIN 2000 has created in the community, partly indicated by a record number of submissions.

January 2000

Gastón Gonnet
Daniel Panario
Alfredo Viola

The Conference

Invited Speakers

Allan Borodin (Canada)
Philippe Flajolet (France)
Joachim von zur Gathen (Germany)

Yoshiharu Kohayakawa (Brazil)
Andrew Odlyzko (USA)
Prabhakar Raghavan (USA)

Program Committee

Ricardo Baeza-Yates (Chile)
Béla Bollobás (USA)
Felipe Cucker (Hong Kong)
Josep Díaz (Spain)
Esteban Feuerstein (Argentina)
Celina M. de Figueiredo (Brazil)
Gastón Gonnet (Switzerland, Chair)
Jozef Gruska (Czech Republic)
Joos Heintz (Argentina/Spain)
Gérard Huet (France)
Marcos Kiwi (Chile)
Ming Li (Canada)
Cláudio L. Lucchesi (Brazil)
Ron Mullin (Canada)

Ian Munro (Canada)
Daniel Panario (Canada)
Dominique Perrin (France)
Patricio Poblete (Chile)
Bruce Reed (France)
Bruce Richmond (Canada)
Vojtech Rödl (USA)
Imre Simon (Brazil)
Neil Sloane (USA)
Endre Szmerédi (USA)
Alfredo Viola (Uruguay)
Yoshiko Wakabayashi (Brazil)
Siang Wun Song (Brazil)
Nivio Ziviani (Brazil)

Organizing Committee

Ed Coffman Jr.
Cristina Cornes
Javier Molina
Laura Molina
Lucia Moura

Daniel Panario (Co-Chair)
Alberto Pardo
Luis Sierra
Alfredo Viola (Co-Chair)

Local Arrangements

The local arrangements for the conference were handled by IDEAS S.R.L.

Organizing Institutions

Instituto de Computación (Universidad de la República Oriental del Uruguay)
Pedeciba Informática

Sponsors and Supporting Organizations

CLEI (Centro Latinoamericano de Estudios en Informática)
CSIC (Comisión Sectorial de Investigación Científica, Universidad de la República)
CONICYT (Consejo Nacional de Investigaciones Científicas y Técnicas)
UNESCO
Universidad ORT del Uruguay
Tecnología Informática

Referees

Carme Alvarez
Andre Arnold
Juan Carlos Augusto
Valmir C. Barbosa
Alejandro Bassi
Gabriel Baum
Denis Bechet
Leopoldo Bertossi
Ralf Borndoerfer
Claudson Bornstein
Richard Brent
Véronique Bruyère
Héctor Cancela
Rodney Can eld
Jianer Chen
Chirstian Choffrut
José Coelho de Pina Jr.
Ed Coffman Jr.
Don Coppersmith
Cristina Cornes
Bruno Courcelle
Gustavo Crispino
Maxime Crochemore
Diana Cukierman
Joe Culberson
Ricardo Dahab
Célia Picinin de Mello
Erik Demaine
Nachum Dershowitz
Luc Devroye
Volker Diekert
Luis Dissett
Juan V. Echagüe
David Eppstein
Mart´n Farach-Colton
Paulo Feo lof f
Henning Fernau
Cristina G. Fernandes
W. Fernández de la Vega
Carlos E. Ferreira
Marcelo Fr´as
Zoltan Furedi

Joaquim Gabarro
Juan Garay
Mark Giesbrecht
Eduardo Giménez
Bernard Gittenberger
Raúl Gouet
Qian-Ping Gu
Marisa Gutiérrez
Ryan Hayward
Ch´nh T. Hoàng
Delia Kesner
Ayman Khalfalah
Yoshiharu Kohayakawa
Teresa Krick
Eyal Kushilevitz
Anton´n Kucera
Imre Leader
Hanno Lefmann
Sebastian Leipert
Stefano Leonardi
Sachin Lodha
F. Javier López
Hosam M. Mahmoud
A. Marchetti-Spaccamela
Claude Marché
Arnaldo Mandel
Mart´n Matamala
Guillermo Matera
Jacques Mazoyer
Alberto Mendelzon
Ugo Montanari
François Morain
Petra Mutzel
Rajagopal Nagarajan
Gonzalo Navarro
Marden Neubert
Cyril Nicaud
Takao Nishizeki
Johan Nordlander
Alfredo Olivero
Alberto Pardo
Jordi Petit

Wojciech Plandowski
Libor Polák
Pavel Pudlak
Davood Ra ei
Ivan Rapaport
Mauricio G.C. Resende
Celso C. Ribeiro
Alexander Rosa
Salvador Roura
Andrzej Ruciński
Juan Sabia
Philippe Schoebelen
Maria Serna
Oriol Serra
Jiri Sgall
Guillermo R. Simari
José Soares
Pablo Solernó
Doug Stinson
Jorge Stol
Leen Stougie
Jayme Szwarc ter
Prasad Tetali
Dimitrios Thilikos
Soledad Torres
Luca Trevisan
Vilmar Trevisan
Andrew Turpin
Kristina Vusković
Lusheng Wang
Sue Whitesides
Thomas Wilke
Hugh Williams
David Williamson
Fatos Xhafa
Daniel Yankelevich
Sheng Yu
Louxin Zhang
Binhai Zhu

Table of Contents

Random Structures and Algorithms

Algorithms I

Combinatorial Designs

Web Graph, Graph Theory I

Graph Theory II

Competitive Analysis, Complexity

Algorithms II

Computational Number Theory, Cryptography

Analysis of Algorithms I

Algebraic Algorithms

Computability

Automata, Formal Languages

Logic, Programming Theory

Analysis of Algorithms II

Algorithmic Aspects of Regularity

Y. Kohayakawa[1*] and V. Rödl[2]

[1] Instituto de Matemática e Estatística, Universidade de São Paulo,
Rua do Matão 1010, 05508–900 São Paulo, Brazil
yoshi@ime.usp.br
[2] Department of Mathematics and Computer Science,
Emory University, Atlanta, GA, 30322, USA
rodl@mathcs.emory.edu

Abstract. Szemerédi's celebrated regularity lemma proved to be a fundamental result in graph theory. Roughly speaking, his lemma states that any graph may be approximated by a union of a bounded number of bipartite graphs, each of which is 'pseudorandom'. As later proved by Alon, Duke, Lefmann, Rödl, and Yuster, there is a fast deterministic algorithm for finding such an approximation, and therefore many of the existential results based on the regularity lemma could be turned into constructive results. In this survey, we discuss some recent developments concerning the algorithmic aspects of the regularity lemma.

1 Introduction

In the course of proving his well known density theorem for arithmetic progressions [47], Szemerédi discovered a fundamental result in graph theory. This result became known as his *regularity lemma* [48]. For an excellent survey on this lemma, see Komlós and Simonovits [35]. Roughly speaking, Szemerédi's lemma states that any graph may be approximated by a union of a bounded number of bipartite graphs, each of which is 'pseudorandom'.

Szemerédi's proof did not provide an efficient algorithm for finding such an approximation, but it was later proved by Alon, Duke, Lefmann, Rödl, and Yuster [1,2] that such an algorithm does exist. Given the wide applicability of the regularity lemma, the result of [1,2] had many consequences. The reader is referred to [1,2,14,35] for the first applications of the algorithmic version of the regularity lemma. For more recent applications, see [5,6,7,12,17,32,52].

If the input graph G has n vertices, the algorithm of [1,2] runs in time $O(M(n))$, where $M(n) = O(n^{2.376})$ is the time needed to multiply two n by n matrices with $\{0, 1\}$-entries over the integers. In [34], an improvement of this is given: it is shown that there is an algorithm for the regularity lemma that runs in time $O(n^2)$ for graphs of order n.

If one allows randomization, one may do a great deal better, as demonstrated by Frieze and Kannan. In fact, they show in [21,22] that there is a randomized

* Partially supported by FAPESP (Proc. 96/04505–2), by CNPq (Proc. 300334/93–1), and by MCT/FINEP (PRONEX project 107/97).

algorithm for the regularity lemma that runs in time $O(n)$ for n-vertex graphs. Quite surprisingly, they in fact show that one may obtain an implicit description of the required output in *constant time*. The key technique here is *sampling*.

The regularity lemma has been generalized for hypergraphs in a few different ways; see, *e.g.*, [8,18,19,21,22,42]. One of these generalizations admits constructive versions, both deterministic [13] and randomized [21,22]. Again, the consequences of the existence of such algorithms are important. For instance, Frieze and Kannan [21,22] prove that all 'dense' MAXSNP-hard problems admit PTAS, by making use of such algorithms. For other applications of an algorithmic hypergraph regularity lemma, see Czygrinow [11].

Let us discuss the organization of this survey. In Section 2, we state the regularity lemma for graphs and hypergraphs. In Section 3, we discuss a few independent lemmas, each of which allows one to produce algorithms for the regularity lemma. In Section 4.1, we state the main result of [34]. Some results of Frieze and Kannan are discussed in Section 4.2. In Section 5, we discuss a recent result on property testing on graphs, due to Alon, Fischer, Krivelevich, and Szegedy [3,4]. The main result in [3,4] is based on a new variant of the regularity lemma. We close with some final remarks.

2   The Regularity Lemma

In this section we state the regularity lemma and briefly discuss the original proof of Szemerédi. In order to be concise, we shall state a hypergraph version that is a straightforward extension of the classical lemma. We remark that this extension was first considered and applied by Prömel and Steger [42].

2.1   The Statement of the Lemma

Given a set V and a non-negative integer r, we write $[V]^r$ for the collection of subsets of V of cardinality r. An *r-uniform hypergraph* or *r-graph* on the *vertex set V* is simply a collection of r-sets $\mathcal{H} \subset [V]^r$. The elements of $\mathcal{H}$ are the *hyperedges* of $\mathcal{H}$

Let $U_1, \ldots, U_r \subset V$ be pairwise disjoint, non-empty subsets of vertices. The density $d_{\mathcal{H}}(U_1, \ldots, U_r)$ of this r-tuple with respect $\mathcal{H}$ is

$$d_{\mathcal{H}}(U_1, \ldots, U_r) = \frac{e(U_1, \ldots, U_r)}{|U_1| \ldots |U_r|}, \tag{1}$$

where $e(U_1, \ldots, U_r)$ is the number of hyperedges $e \in \mathcal{H}$ with $|e \cap U_i| = 1$ for all $1 \leq i \leq r$. We say that the r-tuple $(U_1, \ldots, U_r)$ is *ε-regular* with respect to $\mathcal{H}$ if, for all choices of subsets $U_i' \subset U_i$ with $|U_i'| \geq \varepsilon |U_i|$ for all $1 \leq i \leq r$, we have

$$|d_{\mathcal{H}}(U_1', \ldots, U_r') - d_{\mathcal{H}}(U_1, \ldots, U_r)| \leq \varepsilon. \tag{2}$$

If for any such U_i' ($1 \leq i \leq r$) we have

$$|d_{\mathcal{H}}(U_1', \ldots, U_r') - \alpha| \leq \delta, \tag{3}$$

we say that the r-tuple $(U_1, \ldots, U_r)$ is (α, δ)-*regular*. Finally, we say that a partition $V = V_0 \cup \cdots \cup V_k$ of the vertex set V of $\mathcal{H}$ is ε-*regular* with respect to $\mathcal{H}$ if

(i) $|V_0| < \varepsilon|V|$,
(ii) $|V_1| = \cdots = |V_k|$,
(iii) at least $(1 - \varepsilon)\binom{k}{r}$ of the r-tuples $(V_{i_1}, \ldots, V_{i_r})$ with $1 \leq i_1 < \cdots < i_r \leq k$ are ε-regular with respect to $\mathcal{H}$.

Often, V_0 is called the *exceptional class* of the partition. For convenience, we say that a partition $(V_i)_{i=0}^{k}$ is (ε, k)-*equitable* if it satisfies (i) and (ii) above. The hypergraph version of Szemerédi's lemma reads as follows.

Theorem 1 *For any integers $r \geq 2$ and $k_0 \geq 1$ and real number $\varepsilon > 0$, there are integers $K = K(r, k_0, \varepsilon)$ and $N = N(r, k_0, \varepsilon)$ such that any r-graph $\mathcal{H}$ on a vertex set of cardinality at least N admits an (ε, k)-equitable ε-regular partition with $k_0 \leq k \leq K$.*

Szemerédi [48] considered the case $r = 2$, that is, the case of *graphs*. However, the proof in [48] generalizes in a straightforward manner to a proof of Theorem 1 above; see Prömel and Steger [42], where the authors prove and apply this result.

2.2   Brief Outline of Proofs

Let $\mathcal{H}$ be an r-graph on the vertex set V with $|V| = n$, and let $\Pi = (V_i)_{i=0}^{k}$ be an (ε, k)-equitable partition of V. A crucial definition in Szemerédi's proof of his lemma is the concept of the *index* $\mathrm{ind}(\Pi)$ of Π, given by

$$\mathrm{ind}(\Pi) = \binom{k}{r}^{-1} \sum d(V_{i_1}, \ldots, V_{i_r})^2, \tag{4}$$

where the sum is taken over all r-tuples $1 \leq i_1 < \cdots < i_r \leq k$. Clearly, we always have $0 \leq \mathrm{ind}(\Pi) \leq 1$. For convenience, if $\Pi' = (W_j)_{j=0}^{\ell}$ is an (ε', ℓ)-equitable partition of V, we say that Π' is a *refinement* of Π if, for any $1 \leq j \leq \ell$, there is $1 \leq i \leq k$ for which we have $W_j \subset V_i$. In words, Π' refines Π if any non-exceptional class of Π' is contained in some non-exceptional class of Π. Now, the key lemma in the proof of Theorem 1 is the following (see [42]).

Lemma 2 *For any integer $r \geq 2$ and real number $\varepsilon > 0$, there exist integers $k_0 = k_0(r, \varepsilon)$ and $n_0 = n_0(r, \varepsilon)$ and a positive number $\vartheta = \vartheta(r, \varepsilon) > 0$ for which the following holds. Suppose we have an r-graph $\mathcal{H}$ on a vertex V, with $n = |V| \geq n_0$, and $\Pi = (V_i)_{i=0}^{k}$ is an (ε, k)-equitable partition of V. Then*

(i)   *either Π is ε-regular with respect to $\mathcal{H}$,*
(ii) *or, else, there is a refinement $\Pi' = (W_j)_{j=0}^{\ell}$ of Π such that*
   (a) $|W_0| \leq |V_0| + n/4^k$,
   (b) $|W_1| = \cdots = |W_\ell|$,

(c) $\ell = k4^{k^{r-1}}$,
(d) $\mathrm{ind}(\Pi') \geq \mathrm{ind}(\Pi) + \vartheta$.

Theorem 1 follows easily from Lemma 2: it suffices to recall that the index can never be larger than 1, and hence if we successively apply Lemma 2, starting from an arbitrary partition Π, we must arrive at an ε-regular partition after a bounded number of steps, because (d) of alternative (ii) guarantees that the indices of the partitions that we generate always increase by a fixed positive amount.

Ideally, to turn this proof into an efficient algorithm, given a partition Π, we would like to have (I) an efficient procedure to check whether (i) applies, and (II) if (i) fails, an efficient procedure for finding Π' as specified in (a)–(d).

It turns out that if we actually have, at hand, witnesses for the failure of ε-regularity of $> \varepsilon\binom{k}{r}$ of the r-tuples $(V_{i_1}, \ldots, V_{i_r})$, where $1 \leq i_1 < \cdots < i_r \leq k$, then Π' may be found easily (see [42] for details). Here, by a *witness* for the ε-irregularity of an r-tuple $(U_1, \ldots, U_r)$ we mean an r-tuple $(U_1', \ldots, U_r')$ with $U_i' \subset U_i$ and $|U_i'| \geq \varepsilon|U_i|$ for all $1 \leq i \leq r$ for which (2) fails. We are therefore led to the following decision problem:

Problem 3 *Given an r-graph $\mathcal{H}$, an r-tuple $(U_1, \ldots, U_r)$ of non-empty, pairwise disjoint sets of vertices of $\mathcal{H}$, and a real number $\varepsilon > 0$, decide whether this r-tuple is ε-regular with respect to $\mathcal{H}$.*

In case the answer to Problem 3 is negative for a given instance, we would like to have a witness for the ε-irregularity of the given r-tuple.

3   Conditions for Regularity

3.1   A Hardness Result

It turns out that Problem 3 is hard, as proved by Alon, Duke, Lefmann, Rödl, and Yuster [1,2].

Theorem 4 *Problem 3 is coNP-complete for $r = 2$.*

Let us remark in passing that Theorem 4 is proved in [1,2] for the case in which $\varepsilon = 1/2$; for a proof for arbitrary $0 < \varepsilon \leq 1/2$, see Taraz [49].

Theorem 4 is certainly discouraging. Fortunately, however, there is a way around. We discuss the graph and hypergraph cases separately.

The graph case. In the case $r = 2$, that is, in the case of graphs, one has the following lemma. Below $\mathbb{R}_+$ denotes the set of positive reals. Moreover, a bipartite graph $B = (U, W; E)$ with vertex classes U and W and edge set E is said to be ε-regular if (U, W) is an ε-regular pair with respect to B. Thus, a witness to the ε-irregularity of B is a pair (U', W') with $U' \subset U$, $W' \subset W$, $|U'|, |W'| \geq \varepsilon n$, and $|d_B(U', W') - d_B(U, W)| > \varepsilon$ (see the paragraph before Problem 3).

Lemma 5 *There is a polynomial-time algorithm $\mathcal{A}$ and a function $\varepsilon'_{\mathcal{A}} \colon \mathbb{R}_+ \to$ $\mathbb{R}_+$ such that the following holds. When $\mathcal{A}$ receives as input a bipartite graph $B = (U, W; E)$ with $|U| = |W| = n$ and a real number $\varepsilon > 0$, it either correctly asserts that B is ε-regular, or else it returns a witness for the $\varepsilon'_{\mathcal{A}}(\varepsilon)$-irregularity of B.*

We remark that Lemma 5 implicitly says that $\varepsilon' = \varepsilon'_{\mathcal{A}}(\varepsilon) \leq \varepsilon$, for otherwise $\mathcal{A}$ would not be able to handle an input graph B that is not ε-regular but is ε'-regular. In fact, one usually has $\varepsilon' \ll \varepsilon$.

Note that Lemma 5 leaves open what the behaviour of $\mathcal{A}$ should be when B is ε-regular but is not ε'-regular. Despite this fact, Lemma 5 does indeed imply the existence of a polynomial-time algorithm for finding ε-regular partitions of graphs. We leave the proof of this assertion as an exercise for the reader.

In Sections 3.2, 3.3, and 3.4 we state some independent results that imply Lemma 5, thus completing the general description of a few distinct ways one may prove the algorithmic version of the regularity lemma for graphs.

The hypergraph case. In the case of r-graphs ($r \geq 3$), we do not know a result similar to Lemma 5. The algorithmic version of Theorem 1 for $r \geq 3$ is instead proved by introducing a modified concept of index and then by proving a somewhat more complicated version of Lemma 5. For lack of space, we shall not go into details; the interested reader is referred to Czygrinow and Rödl [13]. In fact, in the remainder of this survey we shall mostly concentrate on graphs.

Even though several applications are known for the algorithmic version of Theorem 1 for $r \geq 3$ (see, *e.g.*, [11,12,13,21,22]), it should be mentioned that the most powerful version of the regularity lemma for hypergraphs is not the one presented above. Indeed, the regularity lemma for hypergraphs proved in [18] seems to have deeper combinatorial consequences.

3.2   The Pair Condition for Regularity

As mentioned in Section 3.1, Lemma 5 may be proved in a few different ways. One technique is presented in this section. The second, which is in fact a generalization of the methods discussed here, is based on Lemmas 9 and 10. Finally, the third method, presented in Section 3.4, is based on a criterion for regularity given in Lemma 11.

Let us turn to the original approach of [1,2]. The central idea here may be summarized in two lemmas, Lemmas 6 and 7 below; we follow the formulation given in [14] (see also [9,20,50,51] and the proof of the upper bound in Theorem 15.2 in [16], due to J. H. Lindsey). Below, $d(x, x')$ denotes the *joint degree* or *codegree* of x and x', that is, the number of common neighbours of x and x'.

Lemma 6 *Let a constant $0 < \varepsilon < 1$ be given and let $B = (U, W; E)$ be a bipartite graph with $|U| \geq 2/\varepsilon$. Let $\varrho = d_B(U, W)$ and let D be the collection of all pairs $\{x, x'\}$ of vertices of U for which*

(i) $d(x), d(x') > (\varrho - \varepsilon)|W|$,

(ii) $d(x, x') < (\varrho + \varepsilon)^2 |W|.$

Then if $|D| > (1/2)(1 - 5\varepsilon)|W|^2$, the pair (U, W) is $(\varrho, (16\varepsilon)^{1/5})$-regular.

Lemma 7 *Let $B = (U, W; E)$ be a graph with (U, W) a (ϱ, ε)-regular pair and with density $d(U, W) = \varrho$. Assume that $\varrho|W| \geq 1$ and $0 < \varepsilon < 1$. Then*

(i) *all but at most $2\varepsilon|W|$ vertices $x \in W$ satisfy*

$$(\varrho - \varepsilon)|W| < d(x), \; d(x') < (\varrho + \varepsilon)|W|,$$

(ii) *all but at most $2\varepsilon|U|^2$ pairs $\{x, x'\}$ of vertices of A satisfy*

$$d(x, x') < (\varrho + \varepsilon)^2 |W|.$$

It is not difficult to see that Lemmas 6 and 7 imply Lemma 5. Indeed, the main computational task that algorithm $\mathcal{A}$ from Lemma 5 has to perform is to compute the codegrees of all pairs of vertices $(x, x') \in U \times U$. Clearly, this may be done in time $O(n^3)$. Observing that this task may be encoded as the squaring of a certain natural $\{0, 1\}$-matrix over the integers, one sees that there is an algorithm $\mathcal{A}$ as in Lemma 5 with time complexity $O(M(n))$, where $M(n) = O(n^{2.376})$ is the time required to carry out such a multiplication (see [10]).

Before we proceed, let us observe that the pleasant fact here is the following. Although the definition of ε-regularity for a pair (U, W) involves a quantification over exponentially many pairs (U', W'), we may essentially check the validity of this definition by examining all pairs $(x, x') \in U \times U$, of which there are only quadratically many. We refer to the criterion for regularity given by Lemmas 6 and 7 as the *pair condition for regularity*.

3.3   An Optimized Pair Condition for Regularity

Here we state an improved version of the pair condition of Section 3.2. The key idea is that it suffices to control the codegrees of a small, suitably chosen set of pairs of vertices to guarantee the regularity of a bipartite graph; that is, we need not examine all pairs $(x, x') \in U \times U$. As it turns out, it suffices to consider the pairs that form the edge set of a linear-sized expander, which reduces the number of pairs to examine from $\binom{n}{2}$ to $O(n)$. This implies that there is an algorithm $\mathcal{A}$ as in Lemma 5 with time complexity $O(n^2)$.

We start with an auxiliary definition.

Definition 8 *Let $0 < \varrho \leq 1$ and $A > 0$ be given. We say that a graph J on n vertices is (ϱ, A)-uniform if for any pair of disjoint sets $U, W \subset V(J)$ such that $1 \leq |U| \leq |W| \leq r|U|$, where $r = \varrho n$, we have*

$$\left| e_J(U, W) - \varrho|U||W| \right| \leq A\sqrt{r|U||W|}.$$

Thus, a (ϱ, A)-uniform graph is a graph with density $\sim \varrho$ in which the edges are distributed in a random-like manner. One may check that the usual binomial random graph $G(n, p)$ is $(p, 20)$-uniform (see, e.g., [31]). More relevant to us is the fact that there are efficiently constructible graphs J that are $(\varrho, O(1))$-uniform, and have arbitrarily large but constant average degree. Here we have in mind the celebrated graphs of Margulis [41] and Lubotzky, Phillips, and Sarnak [39] (see also [40,46]). For these graphs, we have $A = 2$.

Let us now introduce the variant of the pair condition for regularity that is of interest. Let an n by n bipartite graph $B = (U, W; E)$ be given, and suppose that J is a graph on U. Write $e(J)$ for the number of edges in J and let $p = e(B)/n^2$ be the density of B. We say that B satisfies property $\mathcal{P}(J, \delta)$ if

$$\sum_{\{x,y\} \in E(J)} |d(x, y) - p^2 n| \leq \delta p^2 n e(J). \tag{5}$$

Below, we shall only be interested in the case in which J is (ϱ, A)-uniform for some constant A and $\varrho \asymp 1/n$. The results analogous to Lemmas 6 and 7 involving property $\mathcal{P}$ are as follows.

Lemma 9 *For every $\varepsilon > 0$, there exist $r_0 = r_0(\varepsilon)$, $n_0 = n_0(\varepsilon) \geq 1$, and $\delta = \delta(\varepsilon) > 0$ for which the following holds. Suppose $n \geq n_0$, the graph J is a (ϱ, A)-uniform graph with $\varrho = r/n \geq r_0/n$, and B is a bipartite graph as above. Then, if B has property $\mathcal{P}(J, \delta)$, then B is ε-regular.*

Lemma 10 *For every $\delta > 0$, there exist $r_1 = r_1(\delta)$, $n_1 = n_1(\delta) \geq 1$, and $\varepsilon' = \varepsilon'(\delta) > 0$ for which the following holds. Suppose $n \geq n_1$, the graph J is a (ϱ, A)-uniform graph with $\varrho = r/n \geq r_1/n$, and B is a bipartite graph as above. Then, if B does not satisfy property $\mathcal{P}(J, \delta)$, then B is not ε'-regular. Furthermore, in this case, we can find a pair of sets of vertices (U', W') witnessing the ε'-irregularity of B in time $O(n^2)$.*

Lemmas 9 and 10 show that Lemma 5 holds for an algorithm $\mathcal{A}$ with time complexity $O(n^2)$.

The proof of Lemma 9 is similar in spirit to the proof of Lemma 6, but of course one has to make heavy use of the (ϱ, A)-uniformity of J. The sparse version of the regularity lemma (see, e.g., [33]) may be used to prove the ε'-irregularity of the graph B in Lemma 10. However, proving that a suitable witness may be found in quadratic time requires a different approach. The reader is referred to [34] for details.

3.4   Singular Values and Regularity

We now present an approach due to Frieze and Kannan [23]. Let an m by n real matrix $\mathbf{A}$ be given. The *first singular value* $\sigma_1(\mathbf{A})$ of $\mathbf{A}$ is

$$\sigma_1(\mathbf{A}) = \sup\{|x^\top \mathbf{A} y| : \|x\| = \|y\| = 1\}. \tag{6}$$

Above, we use $\| \ \|$ to denote the 2-norm. In what follows, for a matrix $\mathbf{W} = (w_{i,j})$, we let $\|\mathbf{W}\|_\infty = \max |w_{i,j}|$. Moreover, if I and J are subsets of the index sets of the rows and columns of $\mathbf{W}$, we let

$$\mathbf{W}(I,J) = \sum_{i \in I, j \in J} w_{i,j} = \chi_I^\top \mathbf{W} \chi_J, \tag{7}$$

where χ_I and χ_J are the $\{0,1\}$-characteristic vectors for I and J.

Let $B = (U, W; E)$ be a bipartite graph with density $p = e(B)/|U||W|$, and let $\mathbf{A} = (a_{i,j})_{i \in U, j \in W}$ be the natural $\{0,1\}$-adjacency matrix associated with B. Put $\mathbf{W} = \mathbf{A} - p\mathbf{J}$, where $\mathbf{J}$ is the $n \times n$ matrix with all entries equal to 1. It is immediate to check that the following holds:

(*) B is ε-regular if and only if $|\mathbf{W}(U', W')| \leq \varepsilon|U'||W'|$ for all $U' \subset U$ and $W' \subset W$ with $|U'| \geq \varepsilon|U|$ and $|W'| \geq \varepsilon|W|$.

The regularity condition of Frieze and Kannan [23] is as follows.

Lemma 11 *Let* $\mathbf{W}$ *be a matrix whose entries are index by* $U \times W$*, where* $|U| = |W| = n$*, and suppose that* $\|\mathbf{W}\|_\infty \leq 1$*. Let* $\gamma > 0$ *be given. Then the following assertions hold:*

(i) *If there exist* $U' \subset U$ *and* $W' \subset W$ *such that* $|U'|, |W'| \geq \gamma n$ *and*

$$|\mathbf{W}(U', W')| \geq \gamma|U'||W'|,$$

then $\sigma_1(\mathbf{W}) \geq \gamma^3 n$.

(ii) *If* $\sigma_1(\mathbf{W}) \geq \gamma n$*, then there exist* $U' \subset U$ *and* $W' \subset W$ *such that* $|U'|, |W'| \geq \gamma' n$ *and* $|\mathbf{W}(U', W')| \geq \gamma'|U'||W'|$*, where* $\gamma' = \gamma^3/108$*. Furthermore,* U' *and* W' *may be constructed in polynomial time.*

Now, in view of (*) and the fact that singular values may be computed in polynomial time (see, *e.g.*, [29]), Lemma 11 provides a proof for Lemma 5.

4   The Algorithmic Versions of Regularity

In this section, we discuss the constructive versions of Theorem 1 for the graph case, that is $r = 2$. We discuss deterministic and randomized algorithms separately.

4.1   Deterministic Algorithms

Our deterministic result, Theorem 12 below, asserts the existence of an algorithm for finding regular partitions that is asymptotically faster than the algorithm due to Alon, Duke, Lefmann, Rödl, and Yuster [1,2].

Theorem 12 *There is an algorithm $\mathcal{A}$ that takes as input an integer $k_0 \geq 1$, an $\varepsilon > 0$, and a graph G on n vertices and returns an ε-regular (ε, k)-equitable partition for G with $k_0 \leq k \leq K$, where K is as Theorem 1. Algorithm $\mathcal{A}$ runs in time $\leq Cn^2$, where $C = C(\varepsilon, k_0)$ depends only on ε and k_0.*

Theorem 12 follows from the considerations in Sections 3.1 and 3.3 (see [34] for details).

It is easy to verify that there is a constant $\varepsilon' = \varepsilon'(\varepsilon, k_0)$ such that if $e(G) < \varepsilon' n^2$, then *any* (ε, k_0)-equitable partition is ε-regular. Clearly, as the time required to read the input is $\Omega(e(G))$ and, as observed above, we may assume that $e(G) \geq \varepsilon' n^2$, algorithm $\mathcal{A}$ in Theorem 12 is optimal, apart from the value of the constant $C = C(\varepsilon, k_0)$.

A typical application of the algorithmic regularity lemma of [1,2] asserts the existence of a fast algorithm for a graph problem. As it turns out, the running time of such an algorithm is often dominated by the time required for finding a regular partition for the input graph, and hence the algorithm has time complexity $O(M(n))$. In view of Theorem 12, the existence of quadratic algorithms for these problems may be asserted. Some examples are given in [34].

4.2   Randomized Algorithms

We have been discussing deterministic algorithms so far. If we allow randomization, as proved by Frieze and Kannan, a great deal more may be achieved in terms of efficiency [21,22]. The model we adopt is as follows. We assume that sampling a random vertex from G as well as checking an entry of the adjacency matrix of G both have unit cost.

Theorem 13 *There is a randomized algorithm $\mathcal{A}_{\mathrm{FK}}$ that takes as input an integer $k_0 \geq 1$, an $\varepsilon > 0$, a $\delta > 0$, and a graph G on n vertices and returns, with probability $\geq 1 - \delta$, an ε-regular (ε, k)-equitable partition for G with $k_0 \leq k \leq K$, where K is as Theorem 1. Moreover, the following assertions hold:*

(i)   Algorithm $\mathcal{A}_{\mathrm{FK}}$ runs in time $\leq Cn$, where $C = C(\varepsilon, \delta, k_0)$ depends only on ε, δ, and k_0.

(ii)   In fact, $\mathcal{A}_{\mathrm{FK}}$ first outputs a collection of vertices of G of cardinality $\leq C'$, where $C' = C'(\varepsilon, \delta, k_0)$ depends only on ε, δ, and k_0, and the above ε-regular partition for G may be constructed in linear time from this collection of vertices.

The fact that one is able to construct a regular partition for a graph on n vertices in randomized time $O(n)$ is quite remarkable; this is another evidence of the power of randomization. However, even more remarkable is what (ii) implies: given $\varepsilon > 0$, $\delta > 0$, and $k_0 \geq 1$, there is a uniform bound $C' = C'(\varepsilon, \delta, k_0)$ on the number of randomly chosen vertices that will implicitly define for us a suitable ε-regular partition of the given graph, *no matter how large this input graph is*.

Let us try to give a feel on how one may proceed to prove Theorem 13. We shall discuss a lemma that is central in the Frieze–Kannan approach. The idea is that, if a bipartite graph $B = (U, W; E)$ is not ε-regular, then this may be detected, with high probability, by sampling a bounded number of vertices of B. The effectiveness of sampling for detecting "dense spots" in graphs appears already in Goldreich, Goldwasser, and Ron [25,26], where many *constant time* algorithms are developed (we shall discuss such matters in Section 5 below).

Let us state the key technical lemma in the proof of Theorem 13. Let an n by n bipartite graph $B = (U, W; E)$ be given. Our aim is to check whether there exist $U' \subset U$ and $W' \subset W$ such that $|U'|, |W'| \geq \varepsilon n$ and

$$|d_B(U', W') - d_B(U, W)| > \varepsilon.$$

Note that this last inequality is equivalent to $\big|e(U', W') - p|U'||W'|\big| > \varepsilon|U'||W'|$, where $p = e(B)/n^2$ is the density of B. In fact, if

$$\big|e(U', W') - p|U'||W'|\big| > \gamma n^2$$

holds and $\gamma \geq \varepsilon$, then (U', W') must be a witness for the ε-irregularity of B.

We are now ready to state the result of Frieze and Kannan that allows one to prove a randomized version of Lemma 5. The reader may wish to compare this result with Lemmas 7 and 10.

Lemma 14 *There is a randomized algorithm $\mathcal{A}$ that behaves as follows. Let $B = (U, W; E)$ be as above and let γ and $\delta > 0$ be positive reals. Suppose there exist $U' \subset U$ and $W' \subset W$ for which $e(U', W') > p|U'||W'| + \gamma n^2$ holds. Then, on input B, $\gamma > 0$, and $\delta > 0$, algorithm $\mathcal{A}$ determines, with probability $\geq 1 - \delta$, implicitly defined sets $U'' \subset U$ and $W'' \subset W$ with*

$$e(U'', W'') > p|U''||W''| + \frac{1}{16}\gamma n^2.$$

The running time of $\mathcal{A}$ is bounded by some constant $C = C(\gamma, \delta)$ that depends only on γ and δ.

A few comments concerning Lemma 14 are in order. As before, the model here allows for the selection of random vertices of B as well as checking whether two given vertices are adjacent in constant time. In order to define the sets U'' and W'', algorithm $\mathcal{A}$ returns two sets $Z_1 \subset U$ and $Z_2 \subset W$, both of cardinality bounded by some constant depending only on γ and $\delta > 0$. Then, U'' is simply the set of vertices $u \in U$ for which $e(u, Z_2) \geq p|Z_2|$. The set W'' is defined analogously.

Algorithm $\mathcal{A}$ of Lemma 14 is extremely simple, and its elegant proof of correctness is based on the linearity of expectation and on well known large deviation inequalities (see Frieze and Kannan [21]).

We close this section mentioning a result complementary to Lemma 14 (see [15] for a slightly weaker statement). Suppose $B = (U, W; E)$ is ε-regular. Then if $U' \subset U$ and $W' \subset W$ are randomly chosen sets of vertices with $|U'| = |W'| \geq M_0 = M_0(\varepsilon', \delta)$, then the bipartite graph $B' = (U', W'; E')$ induced by B on (U', W') is ε'-regular with probability $\geq 1 - \delta$, as long as $\varepsilon \leq \varepsilon_0(\varepsilon', \delta)$. Here again the striking fact is that a bounded sample of vertices forms a good enough picture of the whole graph.

5 Property Testing

The topic of this section is in nature different from the topics discussed so far. We have been considering how to produce algorithms for finding regular partitions of graphs. In this section, we discuss how non-constructive versions of the regularity lemma may be used to prove the correctness of certain algorithms. We shall discuss a recent result to Alon, Fischer, Krivelevich, and Szegedy [3,4]. These authors develop a new variant of the regularity lemma and use it to prove a far reaching result concerning the *testability* of certain graph properties.

5.1 Definitions and the Testability Result

The general notion of property testing was introduced by Rubinfeld and Sudan [45], but in the context of combinatorial testing it is the work of Goldreich and his co-authors [24,25,26,27,28] that are most relevant to us.

Let $\mathcal{G}^n$ be the collection of all graphs on a fixed n-vertex set, say $[n] = \{1, \ldots, n\}$. Put $\mathcal{G} = \bigcup_{n \geq 1} \mathcal{G}^n$. A *property* of graphs is simply a subset $\mathcal{P} \subset \mathcal{G}$ that is closed under isomorphisms. There is a natural notion of distance in each $\mathcal{G}^n$, the *normalized Hamming distance*: the distance $d(G, H) = d_n(G, H)$ between two graphs G and $H \in \mathcal{G}^n$ is $|E(G) \bigtriangleup E(H)| \binom{n}{2}^{-1}$, where $E(G) \bigtriangleup E(H)$ denotes the symmetric difference of the edge sets of G and H.

We say that a graph G is ε-*far* from having property $\mathcal{P}$ if

$$d(G, \mathcal{P}) = \min_{H \in \mathcal{P}} d(G, H) \geq \varepsilon,$$

that is, at least $\varepsilon \binom{n}{2}$ edges have to be added or removed to G to turn it into a graph that satisfies $\mathcal{P}$.

An ε-*test* for a graph property $\mathcal{P}$ is a randomized algorithm $\mathcal{A}$ that receives as input a graph G and behaves as follows: if G has $\mathcal{P}$ then with probability $\geq 2/3$ we have $\mathcal{A}(G) = 1$, and if G is ε-far from having $\mathcal{P}$ then with probability $\geq 2/3$ we have $\mathcal{A}(G) = 0$. The graph G is given to $\mathcal{A}$ through an oracle; we assume that $\mathcal{A}$ is able to generate random vertices from G and it may *query* the oracle whether two vertices that have been generated are adjacent.

We say that a graph property $\mathcal{P}$ is *testable* if, for all $\varepsilon > 0$, it admits an ε-test that makes at most Q queries to the oracle, where $Q = Q(\varepsilon)$ is a constant that depends only on ε. Note that, in particular, we require the number of queries to be independent of the order of the input graph.

Goldreich, Goldwasser, and Ron [25,26], besides showing that there exist NP graph properties that are not testable, proved that a large class of interesting graph properties *are* testable, including the property of being k-colourable, of having a clique with $\geq \varrho n$ vertices, and of having a cut with $\geq \varrho n^2$ edges, where n is the order of the input graph. The regularity lemma is not used in [25,26]. The fact that k-colourability is testable had in fact been proved implicitly in [15], where regularity is used.

We are now ready to turn to the result of Alon, Fischer, Krivelevich, and Szegedy [3,4]. Let us consider properties from the first order theory of graphs.

Thus, we are concerned with properties that may be expressed through quantification of vertices, Boolean connectives, equality, and adjacency. Of particular interest are the properties that may be expressed in the form

$$\exists x_1, \ldots, x_r \; \forall y_1, \ldots, y_s \; A(x_1, \ldots, x_r, y_1, \ldots, y_s),$$

where A is a quantifier-free first order expression. Let us call such properties *of type* $\exists\forall$. Similarly, we define properties *of type* $\forall\exists$. The main result of [3,4] is as follows.

Theorem 15 *All first order properties of graphs that may be expressed with at most one quantifier as well as all properties that are of type* $\exists\forall$ *are testable. Furthermore, there exist properties of type* $\forall\exists$ *that are not testable.*

The first part of the proof of the positive result in Theorem 15 involves the reduction, up to testability, of properties of type $\exists\forall$ to a certain generalized colourability property. A new variant of the regularity lemma is then used to handle this generalized colouring problem.

5.2   A Variant of the Regularity Lemma

In this section we shall state a variant of the regularity lemma proved in [3,4].

Let us say that a partition $\Pi = (V_i)_{i=1}^k$ of a set V is an *equipartition* of V if all the sets V_i ($1 \leq i \leq k$) differ by at most 1 in size. In this section, we shall not have exceptional classes in our partitions. Below, we shall have an equipartition of V

$$\Pi' = \{V_{i,j} : 1 \leq i \leq k, \; 1 \leq j \leq \ell\}$$

that is a refinement of a given partition $\Pi = (V_i)_{i=1}^k$. In this notation, we understand that, for all i, all the $V_{i,j}$ ($1 \leq j \leq \ell$) are contained in V_i.

Theorem 16 *For every integer k_0 and every function $0 < \varepsilon(r) < 1$ defined on the positive integers, there are constants $K = K(k_0, \varepsilon)$ and $N = N(k_0, \varepsilon)$ with the following property. If G is any graph with at least N vertices, then there exist equipartitions $\Pi = (V_i)_{1 \leq i \leq k}$ and $\Pi' = (V_{i,j})_{1 \leq i \leq k, 1 \leq j \leq \ell}$ of $V = V(G)$ such that the following hold:*

(i) $|\Pi| = k \geq k_0$ *and* $|\Pi'| = k\ell \leq K$;

(ii)   *at least* $(1 - \varepsilon(0))\binom{k}{2}$ *of the pairs* $(V_i, V_{i'})$ *with* $1 \leq i < i' \leq k$ *are* $\varepsilon(0)$-*regular*;

(iii)   *for all* $1 \leq i < i' \leq k$, *we have that at least* $(1-\varepsilon(k))\ell^2$ *of the pairs* $(V_{i,j}, V_{i',j'})$ *with* $j, j' \in [\ell]$ *are* $\varepsilon(k)$-*regular*;

(iv)   *for at least* $(1 - \varepsilon(0))\binom{k}{2}$ *of the pairs* $1 \leq i < i' \leq k$, *we have that for at least* $(1 - \varepsilon(0))\ell^2$ *of the pairs* $j, j' \in [\ell]$ *we have*

$$|d_G(V_i, V_{i'}) - d_G(V_{i,j}, V_{i',j'})| \leq \varepsilon(0).$$

Suppose we have partitions Π and Π' as in Theorem 16 above and that $\varepsilon(k) \ll 1/k$. It is not difficult to see that then, for many 'choice' functions $j \colon [k] \to [\ell]$, we have that $\widetilde{\Pi} = (V_{i,j(i)})_{1 \leq i \leq k}$ is an equipartition of an induced subgraph of G such that the following hold:

(a) *all* the pairs $(V_{i,j(i)}, V_{i',j(i')})$ are $\varepsilon(k)$-regular,

(b) for at least $(1 - \varepsilon(0))\binom{k}{2}$ of the pairs $1 \leq i < i' \leq k$, we have

$$|d_G(V_i, V_{i'}) - d_G(V_{i,j(i)}, V_{i',j(i')})| \leq \varepsilon(0).$$

In a certain sense, this consequence of Theorem 16 lets us ignore the irregular pairs in the partition Π, at the expense of dropping down from the V_i to smaller sets $V_{i,j(i)}$ (still all of cardinality $\Omega(n)$), and having most but not necessarily all densities $d_G(V_{i,j(i)}, V_{i',j(i')})$ under tight control.

Let us remark in passing that, naturally, one may ask whether Theorem 1 may be strengthened by requiring that there should be *no* irregular pairs altogether. This question was already raised by Szemerédi in [48]. As observed by Lovász, Seymour, Trotter, and the authors of [2] (see p. 82 in [2]), such an extension of Theorem 1 does not exist. As noted above, Theorem 16 presents a way around this difficulty.

Theorem 16 and its corollary mentioned above are the main ingredients in the proof of the following result (see [3,4] for details).

Theorem 17 *For every $\varepsilon > 0$ and $h \geq 1$, there is $\delta = \delta(\varepsilon, h) > 0$ for which the following holds. Let H be an arbitrary graph on h vertices and let $\mathcal{P} = \mathrm{Forb}_{\mathrm{ind}}(H)$ be the property of not containing H as an induced subgraph. If an n-vertex graph G is ε-far from $\mathcal{P}$, then G contains δn^h induced copies of H.*

The case in which H is a complete graph follows from the original regularity lemma, but the general case requires the corollary to Theorem 16 discussed above. Note that Theorem 17 immediately implies that the property of membership in $\mathrm{Forb}_{\mathrm{ind}}(H)$ (in order words, the property of not containing an induced copy of H) is a testable property for any graph H.

The proof of Theorem 15 requires a generalization of Theorem 17 related to the colouring problem alluded to at the end of Section 5.1. We refer the reader to [3,4]. We close by remarking that Theorem 16 has an algorithmic version, although we stress that this is not required in the proof of Theorem 15.

6 Concluding Remarks

We have not discussed a few recent, important results that relate to the regularity lemma. We single out three topics that the reader may wish to pursue.

6.1 Constants

The constants involved in Theorem 1 are extremely large. The proof in [48] gives that K in Theorem 1 is bounded from above by a tower of 2s of height ε^{-5}.

A recent result of Gowers [30] in fact shows that this cannot be essentially improved. Indeed, it is proved in [30] that there are graphs for which any ε-regular partition must have at least $G(\varepsilon^{-c})$ parts, where $c > 0$ is some absolute constant and $G(x)$ is a tower of 2s of height $\lfloor x \rfloor$.

The size of K is very often not too relevant in applications, but in certain cases essentially better results may be obtained if one is able to avoid the appearance of such huge constants. In view of Gowers's result, this can only be accomplished by modifying the regularity lemma. One early instance in which this carried out appears in [14]; a more recent example is [21].

6.2   Approximation Schemes for Dense Problems

Frieze and Kannan have developed variants of the regularity lemma for graphs and hypergraphs and discuss several applications in [21,22]. The applications are mostly algorithmic and focus on 'dense' problems, such as the design of a PTAS for the max-cut problem for dense graphs. The algorithmic versions of their variants of the regularity lemma play a central rôle in this approach.

For more applications of algorithmic regularity to 'dense' problems, the reader is referred to [11,12,13,32]

6.3   The Blow-Up Lemma

We close with an important lemma due to Komlós, Sárközy, and Szemerédi [37], the so-called *blow-up lemma*. (For an alternative proof of this lemma, see [44].)

In typical applications of the regularity lemma, once a suitably regular partition of some given graph G is found, one proceeds by embedding some 'target graph' H of bounded degree into G. Until recently, the embedding techniques could only handle graphs H with many fewer vertices than G. The blow-up lemma is a novel tool that allows one to embed target graphs H that even have the same number of vertices as G. The combined use of the regularity lemma and the blow-up lemma is a powerful new machinery in graph theory. The reader is referred to Komlós [36] for a discussion on the applications of the blow-up lemma.

On the algorithmic side, the situation is good: Komlós, Sárközy, and Szemerédi [38] have also proved an algorithmic version of the blow-up lemma (see Rödl, Ruciński, and Wagner [43] for an alternative proof).

References

1. N. Alon, R. A. Duke, H. Lefmann, V. Rödl, and R. Yuster, *The algorithmic aspects of the regularity lemma (extended abstract)*, 33rd Annual Symposium on Foundations of Computer Science (Pittsburgh, Pennsylvania), IEEE Comput. Soc. Press, 1992, pp. 473–481.
2. _____, *The algorithmic aspects of the regularity lemma*, Journal of Algorithms **16** (1994), no. 1, 80–109.
3. N. Alon, E. Fischer, M. Krivelevich, and M. Szegedy, *Efficient testing of large graphs*, submitted, 22pp., 1999.

4. _____, *Efficient testing of large graphs (extended abstract)*, 40th Annual Symposium on Foundations of Computer Science (New York City, NY), IEEE Comput. Soc. Press, 1999, pp. 656–666.

5. N. Alon and E. Fischer, *Refining the graph density condition for the existence of almost K-factors*, Ars Combinatoria **52** (1999), 296–308.

6. N. Alon and R. Yuster, *Almost H-factors in dense graphs*, Graphs and Combinatorics **8** (1992), no. 2, 95–102.

7. _____, *H-factors in dense graphs*, Journal of Combinatorial Theory, Series B **66** (1996), no. 2, 269–282.

8. F. R. K. Chung, *Regularity lemmas for hypergraphs and quasi-randomness*, Random Structures and Algorithms **2** (1991), no. 1, 241–252.

9. F. R. K. Chung, R. L. Graham, and R. M. Wilson, *Quasi-random graphs*, Combinatorica **9** (1989), no. 4, 345–362.

10. D. Coppersmith and S. Winograd, *Matrix multiplication via arithmetic progressions.*, Journal of Symbolic Computation **9** (1990), no. 3, 251–280.

11. A. Czygrinow, *Partitioning problems in dense hypergraphs*, submitted, 1999.

12. A. Czygrinow, S. Poljak, and V. Rödl, *Constructive quasi-Ramsey numbers and tournament ranking*, SIAM Journal on Discrete Mathematics **12** (1999), no. 1, 48–63.

13. A. Czygrinow and V. Rödl, *An algorithmic regularity lemma for hypergraphs*, submitted, 1999.

14. R. A. Duke, H. Lefmann, and V. Rödl, *A fast approximation algorithm for computing the frequencies of subgraphs in a given graph*, SIAM Journal on Computing **24** (1995), no. 3, 598–620.

15. R. A. Duke and V. Rödl, *On graphs with small subgraphs of large chromatic number*, Graphs and Combinatorics **1** (1985), no. 1, 91–96.

16. P. Erdős and J. Spencer, *Probabilistic methods in combinatorics*, Akademiai Kiado, Budapest, 1974, 106pp.

17. E. Fischer, *Cycle factors in dense graphs*, Discrete Mathematics **197/198** (1999), 309–323, 16th British Combinatorial Conference (London, 1997).

18. P. Frankl and V. Rödl, *Extremal problems on set systems*, Random Structures and Algorithms, to appear.

19. _____, *The uniformity lemma for hypergraphs*, Graphs and Combinatorics **8** (1992), no. 4, 309–312.

20. P. Frankl, V. Rödl, and R. M. Wilson, *The number of submatrices of a given type in a Hadamard matrix and related results*, Journal of Combinatorial Theory, Series B **44** (1988), no. 3, 317–328.

21. A. Frieze and R. Kannan, *The regularity lemma and approximation schemes for dense problems*, 37th Annual Symposium on Foundations of Computer Science (Burlington, VT, 1996), IEEE Comput. Soc. Press, Los Alamitos, CA, 1996, pp. 12–20.

22. _____, *Quick approximation to matrices and applications*, Combinatorica **19** (1999), no. 2, 175–220.

23. _____, *A simple algorithm for constructing Szemerédi's regularity partition*, Electronic Journal of Combinatorics **6** (1999), no. 1, Research Paper 17, 7 pp. (electronic).

24. O. Goldreich, *Combinatorial property testing (a survey)*, Randomization methods in algorithm design (Princeton, NJ, 1997), Amer. Math. Soc., Providence, RI, 1999, pp. 45–59.

25. O. Goldreich, S. Goldwasser, and D. Ron, *Property testing and its connection to learning and approximation*, 37th Annual Symposium on Foundations of Computer Science (Burlington, VT, 1996), IEEE Comput. Soc. Press, Los Alamitos, CA, 1996, pp. 339–348.

26. _____, *Property testing and its connection to learning and approximation*, Journal of the Association for Computing Machinery **45** (1998), no. 4, 653–750.

27. O. Goldreich and D. Ron, *Property testing in bounded degree graphs*, 29th ACM Symposium on Theory of Computing (El Paso, Texas), 1997, pp. 406–419.

28. _____, *A sublinear bipartiteness tester for bounded degree graphs*, Combinatorica **19** (1999), no. 3, 335–373.

29. G. H. Golub and C. F. van Loan, *Matrix computations*, Johns Hopkins University Press, London, 1989.

30. W. T. Gowers, *Lower bounds of tower type for Szemerédi's uniformity lemma*, Geometric and Functional Analysis **7** (1997), no. 2, 322–337.

31. P. E. Haxell, Y. Kohayakawa, and T. Łuczak, *The induced size-Ramsey number of cycles*, Combinatorics, Probability, and Computing **4** (1995), no. 3, 217–239.

32. P. E. Haxell and V. Rödl, *Integer and fractional packings in dense graphs*, submitted, 1999.

33. Y. Kohayakawa, *Szemerédi's regularity lemma for sparse graphs*, Foundations of Computational Mathematics (Berlin, Heidelberg) (F. Cucker and M. Shub, eds.), Springer-Verlag, January 1997, pp. 216–230.

34. Y. Kohayakawa, V. Rödl, and L. Thoma, *An optimal deterministic algorithm for Szemerédi's regularity lemma*, submitted, 2000.

35. J. Komlós and M. Simonovits, *Szemerédi's regularity lemma and its applications in graph theory*, Combinatorics—Paul Erdős is eighty, vol. 2 (Keszthely, 1993) (D. Miklós, V. T. Sós, and T. Szőnyi, eds.), Bolyai Society Mathematical Studies, vol. 2, János Bolyai Mathematical Society, Budapest, 1996, pp. 295–352.

36. J. Komlós, *The blow-up lemma*, Combinatorics, Probability and Computing **8** (1999), no. 1-2, 161–176, Recent trends in combinatorics (Mátraháza, 1995).

37. J. Komlós, G. N. Sárközy, and E. Szemerédi, *Blow-up lemma*, Combinatorica **17** (1997), no. 1, 109–123.

38. _____, *An algorithmic version of the blow-up lemma*, Random Structures and Algorithms **12** (1998), no. 3, 297–312.

39. A. Lubotzky, R. Phillips, and P. Sarnak, *Ramanujan graphs*, Combinatorica **8** (1988), 261–277.

40. A. Lubotzky, *Discrete groups, expanding graphs and invariant measures*, Birkhäuser Verlag, Basel, 1994, with an appendix by Jonathan D. Rogawski.

41. G. A. Margulis, *Explicit group-theoretic constructions of combinatorial schemes and their applications in the construction of expanders and concentrators*, Problemy Peredachi Informatsii **24** (1988), no. 1, 51–60.

42. H. J. Prömel and A. Steger, *Excluding induced subgraphs III. A general asymptotic*, Random Structures and Algorithms **3** (1992), no. 1, 19–31.

43. V. Rödl, A. Ruciński, and M. Wagner, *An algorithmic embedding of graphs via perfect matchings*, Randomization and approximation techniques in computer science, Lecture Notes in Computer Science, vol. 1518, 1998, pp. 25–34.

44. V. Rödl and A. Ruciński, *Perfect matchings in ε-regular graphs and the blow-up lemma*, Combinatorica **19** (1999), no. 3, 437–452.

45. R. Rubinfeld and M. Sudan, *Robust characterizations of polynomials with applications to program testing*, SIAM Journal on Computing **25** (1996), no. 2, 252–271.

46. P. Sarnak, *Some applications of modular forms*, Cambridge University Press, Cambridge, 1990.

47. E. Szemerédi, *On sets of integers containing no k elements in arithmetic progression*, Acta Arithmetica **27** (1975), 199–245, collection of articles in memory of Juriĭ Vladimirovič Linnik.

48. _____, *Regular partitions of graphs*, Problèmes Combinatoires et Théorie des Graphes (Colloq. Internat. CNRS, Univ. Orsay, Orsay, 1976) (Paris), Colloques Internationaux CNRS n. 260, 1978, pp. 399–401.

49. A. R. Taraz, *Szemerédis Regularitätslemma*, April 1995, Diplomarbeit, Universität Bonn, 83pp.

50. A. G. Thomason, *Pseudorandom graphs*, Random graphs '85 (Poznań, 1985), North-Holland Math. Stud., vol. 144, North-Holland, Amsterdam–New York, 1987, pp. 307–331.

51. _____, *Random graphs, strongly regular graphs and pseudorandom graphs*, Surveys in Combinatorics 1987 (C. Whitehead, ed.), London Mathematical Society Lecture Note Series, vol. 123, Cambridge University Press, Cambridge–New York, 1987, pp. 173–195.

52. A. Tiskin, *Bulk-synchronous parallel multiplication of Boolean matrices*, Automata, languages and programming, Lecture Notes in Computer Science, vol. 1443, 1998, pp. 494–506.

Small Maximal Matchings in Random Graphs

Michele Zito[*]

Department of Computer Science, University of Liverpool, Liverpool L69 7ZF, UK

Abstract. We look at the minimal size of a maximal matching in general, bipartite and d-regular random graphs. We prove that the ratio between the sizes of any two maximal matchings approaches one in dense random graphs and random bipartite graphs. Weaker bounds hold for sparse random graphs and random d-regular graphs. We also describe an algorithm that with high probability finds a matching of size strictly less than $n/2$ in a cubic graph. The result is based on approximating the algorithm dynamics by a system of linear differential equations.

1   Introduction

A matching in a graph is a set of disjoint edges. Several optimisation problems are definable in terms of matchings. If G is a graph and M is a matching in G, we count the number of edges in M and the goal is to maximise this value, then the corresponding problem is that of finding a maximum cardinality matching in G. This problem has a glorious history and an important place among combinatorial problems [2,5,8]. However few other matching problems share its nice combinatorial properties. If $G = (V, E)$ is a graph, a matching $M \subseteq E$ is *maximal* if for every $e \in E \setminus M$, $M \cup e$ is not a matching; $V(M) = \{v : \exists u \{u, v\} \in M\}$. Let $\beta(G)$ denote the minimum cardinality of a maximal matching in G. The minimum maximal matching problem is that of finding a maximal matching in G with $\beta(G)$ edges. The problem is NP-hard [10]. The size of any maximal matching is at most $2\beta(G)$ [6] in general graphs and at most $\left(2 - \frac{1}{d}\right)\beta(G)$ [11] in regular graphs of degree d. Some negative results are known about the approximability of $\beta(G)$ [11].

In this paper we abandon the pessimistic point of view of worst-case algorithmic analysis by assuming that each input graph G occurs with a given probability. Nothing seems to be known about the most likely value of $\beta(G)$ or the effectiveness of any approximation heuristics in this setting. In Section 2 we prove that the most likely value of $\beta(G)$ can be estimated quite precisely, for instance, if G is chosen at random among all graphs with a given number of vertices. Similar results are proved in Section 3 for dense random bipartite graphs. Also, simple algorithms exist which, *with high probability* (w.h.p.), that is with probability approaching one as $n = |V(G)|$ tends to infinity, return matchings of size $\beta(G) + o(n)$. Lower bounds on $\beta(G)$, improving the ones presented above, are proved also in the case when higher probability is given to graphs with few edges. Most of the bounds on $\beta(G)$ are obtained by exploiting a simple relation between maximal matchings and independent sets. In Section 4 we investigate the possibility of applying a similar reasoning if G is a random d-regular graph. After showing a number

---

[*] Supported by EPSRC grant GR/L/77089.

G. Gonnet, D. Panario, and A. Viola (Eds.): LATIN 2000, LNCS 1776, pp. 18–27, 2000.

of lower bounds on $\beta(G)$ for several values of d, we present an algorithm that finds a maximal matching in a d-regular graph. We prove that with high probability it returns a matching of size asymptotically less than $n/2$ if G is a random cubic graph.

In what follows $\mathcal{G}(n, p)$ ($\mathcal{G}(K_{n,n}, p)$) denotes the usual model of random (bipartite) graphs as defined in [1]. Also $\mathcal{G}(n, d\text{-reg})$ denotes the following model for random d-regular graphs [9, Section 4]. Let n urns be given, each containing d balls (with dn even): a set of $dn/2$ pairs of balls (called a *configuration*) is chosen at random among those containing neither pairs with two balls from the same urn nor couples of pairs with balls coming from just two urns. To get a random $G \in \mathcal{G}(n, d\text{-reg})$ let $\{i, j\} \in E(G)$ if and only if there is a pair with one ball belonging to urn i and the other belonging to urn j. If $\mathcal{G}$ is a random graph model, $G \in \mathcal{G}$ means that G is selected with a probability defined by $\mathcal{G}$. The random variable $X = X_k(G)$ counts the number of maximal matchings of size k in G. The meaning of the sentences "*almost always* (a.a.)", "for *almost every* (a.e.) graph" is defined in [1, Ch. II].

2 General Random Graphs

Let $q = 1 - p$. If U is a random indicator $\Pr[U]$ will denote $\Pr[U = 1]$.

Theorem 1. *If $G \in \mathcal{G}(n, p)$ then $\mathrm{E}(X) = \binom{n}{2k} \frac{(2k)!}{k!} \left(\frac{p}{2}\right)^k q^{\binom{n-2k}{2}}$.*

Proof. Let M_i be a set of k independent edges, assume that G is a random graph sampled according to the model $\mathcal{G}(n, p)$ and let $X_{p,k}^i$ be the random indicator equal to one if M_i is a maximal matching in G. $\mathrm{E}(X_{p,k}^i) = \Pr[X_{p,k}^i] = p^k q^{\binom{n-2k}{2}}$. Then by linearity of expectation

$$\mathrm{E}(X) = \sum\nolimits_{|M_i|=k} \mathrm{E}(X_{p,k}^i) = |\{M_i : |M_i| = k\}| \cdot p^k q^{\binom{n-2k}{2}}$$

The number of matchings of size k is equal to the possible ways of choosing $2k$ vertices out of n times the number of ways of connecting them by k independent edges divided by the number of orderings of these chosen edges. □

A lower bound on $\beta(G)$ is obtained by bounding $\mathrm{E}(X)$ and then using the Markov inequality to prove that $\Pr[X > 0]$ approaches zero as the number of vertices in the graph becomes large. Assuming $2k = n - 2\omega$

$$\mathrm{E}(X) \leq \frac{n^{\frac{n}{2}-\omega}}{(2\omega)!} \left(\frac{p}{2}\right)^{\frac{n}{2}-\omega} q^{2\omega^2 - \omega} \leq \left(\frac{pn}{2}\right)^{\frac{n}{2}} \left(\frac{e}{npq\omega}\right)^{\omega} q^{2\omega^2}$$

and this goes to zero only if $\omega = \Omega(\sqrt{n})$. However a different argument gives a considerably better result.

Theorem 2. $\beta(G) > \frac{n}{2} - \frac{\log n}{\log(1/q)}$ *for a.e. $G \in \mathcal{G}(n, p)$ with p constant.*

Proof. If M is a maximal matching in G then $V \setminus V(M)$ is an independent set. Let $Z = Z_{p,2\omega}$ be the random variable counting independent sets of size $2\omega = \frac{2\log n}{\log(1/q)}$ in a random graph G. If X counts maximal matchings of size $k = \frac{n}{2} - \omega$,

$$\Pr[X > 0] = \Pr[X > 0 \mid Z > 0]\Pr[Z > 0] + \Pr[X > 0 \mid Z = 0]\Pr[Z = 0]$$
$$\leq \Pr[X > 0 \mid Z > 0]\Pr[Z > 0] + 0 \cdot 1 \leq \Pr[Z > 0] \to 0$$

The last result follows from a theorem in [4] on the independence number of dense random graphs. Thus $\beta(G) > \frac{n}{2} - \frac{\log n}{\log 1/q}$ for a.e. $G \in \mathcal{G}(n,p)$. □

The argument before Theorem 2 is weak because even if $E(Z_{p,2\omega})$ is small $E(X)$ might be very large. The random graph G might have very few independent sets of size 2ω but many maximal matchings of size $\frac{n}{2} - \omega$.

Results in [4] also have algorithmic consequences. Grimmett and McDiarmid considered the simple greedy heuristic which repeatedly places a vertex v in the independent set I if there is no $u \in I$ with $\{u, v\} \in E(G)$ and removes it from G. It is easily proved that $|I| \sim \frac{\log n}{\log(1/q)}$.

Theorem 3. $\beta(G) < \frac{n}{2} - \frac{\log n}{2\log(1/q)}$ for a.e. $G \in \mathcal{G}(n,p)$ with p constant.

Proof. Let $\mathcal{IS}$ be an algorithm that first finds a maximal independent set I in G using the algorithm above and then looks for a perfect matching in the remaining graph. With probability approaching one $|I| \geq (1 - \delta)\frac{\log n}{\log(1/q)}$ for all $\delta > 0$. Also, $\mathcal{IS}$ does not expose any edge in $G - I$. Hence $G - I$ is a completely random graph on about $n - |I|$ vertices, each edge in it being chosen with constant probability p. Results in [3] imply that a.a. such graphs contain a matching with at most one unmatched vertex. □

Independent sets are useful also for sparse graphs. If $p = \frac{c}{n}$ a lower bound on $\beta(G)$ can be obtained again by studying $\alpha(G)$, the size of a largest independent set of vertices in G.

Theorem 4. $\beta(G) > \frac{n}{2} - \frac{n \log c}{c}$ for a.e. $G \in \mathcal{G}(n, c/n)$, with $c > 2.27$.

Proof. $\alpha(G) < \frac{2n \log c}{c}$ for a.e. $G \in \mathcal{G}_{n,c/n}$ for $c > 2.27$ [1, Theorem XI.22]. The result follows by an argument similar to that of Theorem 2. □

If $p = \frac{c}{n}$ for c sufficiently small, the exact expression for $E(X)$ in Theorem 1 gives an improved lower bound on $\beta(G)$. Roughly, if c is sufficiently small and U is a large independent set in G then the graph induced by $V \setminus U$ very rarely contains a perfect matching.

Theorem 5. $\beta(G) > \frac{n}{3}$ for a.e. $G \in \mathcal{G}_{n,c/n}$, with $c \in (2.27, 16.99]$

Proof. Let $k = \frac{n}{2} - \frac{dn}{c}$. If $d \in \left(\frac{c}{6}, \frac{c}{2}\right)$ then $k < n/3$. Hence $\frac{n!}{(n-2k)! \, k!} \leq n^{2k}/k!$ and

$$E(X) \leq O(1) \cdot \sqrt{\frac{c}{\pi(c-2d)n}} \left(\frac{c^2 e}{(c-2d)n}\right)^{\frac{n}{2} - \frac{dn}{c}} e^{-\frac{2d^2 n}{c} + d}$$

which goes to zero for every d in the given range. The best choice of d is the smallest and the theorem follows by noticing that $\frac{\log c}{c} < \frac{1}{6}$ if $c > 16.9989$. □

3   Bipartite Graphs

The results in the last section can be extended to the case when $G \in \mathcal{G}(K_{n,n}, p)$. Again $\beta(G)$ is closely related to a graph parameter whose value, at least in dense random graphs, can be estimate rather well. Given a bipartite graph $G = (V_1, V_2, E)$ with $|V_1| = |V_2| = n$, a *split independent set* in G is a set of 2ω independent vertices S with $|S \cap V_i| = \omega$. Let $\sigma(G)$ be the size of a largest split independent set in G. If M is a maximal matching in a bipartite graph G then $V \setminus V(M)$ is a split independent set.

Theorem 6. *If $G \in \mathcal{G}(K_{n,n}, p)$ then*

1. $\mathrm{E}(X) = \binom{n}{k}^2 k! p^k q^{(n-k)^2}$.
2. *If $Z = Z_{p,n-k}$ is the random variable counting split independent sets of size $n - k$ and $Y = Y_{p,k}$ is the random variable counting perfect matchings in $H \in \mathcal{G}(K_{k,k}, p)$ then $\mathrm{E}(X) = \mathrm{E}(Z) \cdot \mathrm{E}(Y)$.*

Proof. Let M_i be a set of k independent edges and $G \in \mathcal{G}((K_{n,n}, p)$ and let $X^i_{p,k}$ be the random indicator equal to one if M_i is a maximal matching in G. $\mathrm{E}(X^i_{p,k}) = \Pr[X^i_{p,k}] = p^k q^{(n-k)^2}$. Then

$$\mathrm{E}(X) = \sum_{|M_i|=k} \mathrm{E}(X^i_{p,k}) = |\{M_i : |M_i| = k\}| \cdot p^k q^{(n-k)^2}$$

The number of matchings of size k is given by the possible ways of choosing k vertices out of n on each side times the number of permutations on k elements. $\square$

If p is constant, it is fairly easy to bound the first two moments of Z and get good estimates on the value of $\sigma(G)$.

Theorem 7. $\sigma(G) \sim \frac{4 \log n}{\log 1/q}$ *for a.e. $G \in \mathcal{G}(K_{n,n}, p)$ with p constant.*

Proof. The expected number of split independent sets of size 2ω is $\binom{n}{\omega}^2 q^{\omega^2}$. Hence, by the Markov inequality, and Stirling's approximation to the factorial $\Pr[Z > 0] < \left(\frac{n^\omega}{\omega!}\right)^2 q^{\omega^2}$ and the right side tends to zero as n grows if $2\omega = 2 \left\lceil \frac{2 \log n}{\log 1/q} \right\rceil$.

Let $2\omega = 2 \left\lfloor \frac{2(1-\epsilon) \log n}{\log 1/q} \right\rfloor$ for any $\epsilon > 0$. The event "$Z = 0$" is equivalent to "$\sigma(G) < 2\omega$" because if there is no split independent set of size 2ω then the largest of such sets can only have less than 2ω elements. By the Chebyshev inequality $\Pr[Z = 0] \leq \mathrm{Var}(Z)/\mathrm{E}(Z)^2$. Also $\mathrm{Var}(Z) = \mathrm{E}(Z^2) - \mathrm{E}(Z)^2$. There are $s_\omega = \binom{n}{\omega}^2$ ways of choosing ω vertices from two disjoint sets of n vertices. If Z^i is the random indicator set to one if S^i is a split independent set in G then $Z = \sum Z^i$ and $\mathrm{E}(Z^2) = \sum_{i,j} \Pr[Z^i \wedge Z^j] = \sum_{i,j} \Pr[Z^j] \sum_i \Pr[Z^i \mid Z^j]$ where the sums are over all $i, j \in \{1, \ldots, s_\omega\}$. Finally by symmetry $\Pr[Z^i \mid Z^j]$ does not actually depend on j but only on the amount of intersection between S^i and S^j. Thus, if $S^1 = \{1, \ldots, 2\omega\}$,

$$\mathrm{E}(Z^2) = \left(\sum_j \Pr[Z^j] \right) \left(\sum_i \Pr[Z^i \mid Z^1] \right) = \mathrm{E}(Z) \cdot \mathrm{E}(Z|Z^1).$$

Thus to prove that $\Pr[Z = 0]$ converges to zero it is enough to show that the ratio $\mathrm{E}(Z|Z^1)/\mathrm{E}(Z)$ converges to one. By definition of conditional expectation

$$\mathrm{E}(Z|Z^1) = \sum_{0 \leq l_1, l_2 \leq \omega} \binom{\omega}{l_1} \binom{\omega}{l_2} \binom{n-\omega}{\omega-l_1} \binom{n-\omega}{\omega-l_2} q^{\omega^2 - l_1 l_2}$$

Define T_{ij} (generic term in $\mathrm{E}(Z|Z^1)/\mathrm{E}(Z)$) by

$$T_{ij} \binom{n}{\omega}^2 = \binom{\omega}{i} \binom{\omega}{j} \binom{n-\omega}{\omega-i} \binom{n-\omega}{\omega-j} q^{-ij}$$

Tedious algebraic manipulations prove that $T_{00} \leq 1 - \frac{2\omega^2}{n-\omega+1} + \frac{\omega^3(2\omega-1)}{(n-\omega+1)^2}$, and, for sufficiently large n, $T_{ij} \leq \frac{\omega^2}{n-\omega+1}$ for $i+j = 1$, and $T_{ij} \leq T_{10}$ for all $i, j \in \{1, \ldots, \omega\}$. From these results it follows that

$$\Pr\left[\sigma < 2\left\lfloor \frac{2(1-\epsilon)\log n}{\log 1/q} \right\rfloor\right] \leq T_{00} + T_{10} + T_{01} + \omega^2 T_{10} - 1$$
$$\leq 1 - 2\omega^2/n + \frac{\omega^3(2\omega-1)}{(n-\omega+1)^2} + \frac{2\omega^2}{n} + \frac{\omega^4}{n} - 1$$
$$\leq \frac{\omega^3(2\omega-1)}{(n-\omega+1)^2} + \frac{\omega^4}{n}$$

$\square$

Theorem 8. $\beta(G) > n - \frac{2\log n}{\log 1/q}$ for a.e. $G \in \mathcal{G}(K_{n,n}, p)$ with p constant.

The similarities between the properties of independent sets in random graphs and those of split independent sets in random bipartite graphs have some algorithmic implications. A simple greedy heuristic almost always produces a solution whose cardinality can be predicted quite tightly. Let I be the independent set to be output. Consider the process that visits the vertices of a random bipartite graph $G(V_1, V_2, E)$ in some fixed order. If $V_i = \{v_1^i, \ldots, v_n^i\}$, then the algorithm will look at the pair (v_j^1, v_j^2) during step j. If $\{v_j^1, v_j^2\} \notin E$ and if there is no edge between v_j^i and any of the vertices which are already in I then v_j^1 and v_j^2 are inserted in I. Let $\sigma_g(G) = |I|$.

Theorem 9. $\sigma_g(G) \sim \frac{\log n}{\log 1/q}$ for a.e. $G \in \mathcal{G}(K_{n,n}, p)$ with p constant.

Proof. Suppose that $2(k-1)$ vertices are already in I. The algorithm above will add two vertices v_1 and v_2 as the kth pair if $\{v_1, v_2\} \notin E$ and there is no edge between either v_1 or v_2 and any of the vertices which are already in I. The two events are independent in the given model and their joint probability is $(1-p) \cdot (1-p)^{2(k-1)} = (1-p)^{2k-1}$. Let W_k (for $k \in \mathbb{N}^+$) be the random variable equal to the number of pairs considered before the kth pair is added to I. W_k has geometric distribution with parameter $P_k = (1-p)^{2k-1}$. Moreover the variables $W_1, W_2, \ldots$ are all independent. Let $Y_\omega = \sum_{k=1}^{\omega} W_k$. The event "$Y_\omega < n$" is implied by "$\sigma_g(G) > 2\omega$": if the split independent set returned by the greedy algorithm contains more than 2ω vertices that means that the algorithm finds ω independent pairs in strictly less than n trials. Also if $Y_\omega < n$ then certainly each of the W_k cannot be larger than n. Hence

$$\Pr[Y_\omega < n] \leq \Pr[\cap_{k=1}^{\omega}\{W_k \leq n\}] = \prod_{k=1}^{\omega}\{1 - [1 - (1-p)^{2k-1}]^n\}$$

Let $\omega = \left\lceil \frac{(1+\epsilon)\log n}{2\log 1/q} \right\rceil$ and, given $\epsilon > 0$ and $r \in \mathbb{N}$, choose $m > r/\epsilon$. For sufficiently large n, $\omega - m > 0$. Hence

$$\Pr[Y_\omega < n] \leq \prod_{k=\omega-m}^{\omega}\{1 - [1 - (1-p)^{2k+1}]^n\}$$

that is at most $\{1 - [1 - (1-p)^{2(\omega-m)+1}]^n\}^m$. Since $(1-x)^n \geq 1 - nx$, we also have $\Pr[Y_\omega < n] \leq \{n(1-p)^{2(\omega-m)+1}\}^m = o(n^{-r})$. The event "$Y_\omega > n$" is equivalent to "$\sigma_g(G) < 2\omega$". Let $\omega = \left\lfloor \frac{(1-\epsilon)\log n}{2\log 1/q} \right\rfloor$. If $Y_\omega > n$ then there must be at least one k for which $W_k > n/\omega$. Hence $\Pr[Y_\omega > n] \leq \Pr[\cup_{k=1}^{\omega}\{W_k > n/\omega\}]$ and this is at most

$$\sum_{k=1}^{\omega} \Pr[W_k > n/\omega] \leq \omega[1 - (1-p)^{2\omega-1}]^{\lfloor n/\omega \rfloor}.$$

By the choice of ω, $(1-p)^{2\omega-1} > \frac{n^{-(1-\epsilon)}}{1-p}$. Hence

$$\Pr[Y_\omega > n] \leq \omega \left[1 - \frac{n^{-(1-\epsilon)}}{1-p}\right]^{\lfloor n/\omega \rfloor} \leq \omega \, \exp\left\{-\frac{n^{-(1-\epsilon)}}{1-p}\left\lfloor \frac{n}{\omega}\right\rfloor\right\}$$

Finally $\Pr[Y_\omega > n] \leq \omega \exp\left\{-\frac{n^\epsilon}{(1-p)\omega} - o(1)\right\}$ since $\lfloor n/\omega \rfloor > n/\omega - 1$, and the result follows from the choice of ω.   □

The greedy algorithm analysed in Theorem 9 does not expose any edge between two vertices that are not selected to be in I. Therefore $G - I$ is a random graph. Classical results ensure the existence of a perfect matching in $G - I$, and polynomial time algorithms exist which find one such a matching. We have proved the following.

Theorem 10. $\beta(G) < n - \frac{\log n}{2\log 1/q}$ for a.e. $G \in \mathcal{G}(K_{n,n}, p)$ with p constant.

4   Regular Graphs

In this section we look at the size of the smallest maximal matchings in random regular graphs. Again known upper bounds on the independence number of such graphs imply, in nearly all interesting cases, good lower bounds on $\beta(G)$.

Theorem 11. For each $d \geq 3$ there exists a constant $\gamma(d)$ such that $\beta(G) \geq \gamma(d)n$ for a.e. $G \in \mathcal{G}(n, d\text{-reg})$.

Proof. It is convenient to use the configuration model described in the introduction. Two pairs of balls in a configuration are *independent* if each ball is chosen from a distinct urn. A matching in a configuration is a set of independent pairs. The expected number of maximal matchings of size k in a random configuration is

$$\frac{n! \, d^{2k}}{k! \, (n-2k)!} \, \frac{[2k(d-1)]!}{[k(2d-1)-nd/2]!} \, \frac{(dn/2)!}{(dn)!} \, 2^{d(n-2k)}$$

If $k = \gamma n$, using Stirling's approximation to the factorial, this is at most

$$f(\gamma,d)^n = \left\{\left(\frac{1}{1-2\gamma}\right)\left[\frac{d(1-2\gamma)^2}{\gamma}\right]^\gamma \frac{[\gamma(d-1)]^{2\gamma(d-1)}}{[\gamma(2d-1)-d/2]^{\gamma(2d-1)}}\left[\frac{\gamma(2d-1)-d/2}{d}\right]^{\frac{d}{2}} 2^{\frac{d}{2}-2\gamma}\right\}^n$$

For every d there exists a unique $\gamma_1(d) \in \left(\frac{d}{2(2d-1)}, \frac{1}{2}\right)$ for which $f(\gamma,d) \geq 1$, for $\gamma \in (\gamma_1(d), 0.5)$. Since the probability that a random configuration corresponds to a d-regular graph is bounded (see for example [1, Chap 2]), the probability that a random d-regular graph has a maximal matching of size γn is at most $f(\gamma,d)^n$. If $d > 6$ a better bound is obtained by using $\gamma(d) = (1 - \alpha_3(d))/2$ where $\alpha_3(d)$ is the smallest value in $(0, 1/2)$ such that $\alpha(G) < \alpha_3(d)n$ for a.a. $G \in \mathcal{G}(n, d\text{-reg})$ [7].   □

The relationship between independent sets and maximal matchings can be further exploited also in the case where $G \in \mathcal{G}(n, d\text{-reg})$, but random regular graphs are rather sparse graphs and the approach used in the previous sections cannot be easily applied in this context. However, a simple greedy algorithm which finds a large independent in a d-regular graph can be modified and incorporated in a longer procedure that finds a small maximal matching in a random regular graph. Consider the following algorithm $\mathcal{A}$.

Input: Random d-regular graph with n vertices
(1) $M \leftarrow \emptyset$;
(2)   **while** there is a vertex of degree d **do**
                choose v in V u.a.r. among the vertices of degree d;
                $M \leftarrow M \cup \{v, u_1\}$; /* Assume $N(v) = \{u_1, \ldots, u_{\deg_G v}\}$ */
                $V \leftarrow V \setminus \{v\}$;
(3)   **for** $j = 1$ **to** $d - 1$ **do**
                choose v u.a.r. among the vertices of degree $d - j$ in $V(M)$;
                $V \leftarrow V \setminus \{v\}$;
(4)   find a maximal matching M' in what is left of G;
(5)   make $M \cup M'$ into a maximal matching for G.

Step (2) essentially mimics one of the algorithms presented in [9], with the only difference that instead of selecting an independent set of vertices, the process selects a set of edges. Step (4) can be clearly performed in polynomial time. In general the set $M \cup M'$ is an *edge dominating set* (each edge in G is adjacent to some edge in $M \cup M'$) but it is not necessarily a matching. However [10] any edge dominating set F can be transformed in polynomial time into a maximal matching M of G with $|M| \leq |F|$. Let $D_i = \{v : \deg_G v = i\}$. In the remaining part of this section we will analyse the evolution of $|D_i|$ for $0 \leq i \leq d$, as the algorithm goes through step (2) and (3). Step (3) is performed in a number of iterations. For $j \geq 0$, let $V_i^j(t)$ be the size of D_i at stage t of iteration j, with the convention that iteration 0 refers to the execution of step (2).

Step (2) for d-regular graphs. Theorem 4 in [9] implies that step (2) proceeds for asymptotically $x_1 = \frac{1}{2} - \frac{1}{2}\left(\frac{1}{d-1}\right)^{\frac{2}{d-2}}$ stages, adding an edge to M at every stage. Let $V_i^{j+}(t) = |D_i \cap V(M)|$ at stage t of iteration j and set $V_i^{j-}(t) = V_i^j(t) - V_i^{j+}(t)$. Let $\Delta V_i^{0 \text{ sign}}(t)$ denote the expected change of $V_i^{0 \text{ sign}}(t)$ (with sign $\in \{\text{""}, \text{"+"}, \text{"−"}\}$) moving from stage t to $t + 1$, of step (2), conditioned to the history of the algorithm's execution up to stage t. Let v be the chosen vertex of degree d. We assume a given fixed ordering among the vertices adjacent to v. The edge $\{v, u_1\}$ is added to M and edges $\{v, u_l\}$ (for $l = 2, \ldots, \deg_G v$) are removed from G. Vertex v becomes of degree zero and the expected reduction in the number of vertices of degree i that are (not) in $V(M)$ is $\frac{iV_i^{0+}(t)}{n-2t}$ (resp. $\frac{iV_i^{0-}(t)}{n-2t}$), that is the probability that a vertex in $D_i \cap V(M)$ (resp. $D_i \cap (V \setminus V(M))$) is hit over d trials. The "loss" of a vertex of degree i implies the "gain" of a vertex of degree $i - 1$. Moreover if $u_1 \in D_{i+1} \cap (V \setminus V(M))$ at stage t, then $u_1 \in D_i \cap V(M)$ at stage $t + 1$. Let $\delta_{r,s} = 1$ if $r = s$ and zero otherwise. In what follows $i \in \{1, \ldots, d-1\}$. Also $V_d^{0-}(t) = V_d^0(t)$. We have

$$\Delta V_d^0(t) = -1 - \frac{dV_d^0(t)}{n-2t}$$
$$\Delta V_i^{0+}(t) = -\frac{iV_i^{0+}(t)}{n-2t} + (1 - \delta_{d-1,i})\frac{(i+1)V_{i+1}^{0+}(t)}{n-2t} + \frac{(i+1)V_{i+1}^{0-}(t)}{nd-2dt}$$
$$\Delta V_i^{0-}(t) = -\frac{iV_i^{0-}(t)}{n-2t} + \frac{(i+1)(d-1)V_{i+1}^{0-}(t)}{nd-2dt}$$
$$\Delta V_0^0(t) = 1 + \frac{V_1^{0-}(t)}{n-2t} + \frac{V_1^{0+}(t)}{n-2t}$$

Setting $x = t/n$, $V_i^{1\ \mathrm{sign}}(t) = nv_i^{1\ \mathrm{sign}}(t/n)$, we can consider the following system of differential equations:

$$v_d^{0'}(x) = -1 - \frac{dv_d^0(x)}{1-2x} \qquad v_d^0(0) = 1$$

$$v_i^{0+'}(x) = -\frac{iv_i^{0+}(x)}{1-2x} + (1-\delta_{d-1,i})\frac{(i+1)v_{i+1}^{0+}(x)}{1-2x} + \frac{(i+1)v_{i+1}^{0-}(x)}{d(1-2x)} \qquad v_i^{0+}(0) = 0$$

$$v_i^{0-'}(x) = -\frac{iv_i^{0-}(x)}{1-2x} + \frac{(i+1)(d-1)v_{i+1}^{0-}(x)}{d(1-2x)} \qquad v_i^{0-}(0) = 0$$

$$v_0^{0'}(x) = 1 + \frac{v_1^{0-}(x)}{1-2x} + \frac{v_1^{0+}(x)}{1-2x} \qquad v_0^0(0) = 0$$

In each case $|V_i^{1\ \mathrm{sign}}(t+1) - V_i^{1\ \mathrm{sign}}(t)|$ is bounded by a constant. Also, the system of differential equations above is sufficiently well-behaved, so that the hypotheses of Theorem 1 in [9] are fulfilled and thus for large n, $V_i^{1\ \mathrm{sign}}(t) \sim nv_i^{1\ \mathrm{sign}}(t/n)$ where $v_i^{1\ \mathrm{sign}}(x)$ are the solutions of the system above.

Lemma 1. *For each $d \in \mathbb{N}^+$ and for each $i \in \{1,\ldots,d\}$, there is a number A_i, two sequences of real numbers $\{B_{i,0}^j\}_{j=1,\ldots,\lceil \frac{d-i+1}{2}\rceil}, \{B_{i,1}^j\}_{j=1,\ldots,\lceil \frac{d-i}{2}\rceil}$, and a number C_i such that the system of differential equation above admits the following unique solutions:*

$$v_i^{0-}(x) = A_i(1-2x) + (1-2x)^{\frac{i}{2}}\left[C_i\log(1-2x) + \sum_{j=0}^{\lceil \frac{d-i+1}{2}\rceil} B_{i,0}^j x^j\right]$$

$$\qquad + (1-2x)^{\frac{i+1}{2}}\sum_{j=0}^{\lceil \frac{d-i}{2}\rceil} B_{i,1}^j x^j$$

$$v_i^{0+}(x) = v_i^{0-}(x)\left[\left(\frac{d}{d-1}\right)^{d-i} - 1\right]$$

$$v_0^0(x) = f_0(x) - f_0(0)$$

where $f_0(x) = x + \left(\frac{d}{d-1}\right)^{d-1}\int \frac{v_1^{0-}(x)}{1-2x}\,\mathrm{d}x$.

Proof. We sketch the proof of the first two results (which can be formally carried out by induction on $d - i$). For $i = d$, $v_d^0(x) = v_d^{0-}(x) = -\frac{1}{d-2}(1-2x) + \frac{d-1}{d-2}(1-2x)^{d/2}$. Assuming the result holds for $v_{i+1}^{0-}(x)$ and letting $D_i = (i+1)\left(\frac{d-1}{d}\right)$, we have

$$v_i^{0-}(x) = D_i(1-2x)^{\frac{i}{2}}\int_0^x \frac{v_{i+1}^{0-}(s)}{(1-2s)^{\frac{i}{2}+1}}\,\mathrm{d}s$$

and the result follows by integration (in particular, the logarithmic terms are present only if $i \le 2$).

Let $I_{i+1}^{0-}(x) = \int_0^x \frac{v_{i+1}^{0-}(s)}{(1-2s)^{\frac{i}{2}+1}}\,\mathrm{d}s$; then $v_i^{0-}(x) = D_i I_{i+1}^{0-}(x)$. Therefore

$$v_i^{0+}(x) = (i+1)\int_0^x \frac{v_{i+1}^{0+}(s)}{(1-2s)^{1+\frac{i}{2}}}\,\mathrm{d}s + \frac{i+1}{d}I_{i+1}^{0-}(x)$$

$$= (i+1)\int_0^x \frac{v_{i+1}^{0+}(s)}{(1-2s)^{1+\frac{i}{2}}}\,\mathrm{d}s + \frac{v_i^{0-}(x)}{d-1}$$

$$= (i+1)\left[\left(\frac{d}{d-1}\right)^{d-i-1} - 1\right]\int_0^x \frac{v_{i+1}^{0-}(s)}{(1-2s)^{1+\frac{i}{2}}}\,\mathrm{d}s + \frac{v_i^{0-}(x)}{d-1}$$

$$= (i+1)\left[\left(\frac{d}{d-1}\right)^{d-i-1} - 1\right]\frac{dv_i^{0-}(x)}{(i+1)(d-1)} + \frac{v_i^{0-}(x)}{d-1}$$

$$= \left[\left(\frac{d}{d-1}\right)^{d-i-1} - 1\right]\frac{dv_i^{0-}(x)}{d-1} + \frac{v_i^{0-}(x)}{d-1}$$

The third result follows by replacing the expression for v_i^{0+} in

$$v_0^{0'}(x) = 1 + \frac{v_1^{0-}(x)}{1-2x} + \frac{v_1^{0+}(x)}{1-2x}$$

□

Lemma 2. *Let x_1 be the smallest root of $v_d^0(x) = 0$. After Step (2) is completed the size of M is asymptotically $x_1 n$ for a.e. $G \in \mathcal{G}(n, d\text{-reg})$.*   □

Step (3.j) for cubic graphs. During this step the algorithm chooses a random vertex in $D_{3-j} \cap V(M)$ and removes it from G (all edges incident to it will not be added to M). Let $c_j(3-j)n/2$ be the number of edges left at the beginning of iteration j. If iteration $j-1$ ended at stage $x_j n$, the parameter c_j satisfies the recurrence:

$$c_j = \frac{c_{j-1}(4-j)}{3-j} - \frac{2(4-j)x_j}{3-j} = \left(1 + \frac{1}{3-j}\right)(c_{j-1} - 2x_j)$$

(with $c_0 = 1$) where x_1 has been defined above and x_2 and x_3 will be defined later. For all $i \in \{1, 2, 3\}$ the expected decrease in the number of vertices of degree i in $V(M)$ (resp. not in $V(M)$) is $\frac{iV_i^{j+}(t)}{c_j n - 2t}$ ($\frac{iV_i^{j-}(t)}{c_j n - 2t}$). The following set of equations describes the expected change in the various $V_i^{j \ \text{sign}}(t)$. In what follows $i \in \{1, \ldots, 3-j\}$. Notice that $V_i^{j+}(t) = 0$ for all $i > 3-j$ during iteration j so there are only $3-j$ equations involving $V_i^{j+}(t)$ but there are always two involving $V_i^{j-}(t)$.

$$\Delta V_0^j(t) = 1 + \frac{V_1^{j-}(t)}{c_j n - 2t} + \frac{V_1^{j+}(t)}{c_j n - 2t}$$

$$\Delta V_i^{j+}(t) = -\delta_{3-j,i} - \frac{iV_i^{j+}(t)}{c_j n - 2t} + \frac{(i+1)V_{i+1}^{j+}(t)}{c_j n - 2t}$$

$$\Delta V_i^{j-}(t) = -\frac{iV_i^{j-}(t)}{c_j n - 2t} + (1 - \delta_{2,i})\frac{(i+1)V_{i+1}^{j-}(t)}{c_j n - 2t}$$

Leading to the d.e.'s

$$v_0^{j'}(x) = 1 + \frac{v_1^{j-}(x)}{c_j - 2x} + \frac{v_1^{j+}(x)}{c_j - 2x} \qquad v_0^j(0) = 0$$

$$v_i^{j+'}(x) = -\delta_{3-j,i} - \frac{iv_i^{j+}(x)}{c_j - 2x} + \frac{(i+1)v_{i+1}^{j+}(x)}{c_j - 2x} \qquad v_i^{j+}(0) = v_i^{(j-1)+}(x_j)$$

$$v_i^{j-'}(x) = -\frac{iv_i^{j-}(x)}{c_j - 2x} + (1 - \delta_{2,i})\frac{(i+1)v_{i+1}^{j-}(x)}{c_j - 2x} \qquad v_i^{j-}(0) = v_i^{(j-1)-}(x_j)$$

Theorem 12. *Let x_j be the smallest positive root of $v_{4-j}^{(j-1)+}(x) = 0$, for $j \in \{1, 2, 3\}$. For a.e. $G \in \mathcal{G}(n, 3\text{-reg})$ algorithm $\mathcal{A}$ returns a maximal matching of size at most*

$$\beta_u(G) \sim n\left(x_1 + \frac{v_1^{2-}(x_3) + v_2^{2-}(x_3)}{2}\right)$$

Proof. The result follows again by applying Theorem 1 in [9] to the random variables $V_i^{1 \ \text{sign}}(t)$. Notice that all functions $v_i^{1 \ \text{sign}}(x)$ have a simple expression which can be derived by direct integration and, in particular,

$$x_2 = \frac{c_1}{2}\left[1 - \exp\left(-\frac{2v_2^0(x_1)}{c_1}\right)\right] \qquad x_3 = \frac{c_2}{2}\left[1 - 4\left(1 - \frac{v_1^{1+}(x_2)}{c_2}\right)^2\right]$$

□

5   Conclusions

In this paper we presented a number of results about the minimal size of a maximal matching in several types of random graphs. If the graph G is dense, with high probability $\beta(G)$ is concentrated around $|V(G)|/2$ (both in the general and bipartite case). Moreover simple algorithms return an asymptotically optimal matching. We also gave simple combinatorial lower bounds on $\beta(G)$ if $G \in \mathcal{G}(n, c/n)$. Finally we presented combinatorial bounds on $\beta(G)$ if $G \in \mathcal{G}(n, d\text{-reg})$ and an algorithm that finds a maximal matching of size asymptotically less than $|V(G)|/2$ in G. The complete analysis was presented for the case when $G \in \mathcal{G}(n, 3\text{-reg})$. In such case the bound in Theorem 11 and the algorithmic result in Theorem 12 imply that $0.3158n < \beta(G) < 0.47563n$. Results similar to Theorem 12 can be proved for random d-regular graphs, although some extra care is needed to keep track of the evolving degree sequence. Our algorithmic results exploit a relationship between independent sets and maximal matchings. In all cases the given minimisation problem is reduced to a maximisation one, and the analysis is completed by exploiting a number of techniques available to deal with the maximisation problem. The weakness of our results for sparse graphs and for regular graphs leaves the open problem of finding a more direct approach which might produce better results.

References

1. B. Bollobás. *Random Graphs*. Academic Press, 1985.
2. J. Edmonds. Paths, Trees and Flowers. *Canadian Journal of Math.*, 15:449–467, 1965.
3. P. Erdős and A. Rényi. On the Existence of a Factor of Degree One of a Connected Random Graph. *Acta Mathematica Academiae Scientiarum Hungaricae*, 17(3–4):359–368, 1966.
4. G. R. Grimmett and C. J. H. McDiarmid. On Colouring Random Graphs. *Mathematical Proceedings of the Cambridge Philosophical Society*, 77:313–324, 1975.
5. J. Hopcroft and R. Karp. An $n^{5/2}$ Algorithm for Maximal Matching in Bipartite Graphs. *SIAM Journal on Computing*, 2:225–231, 1973.
6. B. Korte and D. Hausmann. An Analysis of the Greedy Heuristic for Independence Systems. *Annals of Discrete Mathematics*, 2:65–74, 1978.
7. B. D. McKay. Independent Sets in Regular Graphs of High Girth. *Ars Combinatoria*, 23A:179–185, 1987.
8. S. Micali and V. V. Vazirani. An $O(v^{1/2}e)$ Algorithm for Finding Maximum Matching in General Graphs. In *Proceedings of the 21st Annual Symposium on Foundations of Computer Science*, pages 17–27, New York, 1980.
9. N. C. Wormald. Differential Equations for Random Processes and Random Graphs. *Annals of Applied Probability*, 5:1217–1235, 1995.
10. M. Yannakakis and F. Gavril. Edge Dominating Sets in Graphs. *SIAM Journal on Applied Mathematics*, 38(3):364–372, June 1980.
11. M. Zito. *Randomised Techniques in Combinatorial Algorithmics*. PhD thesis, Department of Computer Science, University of Warwick, 1999.

Some Remarks on Sparsely Connected Isomorphism-Free Labeled Graphs

Vlady Ravelomanana[1] and Loÿs Thimonier[1]

LaRIA
5, Rue du Moulin Neuf
80000 Amiens, France
{thimon,vlady}@laria.u-picardie.fr

Abstract. Given a set $\xi = \{H_1, H_2, \cdots\}$ of connected non-acyclic graphs, a ξ-free graph is one which does not contain any member of ξ as induced subgraph. Our first purpose in this paper is to perform an investigation into the limiting distribution of labeled graphs and multigraphs (graphs with possible self-loops and multiple edges), with n vertices and approximately $\frac{1}{2}n$ edges, in which all sparse connected components are ξ-free. Next, we prove that for any *finite* collection ξ of multicyclic graphs *almost all* connected graphs with n vertices and $n + o(n^1/3)$ edges are ξ-free. The same result holds for multigraphs.

1 Introduction

We consider here labeled *graphs*, i.e., graphs with labeled vertices, undirected edges and without self-loops or multiple edges as well as labeled *multigraphs* which are labeled graphs with self-loops and/or multiple edges. A (n, q) graph (resp. multigraph) is one having n vertices and q edges.

On one hand, classical papers, for e.g. [7], [8], [11] and [13], provide algorithms and analysis of algorithms that deal with random graphs or multigraphs generation, estimating relevant characteristics of their evolution. Starting with an initially empty graph of n vertices, we enrich it by successively adding edges. As random graph evolves, it displays a phase transition similar to the typical phenomena observed with percolation process. On the other hand, various authors such as *Wright* [19], [21] or *Bender, Canfield* and *McKay* [3], [4] studied exact enumeration or asymptotic properties of labeled connected graphs.

In recent years, a lot of research was performed for graphs without certain graphs as *induced subgraphs*. Let H be a connected graph and let F be a family of graphs none of which contains a subgraph isomorphic to H. In this case, we say that the family F is H-*free*. Mostly forbidden subgraphs are triangle, ..., C_n, K_n, $K_{p,q}$ graphs or any combination of them. We refer as bicyclic graphs all connected graphs with n vertices and $(n + 1)$ edges and in general $(q + 1)$-*cyclic* graphs are connected $(n, n + q)$ graphs. Also in this case, we say that it is a q-excess graph. In general, we refer as *multicyclic* a connected graph which is not acyclic. The same nomenclature holds for multigraphs. Denote by $\xi = \{H_1, H_2, H_3, ...\}$ a set of connected multicyclic graphs. A ξ-**free** graph is one which does not contain any member H_i of ξ as induced subgraph. Throughout this

G. Gonnet, D. Panario, and A. Viola (Eds.): LATIN 2000, LNCS 1776, pp. 28–37, 2000.

paper, each H_i is a connected *multicyclic* graph. Our goal in this paper is; to extend the study of random $(n, m(n))$ graphs to *random ξ-free $(n, m(n))$ graphs* when the number of edges, added one at time and at random, reach $m(n) \approx 1/2n$ and to compute the asymptotic number of ξ-free connected graphs when ξ is *finite*. To do this, we will rely strongly on the result of [21], in particular, we will investigate ξ-free connected $(n, n+k)$ graphs when $k = o(n^{1/3})$. Note that similar works can be done with multigraphs.

This paper is organized as follows. In Section 2, we recall some useful definitions of the stuff we will encounter throughout the rest of this document. In Section 3, we will work with the example of the enumeration of bicyclic graphs. The enumeration of these graphs was discovered, as far as we know, independently by *Bagaev* [1] and by *Wright* [19]. The purpose of this example is two-fold. First, it brings a simple new combinatorial point of view to the relationship between the generating functions of some *integer partitions*, on one hand, and *graphs* or *multigraphs*, on the other hand. Next, this example gives us ideas, regarding the *simplest complex components*, of what will happen if we force our graphs to contain some specific configurations (especially the form of the generating function). Section 4 is devoted to the computation of the probability of random graphs without isomorphs in the general case. In Section 5, we give asymptotic formula for the number of connected graph with n vertices, $n + k$ edges as $n \to \infty$ and $k \to \infty$ but $k = o(n^{1/3})$ and prove that *almost* $(n, n + o(n^{1/3}))$ connected graphs are ξ-free when ξ is *finite*.

2 Definitions

Powerful tools in all combinatorial approaches, *generating functions* will be used for our concern. If $F(z)$ is a power series, we write $[z^n] F(z)$ for the coefficient of z^n in $F(z)$. We say that $F(z)$ is the *exponential generating function* (EGF for brief) for a collection F of *labeled* objects if $n! [z^n] F(z)$ is the number of ways to attach objects in F that have n elements (see for instance [18] or [12]). The bivariate EGF for labeled rooted trees satisfies

$$T(w, z) = z \exp\left(T(w, z)\right) = \sum_{n>0} (wn)^{n-1} \frac{z^n}{n!} \,, \tag{1}$$

where the variable w is the variable for edges and z is the variable for vertices. Without ambiguity, one can also associate a given configuration of labeled graph or multigraph with its EGF. For instance, a triangle can be labeled in only one way. Thus,

$$C_3 \to C_3(w, z) = \frac{1}{3!} w^3 z^3 \,. \tag{2}$$

We will denote by W_k, resp. $\widehat{W}_k$, the EGF for labeled multicyclic connected multigraphs, resp. graphs, with k edges more than vertices. These EGF have been computed in [19] and in [13]. Furthermore, we will denote by $W_{k,H}$ and $\widehat{W}_{k,H}$ the EGF of multicyclic H-free multigraphs and graphs with k edges more than vertices. In these notations, the second indice corresponds to the forbidden configuration(s). Recall that a smooth graph or multigraph is one with all vertices of degree ≥ 2 (see [20]). Throughout the rest of this paper, the "*widehat*" notation will be used for EGF of graphs and "*underline*" notation corresponds to the *smoothness* of the species. For example, $\widehat{W_k}$, resp. $\underline{W_k}$ are EGF for respectively connected $(n, n + k)$ smooth graphs and smooth multigraphs.

3   The Link between the EGF of Bicyclic Graphs and Integer Partitions

After the different proofs for trees (see [14] and [9]), *Rényi* [16] found the formula to enumerate *unicyclic graphs* which can be expressed in terms of the generating function of rooted labeled trees

$$\widehat{V}(z) = \frac{1}{2} \ln \frac{1}{1 - T(z)} - \frac{T(z)}{2} - \frac{T(z)^2}{4}. \tag{3}$$

It may be noted that in some connected graphs, as well as multigraphs, the number of edges exceeding the number of vertices can be seen as useful enumerating parameter. The term *bicyclic* graphs, appeared first in the seminal paper of *Flajolet et al.* [11] followed few years later by the huge one of *Janson et al.* [13] and was concerned with all connected graphs with $(n + 1)$ edges and n vertices. *Wright* [19] found recurrent formula well adapted for formal calculation to compute the number of all connected graphs with k edges more than their proper number of vertices for *general k*. Our aim in this section is to show that the problem of the enumeration of *bicyclic graphs* can also be solved with techniques involving integer partitions.

There exist two types of graphs which are connected and have $(n + 1)$ edges as shown by the figures below.

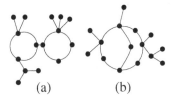

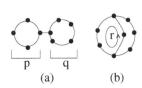

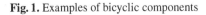

Fig. 1. Examples of bicyclic components

Fig. 2. Smooth bicyclic components without symmetry

Wright [19] showed with his *reduction* method that the EGF of all multicyclic graphs, namely bicyclic graphs, can be expressed in term of the EGF of labeled rooted trees. In order to count the number of ways to label a graph, we can repeatedly *prune* it by suppressing recursively any vertex of degree 1. We then remove as many vertices as edges. As these structures present many symmetries, our experiences suggest so far that we ought to look at our previously described object without symmetry and without the possible rooted subtrees.

There are $\binom{n}{p}\binom{n-p}{q}\frac{(p-1)!}{2}p\frac{(q-1)!}{2}q(n-p-q)! = \frac{n!}{4}$ ways to label the graph represented by the figure 2a $(p \neq q)$ and $2\binom{n}{r}\frac{(r-1)!}{2}r\frac{(n-r)!}{2} = \frac{n!}{2}$ ways for the graph of the figure 2b. Note that the results are independent from the size of the subcycles. One can obtain all smooth bicyclic graphs after considering possible symmetry criterions. In 2a, if the subcycles have the same length, $p = q$, a factor $1/2$ must be considered and we have $n!/8$ ways to label the graph. Similarly, the graph of 2b can have the 3 *arcs* with the same number of vertices. In this case, a factor $1/6$ is introduced. If only two arcs have the same number of vertices, we need a symmetrical factor $1/2$. Thus, the enumeration of smooth bicyclic graphs can be viewed as specific problem of integer partitioning into 2 or 3 parts following the dictates of the basic graphs of the figure 3.

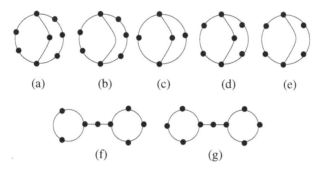

(a)   (b)   (c)   (d)   (e)

(f)   (g)

Fig. 3. The different basic smooth bicyclic graphs

With the same notations as in [6], denote by $P_i(t)$, respectively $Q_i(t)$, the generating functions of the number of partitions of an integer in i parts, respectively in i different parts. Let $\widehat{W_1}(z)$ be the univariate EGF for smooth bicyclic graphs, then we have $\widehat{W_1}(t) = f(P_2(t), P_3(t), Q_2(t), Q_3(t))$. A bit of algebra leads to

$$\widehat{W_1}(z) = \frac{z^4}{24}\frac{(6-z)}{(1-z)^3} . \tag{4}$$

In this formula, the denominator $\frac{1}{(1-z)^3}$ denotes the fact that there is at most 3 arcs or 3 *degrees of liberty* of integer partitions of the vertices in a bicyclic graph. The same remark holds for the denominators $\frac{1}{(1-T(z))^{3k}}$ in Wright's formulae [19], for all $(k+1)$-cyclic connected labeled graphs. The EGF of labeled rooted trees, $T(z)$, is introduced here when *re-expanding* the reduced vertices of some smooth graph. The main consequence of the relation between integer partitions and these EGF is that in any bicyclic graphs containing an induced q-gon as subgraph, the EGF is of the form $\frac{\text{Polynomial in } T(z)}{(1-T(z))^2}$. The form of these EGF is important for the study of the asymptotic behaviour of random graphs or multigraphs. The key point of the study of their characteristics is the analytical properties of *tree polynomial* $t_n(y)$ defined as follow

$$\frac{1}{(1-T(z))} = \sum_{n \geq 0} t_n(y)\frac{z^n}{n!} , \tag{5}$$

where $t_n(y)$ is a polynomial of degree n in y. *Knuth* and *Pittel* [10] studied their properties. For fixed y and $n \to \infty$, we have

$$t_n(y) = \frac{\sqrt{2\pi}n^{(n-1/2+y/2)}}{2^{y/2}\Gamma(y/2)} + O(n^{n-1+y/2}) . \tag{6}$$

This equation tells us that in the EGF of bicyclic graphs

$$\widehat{W_1}(z) = \frac{T(z)^4}{24} \frac{(6 - T(z))}{(1 - T(z))^3} = \frac{5}{24} \frac{1}{(1 - T(z))^3} - \frac{19}{24} \frac{1}{(1 - T(z))^2} + \ldots \tag{7}$$

only the coefficient $\frac{5}{24}$ of $t_n(3)$ is asymptotically significant. Thus in [13, Theorem 5], the authors proved that only leading coefficients of $t_n(3k)$ are used to compute the probability of random graphs or multigraphs. As already said, these coefficients change only slightly in the study of random graphs or multigraphs without forbidden configurations. Denote respectively by V_{C_3} and $\widehat{V_{C_3}}$ the EGF for acyclic multigraphs and graphs without triangle (C_3), we have

$$V_{C_3}(z) = \frac{1}{2} \ln \frac{1}{1 - T(z)} - \frac{T(z)^3}{6} , \tag{8}$$

and

$$\widehat{V_{C_3}}(z) = \frac{1}{2} \ln \frac{1}{1 - T(z)} - \frac{T(z)}{2} - \frac{T(z)^2}{4} - \frac{T(z)^3}{6} . \tag{9}$$

For bicyclic components without triangle, we have respectively for multigraphs and graphs

$$W_{1,C_3}(z) = \frac{T(z)}{24} \frac{(3 + 2T(z))}{(1 - T(z))^3} \text{ and } \widehat{W_{1,C_3}}(z) = \frac{T(z)^5}{24} \frac{(2 + 6T(z) - 3T(z)^2)}{(1 - T(z))^3} . \tag{10}$$

The decompositions of formulae (10), using the tree polynomials described by (5), lead respectively to

$$W_{1,C_3}(z) = \sum_{n \geq 0} \left(\frac{5}{24} t_n(3) - \frac{7}{24} t_n(2) + \frac{1}{12} t_n(1) \right) \frac{z^n}{n!} , \tag{11}$$

$$\widehat{W_{1,C_3}}(z) = \sum_{n \geq 0} \left(\frac{5}{24} t_n(3) - \frac{25}{24} t_n(2) + \frac{47}{24} t_n(1) - \frac{35}{24} - \frac{5}{24} t_n(-1) \right. \\ \left. + \frac{25}{24} t_n(-2) - \frac{5}{8} t_n(-3) + \frac{1}{8} t_n(-4) \right) \frac{z^n}{n!} . \tag{12}$$

Lemma 1. *If $\xi = \{C_k, k \in \Omega\}$ where Ω is a finite set of integers greater to or less than 3, the probability that a random graph or multigraph with has n vertices and $1/2n$ edges only acyclic, unicyclic, bicyclic components all C_k-free, $k \in \Omega$, is*

$$\sqrt{\frac{2}{3}} \cosh \left(\sqrt{\frac{5}{18}} \right) e^{-\sum_{k \in \Omega} \frac{1}{2k}} + O(n^{-1/3}) . \tag{13}$$

$\square$

Proof. This is a corollary of [13, eq (11.7)] using the formulae (8), (9), (11) and (12). Incidentally, random graphs and multigraphs have the same asymptotic behavior as shown by the proof of [13, Theorem 4]. As *multigraphs are graphs without cycles of length* 1 *and* 2, the forbidden cycles of length 1 and 2 bring a factor $e^{-3/4}$ which is cancelled by a factor $e^{+3/4}$ because of the ratio between weighting functions that convert the EGF of graphs and multigraphs into probabilities

$$
\begin{pmatrix} \binom{n}{2} \\ m \end{pmatrix} = \left(\frac{n^{2m}}{2^m m!} \right) \exp\left(-\frac{m}{n} - \frac{m^2}{n^2} + O(\frac{m}{n^2}) + O(\frac{m^3}{n^4}) \right), \ m \le \binom{n}{2}. \quad (14)
$$

The situation changes radically when cycles of length greater to or less than 3 are forbidden. Equations (8), (9) and the "significant coefficient" $\frac{5}{24}$ of $t_n(3)$ in (11) and in (12) and the demonstration of [13, Lemma 3] show us that the term $-\frac{T(z)k}{2k}$, introduced in (8) and (9) for each forbidden k-gon, simply changes the result by a factor of $e^{-1/2k} + O(n^{-1/3})$.

The example of forbidden k-gon suggests itself for a generalization.

4   Random Isomorphism-Free Graphs

The probabilistic results on random H-free graphs/multigraphs can be obtained when looking at the form of the decompositions of their EGF into tree polynomials.

Lemma 2. *Let H be a connected $(n, n + p)$ graph or multigraph. Let $\widehat{R_{q,H}}(w, z)$, resp. $R_{q,H}$, be the bivariate EGF of all connected q-excess graphs, resp. multigraphs, containing at least one subgraph isomorphic to H. Then, $\widehat{R_{q,H}}(w, z)$ is of the form*

$$
\widehat{R_{q,H}}(w, z) = w^q \frac{P(T(wz))}{(1 - T(wz))^k}, \quad (15)
$$

where $k < 3q$ and P is a polynomial. Similar formula holds for multigraphs. □

Proof. The more cycles H has, the more the degree of the denominator of the EGF multicyclic graphs or multigraphs containing subgraph isomorphic to H diminishes. This follows from the fact that EGF of $(q + 1)$-cyclic graphs or multigraphs are simply combination of integer partitions functions up to $3q$. If we force our structures to contain some specific multicyclic subgraphs, some parts are fixed and we strictly diminish the number of parts of integers needed to reconstruct our graphs or multigraphs.

Lemma 3. *Let H be a connected multicyclic graph or multigraph with n vertices and $(n + p)$ edges with $p > 0$ and let $W_{q,H}$, respectively $\widehat{W_{q,H}}$, be the generating functions of connected multicyclic H-free multigraphs, respectively graphs, with q edges, $(q \ge p)$, more than vertices. If $W_q(z) = \frac{c_q}{(1-T(z))^{3q}} + \sum_{i \ge 1} \frac{d_i}{(1-T(z))^{3q-i}}$ is the Wright's EGF of q-excess multigraphs rewritten with tree polynomials then $W_{q,H}(z) = \frac{c_q}{(1-T(z))^{3q}} +$*

$\sum_{i \geq 1} \frac{d_i'}{(1-T(z))^{3q-i}}$. In these formulae the leading coefficient c_q is the same for W_q and $\widehat{W_q}$, defined in [13, equation (8.6)]. Analogous result holds for multigraphs with the EGF $\widehat{W_q}$ and $\widehat{W_{q,H}}$. □

Proof. One can write $W_q(z) = \frac{P_q(T(z))}{(1-T(z))^{3q}}$ where P_q is a polynomial. Then, we can express $P_q(x)$ in term of successive powers of $(1-x)^i$ and c_q equals simply $P_q(1)$. We have $W_{q,H}(w,z) = \frac{w^q c_q}{(1-T(wz))^{3q}} + w^q \sum_{i<3q} d_i' t_n(i) \frac{w^n z^n}{n!}$ because $W_q(w,z) = W_{q,H}(w,z) + R_{q,H}(w,z)$, where $R_{q,H}(w,z)$ is the bivariate EGF of multicyclic connected graphs with q edges more than vertices. As shown by (15), the denominator of $R_{q,H}(w,z)$ is strictly less than $3q$.

We are now ready to state the following result.

Theorem 1. *Let $\xi = \{H_1, H_2, H_3, ... H_m\}$ be a finite collection of multicyclic connected graphs or multigraphs. Then the probability that a random graph with n vertices and $\frac{1}{2}n + O(n^{-\frac{1}{3}})$ edges has r_1 bicyclic components, r_2 tricyclic components, ..., $(k+1)$-cyclic components, all components $\{H_1, H_2, H_3, ... H_m\}$-free and no components of higher cyclic order is*

$$\left(\frac{4}{3}\right)^r \exp\left(-\sum_{p \in \Omega} \frac{1}{2p}\right) \sqrt{\frac{2}{3}} \frac{c_1^{r_1}}{r_1!} \frac{c_2^{r_2}}{r_2!} \cdots \frac{c_k^{r_k}}{r_k!} \frac{r!}{(2r)!} + O(n^{-1/3}) \tag{16}$$

where $\Omega = \{p \geq 3, \exists i \in [1, m] \text{ such that } H_i \text{ is a } p\text{-gon}\}$. □

The theorem 2 below shows that a necessary and sufficient condition to change a coefficient c_i of (16) is that ξ must contain all graphs *contractible* to a certain i-excess graph H_i.

Theorem 2. *Let H be a k-excess multicyclic graph (resp. multigraph) with $k > 0$. Suppose that H has n vertices, $n + k$ edges and $c(H) n!$ is the number of ways to label H (for example $c(K_4) = 1/24$). Denote by $\xi_k(H)$ the set of all k-excess graphs contractible to H. Then the probability that a random graph (resp. multigraph) with n vertices and $m(n) = 1/2n + O(n^{-1/3})$ edges has r_1 bicyclic, r_2 tricyclic, ..., r_p $p + 1$-cyclic components, all without component isomorphic to any member of the set $\xi_k(H)$ is*

$$\left(\frac{4}{3}\right)^r \sqrt{\frac{2}{3}} \frac{c_1^{r_1}}{r_1!} \frac{c_2^{r_2}}{r_2!} \cdots \frac{c_{k-1}^{r_{k-1}}}{r_{k-1}!} \frac{(c_k - c(H))^{r_k}}{r_k!} \frac{c_{k+1}^{r_{k+1}}}{r_{k+1}!} \cdots \frac{c_p^{r_p}}{r_p!} \frac{r!}{(2r)!} + O(n^{-1/3}). \tag{17}$$

□

Proof. The EGF of $\xi_k(H)$ is simply

$$\xi_k(H)(w,z) = w^k c(H) \frac{T(wz)^n}{(1 - T(wz))^{3k}}. \tag{18}$$

Thus in (16) if we want to avoid all graphs *contractible* to H, we have to substract (18) to the EGF of connected k-excess graphs. Lemma 3 shows us that the other coefficients, i.e., c_i for all $i > k$ remain unchanged.

5   Asymptotic Numbers

Denote by $c(n, n+k)$ the number of connected $(n, n+k)$ graphs. Similarly, let $c_\xi(n, n+k)$ be the number of connected $(n, n+k)$ ξ-free graphs. *Wright* [21, Theorem 4] state the following result:

Theorem 3. *If* $k = o(n^{1/3})$, *but* $k \to \infty$ *as* $n \to \infty$ *then*

$$c(n, n+k) = d\,(3\pi)^{1/2}(e/12k)^{k/2}n^{n+1/2(3k-1)}(1+O(k^{-1})+O(k^{3/2}/n^{1/2}))\,. \quad (19)$$

□

Note that later *Voblyi* [17] proved $d_k \to \frac{1}{2\pi}$ as $k \to \infty$. We prove here that a very similar result holds for $c_\xi(n, n+k)$, i.e., for ξ-free connected graphs when ξ is finite and $k = o(n^{1/3})$.

If $X(z)$ and $Y(z)$ are 2 EGF, we note here that $X \geq Y$ iff $\forall n, [z^n]\,X(z) \geq [z^n]\,Y(z)$. Denote by $\widehat{W}_k$, $k \geq 0$ the set of k-excess graphs and $\widehat{W}_k(w, z)$ their bivariate exponential generating function. Thus, $\widehat{W}_k(w, z), k \geq 1$ has the following form

$$\widehat{W}_k(w, z) = \frac{b_k}{(1 - T(wz))^{3k}} - \frac{c_k}{(1 - T(wz))^{3k-1}} + \sum_{s \geq 3k-2} \frac{c_{k,s}}{(1 - T(wz))^s} \quad (20)$$

and $\widehat{W}_0(w, z) = \widehat{V}(w, z)$ as in eq. (3).

Furthermore, denote by $\widehat{W}_{k,\xi}$ the set of connected k-excess ξ-free graphs.

Lemma 4. *If ξ is finite,* $\widehat{W}_{k,\xi}(w, z), k > 0$ *has the following form*

$$\widehat{W}_{k,\xi}(w, z) = \frac{b_k}{(1 - T(wz))^{3k}} - \frac{(c_k + \alpha_k)}{(1 - T(wz))^{3k-1}} + \sum_{s \leq 3k-2} \frac{(c_{k,s} + \alpha k, s)}{(1 - T(wz))^s} \quad (21)$$

where $\alpha_k = 0$ if ξ does not contain a p-gon.

□

Proof. Denote respectively by $S_{k,\xi}$ and $J_{k,\xi}$ the EGF for k-excess graphs containing exactly one occurrence of a member of ξ and k-excess with many occurrences of member of ξ but necessarily juxtaposed, i.e. the deletion of one edge will delete any occurrences of any member of ξ. For example if C_3 and C_4 are in ξ, a "*house*" is a juxtaposition of them. Then, remember that $\widehat{W}_{k,\xi}$ satisfy the recurrence (see also [15])

$$V_w \widehat{W}_{k+1,\xi} + O(S_{k+1,\xi}) + O(J_{k+1,\xi}) = \left(\frac{V_z{}^2 - V_z}{2} - V_w \right) \widehat{W}_{k,\xi}$$

$$+ \sum_{p+q=k} \frac{(V_z \widehat{W}_{p,\xi})\,(V_z \widehat{W}_{q,\xi})}{1 + \delta_{p,q}} \quad (22)$$

where $V_x = x\frac{\partial}{\partial x}$ (see [12]). Lemma 4 follows from the fact that $S_{k,\xi}$ and $J_{k,\xi}$ are of the form described by lemma 2. We have, for $k > 0$ the formula for smooth $(n, n+k)$ graphs with exactly one triangle

$$S_{k,C_3}(z) = \frac{1}{1 - z} \left(V_z \frac{1}{6} z^3 \right) \left(V_z \widehat{W}_{k-1,C_3}(z) \right) + \sum_{i \leq 3k-2} \frac{s_{k,i}}{(1 - z)^i} \,. \quad (23)$$

Thus,

$$S_{k,C_3}(z) = \frac{3}{2}(k-1)\frac{b_{k-1}}{(1-T(z))^{3k-1}} + \sum_{x<3k-1}\frac{s_{k,x}}{(1-T(z))^x} \qquad (24)$$

and in this case α_k of (21) equals $\frac{3}{2}(k-1)b_{k-1}$.

Lemma 5.

$$\frac{b_k}{(1-T(z))^{3k}} - \frac{(c_k+\alpha_k)}{(1-T(z))^{3k-1}} \le \widehat{W}_{k,\xi}(z) \le \frac{b_k}{(1-T(z))^{3k}} \qquad (25)$$

□

Proof. We have to prove only $\widehat{W}_{k,\xi}(z) - \frac{b_k}{(1-T(z))^{3k}} + \frac{(c_k+\alpha_k)}{(1-T(z))^{3k-1}} \ge 0$, since *Wright* [21] show $\widehat{W}_k(z) \le \frac{b_k}{(1-T(z))^{3k}}$ and *a fortiori*, we have $\widehat{W}_{k,\xi}(z) \le \frac{b_k}{(1-T(z))^{3k}}$. Substituting (21) in (22) leads to

$$2(k+1)b_{k+1} = 3k(k+1)b_k + 3\sum_{t=1}^{k-1} t(k-t)b_t b_{k-t} \qquad (26)$$

and

$$\begin{aligned} 2(3k+2)(c_{k+1}+\alpha_{k+1}) &= 8(k+1)b_{k+1} + 3kb_k + (3k+2)(3k-1)(c_k+\alpha_k) \\ &+ 6\sum_{t=1}^{k-1} t(3k-3t-1)b_t(c_{k-t}+\alpha_{k-t}) \end{aligned} \qquad (27)$$

Then we have also as in [21, Lemma 5], $kb_k \le c_k + \alpha_k \le \frac{(c_1+\alpha_1)}{b_1}kb_k$. Still using similar arguments to those of [21], the equivalent of [21, Lemmas 6, 7, 8, 9, 10] can also be obtained here (with the coefficients $c_k + \alpha_k$ instead of c_k) to prove lemma 5 by induction on k.

Theorem 4. *Given a finite collection ξ of multicyclic graphs, almost all connected $(n, n+k)$ graphs are ξ-free when $k = o(n^{1/3})$.* □

Proof. By lemma 5, [21, eq (5.2), (5.3), Theorem 2] and (26).

References

1. Bagaev, G.N.: Random graphs with degree of connectedness equal 2. Discrete Analysis **22** (1973) 3–14, (in Russian).
2. Bagaev, G.N., Voblyi, V.A.: The shrinking-and-expanding method for the graph enumeration. Discrete Mathematics and Applications **8** (1998) 493–498.
3. Bender, E.A., Canfield, E.R., McKay, B.D.: The asymptotic number of labeled connected graphs with a given number of vertices and edges. Random Structures and Algorithms **1** (1990) 127–169.
4. Bender, E.A., Canfield, E.R., McKay, B.D.: Asymptotic Properties of Labeled Connected Graphs. Random Structures and Algorithms **3**, No. 2 (1992) 183–202.

5. Cayley, A.: A Theorem on Trees. Quart. J. Math. Oxford Ser. **23** (1889) 376–378.
6. Comtet, L.: Analyse Combinatoire. Presses Universitaires de France (1970).
7. Erdös, P., Rényi, A.: On random graphs I. Publ. Math. Debrecen **6** (1959) 290–297.
8. Erdös, P., Rényi, A.: On the evolution of random graphs. Magyar Tud. Akad. Mat. Kut. Int. Kzl. **5** (1960) 17–61.
9. Knuth, D.E.: The Art Of Computing Programming, v.1, "Fundamental Algorithms". 2nd Edition, Addition-Wesley, Reading (1973).
10. Knuth, D.E, Pittel, B.: A recurrence related to trees. Proc. Am. Math. Soc. **105** (1989) 335–349.
11. Flajolet, P., Knuth, D.E.,Pittel, B.: The First Cycles in an Evolving Graph. Discrete Mathematics **75** (1989) 167–215.
12. Flajolet, P., Zimmerman, P.,Van Cutsem: A calculus for the random generation of labelled combinatorial structures. Theoretical Computer Sciences **132** (1994) 1–35.
13. Janson, S.,Knuth, D.E., Luczak, T., Pittel, B.: The Birth of the Giant Component. Random Structures and Algorithms **4** (1993) 233–358.
14. Moon, J.W.: Various proofs of Cayley's formula for counting trees. In: Harary, F. (ed.): A seminar on graph theory. New York (1967) 70–78.
15. Ravelomanana, V., Thimonier, L.: Enumeration and random generation of the first multicyclic isomorphism-free labeled graphs. submitted, (1999).
16. Rényi, A.: On connected graphs I. Publ. Math. Inst. Hungarian Acad. Sci. **4** (1959) 385–388.
17. Voblyi, V.A.: Wright and Stepanov-Wright coefficients. Math. Notes **42** (1987) 969–974.
18. Wilf, H.S.: Generatingfunctionology. Academic Press, New-York (1990).
19. Wright, E.M.: The Number of Connected Sparsely Edged Graphs. Journal of Graph Theory **1** (1977) 317–330.
20. Wright, E.M.: The Number of Connected Sparsely Edged Graphs. II. Smooth graphs and blocks. Journal of Graph Theory **2** (1978) 299–305.
21. Wright, E.M.: The Number of Connected Sparsely Edged Graphs. III. Asymptotic results. Journal of Graph Theory **4** (1980) 393–407.

Analysis of Edge Deletion Processes on Faulty Random Regular Graphs

Andreas Goerdt[*1] and Mike Molloy[2]

[1] Fakultät für Informatik, TU Chemnitz, 09111 Chemnitz, Germany
`goerdt@informatik.tu-chemnitz.de`
[2] Department of Computer Science, University of Toronto, Toronto, Canada
`molloy@cs.toronto.edu`

Abstract. Random regular graphs are, at least theoretically, popular communication networks. The reason for this is that they combine low (that is constant) degree with good expansion properties crucial for efficient communication and load balancing. When any kind of communication network gets large one is faced with the question of fault tolerance of this network. Here we consider the question: Are the expansion properties of random regular graphs preserved when each edge gets faulty independently with a given fault probability? We improve previous results on this problem: Expansion properties are shown to be preserved for much higher fault probabilities and lower degrees than was known before. Our proofs are much simpler than related proofs in this area.

Introduction

A natural question in the theory of fault tolerance of communication networks reads: Is it possible to simulate the non-faulty network on the faulty one with a well determined slowdown? Here one assumes that the network proceeds in synchronous steps and in each step each processor (= node of the network) performs some local computation and some communication steps. Ideally one would like to simulate the non-faulty network in such a way that the simulation is slower only by a *constant* factor showing that the time is essentially unchanged. Whereas such efficient simulations are known for networks with unbounded degree, like the hypercube, it is still an important question whether they exist for bounded degree networks like the butterfly [3]. Note that all of this paper refers to *random* faults, that is each component (normally edge or node) gets faulty independently with a given fault probability and the results only hold with high probability meaning with probability going to 1 when the network gets large.

Random regular graphs with given degree $d \geq 3$ are well known to be expander graphs (with high probability) [2]: There is a constant $C (< 1)$ such that each subset X of nodes has $\geq C \cdot |X|$ neighbours adjacent to X but not belonging to X (provided X contains at most half of all vertices). If we ever were

---

[*] Author's work in part performed at the University of Toronto, supported by a grant obtained through Alasdair Urquhart.

G. Gonnet, D. Panario, and A. Viola (Eds.): LATIN 2000, LNCS 1776, pp. 38–47, 2000.

to simulate computation on a random regular graph with slowdown only by a constant factor on the faulty graph we would need a linear size expander inside the faulty graph.

The investigation of random regular graphs with edge faults starts with the paper [9]. In the succeeding paper [10] attention is drawn to the preservation of expansion properties. Some sufficient conditions are given. In work by the first author [4] a threshold result for the existence of a linear size component is proved. In [5] we give a sufficient condition on fault probability and degree such that we can *find* a linear size expander efficiently – a question not treated in the initial work on expansion [10]. Crucial to our result is the notion of a k-core: The k-core of a given graph is the (unique) maximal subgraph where each node has degree at least k. In [5] we first observe that the 3−core of a faulty random regular graph is an expander (this follows simply from randomness properties of the 3−core). Second, we present a simple edge deletion algorithm which is shown to find a 3−core of linear size when $d \geq 42$ and each edge is non-faulty with probability at least $20/d$.

The present paper improves considerably on these results: We give a precise threshold on the fault probability for the existence of a linear size k-core for any $d > k \geq 3$. Thus improving the previous bounds for the existence of an expanding subgraph. For example when the degree is as low as 4 and each edge is faulty with probability $< 1/9$ we have a linear size 3−core and thus an expanding subgraph.

Our proof uses a proof technique originally developed for [7]. It is technically quite simple. This is in sharp constrast to the previous proofs of the weaker results mentioned above relying on technically advanced probability theoretic tools. This technique applies to a wide range of similar problems (see [8]). The technique was inspired by the original (more involved) proof of the k-core threshold for $G_{n,p}$ given in [6].

1 Outline

We will study random regular graphs with edge faults by focussing on the configuration model (cf.[1]). It is well known that properties which hold a.s. (almost surely) for a uniformly random d-regular configuration also hold a.s. for a uniformly random d-regular simple graph. For the configuration model, we consider n disjoint d−element sets called *classes*; the elements of these classes are called *copies*. A configuration is a partition of the set of all copies into 2−element sets, which are edges. Identifying classes with vertices, configurations determine multigraphs and standard graph theoretic terminology can be applied to configurations. More details can be found in [1]. We fix the degree d and the probability p for the rest of this paper and consider probability spaces Con(n, d, p) of random configurations where each edge is present with probability p or absent with fault probability $f = 1 - p$. We call this space the space of *faulty configurations*. An element of this space of is best considered as being generated by the following probabilistic experiment consisting of two stages:

(1) Draw randomly a configuration $\Phi = (\mathcal{W}, E)$ where $\mathcal{W} = W_1 \mathbin{\dot{\cup}} \ldots \mathbin{\dot{\cup}} W_n$ and $|W_i| = d$. (2) Delete each edge of Φ, along with its end-copies, independently with fault probability f.

The probability of a fixed faulty configuration with k edges is $(n \cdot d - 2 \cdot k)!! \cdot (1 - p)^{(n \cdot d/2) - k} \cdot p^k$. Given k, each set of k edges is equally likely to occur. The degree of a class W with respect to a faulty configuration Φ, $\mathrm{Deg}_\Phi(W)$, is the number of copies of W which were not deleted. Note that edges $\{x, y\}$ with $x, y \in W$ contribute with two to the degree. The degree of a copy x, $\mathrm{Deg}_\Phi(x)$, is the degree of the class to which x belongs. The k-core of a faulty configuration is the maximal subconfiguration of the faulty configuration in which each class has a degree $\geq k$. We call classes of degree less than k *light* whereas classes of degree at least k are *heavy*. By $Bin(m, \lambda)$ we denote the binomial distribution with parameters m and success probability λ.

We now give an overview of the proof of the following theorem which is the main result of this paper. For $d > k \geq 3$ we consider the real valued function $L(\lambda) = \lambda / Pr[Bin(d-1, \lambda) \geq k-1]$ which we define for $0 < \lambda \leq 1$. $L(1) = 1$ and $L(\lambda)$ goes to infinity for λ approaching 0. Moreover $L(\lambda)$ has a unique minimum for $1 \geq \lambda > 0$. Let $r(k, d) = \min\{L(\lambda) | 1 \geq lambda > 0\}$. For example we have that $r(3, 4) = 8/9$. The definition of $r(k, d)$ is, no doubt, mysterious at this point, but we will see that it has a very natural motivation.

Theorem 1. *(a) If $p > r(k, d)$ then a random $\Phi \in Con(n, d, p)$ has a k-core of linear size with high probability.*
(b) If $p < r(k, d)$ then a random $\Phi \in Con(n, d, p)$ has only the empty k-core with high probability.

Theorem 1 implies that the analogous result holds for the space of faulty random regular graphs (obtained as: first draw a graph, second delete the faulty edges). The following algorithm which can easily be executed in the faulty network itself is at the heart of our argument.

Algorithm 2 The Global Algorithm
Input: A faulty configuration Φ, output: The k-core of Φ.

while Φ has light classes do
    Φ := the modification of Φ where all light classes are deleted.
od. Output Φ.

Specifically, when we delete a class, W, we delete (i) all copies within W, (ii) all copies of other classes which are paired with copies of W, (iii) W itself. Note that it is possible for W itself to still be undeleted but to contain no copies as they were all deleted as a result of neighbouring classes being deleted, or faulty edges. In this case, of course, W is light and so it will be deleted on the next iteration. At the end of the algorithm Φ has only classes of degree $\geq k$, which form the k-core. The following notion will be used later on: A class W of the faulty configuration Φ *survives* j $(j \geq 0)$ rounds of the global algorithm with degree t iff W has not yet been deleted and has degree t after the j'th execution of the while-loop of the algorithm with input Φ. A class simply survives if it has not yet been deleted.

In section 2 we analyze this algorithm when run for $j - 1$ executions of the loop where we set $j = j(n) = \sqrt{\log_d n}$ throughout. We prove that the number of classes surviving $j - 1$ rounds with degree $t \geq k$ is linear in n with high probability when $p > r(k, d)$ whereas the number of light classes is $o(n)$. (Initially this number is linear in n.) An extra argument presented in section 4 will show how to get rid of these few light classes leaving us with a linear size k-core provided $p > r(k, d)$. On the other hand, if $p < r(k, d)$ then we show that the expected number of classes surviving $j - 1$ rounds with any degree is $o(n)$ and that we have no longer enough classes to form a k-core. This is shown in section 3.

2   Reduction of the Number of Light Classes

For $d \geq t \geq 0$, and for a particular integer j, we let

$$X_t : \mathrm{Con}(n, d, p) \to \mathcal{N} \tag{1}$$

be the number of classes surviving $j - 1$ rounds of the global algorithm with degree equal to t. As usual we can represent $X = X_t$ as a sum of indicator random variables

$$X = X_{W_1} + \cdots + X_{W_n}, \tag{2}$$

where X_W assumes the value 1 when the class W survives $j - 1$ rounds with degree equal to t and 0 when this is not the case. Then $EX = n \cdot E[X_W] = n \cdot Pr[W \text{ survives with degree } t]$ for W arbitrary. We determine $Pr[W \text{ survives with degree } t]$ approximately, that is an interval of width $o(1)$ which includes the probability. The probability of the event: W survives $j - 1$ rounds with degree t, turns out to depend only on the j−environment of W defined as: For a class W the j−environment of W, $j - \mathrm{Env}_\Phi(W)$, is that subconfiguration of Φ which has as classes the classes whose distance from W is at most j. Here distance means the number of edges in a shortest path. The edges of $j - \mathrm{Env}_\Phi(W)$ are those induced from Φ.

The proof of the following lemma follows with standard conditioning techniques observing that the j−environment of a class W in a random configuration can be generated by a natural probabilistic breadth first generation process (cf. [4] for details on this.) Here it is important that j only slowly goes to infinity.

Lemma 1. *Let W be a fixed class then $Pr\{j - Env_\Phi(W) \text{ is a tree}\} \geq 1 - o(1)$.*

Note that the lemma does not mean: Almost always the j-environment of *all* classes is a tree. The definition of j-environment extends to faulty configurations in the obvious manner. Focussing on a j-environment which is a tree is very convenient since in a faulty configuration, it can be thought of as a branching process whereby the number of children of the root is distributed as $Bin(d,p)$, and the number of children of each non-root as $Bin(d-1,p)$.

The following algorithm approximates the effect the global algorithm has on a fixed class W, provided the j-environment of W is a tree.

Algorithm 3 The Local Algorithm.
Input: A (sub-)configuration Γ, which is a j-environment of a class W in a faulty configuration. Γ is a tree with root W.
$\Phi := \Gamma$
for $i = j-1$ downto 0 do
  Modify Φ as follows: Delete all light classes in depth i of the tree Φ.
od.
The output is "W survives with degree t" if W is not deleted and has final degree t. If W is deleted then the output is "W does not survive".

Note that it is not possible for W to survive with degree less than k. By round l of the algorithm we mean an execution of the loop with $i = j-l$ where $1 \leq l \leq j$. A class in depth i where $0 \leq i \leq j$ survives with degree t iff it is not deleted and has degree t after round $j-i$ of the algorithm. Note that classes in depth j are never deleted and so they are considered to always survive. The next lemma states in which respect the local algorithm approximates the global one. The straightforward formal proof is omitted in this abridged version.

Lemma 2. *Let $j \geq 1$. For each class W and each faulty configuration Φ where $j - Env_\Phi(W)$ is a tree we have: After $j-1$ rounds of the global algorithm with Φ the class W survives with degree $t \geq k$. $\Leftrightarrow$ After running the local algorithm with $j - Env_\Phi(W)$ the class W survives with degree $t \geq k$.*

Note that W either survives $j-1$ rounds of the global algorithm and the whole local algorithm with the same degree $t \geq k$ or does not survive the local algorithm in which case it does or does not survive $j-1$ global rounds, but does certainly not survive j global rounds.

We condition the following considerations on the almost sure event that for $j = j(n)$ the j-environment of the class W in the underlying fault free configuration is a tree (cf. Lemma 1). We denote this environment in a random faulty configuration by Γ. We turn our attention to the calculation of the survival probability with the *local* algorithm.

For i with $0 \leq i \leq j-1$ let ϕ_i be the probability that a class in level (=depth) $j-i$ of Γ survives the local algorithm applied to Γ. As the j-enviroment in the underlying fault-free configuration is a tree, the survival events of the children of given class are independent. Therefore:

$$\phi_0 = 1 \text{ and } \phi_i = Pr[Bin(d-1, p \cdot \phi_{i-1}) \geq k-1]. \qquad (3)$$

And furthermore, considering now the root W of the j-environment, we get for $t \geq k$ by analogous considerations:
$Pr[W$ survives the local algorithm with degree $t.] = Pr[Bin(d, p \cdot \phi_{j-1}) = t]$.

We have that the sequence of the ϕ_i's is monotonically decreasing and in the interval $[0, 1]$. Hence $\phi = \phi(p) = \lim_{i \to \infty} \phi_i$ is well defined and as all functions involved are continuous we get: $\phi = Pr[Bin(d-1, p \cdot \phi) \geq k-1]$. (Note that this is no definition of ϕ, the equation is always satisfied by $\phi = 0$.)

Two further notations for subsequent usage: $\lambda_{t,i} = \lambda_{t,i}(p) = Pr[Bin(d, p \cdot \phi_{i-1}) = t]$ for $i \geq 1$. Again we have that the $\lambda_{t,i}$'s are monotonically decreasing and between 0 and 1 and $\lambda_t = \lambda_t(p) = \lim_{i \to \infty} \lambda_{t,i}$. exists. Hence for our fixed class W, considering $j \to \infty$, we get:

$$Pr[W \text{ survives the local algorithm with degree } t.] = \lambda_{t,j} = \lambda_t + o(1). \quad (4)$$

Here is where our formula for $r(k,d)$ comes from:

Lemma 3. $\phi > 0$ iff $p > r(k, d)$.

Proof. First let $\phi > 0$. As stated above we have $\phi = Pr[Bin(d-1, p\phi) \geq k-1]$. Therefore $Pr[Bin(d-1, p\phi) \geq k-1] > 0$ and setting $\lambda = p \cdot \phi$, we get $\lambda/p = Pr[Bin(d-1, \lambda) \geq k-1]$ and so $p = \lambda/Pr(Bin(d-1, \lambda) \geq k-1) = L(\lambda)$ and the result follows.

Now let $p > r(k, d)$. Let λ_0 be such that $r(k, d) = L(\lambda_0)$. We show by induction on i that $p \cdot \phi_i \geq \lambda_0$. For the induction base we get: $p \cdot \phi_0 = p > r(k, d) \geq \lambda_0$ where the last estimate holds because the denominator in the definition of $L(\lambda_0)$ always is ≤ 1. For the induction step we get: $p \cdot \phi_{i+1} = p \cdot Pr[Bin(d-1, p \cdot \phi_i) \geq k-1] \geq p \cdot Pr[Bin(d-1, \lambda_0) \geq k-1] > \lambda_0$ where the last but one estimate uses the induction hypothesis and the last one follows from the assumption. □

We now return to the analysis of the *global* algorithm. The next corollary follows directly with Lemma 1, Lemma 2, and (4).

Corollary 1. *Let W be a fixed class, $t \geq k$ and let $j = j(n) = \sqrt{\log_d n}$. In the space of faulty configurations we have (cf.(2)):*

$$Pr[X_W = 1] = Pr\{W \text{ survives } j(n) - 1 \text{ global rounds with degree } t\}$$
$$= \lambda_t + o(1).$$

Next the announced concentration result:

Theorem 4. *Let $t \geq k$, $X = X_t$ be the random variable defined as in (1), and let $\lambda = \lambda_t$; then we have:*

(1) $EX = \lambda \cdot n + o(n)$.   (2) Almost surely $|X - \lambda \cdot n| \leq o(n)$.

Proof. (1) The claim follows from the representation of X as a sum of indicator random variables (cf. (2)) and with Corollary 1.

(2) We show that $VX = o(n^2)$. This implies the claim with an application of Tschebycheff's inequality. We have $X = X_{W_1} + X_{W_2} + \ldots + X_{W_n}$ (cf. (2)). This and (1) of the present theorem implies $VX = E[X^2] - (EX)^2 = E[X^2] - (\lambda^2 \cdot n^2 + o(n) \cdot n)$. Moreover, $E[X^2] = EX + n \cdot (n-1) \cdot E[X_U \cdot X_W] = \lambda \cdot n + o(n) + n \cdot (n-1) \cdot E[X_U \cdot X_W]$, where U and W are two arbitrary distinct

classes. We need to show that $E[X^2] = \lambda^2 \cdot n^2 + o(n^2)$. This follows from $E[X_U \cdot X_W] = \lambda^2 + o(1)$ showing that the events $X_U = 1$ and $X_W = 1$ are asymptotically independent. This follows by conditioning on the event that the j−environments of U and W are disjoint trees and analyzing the breadth first generation procedure for the $j−$ environment of a given class. Again we need that j goes only slowly to infinity.   □

3   When There is no k-Core

The proof of Theorem 1(b) is now quite simple. First we need the following fact:

Lemma 4. *A.s. a random member of $Con(n, d, p)$ has no k-core on $o(n)$ vertices.*

Proof. The lemma follows from the fact that a random member of $Con(n, d)$ a.s. has no subconfiguration with average degree at least 3 on at most ϵn vertices, where $\epsilon = \epsilon(d)$ is a small positive constant. Consider any $s \leq \epsilon n$. The number of choices for s classes, $1.5s$ edges from amongst those classes, and copies for the endpoint of each edge, is at most:

$$\binom{n}{s}\binom{\binom{s}{2}}{1.5s}d^{3s}.$$

Setting $M(t) = t!/(2^{t/2}(t/2)!)$ to be the number of ways of pairing t copies, we have that for any such collection, the probability that those pairs lie in our random member of $Con(n, d)$ is

$$M((d-3s)n)/M(dn) < (\frac{e}{n})^{1.5s}.$$

Therefore, the expected number of such subconfigurations is at most:

$$\binom{n}{s}\binom{\binom{s}{2}}{1.5s}d^{3s}(\frac{e}{n})^{1.5s} < (\frac{en}{s})^s(\frac{e(s^2/2)}{1.5s})^{1.5s}(\frac{e}{n})^{1.5s}$$

$$\leq (\frac{20d^6 s}{n})^{.5s} = f(s).$$

Therefore, if $\epsilon = 1/40d^6$ then the expected number of such subconfigurations is less than $\sum_{s=2}^{n} f(s)$ which is easily verified to be $o(1)$.   □

Now, by Lemma 3, we have for $p < r(k, d)$ that $\phi = 0$. Therefore, as j goes to infinity the expected number of classes surviving j rounds with degree at least k is $o(n)$ and so almost surely is $o(n)$. With the last lemma we get Theorem 1 (b).

4   When There is a k-Core

In this section, we prove Theorem 1(a). So we assume that $p > r(k, d)$. We start by showing that almost surely very few light clauses survive the first $j(n) - 1$ iterations:

Lemma 5. *In $Con(n, d, p)$ almost surely: The number of light classes after $j(n) - 1 = \sqrt{\log_d n} - 1$ rounds of the global algorithm is reduced to $o(n)$.*

Proof. The proof follows with Theorem 4 applied to $j - 2$ and $j - 1$ (which both go to infinity). $\qquad\square$

In order to eliminate the light classes still present after $j(n) - 1$ global rounds, we need to know something about the distribution of the configurations after $j(n) - 1$ rounds. As usual in similar situations the uniform distribution needs to be preserved. For $\bar{n} = (n_0, n_1, n_2, \dots, n_d)$ where the sum of the n_i is at most n we let $Con(\bar{n})$ be the space of all configurations with n_i classes consisting of i copies. Each configuration is equally likely. The following lemma is proved in [5].

Lemma 6. *Conditioning the space $Con(n, d, p)$ on those configuration which give a configuration in $Con(\bar{n})$ after i global rounds, each configuration from $Con(\bar{n})$ has the same probability to occur after i global rounds.*

After running the global algorithm for $j(n) - 1$ rounds we get by Lemma 5 a configuration uniformly distributed in $Con(\bar{n})$ where $n_1 + n_2 + \dots + n_{k-1} = o(n)$ and $|n_t - \lambda_t \cdot n| \le o(n)$ for $t \ge k$ with high probability. A probabilistic analysis of the following algorithm eliminating the light classes one by one shows that we obtain a linear size k-core with high probability.

Algorithm 5
Input: A faulty configuration Φ.
Output: The k-core of Φ.
while There exist light classes in Φ do
   Choose uniformly at random a light class W from all light classes
   and delete W and the edges incident with W.
od. The classes of degree $\ge k$ are the k-core of Φ.

In order to perform a probabilistic analysis of this algorithm it is again important that the uniform distribution is preserved. A similar result is Proposition 1 in [6] (for the case of graphs instead of configurations).

Lemma 7. *If we apply the algorithm above to a uniformly random $\Phi \in Con(\bar{n})$, ($\bar{n}$ fixed) for a given number of iterations we get: Conditional on the event (in $Con(\bar{n})$) that the configuration obtained, Ψ, is in $Con(n_0', n_1' n_2', n_3', \dots, n_d')$ the configuration Ψ is a uniformly random configuration from this space.*

Lemma 8. *We consider probability spaces $Con(\bar{n})$ where the number of heavy vertices is $\geq \delta \cdot n$. In one round of Algorithm 5 one light class disappears and we get $\leq k - 1$ new light classes. Let $Y : Con(\bar{n}) \rightarrow \mathcal{N}$ be the number of new light classes after one round of Algorithm 5. Let $\nu = \sum_i i \cdot n_i$ and $\pi = (k \cdot n_k)/\nu$. Thus π is the probability to pick a copy of degree k when picking uniformly at random from all copies belonging to edges. Then:*

(a) $Pr[Y = l] = Pr[Bin(deg(W), \pi) = l] + o(1)$.

(b) $EY \leq (k - 1) \cdot \pi + o(1)$.

The straightforward proof of this lemma is omitted due to lack of space. Our next step is to bound π.

Lemma 9. $\pi \leq (1 - \epsilon)/(k - 1)$ *for some $\epsilon > 0$.*

Proof. We will prove that when $p = r(k, d)$ then $\pi = 1/(k - 1)$. Since π is easily shown to be decreasing in p, this proves our lemma. Recall that $r(k, d)$ is defined to be the minimum of the function $L(\lambda)$. Therefore, at $L(\lambda) = r(k, d)$, we have $L'(\lambda) = 0$. Differentiating L, we get:

$$\sum_{i=k-1}^{d-1} \binom{d-1}{i} \lambda^i (1-\lambda)^{d-1-i} = \sum_{i=k-1}^{d-1} \binom{d-1}{i} \lambda^i (1-\lambda)^{d-2-i}(i - (d-1)\lambda). \quad (5)$$

A simple inductive proof shows that the RHS of (5) is equal to

$$(k-1)\binom{d-1}{k-1} \lambda^{k-1} (1-\lambda)^{d-k}. \quad (6)$$

Indeed, it is trivially true for $k = d$, and if it is true for $k = r + 1$ then for $k = r$ the RHS is equal to

$$\binom{d-1}{r-1} \lambda^{r-1} (1-\lambda)^{d-1-r} (r - 1 - (d-1)\lambda) + r\binom{d-1}{r} \lambda^r (1-\lambda)^{d-1-r}$$

$$= (r-1)\binom{d-1}{r-1} \lambda^{r-1} (1-\lambda)^{d-r}$$

Setting $j = i + 1$, and multiplying by λd, the LHS of (5) comes to:

$$\sum_{j=k}^{d} d\binom{d-1}{j-1} \lambda^j (1-\lambda)^{d-j} = \sum_{j=k}^{d} j\binom{d}{j} \lambda^j (1-\lambda)^{d-j},$$

and (6) comes to

$$d(k-1)\binom{d-1}{k-1} \lambda^k (1-\lambda)^{d-k} = k(k-1)\binom{d}{k} \lambda^{k-1} (1-\lambda)^{d-k}.$$

Now, since

$$\pi = \frac{k\binom{d}{k} \lambda^{k-1}(1-\lambda)^{d-k}}{\sum_{j=k}^{d} j\binom{d}{j} \lambda^j (1-\lambda)^{d-j}} + o(1),$$

this establishes our lemma. □

Lemma 10. *Algorithm 5 stops after $o(n)$ rounds of the while loop with a linear size k-core with high probability (with respect to $Con(\bar{n})$).*

Proof. We define Y_i to be the number of light classes remaining after i steps of Algorithm 5. By assumption, $Y_0 = o(n)$. Furthermore, by Lemmas 8 and 9, we have $EY_1 \leq Y_0 - 1 + (k-1)\pi < Y_0 - \epsilon$. Furthermore, it is not hard to verify that, since there are $\Theta(n)$ classes of degree k, then so long as $i = o(n)$ we have

$$EY_{i+1} \leq Y_i - \frac{1}{2}\epsilon,$$

and in particular, the probability that at least ℓ new light vertices are formed during step i is less than the probability that the binomial variable $Bin(k-1, \pi)$ is at least ℓ.

Therefore, for any $t = o(n)$, $Y_0, Y_1, ..., Y_t$ is statistically dominated by a random walk defined as:

$$Z_0 = Y_0; Z_{i+1} = Z_i - 1 + Bin(k-1, \frac{1-\frac{1}{2}\epsilon}{k-1}).$$

Since Z_i has a drift of $-\frac{1}{2}\epsilon$, it is easy to verify that with high probability, $Z_t = 0$ for some $t = o(n)$, and thus with high probability $Y_t = 0$ as well.

If $Y_t = 0$ then we are left with a k-core of linear size. $\square$

Clearly Lemma 10 implies Theorem 1(a).

References

1. Bela Bollobas. Random Graphs. Academic Press. 1985.
2. –.The isoperimetric number of random regular graphs. European Journal of Combinatorics. 1988, 9, 241-244.
3. Richard Cole, Bruce Maggs, Ramesh Sitaraman. Routing on Butterfly networks with random faults. In Proceedings FoCS 1995. IEEE. 558-570.
4. Andreas Goerdt. The giant component threshold for random regular graphs with edge faults. In Proceedings MFCS 1997. LNCS 1295. 279-288.
5. –. Random regular graphs with edge faults: expansion through cores. In Proceedings ISAAC 1998. LNCS 1533, 219-228.
6. Boris Pittel, Joel Spencer, Nicholas Wormald. Sudden emergence of a giant k-core in a random graph. Journal of Combinatorial Theory B 67,1996,111-151.
7. Mike Molloy and Boris Pittel. Subgraphs with average degree 3 in a random graph. In preparation.
8. Mike Molloy and Nick Wormald.In preparation.
9. S. Nikoletseas, K. Palem, P. Spirakis, M. Yung. Vertex disjoint paths and multconnectivity in random graphs: Secure network computing. In Proceedings ICALP 1994. LNCS 820. 508-519.
10. Paul Spirakis and S. Nikoletseas. Expansion properties of random regular graphs with edge faults. In Proceedings STACS 1995. LNCS 900. 421-432.

Equivalent Conditions for Regularity
(Extended Abstract)

Y. Kohayakawa[1*], V. Rödl[2], and J. Skokan[2]

[1] Instituto de Matemática e Estatística, Universidade de São Paulo,
Rua do Matão 1010, 05508–900 São Paulo, Brazil
`yoshi@ime.usp.br`
[2] Department of Mathematics and Computer Science,
Emory University, Atlanta, GA, 30322, USA
`{rodl,jozef}@mathcs.emory.edu`

Abstract. Haviland and Thomason and Chung and Graham were the first to investigate systematically some properties of quasi-random hypergraphs. In particular, in a series of articles, Chung and Graham considered several quite disparate properties of random-like hypergraphs of density $1/2$ and proved that they are in fact equivalent. The central concept in their work turned out to be the so called *deviation* of a hypergraph. Chung and Graham proved that having small deviation is equivalent to a variety of other properties that describe quasi-randomness. In this note, we consider the concept of *discrepancy* for k-uniform hypergraphs with an arbitrary constant density d $(0 < d < 1)$ and prove that the condition of having asymptotically vanishing discrepancy is equivalent to several other quasi-random properties of $\mathcal{H}$, similar to the ones introduced by Chung and Graham. In particular, we give a proof of the fact that having the correct 'spectrum' of the s-vertex subhypergraphs is equivalent to quasi-randomness for any $s \geq 2k$. Our work can be viewed as an extension of the results of Chung and Graham to the case of an arbitrary constant valued density. Our methods, however, are based on different ideas.

1 Introduction and the Main Result

The usefulness of random structures in theoretical computer science and in discrete mathematics is well known. An important, closely related question is the following: which, if any, of the almost sure properties of such structures suffice for a deterministic object to have to be as useful or relevant?

Our main concern here is to address the above question in the context of hypergraphs. We shall continue the study of quasi-random hypergraphs along the lines initiated by Haviland and Thomason [7,8] and especially by Chung [2], and Chung and Graham [3,4]. One of the central concepts concerning hypergraph quasi-randomness, the so called *hypergraph discrepancy*, was investigated by Babai, Nisan, and Szegedy [1], who found a connection between communication complexity and hypergraph discrepancy. This

* Partially supported by FAPESP (Proc. 96/04505–2), by CNPq (Proc. 300334/93–1), and by MCT/FINEP (PRONEX project 107/97).

G. Gonnet, D. Panario, and A. Viola (Eds.): LATIN 2000, LNCS 1776, pp. 48–57, 2000.

connection was further studied by Chung and Tetali [5]. Here, we carry out the investigation very much along the lines of Chung and Graham [3,4], except that we focus on hypergraphs of arbitrary constant density, making use of different techniques.

In the remainder of this introduction, we carefully discuss a result of Chung and Graham [3] and state our main result, Theorem 3 below.

1.1 The Result of Chung and Graham

We need to start with some definitions. For a set V and an integer $k \geq 2$, let $[V]^k$ denote the system of all k-element subsets of V. A subset $\mathcal{G} \subset [V]^k$ is called a k-*uniform hypergraph*. If $k = 2$, we have a *graph*. We sometimes use the notation $\mathcal{G} = (V(\mathcal{G}), E(\mathcal{G}))$. If there is no danger of confusion, we shall identify the hypergraphs with their edge sets. Throughout this paper, the integer k is assumed to be a fixed constant.

For any l-uniform hypergraph $\mathcal{G}$ and $k \geq l$, let $\mathcal{K}_k(\mathcal{G})$ be the set of all k-element sets that span a clique $K_k^{(l)}$ on k vertices. We also denote by $K_k(2)$ the complete k-partite k-uniform hypergraph whose every partite set contains precisely two vertices. We refer to $K_k(2)$ as the *generalized octahedron*, or, simply, the *octahedron*.

We also consider a function $\mu_{\mathcal{H}} \colon [V]^k \to \{-1, 1\}$ such that, for all $e \in [V]^k$, we have

$$\mu_{\mathcal{H}}(e) = \begin{cases} -1, & \text{if } e \in \mathcal{H} \\ 1, & \text{if } e \notin \mathcal{H}. \end{cases}$$

Let $[k] = \{1, 2, \ldots, k\}$, let V^{2k} denote the set of all $2k$-tuples $(v_1, v_2, \ldots, v_{2k})$, where $v_i \in V$ $(1 \leq i \leq 2k)$, and let $\Pi_{\mathcal{H}}^{(k)} \colon V^{2k} \to \{-1, 1\}$ be given by

$$\Pi_{\mathcal{H}}^{(k)}(u_1, \ldots, u_k, v_1, \ldots, v_k) = \prod_{\varepsilon} \mu_{\mathcal{H}}(\varepsilon_1, \ldots, \varepsilon_k),$$

where the product is over all vectors $\varepsilon = (\varepsilon_i)_{i=1}^k$ with $\varepsilon_i \in \{u_i, v_i\}$ for all i and we understand $\mu_{\mathcal{H}}$ to be 1 on arguments with repeated entries.

Following Chung and Graham (see, e.g., [4]), we define the *deviation* dev$(\mathcal{H})$ of $\mathcal{H}$ by

$$\text{dev}(\mathcal{H}) = \frac{1}{m^{2k}} \sum_{u_i, v_i \in V, \ i \in [k]} \Pi_{\mathcal{H}}^{(k)}(u_1, \ldots, u_k, v_1, \ldots, v_k).$$

For two hypergraphs $\mathcal{G}$ and $\mathcal{H}$, we denote by $\binom{\mathcal{H}}{\mathcal{G}}$ the set of all induced subhypergraphs of $\mathcal{H}$ that are isomorphic to $\mathcal{G}$. We also write $\binom{\mathcal{H}}{\mathcal{G}}^{\text{w}}$ for the number of *weak* (i.e., not necessarily induced) subhypergraphs of $\mathcal{H}$ that are isomorphic to $\mathcal{G}$. Furthermore, we need the notion of the *link* of a vertex.

Definition 1 *Let $\mathcal{H}$ be a k-uniform hypergraph and $x \in V(\mathcal{H})$. We shall call the $(k-1)$-uniform hypergraph*

$$\mathcal{H}(x) = \{e \setminus \{x\} \colon e \in \mathcal{H}, \ x \in e\}$$

the link of the vertex x in $\mathcal{H}$. For a subset $W \subset V(\mathcal{H})$, the joint W-link is $\mathcal{H}(W) = \bigcap_{x \in W} \mathcal{H}(x)$. For simplicity, if $W = \{x_1, \ldots, x_k\}$, we write $\mathcal{H}(x_1, \ldots, x_k)$.

Observe that if $\mathcal{H}$ is k-partite, then $\mathcal{H}(x)$ is $(k-1)$-partite for every $x \in V$. Furthermore, if $k = 2$, then $\mathcal{H}(x)$ may be identified with the ordinary graph neighbourhood of x. Moreover, $\mathcal{H}(x, x')$ may be thought of as the 'joint neighbourhood' of x and x'.

In [3], Chung and Graham proved that if the density of an m-vertex k-uniform hypergraph $\mathcal{H}$ is $1/2$, i.e., $|\mathcal{H}| = (1/2 + o(1))\binom{m}{k}$, where $o(1) \to 0$ as $m \to \infty$, then the following statements are equivalent:

$(Q_1(s))$ for all k-uniform hypergraphs $\mathcal{G}$ on $s \geq 2k$ vertices and automorphism group $\mathrm{Aut}(G)$,

$$\left| \binom{\mathcal{H}}{\mathcal{G}} \right| = (1 + o(1)) \binom{m}{s} 2^{-\binom{s}{k}} \frac{s!}{|\mathrm{Aut}(\mathcal{G})|},$$

(Q_2) for all k-uniform hypergraphs $\mathcal{G}$ on $2k$ vertices and automorphism group $\mathrm{Aut}(G)$, we have

$$\left| \binom{\mathcal{H}}{\mathcal{G}} \right| = (1 + o(1)) \binom{m}{2k} 2^{-\binom{2k}{k}} \frac{(2k)!}{|\mathrm{Aut}(G)|},$$

(Q_3) $\mathrm{dev}(\mathcal{H}) = o(1)$,

(Q_4) for almost all choices of vertices x, $y \in V$, the $(k-1)$-uniform hypergraph $\mathcal{H}(x) \triangle \mathcal{H}(y)$, that is, the *complement* $[V]^{k-1} \setminus (\mathcal{H}(x) \triangle \mathcal{H}(y))$ of the symmetric difference of $\mathcal{H}(x)$ and $\mathcal{H}(y)$, satisfies Q_2 with k replaced by $k-1$,

(Q_5) for $1 \leq r \leq 2k - 1$ and almost all x, $y \in V$,

$$\left| \binom{\mathcal{H}(x, y)}{K_r^{(k-1)}} \right| = (1 + o(1)) \binom{m}{r} 2^{-\binom{r}{k-1}}.$$

The equivalence of these properties is to be understood in the following sense. If we have two properties $P = P(o(1))$ and $P' = P'(o(1))$, then "$P \Rightarrow P'$" means that for every $\varepsilon > 0$ there is a $\delta > 0$ so that any k-uniform hypergraph $\mathcal{H}$ on m vertices satisfying $P(\delta)$ must also satisfy $P'(\varepsilon)$, provided $m > M_0(\varepsilon)$.

In [3] Chung and Graham stated that "it would be profitable to explore quasi-randomness extended to simulating random k-uniform hypergraphs $G_p(n)$ for $p \neq 1/2$, or, more generally, for $p = p(n)$, especially along the lines carried out so fruitfully by Thomason [13,14]." Our present aim is to explore quasi-randomness from this point of view. In this paper, we concentrate on the case in which p is an arbitrary constant. In certain crucial parts, our methods are different from the ones of Chung and Graham. Indeed, it seems to us that the fact that the density of $\mathcal{H}$ is $1/2$ is essential in certain proofs in [3] (especially those involving the concept of deviation).

1.2   Discrepancy and Subgraph Counting

The following concept was proposed by Frankl and Rödl and later investigated by Chung [2] and Chung and Graham in [3,4]. For an m-vertex k-uniform hypergraph $\mathcal{H}$ with vertex set V, we define the *density* $d(\mathcal{H})$ and the *discrepancy* $\mathrm{disc}_{1/2}(\mathcal{H})$ of $\mathcal{H}$ by letting $d(\mathcal{H}) = |\mathcal{H}| \binom{m}{k}^{-1}$ and

$$\mathrm{disc}_{1/2}(\mathcal{H}) = \frac{1}{m^k} \max_{\mathcal{G} \subset [V]^{k-1}} \left| |\mathcal{H} \cap \mathcal{K}_k(\mathcal{G})| - |\bar{\mathcal{H}} \cap \mathcal{K}_k(\mathcal{G})| \right|, \tag{1}$$

where the maximum is taken over all $(k-1)$-uniform hypergraphs $\mathcal{G}$ with vertex set V, and $\bar{\mathcal{H}}$ is the complement $[V]^k \setminus \mathcal{H}$ of $\mathcal{H}$.

To accommodate arbitrary densities, we extend the latter concept as follows.

Definition 2 *Let $\mathcal{H}$ be a k-uniform hypergraph with vertex set V with $|V| = m$. We define the discrepancy* $\operatorname{disc}(\mathcal{H})$ *of $\mathcal{H}$ as follows:*

$$\operatorname{disc}(\mathcal{H}) = \frac{1}{m^k} \max_{\mathcal{G} \subset [V]^{k-1}} \left| |\mathcal{H} \cap \mathcal{K}_k(\mathcal{G})| - d(\mathcal{H})|\mathcal{K}_k(\mathcal{G})| \right|, \tag{2}$$

where the maximum is taken over all $(k-1)$-uniform hypergraphs $\mathcal{G}$ with vertex set V.

Observe that if $d(\mathcal{H}) = 1/2$, then $\operatorname{disc}(\mathcal{H}) = (1/2)\operatorname{disc}_{1/2}(\mathcal{H})$, so both notions are equivalent. Following some initial considerations by Frankl and Rödl, Chung and Graham investigated the relation between discrepancy and deviation. In fact, Chung [2] succeeded in proving the following inequalities closely connecting these quantities:

(i) $\operatorname{dev}(\mathcal{H}) < 4^k (\operatorname{disc}_{1/2}(\mathcal{H}))^{1/2^k}$,

(ii) $\operatorname{disc}_{1/2}(\mathcal{H}) < (\operatorname{dev}(\mathcal{H}))^{1/2^k}$.

For simplicity, we state the inequalities for the density $1/2$ case. For the general case, see Section 5 of [2].

Before we proceed, we need to introduce a new concept. If the vertex set of a hypergraph is totally ordered, we say that we have an *ordered* hypergraph. Given two ordered hypergraphs $\mathcal{G}_{\leq}$ and $\mathcal{H}_{\leq'}$, where $\leq$ and $\leq'$ denote the orderings on the vertex sets of $\mathcal{G} = \mathcal{G}_{\leq}$ and $\mathcal{H} = \mathcal{H}_{\leq'}$, we say that a function $f \colon V(\mathcal{G}) \to V(\mathcal{H})$ is an *embedding of ordered hypergraphs* if (i) it is an injection, (ii) it respects the orderings, i.e., $f(x) \leq' f(y)$ whenever $x \leq y$, and (iii) $f(g) \in \mathcal{H}$ if and only if $g \in \mathcal{G}$, where $f(g)$ is the set formed by the images of all the vertices in g. Furthermore, if $\mathcal{G} = \mathcal{G}_{\leq}$ and $\mathcal{H} = \mathcal{H}_{\leq'}$, we write $\binom{\mathcal{H}}{\mathcal{G}}_{\mathrm{ord}}$ for the number of such embeddings.

As our main result, we shall prove the following extension of Chung and Graham's result.

Theorem 3 *Let $\mathcal{H} = (V, E)$ be a k-uniform hypergraph of density $0 < d < 1$. Then the following statements are equivalent:*

(P_1) $\operatorname{disc}(\mathcal{H}) = o(1)$,

(P_2) $\operatorname{disc}(\mathcal{H}(x)) = o(1)$ *for all but $o(m)$ vertices $x \in V$ and $\operatorname{disc}(\mathcal{H}(x,y)) = o(1)$ for all but $o(m^2)$ pairs $x, y \in V$,*

(P_3) $\operatorname{disc}(\mathcal{H}(x,y)) = o(1)$ *for all but $o(m^2)$ pairs $x, y \in V$,*

(P_4) *the number of octahedra $K_k(2)$ in $\mathcal{H}$ is asymptotically minimized among all k-uniform hypergraphs of density d; indeed,*

$$\left| \left(\frac{\mathcal{H}}{K_k(2)} \right)^{\mathrm{w}} \right| = (1 + o(1)) \frac{m^{2k}}{2^k k!} d^{2^k},$$

(P_5) *for any $s \geq 2k$ and any k-uniform hypergraph $\mathcal{G}$ on s vertices with $e(\mathcal{G})$ edges and automorphism group $\operatorname{Aut}(\mathcal{G})$,*

$$\left| \binom{\mathcal{H}}{\mathcal{G}} \right| = (1 + o(1)) \binom{m}{s} d^{e(\mathcal{G})} (1-d)^{\binom{s}{k} - e(\mathcal{G})} \frac{s!}{|\operatorname{Aut}(\mathcal{G})|},$$

(P_5') *for any ordering $\mathcal{H}_\le$ of $\mathcal{H}$ and for any fixed integer $s \ge 2k$, any ordered k-uniform hypergraph $\mathcal{G}_\le$ on s vertices with $e(\mathcal{G})$ edges is such that*

$$\left| \binom{\mathcal{H}}{\mathcal{G}}_{\mathrm{ord}} \right| = (1 + o(1)) \binom{m}{s} d^{e(\mathcal{G})} (1 - d)^{\binom{s}{k} - e(\mathcal{G})},$$

(P_6) *for all k-uniform hypergraphs $\mathcal{G}$ on $2k$ vertices with $e(\mathcal{G})$ edges and automorphism group $\mathrm{Aut}(\mathcal{G})$,*

$$\left| \binom{\mathcal{H}}{\mathcal{G}} \right| = (1 + o(1)) \binom{m}{2k} d^{e(\mathcal{G})} (1 - d)^{\binom{2k}{k} - e(\mathcal{G})} \frac{(2k)!}{|\mathrm{Aut}(\mathcal{G})|}.$$

(P_6') *for any ordering $\mathcal{H}_\le$ of $\mathcal{H}$, any ordered k-uniform hypergraph $\mathcal{G}_\le$ on $2k$ vertices with $e(\mathcal{G})$ edges is such that*

$$\left| \binom{\mathcal{H}}{\mathcal{G}}_{\mathrm{ord}} \right| = (1 + o(1)) \binom{m}{2k} d^{e(\mathcal{G})} (1 - d)^{\binom{2k}{k} - e(\mathcal{G})}.$$

Some of the implications in Theorem 3 are fairly easy or are by now quite standard. There are, however, two implications that appear to be more difficult.

The proof of Chung and Graham that $\mathrm{dev}_{1/2}(\mathcal{H}) = o(1)$ implies P_5 (the 'subgraph counting formula') is based on an approach that has its roots in a seminal paper of Wilson [15]. This beautiful proof seems to make non-trivial use of the fact that $d(\mathcal{H}) = 1/2$. Our proof of the implication that small discrepancy implies the subgraph counting formula ($P_1 \Rightarrow P_5'$) is based on a different technique, which works well in the arbitrary constant density case (see Section 2.2).

Our second contribution, which is somewhat more technical in nature, lies in a novel approach for the proof of the implication $P_2 \Rightarrow P_1$. Our proof is based on a variant of the Regularity Lemma of Szemerédi [12] for hypergraphs [6] (see Section 2.1).

2   Main Steps in the Proof of Theorem 3

2.1   The First Part

The first part of the proof of Theorem 3 consists of proving that properties $P_1, \ldots, P_4$ are mutually equivalent. As it turns out, the proof becomes more transparent if we restrict ourselves to k-partite hypergraphs. In the next paragraph, we introduce some definitions that will allow us to state the k-partite version of $P_1, \ldots, P_4$ (see Theorem 15). We close this section introducing the main tool in the proof of Theorem 15, namely, we state a version of the Regularity Lemma for hypergraphs (see Lemma 20).

Definitions for Partite Hypergraphs.   For simplicity, we first introduce the term *cylinder* to mean partite hypergraphs.

Definition 4 *Let $k \ge l \ge 2$ be two integers. We shall refer to any k-partite l-uniform hypergraph $\mathcal{H} = (V_1 \cup \ldots \cup V_k, E)$ as a k-partite l-cylinder or (k, l)-cylinder. If $l = k - 1$, we shall often write $\mathcal{H}_i$ for the subhypergraph of $\mathcal{H}$ induced on $\bigcup_{j \ne i} V_j$. Clearly, $\mathcal{H} = \bigcup_{i=1}^k \mathcal{H}_i$. We shall also denote by $K_k^{(l)}(V_1, \ldots, V_k)$ the complete (k, l)-cylinder with vertex partition $V_1 \cup \ldots \cup V_k$.*

Definition 5 *For a (k, l)-cylinder $\mathcal{H}$, we shall denote by $\mathcal{K}_j(\mathcal{H})$, $l \leq j \leq k$, the (k, j)-cylinder whose edges are precisely those j-element subsets of $V(\mathcal{H})$ that span cliques of order j in $\mathcal{H}$.*

When we deal with cylinders, we have to measure density according to their natural vertex partitions.

Definition 6 *Let $\mathcal{H}$ be a (k, k)-cylinder with k-partition $V = V_1 \cup \ldots \cup V_k$. We define the k-partite density or simply the density $d(\mathcal{H})$ of $\mathcal{H}$ by*

$$d(\mathcal{H}) = \frac{|\mathcal{H}|}{|V_1| \ldots |V_k|}.$$

To be precise, we should have a distinguished piece of notation for the notion of k-partite density. However, the context will always make clear which notion we mean when we talk about the density of a (k, k)-cylinder.

We should also be careful when we talk about the discrepancy of a cylinder.

Definition 7 *Let $\mathcal{H}$ be a (k, k)-cylinder with vertex set $V = V_1 \cup \ldots \cup V_k$. We define the discrepancy $\mathrm{disc}(\mathcal{H})$ of $\mathcal{H}$ as follows:*

$$\mathrm{disc}(\mathcal{H}) = \frac{1}{|V_1| \ldots |V_k|} \max_{\mathcal{G}} \left| |\mathcal{H} \cap \mathcal{K}_k(\mathcal{G})| - d(\mathcal{H})|\mathcal{K}_k(\mathcal{G})| \right|, \tag{3}$$

where the maximum is taken over all $(k, k-1)$-cylinders $\mathcal{G}$ with vertex set $V = V_1 \cup \ldots \cup V_k$.

We now introduce a simple but important concept concerning the "regularity" of a (k, k)-cylinder.

Definition 8 *Let $\mathcal{H}$ be a (k, k)-cylinder with k-partition $V = V_1 \cup \ldots \cup V_k$ and let $\delta < \alpha$ be two positive real numbers. We say that $\mathcal{H}$ is (α, δ)-regular if the following condition is satisfied: if $\mathcal{G}$ is any $(k, k-1)$-cylinder such that $|\mathcal{K}_k(\mathcal{G})| \geq \delta |V_1| \ldots |V_k|$, then*

$$(\alpha - \delta)|\mathcal{K}_k(\mathcal{G})| \leq |\mathcal{H} \cap \mathcal{K}_k(\mathcal{G})| \leq (\alpha + \delta)|\mathcal{K}_k(\mathcal{G})|. \tag{4}$$

Lemma 9 *Let $\mathcal{H}$ be an (α, δ)-regular (k, k)-cylinder. Then $\mathrm{disc}(\mathcal{H}) \leq 2\delta$.*

Lemma 10 *Suppose $\mathcal{H}$ is a (k, k)-cylinder with k-partition $V = V_1 \cup \ldots \cup V_k$. Put $\alpha = d(\mathcal{H})$ and assume that $\mathrm{disc}(\mathcal{H}) \leq \delta$. Then $\mathcal{H}$ is $(\alpha, \delta^{1/2})$-regular.*

The k-Partite Result. Suppose $\mathcal{H}$ is a k-uniform hypergraph and let $\mathcal{H}'$ be a 'typical' k-partite spanning subhypergraph of $\mathcal{H}$. In this section, we relate the discrepancies of $\mathcal{H}$ and $\mathcal{H}'$.

Definition 11 *Let $\mathcal{H} = (V, E)$ be a k-uniform hypergraph with m vertices and let $\mathcal{P} = (V_i)_1^k$ be a partition of V. We denote by $\mathcal{H}_\mathcal{P}$ the (k, k)-cylinder consisting of the edges $h \in \mathcal{H}$ satisfying $|h \cap V_i| = 1$ for all $1 \leq i \leq k$.*

The following lemma holds.

Lemma 12 *For any partition* $\mathcal{P} = (V_i)_1^k$ *of* V, *we have*

(*i*) $\mathrm{disc}(\mathcal{H}) \geq |d(\mathcal{H}_\mathcal{P}) - d(\mathcal{H})||V_1| \ldots |V_k|/m^k$,
(*ii*) $\mathrm{disc}(\mathcal{H}_\mathcal{P}) \leq 2\,\mathrm{disc}(\mathcal{H})m^k/|V_1| \ldots |V_k|$.

An immediate consequence of the previous lemma is the following.

Lemma 13 *If* $\mathrm{disc}(\mathcal{H}) = o(1)$, *then* $\mathrm{disc}(\mathcal{H}_\mathcal{P}) = o(1)$ *for* $(1 - o(1))k^m$ *partitions* $\mathcal{P} = (V_i)_1^k$ *of* V.

With some more effort, one may prove a converse to Lemma 13.

Lemma 14 *Suppose there exists a real number* $\gamma > 0$ *such that* $\mathrm{disc}(\mathcal{H}_\mathcal{P}) = o(1)$ *for* γk^m *partitions* $\mathcal{P} = (V_i)_1^k$ *of* V. *Then* $\mathrm{disc}(\mathcal{H}) = o(1)$.

We now state the k-partite version of a part of our main result, Theorem 3.

Theorem 15 *Suppose* $V = V_1 \cup \ldots \cup V_k$, $|V_1| = \ldots = |V_k| = n$, *and let* $\mathcal{H} = (V, E)$ *be a* (k, k)-*cylinder with* $|\mathcal{H}| = dn^k$. *Then the following four conditions are equivalent:*

(C_1) $\mathcal{H}$ *is* $(d, o(1))$-*regular;*
(C_2) $\mathcal{H}(x)$ *is* $(d, o(1))$-*regular for all but* $o(n)$ *vertices* $x \in V_k$ *and* $\mathcal{H}(x, y)$ *is* $(d^2, o(1))$-*regular for all but* $o(n^2)$ *pairs* $x, y \in V_k$;
(C_3) $\mathcal{H}(x, y)$ *is* $(d^2, o(1))$-*regular for all but* $o(n^2)$ *pairs* $x, y \in V_k$;
(C_4) *the number of copies of* $K_k(2)$ *in* $\mathcal{H}$ *is asymptotically minimized among all such* (k, k)-*cylinders of density* d, *and equals* $(1 + o(1))n^{2k}d^{2^k}/2^k$.

Remark 1. The condition $|V_1| = \ldots = |V_k| = n$ in the result above has the sole purpose of making the statement more transparent. The immediate generalization of Theorem 15 for $V_1, \ldots, V_k$ of arbitrary sizes holds.

Remark 2. The fact that the minimal number of octahedra in a (k, k)-cylinder is asymptotically $(1 + o(1))n^{2k}d^{2^k}/2^k$ is not difficult to deduce from a standard application of the Cauchy–Schwarz inequality for counting "cherries" (paths of length 2) in bipartite graphs.

We leave the derivation of the equivalence of properties $P_1, \ldots, P_4$ from Theorem 15 to the full paper.

A Regularity Lemma. The hardest part in the proof of Theorem 15 is the implication $C_2 \Rightarrow C_1$. In this paragraph, we discuss the main tool used in the proof of this implication. It turns out that, in what follows, the notation is simplified if we consider $(k + 1)$-partite hypergraphs.

Throughout this paragraph, we let $\mathcal{G}$ be a fixed $(k + 1, k)$-cylinder with vertex set $V(\mathcal{G}) = V_1 \cup \ldots \cup V_{k+1}$. Recall that $\mathcal{G} = \bigcup_{i=1}^{k+1} \mathcal{G}_i$, where $\mathcal{G}_i$ is the corresponding (k, k)-cylinder induced on $\bigcup_{j \neq i} V_j$. In this section, we shall focus on "regularizing" the (k, k)-cylinders $\mathcal{G}_1, \ldots, \mathcal{G}_k$, ignoring $\mathcal{G}_{k+1}$. Alternatively, we may assume that $\mathcal{G}_{k+1} = \emptyset$.

Definition 16 *Let* $\mathcal{F} = \bigcup_{i=1}^{k} \mathcal{F}_i$ *be a* $(k, k-1)$-*cylinder with vertex set* $V_1 \cup \ldots \cup V_k$. *For a vertex* $v \in V_{k+1}$, *we define the* $\mathcal{G}$-*link* $\mathcal{G}_{\mathcal{F}}(x)$ *of* x *with respect to* $\mathcal{F}$ *to be the* $(k, k-1)$-*cylinder* $\mathcal{G}_{\mathcal{F}}(x) = \mathcal{G}(x) \cap \mathcal{F}$.

Definition 17 *Let* $W \subset V_{k+1}$ *and let* $\mathcal{F} = \bigcup_{i=1}^{k} \mathcal{F}_i$ *be as above. We shall say that the pair* $(\mathcal{F}, W)$ *is* (ε, d)-*regular if*

$$\left| \frac{|\mathcal{K}_k(\mathcal{G}_{\mathcal{F}}(x))|}{|\mathcal{K}_k(\mathcal{F})|} - d \right| < \varepsilon \tag{5}$$

for all but at most $\varepsilon|W|$ *vertices* $x \in W$, *and*

$$\left| \frac{|\mathcal{K}_k(\mathcal{G}_{\mathcal{F}}(x)) \cap \mathcal{K}_k(\mathcal{G}_{\mathcal{F}}(y))|}{|\mathcal{K}_k(\mathcal{F})|} - d^2 \right| < \varepsilon \tag{6}$$

for all but at most $\varepsilon|W|^2$ *pairs* $x, y \in W$.

Definition 18 *Let* t *be a positive integer and let* $V_{k+1} = W_1 \cup \ldots \cup W_t$ *be an arbitrary partition of* V_{k+1}. *For every* $i \in [k]$, *consider a* t-*partition* $P_i^{(t)} = \{\mathcal{E}_1^{(i)}, \ldots, \mathcal{E}_t^{(i)}\}$ *of* $V_1 \times \ldots \times V_{i-1} \times V_{i+1} \times \ldots \times V_k = \bigcup_{\alpha=1}^{t} \mathcal{E}_\alpha^{(i)}$. *Put* $P^{(t)} = (P_1^{(t)}, \ldots, P_k^{(t)})$. *We shall write* $\mathcal{E}(P^{(t)})$ *for the collection of all* $(k, k-1)$-*cylinders* $\mathcal{E}$ *of the form* $\mathcal{E}_{\alpha_1}^{(1)} \cup \ldots \cup \mathcal{E}_{\alpha_k}^{(k)}$, *where* $\mathcal{E}_{\alpha_i}^{(i)} \in P_i^{(t)}$ *for all* $1 \leq i \leq k$.

Clearly, with the notation as above, we have $|\mathcal{E}(P^{(t)})| = t^k$. Moreover, observe that each of the t^{k+1} pairs $(\mathcal{E}, W_i)$, where $\mathcal{E} \in \mathcal{E}(P^{(t)})$ and $1 \leq i \leq t$, may be classified as ε-regular or ε-irregular (i.e., not ε-regular), according to Definition 17. Also, notice that each $v = (v_1, \ldots, v_{k+1}) \in V_1 \times \ldots \times V_{k+1}$ is 'covered' by exactly one such pair, that is, $v \in \mathcal{K}_k(\mathcal{E}) \times W_i$ for a unique pair $(\mathcal{E}, W_i)$.

Definition 19 *Let* $P^{(t)} = \left(P_i^{(t)}\right)_1^k$ *and* $\left(W_i\right)_1^t$ *be as in Definition 18. We shall say that the system of partitions* $\left\{P_1^{(t)}, \ldots, P_k^{(t)}, \{W_1, \ldots, W_t\}\right\}$ *is* ε-*regular if the number of* $(k+1)$-*tuples* $(v_1, \ldots, v_{k+1}) \in V_1 \times \ldots \times V_{k+1}$ *that are not covered by the family of* ε-*regular pairs* $(\mathcal{E}, W_i)$ *with* $\mathcal{E} \in \mathcal{E}(P^{(t)})$ *and* $1 \leq i \leq t$ *is at most* $\varepsilon|V_1| \ldots |V_{k+1}|$.

The main tool in the proof of $C_2 \Rightarrow C_1$ is the following result (see [9] for the details).

Lemma 20 *For every* $\varepsilon > 0$ *and* $t_0 \geq 1$, *there exist integers* n_0 *and* T_0 *such that every* $(k+1, k)$-*cylinder* $\mathcal{G} = \bigcup_{i=1}^{k+1} \mathcal{G}_i$ *with vertex set* $V_1 \cup \ldots \cup V_{k+1}$, *where* $|V_i| \geq n_0 \ \forall i$, $1 \leq i \leq k+1$, *admits an* ε-*regular system of partitions* $\{P_1^{(t)}, \ldots, P_k^{(t)}, \{W_1, \ldots, W_t\}\}$ *with* $t_0 < t < T_0$.

2.2 The Subgraph Counting Formula

In this section, we shall state the main result that may be used to prove the implication $P_1 \Rightarrow P_5'$. To this end, we need to introduce some notation. Throughout this section, $s \geq 2k$ is some fixed integer.

If $\mathcal{H}$ and $\mathcal{G}$ are, respectively, k-uniform and ℓ-uniform ($k \geq \ell$), then we say that $\mathcal{H}$ is *supported* on $\mathcal{G}$ if $\mathcal{H} \subset \mathcal{K}_k(\mathcal{G})$.

Suppose we have pairwise disjoint sets $W_1, \ldots, W_s$, with $|W_i| = n$ for all i. Suppose further that we have a sequence $\mathcal{G}^{(2)}, \ldots, \mathcal{G}^{(k)}$ of s-partite cylinders on $W_1 \cup \ldots \cup W_s$, with $\mathcal{G}^{(i)}$ an (s, i)-cylinder and, moreover, such that $\mathcal{G}^{(i)}$ is supported on $\mathcal{G}^{(i-1)}$ for all $3 \leq i \leq k$. Suppose also that, for all $2 \leq i \leq k$ and for all $1 \leq j_1 < \ldots < j_i \leq s$, the (i, i)-cylinder $\mathcal{G}[j_1, \ldots, j_i] = \mathcal{G}^{(i)}[W_{j_1} \cup \ldots \cup W_{j_i}]$ is (γ_i, δ)-*regular with respect to* $\mathcal{G}^{(i-1)}[j_1, \ldots, j_i] = \mathcal{G}^{(i-1)}[W_{j_1} \cup \ldots \cup W_{j_i}]$, that is, whenever $\mathcal{G} \subset \mathcal{G}^{(i-1)}[j_1, \ldots, j_i]$ is such that $|\mathcal{K}_i(\mathcal{G})| \geq \delta|\mathcal{K}_i(\mathcal{G}^{(i-1)}[j_1, \ldots, j_i])|$, we have

$$(\gamma_i - \delta)|\mathcal{K}_i(\mathcal{G})| \leq |\mathcal{G}[j_1, \ldots, j_i] \cap \mathcal{K}_i(\mathcal{G})| \leq (\gamma_i + \delta)|\mathcal{K}_i(\mathcal{G})|.$$

Finally, let us say that a copy of $K_s^{(k)}$ in $W_1 \cup \ldots \cup W_s$ is *transversal* if $|V(K_s^{(k)}) \cap W_i| = 1$ for all $1 \leq i \leq s$.

Our main result concerning counting subhypergraphs is then the following.

Theorem 21 *For any $\varepsilon > 0$ and any $\gamma_2, \ldots, \gamma_k > 0$, there is $\delta_0 > 0$ such that if $\delta < \delta_0$, then the number of transversal $K_s^{(k)}$ in $\mathcal{G}^{(k)}$ is $(1 + O(\varepsilon))\gamma_k^{\binom{s}{k}} \ldots \gamma_2^{\binom{s}{2}} n^s$.*

Theorem 21 above is an instance of certain counting lemmas developed by Rödl and Skokan for such *complexes* $\mathcal{G} = \left(\mathcal{G}^{(i)}\right)_{2 \leq i \leq k}$ (see, e.g., [11]).

3   Concluding Remarks

We hope that the discussion above on our proof approach for Theorem 3 gives some idea about our methods and techniques. Unfortunately, because of space limitations and because we discuss the motivation behind our work in detail, we are unable to give more details. We refer the interested reader to [9].

It is also our hope that the reader will have seen that many interesting questions remain. Probably, the most challenging of them concerns developing an applicable theory of sparse quasi-random hypergraphs. Here, we have in mind such lemmas for sparse quasi-random graphs as the ones in [10].

References

1. L. Babai, N. Nisan, and M. Szegedy, *Multiparty protocols, pseudorandom generators for logspace, and time-space trade-offs*, J. Comput. System Sci. **45** (1992), no. 2, 204–232, Twenty-first Symposium on the Theory of Computing (Seattle, WA, 1989).
2. F.R.K. Chung, *Quasi-random classes of hypergraphs*, Random Structures and Algorithms **1** (1990), no. 4, 363–382.
3. F.R.K. Chung and R.L. Graham, *Quasi-random hypergraphs*, Random Structures and Algorithms **1** (1990), no. 1, 105–124.
4. _____, *Quasi-random set systems*, Journal of the American Mathematical Society **4** (1991), no. 1, 151–196.
5. F.R.K. Chung and P. Tetali, *Communication complexity and quasi randomness*, SIAM J. Discrete Math. **6** (1993), no. 1, 110–123.

6. P. Frankl and V. Rödl, *The uniformity lemma for hypergraphs*, Graphs and Combinatorics **8** (1992), no. 4, 309–312.
7. J. Haviland and A.G. Thomason, *Pseudo-random hypergraphs*, Discrete Math. **75** (1989), no. 1–3, 255–278, Graph theory and combinatorics (Cambridge, 1988).
8. _____ , *On testing the "pseudo-randomness" of a hypergraph*, Discrete Math. **103** (1992), no. 3, 321–327.
9. Y. Kohayakawa, V. Rödl, and J. Skokan, *Equivalent conditions for regularity*, in preparation, 1999.
10. Y. Kohayakawa, V. Rödl, and E. Szemerédi, *The size-Ramsey number of graphs of bounded degree*, in preparation, 1999.
11. V. Rödl and J. Skokan, *Uniformity of set systems*, in preparation, 1999.
12. E. Szemerédi, *Regular partitions of graphs*, Problèmes Combinatoires et Théorie des Graphes (Colloq. Internat. CNRS, Univ. Orsay, Orsay, 1976) (Paris), Colloques Internationaux CNRS n. 260, 1978, pp. 399–401.
13. A.G. Thomason, *Pseudorandom graphs*, Random graphs '85 (Poznań, 1985), North-Holland Math. Stud., vol. 144, North-Holland, Amsterdam-New York, 1987, pp. 307–331.
14. _____ , *Random graphs, strongly regular graphs and pseudorandom graphs*, Surveys in Combinatorics 1987 (C. Whitehead, ed.), London Mathematical Society Lecture Note Series, vol. 123, Cambridge University Press, Cambridge–New York, 1987, pp. 173–195.
15. R.M. Wilson, *Cyclotomy and difference families in elementary abelian groups*, J. Number Theory **4** (1972), 17–47.

Cube Packing

F.K. Miyazawa[1]* and Y. Wakabayashi[2]*

[1] Instituto de Computação — Universidade Estadual de Campinas
Caixa Postal 6176 — 13083-970 — Campinas–SP — Brazil
fkm@dcc.unicamp.br
[2] Instituto de Matemática e Estatística — Universidade de São Paulo
Rua do Matão, 1010 — 05508-900 — São Paulo–SP — Brazil
yw@ime.usp.br

Abstract. The Cube Packing Problem (CPP) is defined as follows. Find a packing of a given list of (small) cubes into a minimum number of (larger) identical cubes. We show first that the approach introduced by Coppersmith and Raghavan for general online algorithms for packing problems leads to an online algorithm for CPP with asymptotic performance bound 3.954. Then we describe two other offline approximation algorithms for CPP: one with asymptotic performance bound 3.466 and the other with 2.669. A parametric version of this problem is defined and results on online and offline algorithms are presented. We did not find in the literature offline algorithms with asymptotic performance bounds as good as 2.669.

1 Introduction

The *Cube Packing Problem* (CPP) is defined as follows. Given a list L of n cubes (of different dimensions) and identical cubes, called *bins*, find a packing of the cubes of L into a minimum number of bins. The packings we consider are all orthogonal. That is, with respect to a fixed side of the bin, the sides of the cubes must be parallel or orthogonal to it.

CPP is a special case of the *Three-dimensional Bin Packing Problem* (3BP). In this problem the list L consists of rectangular boxes and the bins are also rectangular boxes. Here, we may assume that the bins are cubes, since otherwise we can scale the bins and the boxes in L correspondingly.

In 1989, Coppersmith and Raghavan [6] presented an online algorithm for 3BP, with asymptotic performance bound 6.25. Then, in 1992, Li and Cheng [11] presented an algorithm with asymptotic performance bound close to 4.93. Improving the latter result, Csirik and van Vliet [7], and also Li and Cheng [10] designed algorithms for 3BP with asymptotic performance bound 4.84 (the best bound known for this problem). Since CPP is a special case of 3BP, these

---

* This work has been partially supported by Project ProNEx 107/97 (MCT/FINEP), FAPESP (Proc. 96/4505–2), and CNPq individual research grants (Proc. 300301/98-7 and Proc. 304527/89-0).

algorithms can be used to solve it. Our aim is to show that algorithms with better asymptotic performance bounds can be designed.

Results of this kind have already been obtained for the 2-dimensional case, more precisely, for the *Square Packing Problem* (SPP). In this problem we are given a list of squares and we are aked to pack them into a minimum number of square bins. In [6], Coppersmith and Raghavan observe that their technique leads to an online algorithm for SPP with asymptotic performance bound 2.6875. They also proved that any online algorithm for packing d-dimensional squares, $d \geq 2$, must have asymptotic performance bound at least $4/3$. Ferreira, Miyazawa and Wakabayashi [9] presented an offline algorithm for SPP with asymptotic performance bound 1.988. For the more general version of the 2-dimensional case, where the items of L are rectangles (instead of squares), Chung, Garey and Johnson [2] designed an algorithm with asymptotic performance bound 2.125.

For more results on packing problems the reader is referred to [1,3,4,5,8].

The remainder of this paper is organized as follows. In Section 2 we present some notation and definitions. In Section 3 we describe an online algorithm for CPP that uses an approach introduced by Coppersmith and Raghavan [6], showing that its asymptotic performance bound is at most 3.954. In Section 4 we present an offline algorithm with asymptotic performance bound 3.466. We mention a parametric version for these algorithms and derive asymptotic performance bounds. In Section 5 we present an improved version of the offline algorithm described in Section 4. We show that this algorithm has asymptotic performance bound 2.669. Finally, in Section 6 we present some concluding remarks.

2   Notation and Definitions

The reader is referred to [14] for the basic concepts and terms related to packing. Without loss of generality, we assume that the bins have unit dimensions, since otherwise we can scale the cubes of the instance to fulfill this condition.

A rectangular box b with length x, width y and height z is denoted by a triplet $b = (x, y, z)$. Thus, a cube is simply a triplet of the form (x, x, x). The *size* of a cube $c = (x, x, x)$, denoted by $s(c)$, is x. Here we assume that every cube in the input list L has size at most 1. The *volume* of a list L, denoted by $V(L)$, is the sum of the volumes of the items in L.

For a given list L and algorithm $\mathcal{A}$, we denote by $\mathcal{A}(L)$ the number of bins used when algorithm $\mathcal{A}$ is applied to list L, and by $\mathrm{OPT}(L)$ the optimum number of bins for a packing of L. We say that an algorithm $\mathcal{A}$ has an *asymptotic performance bound* α if there exists a constant β such that

$$\mathcal{A}(L) \leq \alpha \cdot \mathrm{OPT}(L) + \beta, \quad \text{for all input list } L.$$

If $\beta = 0$ then we say that α is an *absolute performance bound* for algorithm $\mathcal{A}$.

If $\mathcal{P}$ is a packing, then we denote by $\#(\mathcal{P})$ the number of bins used in $\mathcal{P}$.

An algorithm to pack a list of items $L = (c_1, \ldots, c_n)$ is said to be *online* if it packs the items in the order given by the list L, without knowledge of the subsequent items on the list. An algorithm that is not online is said to be *offline*.

We consider here a *parametric* version of CPP, denoted by CPP_m, where m is a natural number. In this problem, the instance L consists of cubes with size at most $1/m$. Thus CPP_1 and CPP are the same problem.

3   The Online Algorithm of Coppersmith and Raghavan

In 1989, Coppersmith and Raghavan [6] introduced an online algorithm for the multidimensional bin packing problem. In this section we describe a specialized version of this algorithm for CPP. Our aim is to derive an asymptotic performance bound for this algorithm (not explicitly given in the above paper).

The main idea of the algorithm is to round up the dimensions of the items in L using a rounding set $S = \{1 = s_0, s_1, \ldots, s_i, \ldots\}$, $s_i > s_{i+1}$. The first step consists in rounding up each item size to the nearest value in S. The rounding set S for CPP is $S := S_1 \cup S_2 \cup S_3$, where

$$S_1 = \{1\}, \quad S_2 = \{1/2, 1/4, \ldots, 1/2^k, \ldots\}, \quad S_3 = \{1/3, 1/6, \ldots, 1/(3 \cdot 2^k), \ldots\}.$$

Let $\overline{x}$ be the value obtained by rounding up x to the nearest value in S. Given a cube $c = (x, x, x)$, define $\overline{c}$ as the cube $\overline{c} := (\overline{x}, \overline{x}, \overline{x})$. Let $\overline{L}$ be the list obtained from L by rounding up the sizes of the cubes to the values in the rounding set S. The idea is to pack the cubes of the list $\overline{L}$ instead of L, so that the packing of each cube $\overline{c} \in \overline{L}$ represents the packing of $c \in L$. The packing of $\overline{L}$ is generated into bins belonging to three different groups: G_1, G_2 and G_3. Each group G_i contains only bins of dimensions $(x, x, 1)$, $x \in S_i$, $i = 1, 2, 3$. A bin of dimension $(x, x, 1)$ will have only cubes $c = (x, x, x)$ packed into it. We say that a cube $c = (x, x, x)$ is of *type i*, if $\overline{x} \in S_i$, $i = 1, 2, 3$.

To pack the next unpacked cube $\overline{c} \in \overline{L}$ with size $x \in S_i$, we proceed as follows.

1. Let $B \in G_i$ be the first bin $B = (x, x, 1)$, such that $\sum_{b \in B} s(b) + x \leq 1$ (if there exists such a bin B).
2. If there is a bin B as in step 1, pack $\overline{c}$ in a Next Fit manner into B.
3. Otherwise,
   a) take the first empty bin $C = (y, y, 1)$, $y \in S_i$, with $y > x$ and y as small as possible. If there is no such bin C, take a new bin $(1, 1, 1)$ and replace it by i^2 bins of dimensions $(\frac{1}{i}, \frac{1}{i}, 1)$ and let $C = (y, y, 1)$ be the first of these i^2 bins.
   b) If $y > x$, then replace C by other four bins of dimensions $(\frac{y}{2}, \frac{y}{2}, 1)$. Continue in this manner replacing one of these new bins by four bins, until there is a bin C of dimension $C = (\frac{y}{2^m}, \frac{y}{2^m}, 1)$ with $\frac{y}{2^m} = x$.
   c) Pack $\overline{c}$ in a Next Fit manner into the first bin C.
4. Update the group G_i.

Let us now analyse the asymptotic performance of the algorithm we have described. Consider $\mathcal{P}$ the packing of L generated by this algorithm, L_i the set of all cubes of type i in L, and $\mathcal{P}_i$ the set of bins of $\mathcal{P}$ having only cubes of type i. Now, let us consider the bins $B = (x, x, 1)$, $x \in S_i$, and compute the volume occupied by the cubes of $\overline{L}$ that were packed into these bins. All bins in the group G_1 are completely filled. Thus, $\#(\mathcal{P}_1) = V(\overline{L}_1)$. For the bins in the groups G_2 and G_3 the unoccupied volume is at most 1 for each group. Therefore, we have $\#(\mathcal{P}_i) \leq V(\overline{L}_i) + 1$, $\quad i = 2, 3$.

Now, let us consider the volume we increased because of the rounding process. Each cube $c \in L_1$ has volume at least $\frac{1}{8}$ of $\overline{c}$, and each cube $c \in L_2 \cup L_3$ has volume at least $\frac{8}{27}$ of $\overline{c}$. Hence, we have the following inequalities:

$$\#(\mathcal{P}_1) \leq \frac{1}{1/8} V(L_1), \text{ and } \#(\mathcal{P}_2 \cup \mathcal{P}_3) \leq \frac{1}{8/27} V(L_2 \cup L_3) + 2.$$

Let $n_1 := \#(\mathcal{P}_1)$ and $n_{23} := \#(\mathcal{P}_2 \cup \mathcal{P}_3) - 2$. Thus, using the inequalities above and the fact that the volume of the cubes in L is a lower bound for the optimum packing, we have $\mathrm{OPT}(L) \geq V(L) \geq \frac{1}{8} n_1 + \frac{8}{27} n_{23}$.

Since $\mathrm{OPT}(L) \geq n_1$, it follows that $\mathrm{OPT}(L) \geq \max\{n_1, \frac{1}{8} n_1 + \frac{8}{27} n_{23}\}$. Now using the fact that $\#(\mathcal{P}) = \#(\mathcal{P}_1) + \#(\mathcal{P}_2 \cup \mathcal{P}_3) = n_1 + n_{23} + 2$, we have

$$\#(\mathcal{P}) \leq \alpha \cdot \mathrm{OPT}(L) + 2,$$

where $\alpha = (n_1 + n_{23})/(\max\{n_1, \frac{1}{8} n_1 + \frac{8}{27} n_{23}\})$. Analysing the two possible cases for the denominator, we obtain $\alpha \leq 3.954$.

The approach used above can also be used to develop online algorithms for the parametric version CPP_m. In this case we partition the packing into two parts. One part is an optimum packing with all bins, except perhaps the last (say n' bins), filled with m^3 cubes of volume at least $(1/(m+1))^3$ each. The other part is a packing with all bins, except perhaps a fixed number of them (say n'' bins), having an occupied volume of at least $((m+1)/(m+2))^3$.

It is not difficult to show that the asymptotic performance bound α_m of CPP_m is bounded by $(n'+n'')/(\max\{n', (m/(m+1))^3 n' + ((m+1)/(m+2))^3 n''\})$. For $m = 2$ and $m = 3$ these values are at most 2.668039 and 2.129151, respectively.

4   An Offline Algorithm

Before we present our first offline algorithm for CPP, let us describe the algorithm NFDH (Next Fit Descreasing Height), which is used as a subroutine.

NFDH first sorts the cubes of L in nonincreasing order of their size, say $c_1, c_2, \ldots, c_n$. The first cube c_1 is packed in the position $(0, 0, 0)$, the next one is packed in the position $(s(c_1), 0, 0)$ and so on, side by side, until a cube that does not fit in this layer is found. At this moment the next cube c_k is packed in the position $(0, s(c_1), 0)$. The process continues in this way, layer by layer, until a cube that does not fit in the first level is found. Then the algorithm packs this cube in a new level at height $s(c_1)$. When a cube cannot be packed in a bin, it is packed in a new bin. The process proceeds in this way until all cubes of L have been packed.

The following results will be used in the sequel. The proof of Lemma 1 is left to the reader. Theorem 2 follows immediately from Lemma 1.

Theorem 1 (Meir and Moser [12]). *Any list L of k-dimensional cubes, with sizes $x_1 \geq x_2 \geq \cdots \geq x_n \geq \cdots$, can be packed by algorithm NFDH into only one k-dimensional rectangular parallelepiped of volume $a_1 \times a_2 \times \ldots \times a_k$ if $a_j > x_1$ $(j = 1, \ldots, k)$ and $x_1^k + (a_1 - x_1)(a_2 - x_1) \cdots (a_k - x_1) \geq V(L)$.*

Lemma 1. *For any list of cubes $L = (c_1, \ldots, c_n)$ such that $x(c_i) \leq \frac{1}{m}$, the following holds for the packing of L into unit bins:*

$$\mathrm{NFDH}(L) \leq ((m+1)/m)^3 V(L) + 2.$$

Theorem 2. *For any list of cubes $L = (b_1, \ldots, b_n)$ such that $x(b_i) \leq \frac{1}{m}$, the following holds for the packing of L into unit bins:*

$$\mathrm{NFDH}(L) \leq ((m+1)/m)^3 \mathrm{OPT}(L) + 2.$$

Before presenting the first offline algorithm, called CUBE, let us introduce a definition and the main ideas behind it.

If a packing $\mathcal{P}$ of a list L satisfies the inequality $\#(\mathcal{P}) \leq \frac{V(L)}{v} + C$, where v and C are constants, then we say that v is a *volume guarantee* of the packing $\mathcal{P}$ (for the list L). Algorithm CUBE uses an approach, which we call *critical set combination* (see [13]), based on the following observation.

Recall that in the analysis of the performance of the algorithm presented in Section 3 we considered the packing divided into two parts. One optimum packing, of the list L_1, with a volume guarantee $\frac{1}{8}$, and the other part, of the list $L_{23} = L_2 \cup L_3$, with a volume guarantee $\frac{8}{27}$. If we consider this partition of L, the volume we can guarantee in each bin is the best possible, as we can have cubes in L_1 with volume very close to $\frac{1}{8}$, and cubes in L_{23} for which we have a packing with volume occupation in each bin very close to $\frac{8}{27}$. In the critical set combination approach, we first define some subsets of cubes in L_1 and L_{23} with small volumes as the critical sets. Then we combine the cubes in these critical sets obtaining a partial packing that is part of an optimum packing and has volume occupation in each bin better than $\frac{1}{8}$. That is, sets of cubes that would lead to small volume occupation are set aside and they are combined appropriately so that the resulting packing has a better volume guarantee.

Theorem 3. *For any list L of cubes for CPP, we have*

$$\mathrm{CUBE}(L) \leq 3.466 \cdot \mathrm{OPT}(L) + 4.$$

Proof. First, consider the packing $\mathcal{P}_{AB}$. Since each bin of $\mathcal{P}_{AB}$, except perhaps the last, contains one cube of L_A and seven cubes of L_B, we have

$$\#(\mathcal{P}_{AB}) \leq \frac{1}{83/216} V(L_{AB}) + 1, \tag{1}$$

where L_{AB} is the set of cubes of L packed in $\mathcal{P}_{AB}$.

Algorithm CUBE

> // To pack a list of cubes L into unit bins $B = (1, 1, 1)$.

1 Let $p = 0.354014$; and L_A, L_B be sublists of L defined as follows.

$$L_A \leftarrow \{c \in L_1 : \tfrac{1}{2} < s(c) \leq (1 - p)\}, \quad L_B \leftarrow \{c \in L_2 : \tfrac{1}{3} < s(c) \leq p\}.$$

2 Generate a partial packing $\mathcal{P}_{AB}$ of $L_A \cup L_B$, such that $\mathcal{P}_{AB}$ is the union of packings $\mathcal{P}_{AB}^1, \ldots, \mathcal{P}_{AB}^k$, where $\mathcal{P}_{AB}^i$ is a packing generated for one bin, consisting of one cube of L_A and seven cubes of L_B, except perhaps the last (that can have fewer cubes of L_B). [The packing $\mathcal{P}_{AB}$ will contain all cubes of L_A or all cubes of L_B.] Update the list L by removing the cubes packed in $\mathcal{P}_{AB}$.

3 $\mathcal{P}^{\cdot} \leftarrow \mathrm{NFDH}(L)$;

4 Return $\mathcal{P}^{\cdot} \cup \mathcal{P}_{AB}$.

end algorithm.

Now consider a partition of $\mathcal{P}'$ into three partial packings $\mathcal{P}_1$, $\mathcal{P}_2$ and $\mathcal{P}_3$, defined as follows. The packing $\mathcal{P}_1$ has the bins of $\mathcal{P}'$ with at least one cube of size greater than $\tfrac{1}{2}$. The packing $\mathcal{P}_2$ has the bins of $\mathcal{P}' \setminus \mathcal{P}_1$ with at least one cube of size greater than $\tfrac{1}{3}$. The packing $\mathcal{P}_3$ has the remaining bins, *i.e.*, the bins in $\mathcal{P}' \setminus (\mathcal{P}_1 \cup \mathcal{P}_2)$. Let L_i be the set of cubes packed in $\mathcal{P}_i$, $i = 1, 2, 3$.

Since all cubes of L_3 have size at most $1/3$, and they are packed in $\mathcal{P}_3$ with algorithm NFDH, by Lemma 1, we have

$$\#(\mathcal{P}_3) \leq \frac{1}{27/64} V(L_3) + 2. \tag{2}$$

Case 1. L_B is totally packed in $\mathcal{P}_{AB}$.

In this case, every cube of L_1 has volume at least $\tfrac{1}{8}$. Therefore

$$\#(\mathcal{P}_1) \leq \frac{1}{1/8} V(L_1). \tag{3}$$

Now, since every cube of L_2 has size at least p and each bin of packing $\mathcal{P}_2$ has at least 8 cubes of L_2, we have

$$\#(\mathcal{P}_2) \leq \frac{1}{8p^3} V(L_2) + 1. \tag{4}$$

Since $8p^3 = \min\{8p^3, \frac{83}{216}, \frac{27}{64}\}$, using (1), (2) and (4), and setting $\mathcal{P}_{aux} := \mathcal{P}_{AB} \cup \mathcal{P}_3 \cup \mathcal{P}_2$, we have

$$\#(\mathcal{P}_{aux}) \leq \frac{1}{8p^3} V(L_{aux}) + 4. \tag{5}$$

Clearly, $\mathcal{P}_1$ is an optimum packing of L_1, and hence

$$\#(\mathcal{P}_1) \leq \mathrm{OPT}(L). \tag{6}$$

Defining $h_1 := \#(\mathcal{P}_1)$ and $h_2 := \#(\mathcal{P}_{aux}) - 4$, and using inequalities (3), (5) and (6) we have

$$\text{CUBE}(L) \leq \alpha' \cdot \text{OPT}(L) + 4,$$

where $\alpha' = (h_1 + h_2)/(\max\{h_1, \frac{1}{8}h_1 + 8p^3 h_2\}) \leq 3.466$.

Case 2. L_A is totally packed in $\mathcal{P}_{AB}$.

In this case, the volume guarantee for the cubes in L_1 is better than the one obtained in Case 1. Each cube of L_1 has size at least $1 - p$. Thus, $\#(\mathcal{P}_1) \leq \frac{1}{(1-p)^3} V(L_1)$. For the packing $\mathcal{P}_2$, we obtain a volume guarantee of at least $\frac{8}{27}$, and the same holds for the packings $\mathcal{P}_3$ and $\mathcal{P}_{AB}$. Thus, for $\mathcal{P}_{aux}$ as above, $\#(\mathcal{P}_{aux}) \leq \frac{1}{8/27} V(L_{aux}) + 4$.

Since $\#(\mathcal{P}_1) \leq \text{OPT}(L)$, combining the previous inequalities and proceeding as in Case 1, we have

$$\text{CUBE}(L) \leq \alpha'' \cdot \text{OPT}(L) + 4,$$

where $\alpha'' = (h_1 + h_2)/(\max\{h_1, (1-p)^3 h_1 + \frac{8}{27}h_2\}) \leq 3.466$.

The proof of the theorem follows from the results obtained in Case 1 and Case 2. We observe that the value of p was obtained by imposing equality for the values of α' and α''. $\square$

Algorithm CUBE can also be generalized for the parametric problem CPP_m. The idea is the same as the one used in algorithm CUBE. The input list is first subdivided into two parts, P_1 and P_2. Part P_1 consists of those cubes with size in $\left(\frac{1}{m+1}, \frac{1}{m}\right]$, and part P_2 consists of the remaining cubes. The critical cubes in each part are defined using an appropriate value of $p = p(m)$, and then combined. The analysis is also divided into two parts, according to which critical set is totally packed in the combined packing. It is not difficult to derive the bounds $\alpha(\text{CUBE}_m)$ that can be obtained for the corresponding algorithms. For $m = 2$ and $m = 3$ the values of $\alpha(\text{CUBE}_m)$ are at most 2.42362852 ($p = 0.26355815$) and 1.98710756 ($p = 0.20916664$), respectively.

5   An Improved Algorithm for CPP

We present in this section an algorithm for the cube packing problem that is an improvement of algorithm CUBE described in the previous section. For that, we consider another restricted version of CPP, denoted by CPP^k, where k is an integer greater than 2. In this problem the instance is a list L consisting of cubes of size greater than $\frac{1}{k}$. We use in the sequel the following result for CPP^3.

Lemma 2. *There is a polynomial time algorithm to solve CPP^3.*

Proof. Let $L_1 = \{c \in L : s(c) > \frac{1}{2}\}$ and $L_2 = L \setminus L_1$. Without loss of generallity, consider $L_1 = (c_1, \ldots, c_k)$. Pack each cube $c_i \in L_1$ in a unit bin C_i at the corner $(0,0,0)$. Note that it is possible to pack seven cubes with size at most $1 - s(c_i)$

in each bin C_i. Now, for each bin C_i, consider seven other smaller bins $C_i^{(j)}$, $j = 1, \ldots, 7$, each with size $1 - s(c_i)$. Consider a bipartite graph G with vertex set $X \cup Y$, where X is the set of the small bins, and Y is precisely L_2. In G there is an edge from a cube $c \in L_2$ to a bin $C_i^{(j)} \in X$ if and only if c can be packed into $C_i^{(j)}$. Clearly, a maximum matching in G corresponds to a maximum packing of the cubes of L_2 into the bins occupied by the cubes of L_1. Denote by $\mathcal{P}_{12}$ the packing of L_1 combined with the cubes of L_2 packed with the matching strategy. The optimum packing of L can be obtained by adding to the packing $\mathcal{P}_{12}$ the bins packed with the remaining cubes of L_2 (if existent), each with 8 cubes, except perhaps the last. $\square$

We say that a cube c is of type G, resp. M, if $s(c) \in \left(\frac{1}{2}, 1\right]$, resp. $s(c) \in \left(\frac{1}{3}, \frac{1}{2}\right]$.

Lemma 3. *It is possible to generate an optimum packing of an instance of* CPP^3 *such that each bin, except perhaps one, has one of the following configurations:*
(a) C1: configuration consisting of 1 cube of type G and 7 cubes of type M;
(b) C2: configuration consisting of exactly 1 cube of type G; and
(c) C3: configuration consisting of 8 cubes of type M.

Lemma 2 shows the existence of a polynomial time optimum algorithm for CPP^3. In fact, it is not difficult to design a greedy-like algorithm to solve CPP^3 in time $O(n \log n)$. Such an algorithm is given in [9] for SPP^3 (defined analogously with respect to SPP).

We are now ready to present the improved algorithm for the cube packing problem, which we call ICUBE (Improved CUBE).

Algorithm ICUBE

   // *To pack a list of cubes L into unit bins $B = (1, 1, 1)$.*

1. Let $L_1^{\cdot} \leftarrow \{q \in L : \frac{1}{3} < s(q) \leq 1\}$.
2. Generate an optimum packing $\mathcal{P}_1^{\cdot}$ of $L_1^{\cdot}$ (in polynomial time), with bins as in Lemma 3. That is, solve CPP^3 with input list $L_1^{\cdot}$.
3. Let $\mathcal{P}_A$ be the set of bins $B \in \mathcal{P}_1^{\cdot}$ having configuration $C2$ with a cube $q \in B$ with $s(q) \leq \frac{2}{3}$; let L_A be the set of cubes packed in $\mathcal{P}_A$.
4. Let $L_B \leftarrow \{q \in L : 0 < s(q) \leq \frac{1}{3}\}$.
5. Generate a packing $\mathcal{P}_{AB}$ filling the bins in $\mathcal{P}_A$ with cubes of L_B (see below).
6. Let L_1 be the set of all packed cubes, and $\mathcal{P}_1$ the packing generated for L_1.
7. Let $\mathcal{P}_2$ be the packing of the unpacked cubes of L_B generated by NFDH.
8. Return the packing $\mathcal{P}_1 \cup \mathcal{P}_2$.
end algorithm

To generate the packing $\mathcal{P}_{AB}$, in step 5 of algorithm ICUBE, we first partition the list L_B into 5 lists, $L_{B,3}$, $L_{B,4}$, $L_{B,5}$, $L_{B,6}$, $L_{B,7}$, defined as follows. $L_{B,i} = \{c \in L_B : \frac{1}{i+1} < s(c) \leq \frac{1}{i}\}$, $i = 3, \ldots, 6$ and $L_{B,7} = \{c \in L_B : s(c) \leq \frac{1}{7}\}$. Then we combine the cubes in each of these lists with the packing $\mathcal{P}_A$ generated in step 3.

Now consider the packing of cubes of $L_{B,3}$ into bins of $\mathcal{P}_A$. Since we can pack 19 cubes of $L_{B,3}$ into each of these bins, we generate such a packing until all cubes of $L_{B,3}$ have been packed, or until there are no more bins in $\mathcal{P}_A$. We generate similar packings combining the remaining bins of $\mathcal{P}_A$ with the lists $L_{B,4}$, $L_{B,5}$ and $L_{B,6}$. To pack the cubes of $L_{B,7}$ into bins of $\mathcal{P}_A$ we consider the empty space of the bin divided into three smaller bins of dimensions $(1, 1, \frac{1}{3})$, $(1, \frac{1}{3}, \frac{1}{3})$ and $(\frac{1}{3}, \frac{1}{3}, \frac{1}{3})$. Then use NFDH to pack the cubes in $L_{B,7}$ into these smaller bins. We continue the packing of $L_{B,7}$ using other bins of $\mathcal{P}_A$ until there are no more unpacked cubes of $L_{B,7}$, or all bins of $\mathcal{P}_A$ have been considered.

Theorem 4. *For any instance L of* CPP, *we have*

$$\text{ICUBE}(L) \leq 2.669 \cdot \text{OPT}(L) + 7.$$

Proof. (Sketch) Let $\mathcal{C}_1', \mathcal{C}_2'$ and $\mathcal{C}_3'$ be the set of bins used in $\mathcal{P}_1'$ with configurations C1, C2 and C3, respectively. Considering the volume guarantees of $\mathcal{C}_1', \mathcal{C}_2'$ and $\mathcal{C}_3'$, we have $\#(\mathcal{C}_1') \leq \frac{1}{1/8+7/27}V(\mathcal{C}_1') + 1$, $\#(\mathcal{C}_2') \leq \frac{1}{1/8}V(\mathcal{C}_2') + 1$, and $\#(\mathcal{C}_3') \leq \frac{1}{8/27}V(\mathcal{C}_3') + 1$.

We call L_A the set of cubes packed in $\mathcal{C}_2'$, and consider it a *critical set* ($L_A := \{q \in L : \frac{1}{2} < s(q) \leq \frac{2}{3}\}$). The bins of $\mathcal{C}_2'$ are additionally filled with the cubes in L_1', defined as L_B, until possibly all cubes of L_B have been packed ($L_B := \{q \in L : 0 < s(q) \leq \frac{1}{3}\}$). We have two cases to analyse.

Case 1: All cubes of L_B have been packed in $\mathcal{P}_{AB}$.
The analysis of this case is simple and will be omitted.

Case 2: There are cubes of L_B not packed in $\mathcal{P}_{AB}$.
Note that the volume occupation in each bin with configuration C1 or C3 is at least $\frac{8}{27}$. For the bins with configuration C2, we have a volume occupation of $\frac{1}{8}$. In step 5, the bins with configuration C2 are additionaly filled with cubes of L_B generating a combined packing $\mathcal{P}_{AB}$.

In this case, all cubes of L_A have been packed with cubes of L_B. Thus, each bin of $\mathcal{P}_{AB}$ has a volume ocupation of at least $\frac{8}{27}$. The reader can verify this fact by adding up the volume of these cubes in L_A and the cubes of $L_{B,i}$, $i = 3, \ldots, 6$. For bins combining cubes of L_A with $L_{B,7}$, we use Theorem 1 to guarantee this minimum volume ocupation for the resulting packed bins. Therefore, we have an optimum packing of L_1 with volume guarantee at least $\frac{8}{27}$. Thus we have $\#(\mathcal{P}_1) \leq \text{OPT}(L)$, and $\#(\mathcal{P}_1) \leq \frac{V(L)}{8/27} + 6$.

The packing $\mathcal{P}_2$ is generated by algorithm NFDH for a list of cubes with size not greater than $\frac{1}{3}$. Therefore, by Lemma 1, we have $\#(\mathcal{P}_2) \leq \frac{V(L)}{27/64} + 2$.

Now, proceeding as in the proof of Theorem 3, we obtain

$$\text{ICUBE}(L) \leq \alpha \cdot \text{OPT}(L) + 8,$$

where $\alpha = \frac{1945}{729} \leq 2.669$. $\square$

6   Concluding Remarks

We have described an online algorithm for CPP that is a specialization of an approach introduced by Coppersmith and Raghavan [6] for a more general setting.

Our motivation in doing so was to obtain the asymptotic performance bound (3.954) of this algorithm, so that we could compare it with the bounds of the offline algorithms presented here.

We have shown a simple offline algorithm for CPP with asymptotic performance bound 3.466. Then we have designed another offline algorithm that is an improvement of this algorithm, with asymptotic performance bound 2.669. This result can be generalized to k-dimensional cube packing, for $k > 3$, by making use of the Theorem 1 and generalizing the techniques used in this paper. Both algorithms can be implemented to run in time $O(n \log n)$, where n is the number of cubes in the list L. We have also shown that if the instance consists of cubes with size greater than $1/3$ there is a polynomial exact algorithm.

We did not find in the literature offlines algorithms for CPP with asymptotic performance bound as good as 2.669.

References

1. B. S. Baker, A. R. Calderbank, E. G. Coffman Jr., and J. C. Lagarias. Approximation algorithms for maximizing the number of squares packed into a rectangle. *SIAM J. Algebraic Discrete Methods*, 4(3):383–397, 1983.
2. F. R. K. Chung, M. R. Garey, and D. S. Johnson. On packing two-dimensional bins. *SIAM J. Algebraic Discrete Methods*, 3:66–76, 1982.
3. E. G. Coffman, Jr., M. R. Garey, and D. S. Johnson. Approximation algorithms for bin packing – an updated survey. In G. Ausiello et al. (eds.) *Algorithms design for computer system design*, 49–106. Springer-Verlag, New York, 1984.
4. E. G. Coffman, Jr., M. R. Garey, and D. S. Johnson. Approximation algorithms for bin packing – a survey. In D. Hochbaum (ed.) *Approximation algorithms for NP-hard problems*, 46–93, PWS, 1997.
5. E. G. Coffman Jr. and J. C. Lagarias. Algorithms for packing squares: A probabilistic analysis. *SIAM J. Comput.*, 18(1):166–185, 1989.
6. D. Coppersmith and P. Raghavan. Multidimensional on-line bin packing: algorithms and worst-case analysis. *Oper. Res. Lett.*, 8(1):17–20, 1989.
7. J. Csirik and A. van Vliet. An on-line algorithm for multidimensional bin packing. *Operations Research Letters*, 13:149–158, 1993.
8. H. Dyckhoff, G. Scheithauer, and J. Terno. Cutting and packing. In F. Maffioli, M. Dell'Amico and S. Martello (eds.) *Annotated Bibliographies in Combinatorial Optimization*, Chapter 22, 393–412. John Wiley, 1997.
9. C. E. Ferreira, F. K. Miyazawa, and Y. Wakabayashi. Packing squares into squares. *Pesquisa Operacional*, a special volume on cutting and packing. To appear.
10. K. Li and K-H. Cheng. A generalized harmonic algorithm for on-line multidimensional bin packing. TR UH-CS-90-2, University of Houston, January 1990.
11. K. Li and K-H. Cheng. Generalized first-fit algorithms in two and three dimensions. *Int. J. Found. Comput Sci.*, 1(2):131–150, 1992.
12. A. Meir and L. Moser. On packing of squares and cubes. *J. Combinatorial Theory Ser. A*, 5:116–127, 1968.
13. F. K. Miyazawa and Y. Wakabayashi. Approximation algorithms for the orthogonal z-oriented three-dimensional packing problem. To appear in *SIAM J. Computing*.
14. F. K. Miyazawa and Y. Wakabayashi. An algorithm for the three-dimensional packing problem with asymptotic performance analysis. *Algorithmica*, 18(1):122–144, 1997.

Approximation Algorithms for Flexible Job Shop Problems

Klaus Jansen[1], Monaldo Mastrolilli[2], and Roberto Solis-Oba[3]

[1] Institut für Informatik und Praktische Mathematik, Universität zu Kiel, Germany
kj@informatik.uni-kiel.de [*]
[2] IDSIA Lugano, Switzerland, monaldo@idsia.ch [*]
[3] Department of Computer Science, The University of Western Ontario, Canada
solis@brown.csd.uwo.ca [*]

Abstract. The Flexible Job Shop Problem is a generalization of the classical job shop scheduling problem in which for every operation there is a group of machines that can process it. The problem is to assign operations to machines and to order the operations on the machines, so that the operations can be processed in the smallest amount of time. We present a linear time approximation scheme for the non-preemptive version of the problem when the number m of machines and the maximum number μ of operations per job are fixed. We also study the preemptive version of the problem when m and μ are fixed, and present a linear time $(2 + \varepsilon)$-approximation algorithm for the problem with migration.

1 Introduction

The job shop scheduling problem is a classical problem in Operations Research [10] in which it is desired to process a set $\mathcal{J} = \{J_1, \ldots, J_n\}$ of n jobs on a group $M = \{1, \ldots, m\}$ of m machines in the smallest amount of time. Every job J_j consists of a sequence of μ operations $O_{1j}, O_{2j}, \ldots, O_{\mu j}$ which must be processed in the given order. Every operation O_{ij} has assigned a unique machine $m_{ij} \in M$ which must process the operation without interruption during p_{ij} units of time, and a machine can process at most one operation at a time.

In this paper we study a generalization of the job shop scheduling problem called the *flexible job shop problem* [1], which models a wide variety of problems encountered in real manufacturing systems [1,13]. In the flexible job shop problem an operation O_{ij} can be processed by any machine from a given group $M_{ij} \subseteq M$. The processing time of operation O_{ij} on machine $k \in M_{ij}$ is p_{ij}^k. The goal is to choose for each operation O_{ij} an eligible machine and a starting time so that the maximum completion time $C_{\max}$ over all jobs is minimized. $C_{\max}$ is called the *makespan* or the *length* of the schedule.

---

[*] This research was done while the author was at IDSIA Lugano, Switzerland.
[*] This author was supported by the Swiss National Science Foundation project 21-55778.98.
[*] This research was done while the author was at MPII Saarbrücken, Germany.

G. Gonnet, D. Panario, and A. Viola (Eds.): LATIN 2000, LNCS 1776, pp. 68–77, 2000.
© Springer-Verlag Berlin Heidelberg 2000

The flexible job shop problem is more complex than the job shop problem because of the need to assign operations to machines. Following the three-field $\alpha|\beta|\gamma$ notation suggested by Vaessens [13] and based on that of [4], we denote our problem as $m1m|chain, op \leq \mu|C_{\max}$. In the first field m specifies that the number of machines is a constant, 1 specifies that any operation requires at most one machine to be processed, and the second m gives an upper bound on the number of machines that can process an operation. The second field states the precedence constraints and the maximum number of operations per job, while the third field specifies the objective function. The following special cases of the problem are already NP-hard (see [13] for a survey): 2 1 2$|chain, n = 3|C_{\max}$, 3 1 2$|chain, n = 2|C_{\max}$, 2 1 2$|chain, op \leq 2|C_{\max}$.

The job shop scheduling problem has been extensively studied. The problem is known to be strongly NP-hard even if each job has at most three operations and there are only two machines [10]. Williamson et al. [14] proved that when the number of machines, jobs, and operations per job are part of the input there does not exist a polynomial time approximation algorithm with worst case bound smaller than $\frac{5}{4}$ unless $P = NP$. On the other hand the preemptive version of the job shop scheduling problem is NP-complete in the strong sense even when $m = 3$ and $\mu = 3$ [3]. Jansen et al. [8] have designed a linear time approximation scheme for the case when m and μ are fixed. When m and μ are part of the input the best known result [2] is an approximation algorithm with worst case bound $O([\log(m\mu)\log(\min\{m\mu, p_{max}\})/\log\log(m\mu)]^2)$, where p_{max} is the largest processing time among all operations.

Scheduling jobs with chain precedence constraints on unrelated parallel machines is equivalent to the flexible job shop problem. For the first problem, Shmoys et al. [12] have designed a polynomial-time randomized algorithm that, with high probability, finds a schedule of length at most $O((\log^2 n/\log\log n)C_{\max}^*)$, where $C_{\max}^*$ is the optimal makespan.

In this work we study the preemptive and non-preemptive versions of the flexible job shop scheduling problem when the number of machines m and the number of operations per job μ are fixed. We generalize the techniques described in [8] for the job shop scheduling problem and design a linear time approximation scheme for the flexible job shop problem. In addition, each job J_j has a *delivery time* q_j. If in a schedule J_j completes its processing at time C_j, then its *delivery completion time* is equal to $C_j + q_j$. The problem now is to find a schedule that minimizes the maximum delivery completion time L_{max}. We notice that by using the same techniques we can also handle the case in which each job J_j has a *release time* r_j when it becomes available for processing and the objective is to minimize the makespan.

Our techniques allow us also to design a linear time approximation scheme for the preemptive version of the flexible job shop problem without migration. No migration means that each operation must be processed by a unique machine. So if an operation is preempted, its processing can only be resumed on the same machine on which it was being processed before the preemption. Due to space limitations we do not describe this algorithm here. We also study the preemptive flexible job shop problem with migration, and present a $(2 + \varepsilon)$-approximation algorithm for it. The last algorithm handles release and delivery times, and both of them produce solutions with only a constant number of preemptions.

2   The Non-preemptive Flexible Job Shop Problem

Consider an instance of the flexible job shop problem with release and delivery times. Let L^*_{max} be the length of an optimum schedule. For every job J_j, let $P_j = \sum_{i=1}^{\mu}[\min_{s \in M_{kj}} p^s_{ij}]$ denote its minimum processing time. Let $P = \sum_{J_j \in \mathcal{J}} P_j$. Let r_j be the release time of job J_j and q_j be its delivery time. We define $t_j = r_j + P_j + q_j$ for all jobs J_j, and we let $t_{max} = \max_j t_j$.

Lemma 1.
$$\max\left\{\frac{P}{m}, t_{max}\right\} \leq L^*_{max} \leq P + t_{max}. \tag{1}$$

We divide all processing, release, and delivery times by $\max\left\{\frac{P}{m}, t_{max}\right\}$, and thus by Lemma 1,
$$1 \leq L^*_{max} \leq m + 1, \text{ and } t_{max} \leq 1. \tag{2}$$

We observe that Lemma 1 holds also for the preemptive version of the problem with or without migration. Here we present an algorithm for the non-preemptive flexible job shop problem that works for the case when all release times are zero. The algorithm works as follows. First we show how to transform an instance of the flexible job shop problem into another instance without delivery times. Then we define a set of time intervals and assign operations to the intervals so that operations from the same job that are assigned to different intervals appear in the correct order, and the total length of the intervals is no larger than the length of an optimum schedule. We perform this step by first fixing the position of the operations of a constant number of jobs (which we call the long jobs), and then using linear programming to determine the position of the remaining operations.

Next we use an algorithm by Sevastianov [11] to find a feasible schedule for the operations within each interval. Sevastianov's algorithm finds for each interval a schedule of length equal to the length of the interval plus $m\mu^3 p_{max}$, where p_{max} is the largest processing time of any operation in the interval. In order to keep this enlargement small, we remove from each interval a subset $\mathcal{V}$ of jobs with large operations before running Sevastianov's algorithm. Those operations are scheduled at the beginning of the solution, and by choosing carefully the set of long jobs we can show that the total length of the operations in $\mathcal{V}$ is very small compared to the overall length of the schedule.

2.1   Getting Rid of the Delivery Times

We use a technique by Hall and Shmoys [6] to transform an instance of the flexible job shop problem into another with only a constant number of different delivery times. Let q_{max} be the maximum delivery time and let $\varepsilon > 0$ be a constant value. The idea is to round each delivery time down to the nearest multiple of $\frac{\varepsilon}{2}q_{max}$ to get at most $1 + 2/\varepsilon$ distinct delivery times. Next, apply a $(1 + \varepsilon/2)$-approximation algorithm for the flexible job shop problem that can handle $1 + 2/\varepsilon$ distinct delivery times (this algorithm is described below). Finally, add $\frac{\varepsilon}{2}q_{max}$ to the completion time of each job; this increases the length of the solution by $\frac{\varepsilon}{2}q_{max}$. The resulting schedule is feasible for the original instance, so this is a $(1 + \varepsilon)$-approximation algorithm for the original problem. In the remainder of this section, we restrict our attention to the problem for which the delivery times $q_1, ..., q_n$ can take only $\chi \leq 1 + \frac{2}{\varepsilon}$ distinct values, which we denote by $\delta_1 > ... > \delta_\chi$.

The delivery time of a job can be interpreted as an additional *delivery operation* that must be processed on a *non-bottleneck machine* after the last operation of the job. A non-bottleneck machine is a machine that can process simultaneously any number of operations. Moreover, every feasible schedule for the jobs $\mathcal{J}$ can be transformed into another feasible schedule, in which all delivery operations finish at the same time, without increasing the length of schedule: simply shift the delivery operations to the end of the schedule. Therefore, we only need to consider a set $\mathcal{D} = \{d_1, ..., d_\chi\}$ of χ different delivery operations, where d_i has processing time δ_i.

2.2   Relative Schedules

Assume that the jobs are indexed so that $P_1 \geq P_2 \geq ... \geq P_n$. Let $\mathcal{L} \subset \mathcal{J}$ be the set formed by the first k jobs, i.e., the k jobs with longest minimum processing time, where k is a constant to be defined later. We call $\mathcal{L}$ the set of *long* jobs. An operation from a long job is called a long operation, regardless of its processing time. Let $\mathcal{S} = \mathcal{J} \setminus \mathcal{L}$ be the set of *short* jobs.

Consider any feasible schedule for the jobs in $\mathcal{J}$. This schedule assigns a machine to every operation and it also defines a relative ordering for the starting and finishing times of the operations. A *relative schedule* R for $\mathcal{L}$ is an assignment of machines to long operations and a relative ordering for the starting and finishing times of the long operations and the delivery operations, such that there is a feasible schedule for $\mathcal{J}$ that respects R. This means that for every relative schedule R there is a feasible schedule for $\mathcal{J}$ that assigns the same machines as R to the long operations and that schedules the long and the delivery operations in the same relative order as R. Since there is a constant number of long jobs, then there is also a constant number of different relative schedules.

Lemma 2. *The number of relative schedules is at most* $m^{\mu k}(2(\mu k + \chi))!$.

If we build all relative schedules, one of them must be equal to the relative schedule defined by some optimum solution. Since it is possible to build all relative schedules in constant time, we might assume without loss of generality that we know how to find a relative schedule R such that some optimum schedule for $\mathcal{J}$ respects R.

Fix a relative schedule R as described above. The ordering of the starting and finishing times of the operations divide the time line into intervals that we call *snapshots*. We can view a relative schedule as a sequence of snapshots $M(1), M(2), \ldots, M(g)$, where $M(1)$ is the unbounded snapshot whose right boundary is the starting time of the first operation according to R, and $M(g)$ is the snapshot bounded on the right by the finishing time of the delivery operations. The number of snapshots g is at most $2\mu k + \chi + 1$ because the starting and finishing times of every operation might bound a snapshot.

2.3   Scheduling the Small Jobs

Given a relative schedule R as described above, to obtain a solution for the flexible job shop problem we need to schedule the small operations within the snapshots defined by R. We do this in two steps. First we use a linear program $LP(R)$ to assign small operations to snapshots and machines, and second, we find a feasible schedule for the small operations within every snapshot.

To formulate the linear program we need first to define some variables. For each snapshot $M(\ell)$ we use a variable t_ℓ to denote its length. For each $J_j \in \mathcal{S}$ we define a set of decision variables $x_{j,(i_1,\ldots,i_\mu),(s_1,\ldots,s_\mu)}$ with the following meaning: $x_{j,(i_1,\ldots,i_\mu),(s_1,\ldots,s_\mu)} = f$ iff for all $q = 1, \ldots, \mu$, an f fraction of the q-th operation of job J_j is completely scheduled in the i_q-th snapshot and on machine s_q.

Let α_j be the snapshot where the delivery operation of job J_j starts. For every variable $x_{j,(i_1,\ldots,i_\mu),(s_1,\ldots,s_\mu)}$ we need $1 \leq i_1 \leq i_2 \leq \cdots \leq i_\mu < \alpha_j$ to ensure that the operations of J_j are scheduled in the proper order. Let $A_j = \{(i,s) \mid i = (i_1,\ldots,i_\mu)\ 1 \leq i_1 \leq \ldots \leq i_\mu < \alpha_j,\ s = (s_1,\ldots,s_\mu)\ s_q \in M_{qj}$ and no long operation is scheduled by R at snapshot i_q on machine s_q, for all $q = 1, \ldots, \mu\}$.

The load $L_{\ell,h}$ on machine h in snapshot $M(\ell)$ is the total processing time of the operations from small jobs assigned to h during $M(\ell)$, i.e.,

$$L_{\ell,h} = \sum_{J_j \in \mathcal{S}} \sum_{(i,s) \in A_j} \sum_{\substack{q=1 \\ i_q=\ell, s_q=h}}^{\mu} x_{jis} p_{qj}^{s_q}, \tag{3}$$

where i_q and s_q are the q-th components of tuples i and s respectively.

For every long operation O_{ij} let α_{ij} and β_{ij} be the indices of the first and last snapshots where the operation is scheduled. Let p_{ij} be the processing time of long operation O_{ij} according to the machine assignment defined by the relative schedule R. We are ready to describe the linear program $LP(R)$ that assigns small operations to snapshots.

Minimize $\sum_{\ell=1}^{g} t_\ell$

s.t.   (1) $\sum_{\ell=\alpha_{ij}}^{\beta_{ij}} t_\ell = p_{ij}$,   for all $J_j \in \mathcal{L}$, $i = 1, \ldots, \mu$,

       (2) $\sum_{\ell=\alpha_j}^{g} t_\ell = \delta_j$,   for all delivery operations d_j ,

       (3) $\sum_{(i,s) \in A_j} x_{jis} = 1$,   for all $J_j \in \mathcal{S}$,

       (4) $L_{\ell,h} \leq t_\ell$,   for all $\ell = 1, \ldots, g$, $h = 1, \ldots, m$.

       (5) $t_\ell \geq 0$,   for all $\ell = 1, \ldots, g$,

(6) $x_{jis} \geq 0$,   for all $J_j \in \mathcal{S}$, $(i, s) \in A_j$.

Lemma 3. *An optimum solution of $LP(R)$ has value no larger than the length of an optimum schedule S^* that respects the relative schedule R.*

One can solve $LP(R)$ optimally in polynomial time and get only a constant number of jobs with fractional assignments, since a basic feasible solution of $LP(R)$ has at most $k\mu + n - k + mg + \chi$ variables with positive value. By constraint (3) every small job has at least one positive variable associated with it, and so there are at most $mg + k\mu + \chi$ jobs with fractional assignments. We show later how to get rid of any constant number of fractional assignments by only slightly increasing the length of the solution.

The drawback of this approach is that solving the linear program might take a very long time. Since we want to get an approximate solution to the flexible job shop problem it is not necessary to find an optimum solution for $LP(R)$, an approximate solution would suffice.

2.4 Approximate Solution of the Linear Program

A *convex block-angular resource sharing problem* has the form:

$$\min\left\{\lambda \;\middle|\; \sum_{k=1}^{K} f_i^k(x^k) \leq \lambda, \text{for all } i = 1, \ldots, N, \text{and } x^k \in B^k, k = 1, \ldots, K\right\}$$

where $f_i^k : B^k \to \Re^+$ are N non-negative continuous convex functions, and B^k are disjoint convex compact nonempty sets called *blocks*, $1 \leq k \leq K$. The Potential Price Directive Decomposition Method of Grigoriadis and Khachiyan [5] can find a $(1 + \rho)$-approximate solution to this problem for any $\rho > 0$. This algorithm needs $O(N(\rho^{-2} \ln \rho^{-1} + \ln N)(N \ln \ln(N/\rho) + KF))$ time, where F is the time needed to find a ρ-approximate solution to the following problem on any block B^k, for some vector $(p_1, \ldots, p_N) \in \Re^N$: $\min\left\{\sum_{i=1}^{N} p_i f_i^k(x^k) \,\middle|\, x^k \in B^k\right\}$.

We can write $LP(R)$ as a convex block-angular resource sharing problem as follows. First we guess the value s of an optimum solution for $LP(R)$, and add the constraint: $\sum_{\ell=1}^{g} t_\ell \leq s$ to the linear program. Note that $s \leq m + 1$. Then we replace constraint (4) by constraint (4'), where λ is a non-negative value:

(4') $L_{\ell,h} - t_\ell + m + 1 \leq \lambda$,   for all $\ell = 1, \ldots, g$, $h = 1, \ldots, m$.

This new linear program, that we denote as $LP(R, s, \lambda)$, has the above block-angular structure. The blocks $B_j = \{x_{jis} \mid \text{constraints (3) and (6) hold}\}$, are $(mg)^\mu$-dimensional *simplices*. The block $B_{|\mathcal{S}|+1} = \{t_\ell \mid \sum_{\ell=1}^{g} t_\ell \leq s$ and constraints (1),(2), and (5) hold$\}$ has also constant dimension. Let $f_{\ell,h} = L_{\ell,h} - t_\ell + m + 1$. Since $t_\ell \leq s \leq m + 1$, these functions are non-negative. Each block B_i has constant dimension, and so the above block optimization problem can be solved in constant time. Therefore the algorithm of [5] finds a $(1 + \rho)$-approximate solution for $LP(R, s, \lambda)$ in $O(n)$ time for any value $\rho > 0$. This gives a feasible solution of $LP(R, s, (m + 1 + \rho'))$ for $\rho = \rho'/(m + 1)$.

Let $L^*_{\max}$ be the length of an optimum schedule and assume that R is a relative schedule for $\mathcal{L}$ in an optimum schedule. We can use binary search on the interval $[1, 1+m]$ to find a value $s \leq (1+\frac{\varepsilon}{8})L^*_{\max}$ such that $LP(R, s, (m+1+\rho'))$ has a solution for $\rho' = \frac{\varepsilon}{8g}$. This search can be performed in $O(\log(\frac{1}{\varepsilon} \log m))$ iterations by performing the binary search only on the following values,

$$(1 + \frac{\varepsilon}{8}), (1 + \frac{\varepsilon}{8})^2, ..., (1 + \frac{\varepsilon}{8})^{b-1}, m + 1 \tag{4}$$

where b is the smallest integer such that $(1 + \varepsilon/8)^b \geq m + 1$. Thus, $b \leq \ln(m + 1)/\ln(1 + \varepsilon/8) + 1 = O(\frac{1}{\varepsilon} \log m)$, since $\ln(1 + \varepsilon/8) \geq \frac{\varepsilon/8}{1+\varepsilon/8}$. To see that this search yields the desired value for s, note that there exists a nonnegative integer $i \leq b$ such that $L^*_{\max} \in [(1+\frac{\varepsilon}{8})^i, (1+\frac{\varepsilon}{8})^{i+1}]$ and therefore with the above search we find a value $s \leq (1 + \frac{\varepsilon}{8})L^*_{\max}$ for which $LP(R, s, m + 1 + \rho')$ has a feasible solution. Linear program $LP(R, s, m + 1 + \rho')$ assumes that the length of each snapshot is increased by ρ', and therefore the total length of the solution is $(1 + \frac{\varepsilon}{8})L^*_{\max} + g\rho' \leq (1 + \frac{\varepsilon}{4})L^*_{\max}$.

Lemma 4. *A solution for $LP(R, s, m + 1 + \rho')$, with $s \leq (1 + \frac{\varepsilon}{8})L^*_{max}$ and $\rho' = \frac{\varepsilon}{8g}$, of value at most $(1 + \frac{\varepsilon}{4})L^*_{\max}$ can be found in linear time.*

By using a similar technique as in [8] we can modify any feasible solution for $LP(R, s, m + 1 + \rho')$ to get a new feasible solution in which all but a constant number of variables x_{jis} have integer values. Moreover we can do this rounding step in linear time.

Lemma 5. *A solution for $LP(R, s, m+1+\rho')$ can be transformed in linear time into another solution for $LP(R, s, m + 1 + \rho')$ in which the set $\mathcal{F}$ of jobs that still have fractional assignments after the rounding procedure has size $|\mathcal{F}| \leq mg$.*

2.5   Generating a Feasible Schedule

To get a feasible schedule from the solution of the linear program we need to remove all jobs $\mathcal{F}$ that received fractional assignment. These jobs are scheduled sequentially at the beginning of the schedule.

For every operation of the small jobs, consider its processing time according to the machine selected for it by the solution of the linear program. Let $\mathcal{V}$ be the set formed by the small jobs containing at least one operation with processing time larger than $\tau = \frac{\varepsilon}{8\mu^3 mg}$. Note that $|\mathcal{V}| \leq \frac{m(m+1)}{\tau} = \frac{8\mu^3 m^2 (m+1)}{\varepsilon} g$. We remove from the snapshots all jobs in $\mathcal{V}$ and place them sequentially at the beginning of the schedule.

Let $O(\ell)$ be the set of operations from small jobs that remain in snapshot $M(\ell)$. Let $p_{max}(\ell)$ be the maximum processing time among the operations in $O(\ell)$. Every snapshot $M(\ell)$ defines an instance of the job shop problem, since the solution of the linear program assigns a unique machine to every operation. Hence we can use Sevastianov's algorithm [11] to find in $O(n^2 \mu^2 m^2)$ time a feasible schedule for the operations $O(\ell)$; this schedule has length at most $\bar{t}_\ell = t_\ell + \rho' + \mu^3 m p_{max}(\ell)$. We must increase the length of every snapshot $M(\ell)$ to $\bar{t}_\ell$ to accommodate the schedule produced by Sevastianov's algorithm. Summing up all these enlargements, we get:

Lemma 6.

$$\sum_{\ell=1}^{g} \mu^3 m p_{max}(\ell) \le \mu^3 m g \tau = \frac{\epsilon}{8} \le \frac{\epsilon}{8} L^*_{max}. \tag{5}$$

The total length of the snapshots $M(\alpha_{ij}), \ldots, M(\beta_{ij})$ that contain a long operation O_{ij} might be larger than p_{ij}. This creates some idle times on machine m_{ij}. We start operations O_{ij} for long jobs $\mathcal{L}$ at the beginning of the enlarged snapshot $M(\alpha_{ij})$. The resulting schedule is clearly feasible. Let $P(J') = \sum_{J_j \in J'} P_j$ be the total processing time of all jobs in some set $J' \subset \mathcal{J}$ when the operations of those jobs are assigned to the machines with the lowest processing times.

Lemma 7. *A feasible schedule for the jobs $\mathcal{J}$ of length at most $(1 + \frac{3}{8}\epsilon)L^*_{max} + P(\mathcal{F} \cup \mathcal{V})$ can be found in $O(n^2)$ time.*

We can choose the number k of long jobs so that $P(\mathcal{F} \cup \mathcal{V}) \le \frac{\epsilon}{8} L^*_{max}$.

Lemma 8. *[7] Let $\{d_1, d_2, \ldots, d_n\}$ be positive values and $\sum_{j=1}^{n} d_j \le m$. Let q be a nonnegative integer, $\alpha > 0$, and $n \ge (q+1)^{\lceil \frac{1}{\alpha} \rceil}$. There exists an integer k such that $d_{k+1} + \ldots + d_{k+qk} \le \alpha m$ and $k \le (q+1)^{\lceil \frac{1}{\alpha} \rceil}$.*

Let us choose $\alpha = \frac{\epsilon}{8m}$ and $q = \left(\left\lfloor \frac{8\mu^3 m(m+1)}{\epsilon} \right\rfloor + 1 \right) m(2\mu + \chi + 1)$. By Lemma 5, $|\mathcal{F} \cup \mathcal{V}| \le mg + \frac{8\mu^3 m^2 (m+1)}{\epsilon} g \le qk$. By Lemma 8 it is possible to choose a value $k \le (q+1)^{\lceil \frac{1}{\alpha} \rceil}$ so that the total processing time of the jobs in $\mathcal{F} \cup \mathcal{V}$ is at most $\frac{\epsilon}{8} \le \frac{\epsilon}{8} L^*_{max}$. This value of k can clearly be computed in constant time. We select the set $\mathcal{L}$ of large jobs as the set consisting of the k jobs with largest processing times $P_j = \sum_{i=1}^{\mu} [\min_{s \in M_{kj}} p_{ij}^s]$.

Lemma 9.

$$P(\mathcal{F} \cup \mathcal{V}) \le \frac{\epsilon}{8} L^*_{max}. \tag{6}$$

Theorem 1. *For any fixed m and μ, there is an algorithm for the flexible job shop scheduling problem that computes for any value $\varepsilon > 0$, a feasible schedule of length at most $(1 + \epsilon)L^*_{max}$ in $O(n)$ time.*

Proof. By Lemmas 7 and 9, the above algorithm finds in $O(n^2)$ a schedule of length at most $(1 + \frac{1}{2}\epsilon)L^*_{max}$. This algorithm can handle $1 + \frac{2}{\epsilon}$ distinct delivery times. By the discussion at the beginning of Section 3.1 it is easy to modify the algorithm so that it handles arbitrary delivery times and it yields a schedule of length at most $(1 + \varepsilon)L^*_{max}$. For every fixed m, μ, and ϵ, all computations can be carried out in $O(n)$ time, with exception of the algorithm of Sevastianov that runs in $O(n^2)$ time. The latter can be sped up to get linear time by "glueing" pairs of small jobs together as described in [8]. $\qquad \square$

3   Preemptive Flexible Job Shop Problem with Migration

Let t a nonnegative variable, with $t \geq \delta_1$, that denotes the length of a schedule (with delivery times). Consider χ time intervals defined as follows, $[0, t - \delta_1], [t - \delta_1, t - \delta_2], ..., [t - \delta_{\chi-1}, t - \delta_\chi]$ where $\delta_1 > ... \delta_\chi$ are the delivery times. First we ignore the release times and select the machines and the time intervals in which the operations of every job are going to be processed. To do this we define a linear program LP (similar to that of Sect. 2.3) that minimize the value of t. The optimum solution of the LP has value no larger than L_{max}^*. Again, by using the Logarithmic Potential Price Directive Decomposition Method [5] and the rounding technique of [8], we can compute in linear time a $(1 + \frac{\varepsilon}{4})$-approximate solution $\tilde{S}$ of the LP such that the size of the set $\mathcal{F}$ of jobs that receive fractional assignments is bounded by $m\chi$.

Let $\mathcal{P}$ denote the set of jobs from $\mathcal{J} \setminus \mathcal{F}$ for which at least one operation has processing time greater than $\frac{\varepsilon \tilde{t}}{4\chi\mu^3 m(1+\frac{\varepsilon}{8})}$, where $\tilde{t}$ is the value of $\tilde{S}$. Let $\mathcal{L} = \mathcal{F} \cup \mathcal{P}$ and $\mathcal{S} = \mathcal{J} \setminus \mathcal{L}$. According to $\tilde{S}$, find a feasible schedule $\sigma_\mathcal{S}$ for the jobs from $\mathcal{S}$ applying Sevastianov's algorithm. We use the algorithm of Sevastianov to find a schedule for the operations assigned to each time interval. The maximum delivery completion time (when release times are ignored) of $\sigma_\mathcal{S}$ is at most $(1 + \varepsilon)L_{max}^*$. By adding release times the length of $\sigma_\mathcal{S}$ is at most $(2 + \varepsilon)L_{max}^*$, since the maximum release time cannot be more than L_{max}^*. Again, the algorithm of Sevastianov can be sped up to take $O(n)$ time, and computing the schedule $\sigma_\mathcal{S}$ takes linear time.

Now, we ignore the delivery times for jobs from $\mathcal{L}$ (they are considered later). We note that the cardinality of set $\mathcal{L}$ is bounded by $O(\frac{\mu^3 m^2}{\varepsilon^2})$. As we did for the delivery times, the release times can be interpreted as additional operations of jobs that have to be processed on a non-bottleneck machine. Because of this interpretation, we can add to the set $\mathcal{O}_\mathcal{L}$ of operations from $\mathcal{L}$ a set $\mathcal{R}$ of release operations O_{0j} with processing times r_j. Each job $J_j \in \mathcal{L}$ has to perform its release operation O_{0j} on a non-bottleneck machine at the beginning. A *relative order* R is an ordered sequence of the starting and finishing times of all operations from $\mathcal{O}_\mathcal{L} \cup \mathcal{R}$, such that there is a feasible schedule for $\mathcal{L}$ that respects R. The ordering of the starting and finishing times of the operations divide the time line into intervals. We observe that the number of intervals g is bounded by $O(\frac{\mu^4 m^2}{\varepsilon^2})$. Note that a relative order is defined without assigning operations of long jobs to machines.

For every relative order R we define a linear program to assign (fractions of) operations to machines and intervals that respects R. We build all relative orders R and solve the corresponding linear programs in constant time. At the end we select a relative schedule R^* with the smallest solution value. We can show that the value of this solution is no larger than the length of an optimum preemptive schedule for $\mathcal{L}$. This solution is in general not a feasible schedule for $\mathcal{L}$ since the order of fractions of operations within an interval could be incorrect. However the set of operations assigned to each interval gives an instance of the preemptive open shop problem, which can be solved exactly in constant time [9].

This can be done without increasing the length of each interval and with in total only a constant number of preemptions (at most $O(m^2)$ preemptions for each interval). By adding the delivery times the length of the computed schedule $\sigma_{\mathcal{L}}$ is at most $2L^*_{\max}$. The output schedule is obtained by appending $\sigma_{\mathcal{S}}$ after $\sigma_{\mathcal{L}}$.

Theorem 2. *For any fixed m, μ, and $\varepsilon > 0$, there is a $(2 + \varepsilon)$-linear-time approximation algorithm for the preemptive flexible job shop scheduling problem with migration.*

References

1. P. Brandimarte, Routing and scheduling in a flexible job shop by tabu search, *Annals of Operations Research*, 22, 158-183, 1993.
2. L.A. Goldberg, M. Paterson, A. Srinivasan, and E. Sweedyk, Better approximation guarantees for job-shop scheduling, *Proceedings of the 8th Symposium on Discrete Algorithms* (SODA 97), 599-608.
3. T. Gonzales and S. Sahni, Flowshop and jobshop schedules: complexity and approximation, *Operations Research* 26 (1978), 36-52.
4. R.L. Graham, E.L. Lawler, J.K. Lenstra, and A.H.G. Rinnoy Kan, Optimization and approximation in deterministic sequencing and scheduling, *Ann. Discrete Math.* 5 (1979), 287-326.
5. M.D. Grigoriadis and L.G. Khachiyan, Coordination complexity of parallel price-directive decomposition, *Mathematics of Operations Research* 21 (1996), 321-340.
6. L.A. Hall and D.B. Shmoys, Approximation algorithms for constrained scheduling problems, *Proceedings of the IEEE 30th Annual Symposium on Foundations of Computer Science* (FOCS 89), 134-139.
7. K. Jansen and L. Porkolab, Linear-time approximation schemes for scheduling malleable parallel tasks, *Proceedings of the 10th Annual ACM-SIAM Symposium on Discrete Algorithms*, (SODA 99), 490-498.
8. K. Jansen, R. Solis-Oba and M.I. Sviridenko, A linear time approximation scheme for the job shop scheduling problem, Proceedings of the Second International Workshop on Approximation Algorithms (APPROX 99), 177-188.
9. E. L. Lawler, J. Labetoulle, On Preemptive Scheduling of Unrelated Parallel Processors by Linear Programming, *Journal of the ACM*, vol. 25, no. 4, pp. 612–619, October 1978.
10. E.L. Lawler, J.K. Lenstra, A.H.G. Rinnooy Kan, and D.B. Shmoys, Sequencing and scheduling: Algorithms and complexity, in: Handbook in Operations Research and Management Science, Vol. 4, North-Holland, 1993, 445-522.
11. S.V. Sevastianov, Bounding algorithms for the routing problem with arbitrary paths and alternative servers, *Cybernetics* 22 (1986), 773-780.
12. D.B. Shmoys, C. Stein, and J. Wein, Improved approximation algorithms for shop scheduling problems, *SIAM Journal on Computing* 23 (1994), 617-632.
13. R.J.M. Vaessens, Generalized job shop scheduling: complexity and local search, Ph.D. thesis (1995), Eindhoven University of Technology.
14. D. Williamson, L. Hall, J. Hoogeveen, C.. Hurkens, J.. Lenstra, S. Sevastianov, and D. Shmoys, Short shop schedules, *Operations Research* 45 (1997), 288-294.

Emerging Behavior as Binary Search Trees Are Symmetrically Updated

Stephen Taylor

College of the Holy Cross, Worcester MA 01610-2395, USA,
staylor@holycross.edu

Abstract. When repeated updates are made to a binary search tree, the expected search cost tends to improve, as observed by Knott. For the case in which the updates use an asymmetric deletion algorithm, the Knott effect is swamped by the behavior discovered by Eppinger. The Knott effect applies also to updates using symmetric deletion algorithms, and it remains unexplained, along with several other trends in the tree distribution. It is believed that updates using symmetric deletion do not cause search cost to deteriorate, but the evidence is all experimental. The contribution of this paper is to model separately several different trends which may contribute to or detract from the Knott effect.

1 Background

A binary search tree (BST) is a tree structure with a key value stored in each node. For each node, the key value is an upper bound on the values of keys in the left subtree, and a lower bound on keys in the right subtree. If there are no duplicate keys, a search of the tree for any given key value involves examining nodes in a single path from the root. An insertion into the tree is made by searching for a candidate key, then placing it as a child of the last node reached in the search, so that an inserted key is always a leaf. Deletions are more complicated, and use one of the algorithms described in section (2.2.)

When a BST with n nodes is grown by random insertions (RI) with no deletions, the average search cost is $O(\log n)$, or equivalently, the total pathlength from the root to every node in the tree (this is the internal pathlength, or IPL) is $O(n \log n)$.

An update consists of deleting the node with some particular key-value, and inserting another, either with the same key, or a different key.

Culberson [CM90] refers to the leftward-only descendents of the root as the *backbone*, and the distances between key-values of the backbone as *intervals*. [Bri86] calls the backbone and the corresponding rightward-only descendents of the root the *shell*. The *length* of the shell is the pathlength from smallest to largest key. *Shell intervals* are defined by the key-values of the shell nodes.

1.1 Related Work

When repeated updates are made to a binary search tree, the expected search cost tends to improve. This *Knott effect* was first reported in [Kno75]. It turns out

G. Gonnet, D. Panario, and A. Viola (Eds.): LATIN 2000, LNCS 1776, pp. 78–87, 2000.

that for updates using asymmetric [Hib62] deletion, the Knott effect is swamped by the *Eppinger effect*. In [Epp83] Eppinger observed that after $O(n^2)$ updates, the tree of size n has expected search time greater than $O(\ln n)$. There is a striking early improvement and search costs drop to about 92% of initial values, and then, after about $n^2/2$ iterations, the cost begins to rise. It levels out after about n^2 iterations. For a tree with 128 nodes, the final search cost levels out to be about the same as for a tree built with insertions only; smaller trees, as conjectured by Knott, fare better; but larger trees do worse. For trees of 2048 nodes, the asymptotic search cost is about 50% greater than for an RI BST.

Culberson [CM90] has given a model which explains this for updates in which the item removed is always the item re-inserted, which he calls the Exact Fit Domain (EFD) model. Culberson's model is based on directed random walks, and finds that the expected search cost is $O(\sqrt{n})$. We call similar undirected random walks in trees updated with symmetrical deletion the *Culberson effect*, although the time scales and results are different.

The Knott effect remains unexplained; and it is not the only unexplained behavior. Simulations reported in Evans and Culberson [EC94] for two symmetric update algorithms show a reduced average pathlength, as predicted by the Knott effect, but also that pathlengths from the root to the largest and smallest leaves (and perhaps to some other, unmeasured subset of nodes) were 1.2 to 1.3 times longer than would be expected in a random binary search tree. We call this the *Evans effect*.

[JK78] demonstrate analytically that Knott's conjecture is correct for trees with three nodes. [BY89] does the same for trees with four nodes. [Mes91] analyzes an update using symmetric deletion for the tree of three nodes.

Martinez and Roura [MR98] provide randomized algorithms which maintain the distribution of trees after update to be the same as a RI binary search tree. Their algorithms are not susceptible to the breakdown caused by sorted input, nor to the Eppinger or Culberson effects. However, they are also immune to the Knott effect, and thus miss any improvements in search times it might provide.

There are several open questions about update with symmetric deletion.

1. Does the Knott conjecture hold if it is revised for symmetric deletion?
2. If so, why? Is there a model which explains why stirring up the tree should result in shorter internal path lengths? The effect is apparent even in Hibbard deletions, before it is overwhelmed by the skewing of the tree.
3. Is there a long-term degeneration in the tree? If so, it must be over a much longer term than the Hibbard deletion degeneration, because Eppinger's and Culberson's simulations did not detect it.

2   Update Methodology

2.1   Exact Fit Domain Model

Culberson [CM89,CM90] proposed the unrealistic Exact Fit Domain (EFD) model to simplify analysis. The assumption is there are only n keys possible and no duplicates in the tree, so that when an update occurs, the new key must be the same as the one which was just deleted. This has the effect of localizing the effects of update operations, making them easier to analyze. We assert without mathematical justification that the EFD model gives qualitatively similar results to a more realistic random update model. Since repeated insertions result in relatively well-understood behavior if they are not in the neighborhood of a deletion, we claim that the EFD model simply telescopes the effect of separated deletions and insertions in the same area. Time scales for emerging behavior may be changed by the EFD model, but perhaps not other measurements. In support of this suggestion, note the graphs of fig. (1.) [Grafting deletion is defined below.]

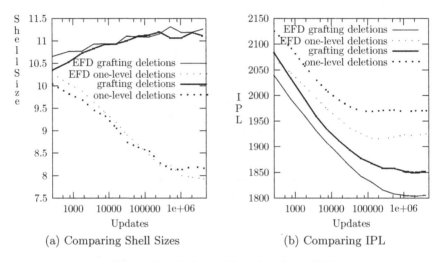

(a) Comparing Shell Sizes   (b) Comparing IPL

Fig. 1. Simulations with and without EFD.

2.2   Deletion Algorithms

Hibbard's Asymmetrical Deletion Algorithm   When Hibbard formulated his deletion algorithm for binary trees in [Hib62], he was aware of the asymmetry. His algorithm has two steps:

1. If the right subtree of the node to be deleted is not empty, replace the key of the deleted node with its *successor*, the left-most node in the right subtree. Then delete the successor.
2. If the right subtree of the node to be deleted is empty, replace the node with its left subtree.

There are two, different, asymmetries: the deleted node is preferentially updated from the right subtree; and the case when the right subtree is empty doesn't have a matching simple case when the left subtree is empty.

Hibbard proved that his asymmetric deletion did not change the distribution of tree shapes. He assumed that this meant that a symmetric algorithm was unnecessary.

Symmetrical Grafting Deletion This algorithm is a combination of the Hibbard deletion algorithm and its mirror image. Whether the right-favored or left-favored version of deletion is used is of equal probability, so the algorithm is symmetrical. In our simulations, we use simple alternation rather than a random number generator to decide which to use, but check for empty subtrees before considering the successor or predecessor key. We call this *grafting* deletion because the subtree is grafted into the place of the deleted node when possible. Most published symmetric deletion algorithms are variants on grafting deletion. Simulations, for example fig. (2) show that one property of grafting deletion is that zero-size subtrees rapidly get less common than in RI BSTs.

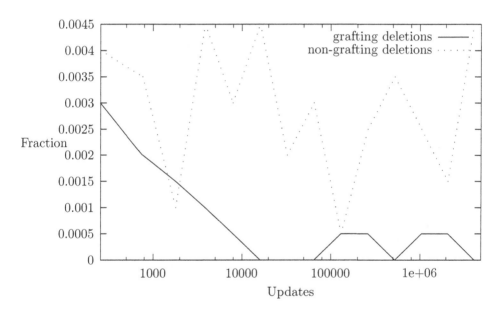

Fig. 2. Zero-size left-subtrees of root in 256 node BST.

Symmetrical Non-Grafting Deletion This is a symmetric deletion algorithm which lacks an optimization for empty subtrees. The algorithm replaces a deleted node with its successor or predecessor in the tree, unless there is none, in which case it is replaced with the predecessor or successor. If there is neither predecessor nor successor (the node to be deleted is a leaf) the node is simply removed.

The algorithm alternates between favoring predecessors and successors, so it is symmetrical. Because it lacks an optimization to reduce the height of the tree by grafting a subtree nearer the root when the other subtree is empty, we might expect that it would produce a distribution of binary search trees which includes rather more zero-sized subtrees than algorithms which include such an optimization. (These include the asymmetrical Hibbard deletion algorithm.)

This algorithm is easier to analyze using a Markov chain, because the state-space of trees of size n can be described by a single variable, the size of the left subtree. In the particularly easy case of the Exact Fit Domain, in which the replacement key in an update is always the same as the key deleted, the size of the subtree can change only if the root is deleted, and only by one.

Assume the BST has n nodes, and $\pi_{k,t}$ be the probability that the left subtree has k nodes. When the root is deleted for time $t > 0$, we have for each t the n simultaneous equations (here we use *Iverson notation* [GKP89]: $[P](term)$ evaluates to *term* if P is true, otherwise to zero:)

$$\pi_{k,t} = \left(1 - \frac{1}{n}\right)\pi_{k,t-1} + [k > 0]\left(\frac{1}{2n}\right)\pi_{k-1,t-1} + [k < n]\left(\frac{1}{2n}\right)\pi_{k+1,t-1} \quad (1)$$

and assuming that there is a steady state, we can rewrite this as

$$\pi_{k,\infty} = \left(1 - \frac{1}{n}\right)\pi_{k,\infty} + [k > 0]\left(\frac{1}{2n}\right)\pi_{k-1,\infty} + [k < n]\left(\frac{1}{2n}\right)\pi_{k+1,\infty} \quad (2)$$

With the additional equation $\sum_k \pi_{k,\infty} = 1$ we can solve the system to find

$$\pi_{0,\infty} = \pi_{n-1,\infty} = \frac{1}{2(n-1)} \quad (3)$$

$$\pi_{k,\infty} = \frac{1}{n-1} \quad [0 < k < n-1] \quad (4)$$

3   Emerging Behavior

3.1   The Knott Effect

The Knott effect is the observed tendency of a binary search tree to become more compact. That is, after a number of deletion and insertion operations are performed on a random binary search tree, the resulting trees have smaller IPL and therefore smaller search times. Knuth speculates that this effect may be due to the tendency of (grafting) delete operations to remove empty subtrees.

A RI BST has one of the two worst possible keys at the root of any subtree with probability 2/|size of subtree|. As a result of updates it evolves toward a steady state in which the probability of zero subtrees is smaller. For the case of update with non-grafting deletion, in the steady state, every subtree size except zero and the largest possible is equally probable, and those two sizes are half as likely as the others, as we have shown in eq. (3) and (4)

If we make the assumption that subtrees have the same distribution as the root, this leads naturally to a recurrence for the IPL of such a tree.

$$f_n = \begin{cases} 0 & n = 0 \cup n = 1 \\ n - 1 + \frac{f_{n-1}}{n-1} + 2\sum_{i=1}^{n-2} \frac{f_i}{n-1} & n > 1 \end{cases} \quad (5)$$

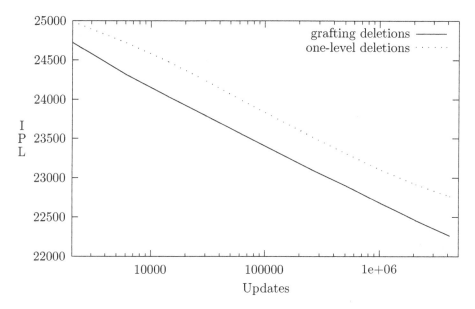

Fig. 3. Internal Pathlength declines as tree updated.

$$f_n = \begin{cases} 0 & n = 0 \cup n = 1 \\ f_{n-1} + \frac{f_{n-2}}{n-1} + \frac{2n-3}{n-1} & n > 1 \end{cases} \tag{6}$$

This can be evaluated numerically and compared with the corresponding recurrence for an RI BST. The comparison shows that for very large values of n, f_n grows quite close to IPL_n; only for small to intermediate values do they diverge.

3.2   The Evans Effect

The Evans effect is reported in [EC94] for search trees updated with symmetric deletion algorithms. They report shells which are 1.2 to 1.3 times as long as those of a RI BST. Presumably there are subtree shells to which the effect would also apply, but clearly not to every path from the root, since the same simulations also showed a Knott effect reduction in average pathlength. Figure (4) shows the Evans effect. Note that the Evans effect doesn't hold for non-grafting deletions; the shell size (which is after all, the sum of two paths) follows the IPL down. For grafting deletions, the shell size gradually rises. This suggests that the Evans effect might be due to the grafting of subshell backbone unto the shell.

We can easily compute the expected size of the initial size of the shell in a RI BST. By symmetry, the size of the shell should be twice the length of the backbone, and this turns out to have a simple recurrence.

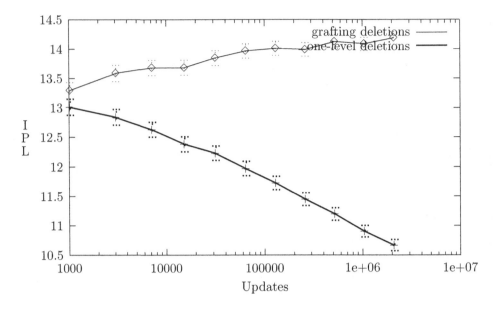

Fig. 4. Changes in shell size as tree updated.

A tree with one node has a backbone length of zero. A tree with n nodes has a left-subtree of size k, $0 \leq k < n$ with probability $\frac{1}{n}$, and so

$$b_n = \begin{cases} 0 & n = 1 \\ \sum_{i=1}^{n-1} \frac{1+b_i}{n} & n > 1, \end{cases} \tag{7}$$

which has the solution $b_n = H_n - 1 \approx \gamma - 1 + \ln n$. The expected size of a RI BST shell is then

$$E(\text{shell}) = 2\gamma - 2 + 2\ln n \approx 2(\ln n) - 0.845568670 \tag{8}$$

The root of a n-node tree which has evolved to a steady state with one-level deletion will have its left or right subtree empty with probability $\frac{1}{2(n-1)}$ and left subtrees of size k, $0 < k < n-1$ with probability $\frac{1}{n-1}$. This leads to a recurrence for the length of the backbone:

$$c_n = \begin{cases} 0 & n = 1 \\ \frac{1+c_{n-1}}{2(n-1)} + \sum_{i=1}^{n-2} \frac{1+c_i}{n-1} & n > 1, \end{cases} \tag{9}$$

This doesn't solve so quickly or easily, but the equivalent form

$$c_{n+1} = \left(1 - \frac{1}{2n}\right) c_n + \frac{c_{n-1}}{2n} + \frac{1}{n} \tag{10}$$

suggests that c_n is smaller but asymptotically grows at almost the same rate as b_n.

Similarly, according to our observation of multi-level delete, the left subtree evolves to be almost never empty. So a recurrence for the backbone for such trees can neglect the zero case.

$$
d_n = \begin{cases} 0 & n = 1 \\ \sum_{i=1}^{n-2} \frac{1+d_i}{n-2} & n > 1, \end{cases}
\tag{11}
$$

3.3   The Culberson Effect

The Culberson Effect, as explained in [CM89] and [CM90] is the tendency of interval endpoints along the shell of the binary search tree to engage in a random walk as time passes. Collisions between endpoints cause adjacent intervals to combine, so that the subsequent expected position and size of the resulting coalesced interval differs from the expected position and size of either of the two original intervals.

In Culberson's formulation for asymmetric update, the random walk is directed; in the case of symmetric deletion, the random walk is undirected, and therefore the effect is more subtle. Figure (5a) shows interval sizes near the root as they evolve with one-level-deletion. Figure (5b) illustrates them for grafting deletion. As each node is deleted (which may occur on any update with a prob-

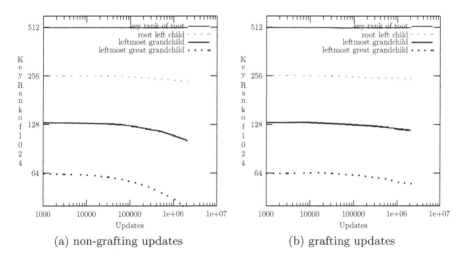

(a) non-grafting updates   (b) grafting updates

Fig. 5. Subtree sizes on shell grow as tree updated.

ability of $1/n$) the key value for that node will take on the value of either the predecessor or the successor node, that is, move either right or left (grafting deletion may cause the keyvalue to skip over several intermediate keys.) After the deletion, one end of the interval defined by the node has moved. The expected position of the node has not changed, but the expected distance from the initial position is one *step*.

An interval is identified by its relative position in the shell, not its bounds. Thus, we can speak of the movement of an interval when deletion of one of its endpoint nodes causes a new key value to appear in the at one end of the interval. This isn't quite Brownian motion, since colliding intervals coalesce instead of rebounding, but after 'long enough' one might expect the intervals nearer the root in the shell to be pushing outward.

There are several special cases. When a shell node deletion occurs through key replacement by the key of its successor or predecessor, which will usually not be a shell node, the interval has moved right (left.) However, if the successor (predecessor) is a shell node with an empty left (right) subtree, there is now one fewer node on the shell, and an interval has disappeared.

Following Culberson, we find that the endpoint of an interval moves either left or right when it is (symmetrically) updated; that is, the key value which defines the endpoint of an interval changes either up or down as the backbone node which holds it is updated. If there were no interval collisions, the expected value of the key would stay constant, while its variance would increase linearly with the number of updates. Since the probability that an update to the tree will involve an endpoint is $1/n$, the expected excursion of a key value is $O(\sqrt{\text{updates}/n})$.

But the size of an interval is bounded below by one. The interval would cease to exist if its endpoints collided. So the expected size of (remaining) intervals will increase as the tree is updated. This effect is clearly visible in fig. (5.) It is much more dramatic for non-grafting deletion, perhaps because in order for a collision to take place the lower endpoint must have a zero-sized subtree, and the grafting deletion algorithm prunes the population of zero-sized subtrees.

The Culberson effect should slightly increase IPL.

4   Recapitulation

Two unrealistic frameworks, *Exact Fit Domain* update, and *non-grafting deletion* are being used to begin understanding three effects in the evolution of binary search trees under update.

Tentatively it appears that the Knott effect may be less significant for a large BST; the effect of fewer zero-size subtrees is predicted to disappear with non-grafting deletion for trees of 100000 or more nodes.

Simulations show that the Culberson effect is still increasing after $n^{\frac{11}{4}}$ deletions. The fact that non-grafting deletion has a stronger Culberson effect needs to be accounted for in modeling. Notice that zero-length subtrees, which figure in the disappearance of intervals in the Culberson model, become quite rare when grafting deletes are used, but are relatively common with non-grafting deletes.

What are the implications of the Evans effect for total IPL? Shell paths seem to be shorter than average paths in the tree, even after Evans stretching, and we would expect that shell lengths would be a less important contributor to IPL in a larger BST.

References

BY89.   Ricardo A. Baeza-Yates. A Trivial Algorithm Whose Analysis Is Not: A Continuation. *BIT*, 29(3):278–394, 1989.

Bri86.   Keith Brinck. On deletion in threaded binary trees. *Journal of Algorithms*, 7(3):395–411, September 1986.

CM89.   Joseph Culberson and J. Ian Munro. Explaining the behavior of binary trees under prolonged updates: A model and simulations. *The Computer Journal*, 32(1), 1989.

CM90.   Joseph Culberson and J. Ian Munro. Analysis of the standard deletion algorithms in exact fit domain binary search trees. *Algorithmica*, 5(3):295–311, 1990.

EC94.   Patricia A. Evans and Joseph Culberson. Asymmetry in binary search tree update algorithms. Technical Report TR 94-09, University of Alberta Department of Computer Science, Edmonton, Alberta, Canada, May 1994.

Epp83.   Jeffrey L. Eppinger. An empirical study of insertion and deletion in binary trees. *CACM*, 26(9):663–669, September 1983.

GBY91.   Gaston H. Gonnet and Ricardo Baeza-Yates. *Handbook of Algorithms and Data Structures*. Addison-Wesley, 2nd edition, 1991.

GKP89.   Ronald L. Graham, Donald E. Knuth, and Oren Patashnik. *Concrete Mathematics*. Addison-Wesley, 1989.

Hib62.   Thomas N. Hibbard. Some combinatorial properties of certain trees with applications to searching and sorting. *JACM*, 9:13–28, 1962.

JK78.   Arne T. Jonassen and Donald E. Knuth. A trivial algorithm whose analysis isn't. *Journal of Computer and System Sciences*, 16:301–322, 1978.

Kno75.   Gary D. Knott. *Deletion in Binary Storage Trees*. PhD thesis, Stanford University, 1975. Available as Tech. Rep. STAN-CS-75-491.

Knu97.   Donald E. Knuth. *The Art of Computer Programming: Volume 3 / Sorting and Searching*. Addison-Wesley, 2nd edition, 1997.

MR98.   Conrado Martinez and Salvador Roura. Randomized binary search trees. *JACM*, 45(2):228–323, March 1998.

Mes91.   Xavier Messeguer. Dynamic behaviour in updating process over BST of size two with probabilistic deletion algorithms. *IPL*, 38:89–100, April 1991.

The LCA Problem Revisited

Michael A. Bender[1][*] and Martín Farach-Colton[2][**]

[1] Department of Computer Science, State University of New York at Stony Brook,
Stony Brook, NY 11794-4400, USA. Email: bender@cs.sunysb.edu.
[2] Department of Computer Science, Rutgers University,
Piscataway, NJ 08855, USA. Email: farach@cs.rutgers.edu.

Abstract. We present a very simple algorithm for the Least Common Ancestors problem. We thus dispel the frequently held notion that optimal LCA computation is unwieldy and unimplementable. Interestingly, this algorithm is a sequentialization of a previously known PRAM algorithm.

1   Introduction

One of the most fundamental algorithmic problems on trees is how to find the *Least Common Ancestor (LCA)* of a pair of nodes. The LCA of nodes u and v in a tree is the shared ancestor of u and v that is located farthest from the root. More formally, the LCA Problem is stated as follows: Given a rooted tree T, how can T be preprocessed to answer LCA queries quickly for any pair of nodes. Thus, one must optimize both the preprocessing time and the query time.

The LCA problem has been studied intensively both because it is inherently beautiful algorithmically and because fast algorithms for the LCA problem can be used to solve other algorithmic problems.

In [HT84], Harel and Tarjan showed the surprising result that LCA queries can be answered in constant time after only linear preprocessing of the tree T. This classic paper is often cited because linear preprocessing is necessary to achieve optimal algorithms in many applications. However, it is well understood that the actual algorithm presented is far too complicated to implement effectively. In [SV88], Schieber and Vishkin introduced a new LCA algorithm. Although their algorithm is vastly simpler than Harel and Tarjan's—indeed, this was the point of this new algorithm—it is far from simple and still not particularly implementable.

The folk wisdom of algorithm designers holds that the LCA problem still has no implementable optimal solution. Thus, according to hearsay, it is better to have a solution to a problem that does not rely on LCA precomputation if possible. We argue in this paper that this folk wisdom is wrong.

In this paper, we present not only a *simplified* LCA algorithm, we present a *simple* LCA algorithm! We devise this algorithm by reëngineering an existing

[*] Supported in part by ISX Corporation and Hughes Research Laboratories.
[**] Supported in part by NSF Career Development Award CCR-9501942, NATO Grant CRG 960215, NSF/NIH Grant BIR 94-12594-03-CONF.

G. Gonnet, D. Panario, and A. Viola (Eds.): LATIN 2000, LNCS 1776, pp. 88–94, 2000.

complicated LCA algorithm: in [BBG+89] a PRAM algorithm was presented that preprocesses and answers queries in $O(\alpha(n))$ time and preprocesses in linear work. Although at first glance, this algorithm is not a promising candidate for implementation, it turns out that almost all of the complications are PRAM induced: when the PRAM complications are excised from this algorithm so that it is lean, mean, and sequential, we are left with an extremely simple algorithm.

In this paper, we present this reëngineered algorithm. Our point is not to present a new algorithm. Indeed, we have already noted that this algorithm has appeared as a PRAM algorithm before. The point is to change the folk wisdom so that researchers are free to use the full power and elegance of LCA computation when it is appropriate.

The remainder of the paper is organized as follows. In Section 2, we provide some definitions and initial lemmas. In Section 3, we present a relatively slow algorithm for LCA preprocessing. In Section 4, we show how to speed up the algorithm so that it runs within the desired time bounds. Finally, in Section 5, we answer some algorithmic questions that arise in the paper but that are not directly related to solving the LCA problem.

2   Definitions

We begin by defining the *Least Common Ancestor (LCA) Problem* formally.

Problem 1. The *Least Common Ancestor (LCA)* problem:

Structure to Preprocess: A rooted tree T having n nodes.
Query: For nodes u and v of tree T, query $\text{LCA}_T(u, v)$ returns the least common ancestor of u and v in T, that is, it returns the node furthest from the root that is an ancestor of both u and v. (When the context is clear, we drop the subscript T on the LCA.)

The *Range Minimum Query (RMQ) Problem*, which seems quite different from the LCA problem, is, in fact, intimately linked.

Problem 2. The *Range Minimum Query (RMQ)* problem:

Structure to Preprocess: A length n array A of numbers.
Query: For indices i and j between 1 and n, query $\text{RMQ}_A(x, y)$ returns the index of the smallest element in the subarray $A[i \ldots j]$. (When the context is clear, we drop the subscript A on the RMQ.)

In order to simplify the description of algorithms that have both preprocessing and query complexity, we introduce the following notation. If an algorithm has preprocessing time $f(n)$ and query time $g(n)$, we will say that the algorithm has complexity $\langle f(n), g(n) \rangle$.

Our solutions to the LCA problem are derived from solutions to the RMQ problem. Thus, before proceeding, we reduce the LCA problem to the RMQ problem. The following simple lemma establishes this reduction.

Lemma 1. *If there is an $\langle f(n), g(n)\rangle$-time solution for RMQ, then there is an $\langle f(2n-1) + O(n), g(2n-1) + O(1)\rangle$-time solution for LCA.*

As we will see, the $O(n)$ term in the preprocessing comes from the time needed to create the soon-to-be-presented length $2n-1$ array, and the $O(1)$ term in the query comes from the time needed to convert the RMQ answer on this array to an LCA answer in the tree.

Proof: Let T be the input tree. The reduction relies on one key observation:

Observation 2 *The LCA of nodes u and v is the shallowest node encountered between the visits to u and to v during a depth first search traversal of T.*

Therefore, the reduction proceeds as follows.

1. Let array $E[1, \ldots, 2n-1]$ store the nodes visited in an Euler Tour of the tree T. [1] That is, $E[i]$ is the label of the ith node visited in the Euler tour of T.
2. Let the *level* of a node be its distance from the root. Compute the Level Array $L[1, \ldots, 2n-1]$, where $L[i]$ is the level of node $E[i]$ of the Euler Tour.
3. Let the *representative* of a node in an Euler tour be the index of first occurrence of the node in the tour[2]; formally, the representative of i is $\mathrm{argmin}_j\{E[j] = i\}$. Compute the Representative Array $R[1, \ldots, n]$, where $R[i]$ is the index of the representative of node i.

Each of these three steps takes $O(n)$ time, yielding $O(n)$ total time. To compute $\mathrm{LCA}_T(x, y)$, we note the following:

- The nodes in the Euler Tour between the first visits to u and to v are $E[R[u], \ldots, R[v]]$ (or $E[R[v], \ldots, R[u]]$).
- The shallowest node in this subtour is at index $\mathrm{RMQ}_L(R[u], R[v])$, since $L[i]$ stores the level of the node at $E[i]$, and the RMQ will thus report the position of the node with minimum level. (Recall Observation 2.)
- The node at this position is $E[\mathrm{RMQ}_L(R[u], R[v])]$, which is thus the output of $\mathrm{LCA}_T(u, v)$.

Thus, we can complete our reduction by preprocessing Level Array L for RMQ. As promised, L is an array of size $2n-1$, and building it takes time $O(n)$. Thus, the total preprocessing is $f(2n-1) + O(n)$. To calculate the query time observe that an LCA query in this reduction uses one RMQ query in L and three array references at $O(1)$ time each. The query thus takes time $g(2n-1) + O(1)$, and we have completed the proof of the reduction. ■

---

[1] The Euler Tour of T is the sequence of nodes we obtain if we write down the label of each node each time it is visited during a DFS. The array of the Euler tour has length $2n-1$ because we start at the root and subsequently output a node each time we traverse an edge. We traverse each of the $n-1$ edges twice, once in each direction.

[2] In fact, any occurrence of i will suffice to make the algorithm work, but we consider the first occurrence for the sake of concreteness.

From now on, we focus only on RMQ solutions. We consider solutions to the general RMQ problem as well as to an important restricted case suggested by the array L. In array L from the above reduction adjacent elements differ by $+1$ or -1. We obtain this ± 1 restriction because, for any two adjacent elements in an Euler tour, one is always the parent of the other, and so their levels differ by exactly one. Thus, we consider the ± 1-RMQ problem as a special case.

2.1 A Naïve Solution for RMQ

We first observe that RMQ has a solution with complexity $\langle O(n^2), O(1) \rangle$: build a table storing answers to all of the n^2 possible queries. To achieve $O(n^2)$ preprocessing rather than the $O(n^3)$ naive preprocessing, we apply a trivial dynamic program. Notice that answering an RMQ query now requires just one array lookup.

3 A Faster RMQ Algorithm

We will improve the $\langle O(n^2), O(1) \rangle$-time brute-force table algorithm for (general) RMQ. The idea is to precompute each query whose length is a power of two. That is, for every i between 1 and n and every j between 1 and $\log n$, we find the minimum element in the block starting at i and having length 2^j, that is, we compute $M[i,j] = \mathrm{argmin}_{k=i\ldots i+2^j-1}\{A[k]\}$. Table M therefore has size $O(n \log n)$, and we fill it in time $O(n \log n)$ by using dynamic programming. Specifically, we find the minimum in a block of size 2^j by comparing the two minima of its two constituent blocks of size 2^{j-1}. More formally, $M[i,j] = M[i,j-1]$ if $A[M[i,j-1]] \leq M[i+2^{j-1}-1,j-1]$ and $M[i,j] = M[i+2^{j-1}-1,j-1]$ otherwise.

How do we use these blocks to compute an arbitrary RMQ(i,j)? We select two overlapping blocks that entirely cover the subrange: let 2^k be the size of the largest block that fits into the range from i to j, that is let $k = \lfloor \log(j-i) \rfloor$. Then RMQ$(i,j)$ can be computed by comparing the minima of the following two blocks: i to $i + 2^k - 1$ ($M(i,k)$) and $j - 2^k + 1$ to j ($M(j - 2^k + 1, k)$). These values have already been computed, so we can find the RMQ in constant time.

This gives the *Sparse Table (ST)* algorithm for RMQ, with complexity $\langle O(n \log n), O(1) \rangle$. Notice that the total computation to answer an RMQ query is three additions, 4 array reference and a minimum, in addition to two other operations: a log and a floor. These can be seen together as the problem of finding the most significant bit of a word. Notice that we must have one such operation in our algorithm, since Harel and Tarjan [HT84] showed that LCA computation has a lower bound of $\Omega(\log \log n)$ on a pointer machine. Furthermore, the most-significant-bit operation has a very fast table lookup solution.

Below, we will use the ST algorithm to build an even faster algorithm for the ± 1RMQ problem.

4 An $\langle O(n), O(1) \rangle$-Time Algorithm for ± 1RMQ

Suppose we have an array A with the ± 1 restriction. We will use a table-lookup technique to precompute answers on small subarrays, thus removing the log factor from the preprocessing. To this end, partition A into blocks of size $\frac{\log n}{2}$. Define an array $A'[1, \ldots, 2n/\log n]$, where $A'[i]$ is the minimum element in the ith block of A. Define an equal size array B, where $B[i]$ is a position in the ith block in which value $A'[i]$ occurs. Recall that RMQ queries return the position of the minimum and that the LCA to RMQ reduction uses the position of the minimum, rather than the minimum itself. Thus we will use array B to keep track of where the minima in A' came from.

The ST algorithm runs on array A' in time $\langle O(n), O(1) \rangle$. Having preprocessed A' for RMQ, consider how we answer any query RMQ(i, j) in A. The indices i and j might be in the same block, so we have to preprocess each block to answer RMQ queries. If $i < j$ are in different blocks, the we can answer the query RMQ(i, j) as follows. First compute the values:

1. The minimum from i forward to the end of its block.
2. The minimum of all the blocks in between between i's block and j's block.
3. The minimum from the beginning of j's block to j.

The query will return the position of the minimum of the three values computed. The second minimum is found in constant time by an RMQ on A', which has been preprocessed using the ST algorithm. But, we need to know how to answer range minimum queries inside blocks to compute the first and third minima, and thus to finish off the algorithm. Thus, the in-block queries are needed whether i and j are in the same block or not.

Therefore, we focus now only on in-block RMQs. If we simply performed RMQ preprocessing on each block, we would spend too much time in preprocessing. If two block were identical, then we could share their preprocessing. However, it is too much to hope for that blocks would be so repeated. The following observation establishes a much stronger shared-preprocessing property.

Observation 3 *If two arrays $X[1, \ldots, k]$ and $Y[1, \ldots, k]$ differ by some fixed value at each position, that is, there is a c such that $X[i] = Y[i] + c$ for every i, then all RMQ answers will be the same for X and Y. In this case, we can use the same preprocessing for both arrays.*

Thus, we can *normalize* a block by subtracting its initial offset from every element. We now use the ± 1 property to show that there are very few kinds of normalized blocks.

Lemma 4. *There are $O(\sqrt{n})$ kinds of normalized blocks.*

Proof: Adjacent elements in normalized blocks differ by $+1$ or -1. Thus, normalized blocks are specified by a ± 1 vector of length $(1/2 \cdot \log n) - 1$. There are $2^{(1/2 \cdot \log n) - 1} = O(\sqrt{n})$ such vectors.   ∎

We are now basically done. We create $O(\sqrt{n})$ tables, one for each possible normalized block. In each table, we put all $(\frac{\log n}{2})^2 = O(\log^2 n)$ answers to all in-block queries. This gives a total of $O(\sqrt{n}\log^2 n)$ total preprocessing of normalized block tables, and $O(1)$ query time. Finally, compute, for each block in A, which normalized block table it should use for its RMQ queries. Thus, each in-block RMQ query takes a single table lookup.

Overall, the total space and preprocessing used for normalized block tables and A' tables is $O(n)$ and the total query time is $O(1)$. We show a complete example below.

4.1   Wrapping Up

We started out by showing a reduction from the LCA problem to the RMQ problem, but with the key observation that the reduction actually leads to a ±1RMQ problem.

We gave a trivial $\langle O(n^2), O(1)\rangle$-time table-lookup algorithm for RMQ, and show how to sparsify the table to get a $\langle O(n\log n), O(1)\rangle$-time table-lookup algorithm. We used this latter algorithm on a smaller summary array A' and needed only to process small blocks to finish the algorithm. Finally, we notice that most of these blocks are the same, from the point of view of the RMQ problem, by using the ±1 assumption given by the original reduction.

5   A Fast Algorithm for RMQ

We have a $\langle O(n), O(1)\rangle$ ±1RMQ. Now we show that the general RMQ can be solved in the same complexity. We do this by reducing the RMQ problem to the LCA problem! Thus, to solve a general RMQ problem, one would convert it to an LCA problem and then back to a ±1RMQ problem.

The following lemma establishes the reduction from RMQ to LCA.

Lemma 5. *If there is a $\langle O(n), O(1)\rangle$ solution for LCA, then there is a $\langle O(n), O(1)\rangle$ solution for RMQ.*

We will show that the $O(n)$ term in the preprocessing comes from the time needed to build the Cartesian Tree of A and the $O(1)$ term in the query comes from the time needed to covert the LCA answer on this tree to an RMQ answer on A.

Proof: Let $A[1,\dots,n]$ be the input array.

The Cartesian Tree of an array is defined as follows. The root of a Cartesian Tree is the minimum element of the array, and the root is labeled with the position of this minimum. Removing the root element splits the array into two pieces. The left and right children of the root are the recursively constructed Cartesian trees of the left and right subarrays, respectively.

A Cartesian Tree can be built in linear time as follows. Suppose C_i is the Cartesian tree of $A[1,\dots,i]$. To build C_{i+1}, we notice that node $i+1$ will belong to the rightmost path of C_{i+1}, so we climb up the rightmost path of C_i until finding the position where $i+1$ belongs. Each comparison either adds an element to the rightmost path or removes one, and each node can only join the rightmost path and leave it once. Thus the total time to build C_n is $O(n)$.

The reduction is as follows.

- Let C be the Cartesian Tree of A. Recall that we associate with each node in C the corresponding corresponding to $A[i]$ with the index i.

Claim. $\text{RMQ}_A(i,j) = \text{LCA}_C(i,j)$.

Proof: Consider the least common ancestor, k, of i and j in the Cartesian Tree C. In the recursive description of a Cartesian tree, k is the first node that separates i and j. Thus, in the array A, element $A[k]$ is between elements $A[i]$ and $A[j]$. Furthermore, $A[k]$ must be the smallest such element in the subarray $A[i, \ldots, j]$ since otherwise, there would be an smaller element k' in $A[i, \ldots, j]$ that would be an ancestor of k in C, and i and j would already have been separated by k'.

More concisely, since k is the first element to split i and j, it is between them because it splits them, and it is minimal because it is the first element to do so. Thus it is the RMQ.   □

We see that we can complete our reduction by preprocessing the Cartesian Tree C for LCA. Tree C takes time $O(n)$ to build, and because C is an n node tree, LCA preprocessing takes $O(n)$ time, for a total of $O(n)$ time. The query then takes $O(1)$, and we have completed the proof of the reduction.   ■

References

BBG+89. O. Berkman, D. Breslauer, Z. Galil, B. Schieber, and U. Vishkin. Highly parallelizable problems. In *Proc. of the 21st Ann. ACM Symp. on Theory of Computing*, pages 309–319, 1989.

HT84. D. Harel and R. E. Tarjan. Fast algorithms for finding nearest common ancestors. *SIAM J. Comput.*, 13(2):338–355, 1984.

SV88. B. Schieber and U. Vishkin. On finding lowest common ancestors: Simplification and parallelization. *SIAM J. Comput.*, 17:1253–1262, 1988.

Optimal and Pessimal Orderings of Steiner Triple Systems in Disk Arrays

Myra B. Cohen and Charles J. Colbourn

Computer Science, University of Vermont, Burlington, VT 05405, USA.
{mcohen,colbourn}@cs.uvm.edu
http://www.cs.uvm.edu/~{mcohen,colbourn}

Abstract. Steiner triple systems are well studied combinatorial designs that have been shown to possess properties desirable for the construction of multiple erasure codes in RAID architectures. The ordering of the columns in the parity check matrices of these codes affects system performance. Combinatorial problems involved in the generation of good and bad column orderings are defined, and examined for small numbers of accesses to consecutive data blocks in the disk array.

1 Background

A *Steiner triple system* is an ordered pair (S, T) where S is a finite set of points or symbols and T is a set of 3-element subsets of S called *triples*, such that each pair of distinct elements of S occurs together in exactly one triple of T. The *order* of a Steiner triple system (S, T) is the size of the set S, denoted $|S|$. A Steiner triple system of order v is often written as $STS(v)$. An $STS(v)$ exists if and only if $v \equiv 1, 3 \pmod 6$ (see [5], for example). We can relax the requirement that every pair occurs exactly once as follows. Let (V, B) be a set V of elements together with a collection B of 3-element subsets of V, so that no pair of elements of V occurs as a subset of more than one $B \in B$. Such a pair (V, B) is an (n, ℓ)-*configuration* when $n = |V|$ and $\ell = |B|$, and every element of V is in at least one of the sets in B.

Let C be a configuration (V, B). We examine the following combinatorial problems. When does there exist a Steiner triple system (S, T) of order v in which the triples can be ordered $T_0, \ldots, T_{b-1}$, so that every ℓ consecutive triples form a configuration isomorphic to C? Such an ordering is a C-*ordering* of the Steiner triple system. When we treat the first triple as following the last (and hence cyclically order the triples), and then enforce the same condition, the ordering is a C-*cyclic ordering*. The presence of configurations in Steiner triple systems has been studied in much detail; see [5] for an extensive survey. Apparently, the presence or absence of configurations among consecutive triples in a triple ordering of an STS has not been previously examined.

Our interest in these problems arises from an application in the design of erasure codes for disk arrays. Prior to examining the combinatorial problems posed, we explore the disk array application. As processor speeds have increased

G. Gonnet, D. Panario, and A. Viola (Eds.): LATIN 2000, LNCS 1776, pp. 95–104, 2000.
© Springer-Verlag Berlin Heidelberg 2000

rapidly in recent years, one method of bridging the Input-Output (I/O) performance gap has been to use redundant arrays of independent disks (RAID) [9]. Individual data reads and writes are *striped* across multiple disks, thereby creating I/O parallelism. Encoding redundant information onto additional disks allows reconstruction of lost information in the presence of disk failures. This creates disk arrays with high throughput and good reliability. However, an array of disks has a substantially greater probability of a disk failure than does an individual disk [8,9]. Indeed, Hellerstein *et al.* [8] have shown that the reliability of an array of 1000 disks which protects against one error, even with periodic daily or weekly repairs, has a lower reliability than an individual disk. Most systems that are available currently handle only one or two disk failures [15]. As arrays grow in size, the need for greater redundancy without a reduction in performance becomes important.

A catastrophic disk failure is an *erasure*. When a disk fails all of the information is lost or erased. Codes that can correct for n erasures are called *n-erasure correcting codes*. The minimum number of additional disks that must be accessed for each write in an n-erasure code, *the update penalty*, has been shown to be n [1,8]. Chee, Colbourn, and Ling [1] have shown that Steiner triple systems possess properties that make them desirable 3-erasure correcting codes with minimal update penalties. The correspondence between Steiner triple systems and parity check matrices is that used by Hellerstein *et al.* [3,8]. Codewords in a binary linear code are viewed as vectors of *information* and *check bits*. The code can then be defined in terms of a $c \times (k + c)$ parity check matrix, $H = [P|I]$ where k is the number of information disks, I is the $c \times c$ identity matrix and P is a $c \times k$ matrix that determines the equations of the check disks. The columns of P are indexed by the k information disks. The columns of I and the rows of H are indexed by the c check disks. A set of disk failures is *recoverable* if and only if the corresponding set of equations in its parity check matrix is linearly independent [1,8]. Any set of t binary vectors is *linearly independent over* GF[2] if and only if the vector sum modulo two of those columns, or any non-empty subset of those columns, is not equal to the zero vector [8].

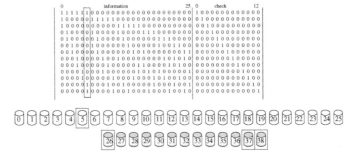

Fig. 1. Steiner (13) Parity Check Matrix: The shaded disks are check disks.

Figure 1 shows a parity check matrix for an STS(13). Cohen and Colbourn [3] examine the ordering of columns in the parity check matrices. This departs from the standard theory of error correcting codes where the order of columns in a parity check matrix is unimportant [16].

One particular class of these codes, anti-Pasch Steiner triple systems, has been shown to correct for all 4-erasures except for *bad erasures* [1]. A bad erasure is one that involves an information disk and all three of its check disks.

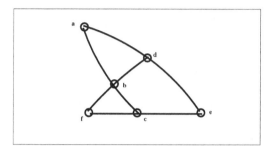

Fig. 2. Pasch Configuration

Figure 2 shows six elements $\{a, b, c, d, e, f\}$ and four triples $\{a, b, c\}, \{a, d, e\}$, $\{f, b, d\}$ and $\{f, c, e\}$. These form a (6,4)-configuration called a *Pasch configuration* or *quadrilateral* [14]. The points represent the check disks (rows of the parity check matrix). Each triple represents an information disk (column of the parity check matrix). If we convert this diagram to a (portion of a) parity check matrix we find that if all four information disks fail there is an irrecoverable loss of information. The resulting four columns in the parity check matrix are linearly dependent and therefore cannot be reconstructed. Anti-Pasch Steiner triple systems yield codes which avoid this configuration. The existence of anti-Pasch $STS(v)$ for all $v \equiv 1$ or 3 (mod 6) except when $v=7$ or 13 was recently solved [7,14].

Cohen and Colbourn [3] examined some of the issues pertaining to encoding Steiner triple systems in a disk array. In a multiple erasure correcting disk array, there may be an overlap among the check disks accessed for consecutive information disks in reads and writes. The number of disks needed in an individual write can therefore be minimized by ordering the columns of this matrix. Using the assumption that the most expensive part of reading or writing in a disk array is the physical read or write to the disk, this overlap can have a significant effect on performance. Cohen and Colbourn [3] describe a write to a triple erasure code as follows. First the information disks are read followed by all of their associated check disks. In the case when check disks overlap, the physical read only takes place once. All of the new parity is computed and then this new parity and the new information is written back to the disks. Once again, the shared check disks are only physically written to one time. Theoretically, the update penalty is the same for all reads and writes in an array. But when more than one information disk in an array shares a common check disk this saves two disk accesses, one read and one write. This finally leads to the questions posed at the outset. In particular, can one ordering be found that optimizes writes of various sizes in such an array?

In order to derive some preliminary results about ordering we have implemented a computer simulation [4,3]. *RaidSim* [9,12,13] is a simulation program written at the University of California at Berkeley [12]. Holland [9] extends it to include declustered parity and online reconstruction. The raidSim program models disk reads and writes and simulates the passage of time. The modified version from [9] is the starting point for our experiments. RaidSim is extended to include mappings for Steiner triple systems and to tolerate multiple disk failures and to detect the existence of unrecoverable four and five erasures [4].

The performance experiments are run with a simulated user concurrency level of 500. Six Steiner triple systems of order 15 are used in these experiments. These are the systems numbered 1, 2, 20, 38, 67 and 80 in [5]. There are 80 non-isomorphic systems of order 15. The number of Pasch configurations in STS(15) range from 105 in STS(15) system one to zero in STS(15) system 80.

2   Pessimal Ordering

A worst triple ordering is one in which consecutive triples are all disjoint. Indeed, if the reads and writes involve at most ℓ consecutive data blocks, a worst triple (or column) ordering is one in which each set of ℓ consecutive triples has all triples disjoint. Let D_ℓ be the $(3\ell, \ell)$-configuration consisting of ℓ disjoint triples. A *pessimal ordering* of an STS is a D_ℓ-ordering. It is easily seen that a $D_{\ell+1}$-ordering is also a D_ℓ-ordering.

The unique STS(7) has no D_2-ordering since every two of its triples intersect. The unique STS(9) has no D_2-ordering, as follows. Consider a triple T. There are exactly two triples, T' and T'' disjoint from T (and indeed T' and T'' are disjoint from each other as well). Without loss of generality, suppose that T' precedes T and T'' follows T in the putative ordering. Then no triple can precede T' or follow T''. These two small cases are, in a sense, misleading. Both STS(13)s and all eighty STS(15)s admit not only a D_2-ordering, but also a D_3-cyclic ordering. This is easily established using a simple backtracking algorithm.

We establish a general result:

Theorem 1. *For* $v \equiv 1, 3 \pmod{6}$ *and* $v \geq 9\ell - 6$, *there exists an* $STS(v)$ *with a* D_ℓ-*ordering.*

Proof. When $v \equiv 3 \pmod{6}$, there is a Steiner triple system $(S, \mathcal{T})$ of order v in which the triples can be partitioned into $(v-1)/2$ classes $R_1, \ldots, R_{(v-1)/2}$, so that within each class all triples are disjoint. Each class R_i contains $v/3$ triples. (This is a *Kirkman triple system*; see [5].) When $v \equiv 1 \pmod{6}$ and $v \geq 19$, there is a Steiner triple system $(S, \mathcal{T})$ of order v in which the triples can be partitioned into $(v+1)/2$ classes $R_1, \ldots, R_{(v+1)/2}$, so that within each class all triples are disjoint. R_1 contains $(v-1)/6$ triples, and each other class contains $(v-1)/3$. (This is a *Hanani triple system*; see [5].)

Our orderings place all triples of R_i before all triples of R_{i+1} for each $1 \leq i < s$. We must order the triples of each class R_i. To do this, we first order the triples of R_1 arbitrarily. Let us then suppose that $R_1, \ldots, R_{i-1}$ have been ordered. To select the jth triple in the ordering of R_i for $1 \leq j < \ell$, we choose a triple which has not already been chosen in R_i, and which does not intersect any of the last $\ell - j$ triples in R_{i-1}. Such a triple exists, since $j - 1$ triples have been chosen, and at most $3(\ell - j)$ triples of R_i intersect any of the last $\ell - j$ triples of R_{i-1}, but $3(\ell - j) + j - 1 < 3\ell - 2 \leq \lfloor v/3 \rfloor$ for all $j \geq 1$.

A similar proof yields D_ℓ-cyclic orderings when v is larger. What is striking about the computational results for order 15 is not that an ordering for some system can be found, but that every system has a D_3-cyclic ordering. This suggests the possibility that for v sufficiently large, every STS(v) admits a D_ℓ-cyclic ordering. To verify this, form the *t-intersection graph* G_t of a triple system $(S, \mathcal{T})$ by including a vertex for each triple in $\mathcal{T}$, and making two vertices adjacent if the corresponding triples share t elements. A D_2-cyclic ordering of $(S, \mathcal{T})$ is equivalent to a Hamilton cycle in G_0. But more is true. A D_ℓ-cyclic ordering of $(S, \mathcal{T})$ is equivalent to the $(\ell - 1)$st power of a Hamilton cycle in G_0. Komlós, Sárközy, and Szemerédi [11] establish that for any $\varepsilon > 0$ and any sufficiently large n-vertex graph G of minimum degree at least $\left(\frac{k}{k+1} + \varepsilon \right) n$, G contains the kth power of a Hamilton cycle. Now G_0 has $v(v-1)/6$ vertices and degree $(v(v-10) + 21)/6$, and so G_0 meets the required conditions. Thus when ℓ is fixed, *every* sufficiently large STS(v) admits a D_ℓ-ordering. We give a direct proof of this, which does not rely upon probabilistic methods.

Theorem 2. *For ℓ a positive integer and $v \geq 81(\ell-1)+1$, every STS(v) admits a D_ℓ-ordering.*

Proof. Let $(S, \mathcal{T})$ be an STS(v). Form the 1-intersection graph G_1 of $(S, \mathcal{T})$. G_1 is regular of degree $3(v-1)/2$, and therefore has a proper vertex coloring in $s \leq 3(v-1)/2$ colors. Form a partition of $\mathcal{T}$, defining classes $R_1, \ldots, R_s$ of triples by placing a triple in the class R_i when the corresponding vertex of G_1 has the ith color. Let us suppose without loss of generality that $|R_i| \leq |R_{i+1}|$ for $1 \leq i < s$. Now if $3|R_1| < |R_s|$, there must be a triple of R_s intersecting no triple of R_1. When this occurs, move such a triple from R_s to R_1. This can be repeated until $3|R_1| \geq |R_s|$. Since $\sum_{i=1}^{s} |R_i| = v(v-1)/6$, we find that $|R_s| \geq \lceil v/9 \rceil$ and thus $|R_1| \geq \lceil v/27 \rceil$. But then $|R_1| > 3\ell - 3$, and we can apply precisely the method in the proof of Theorem 1 to produce the ordering required.

The bound on v can almost certainly be improved upon. Indeed for $\ell = 3$, we expect that every STS(v) with $v > 13$ has a D_3-cyclic ordering.

3   Optimal Ordering

Optimal orderings pose more challenging problems. We wish to minimize rather than maximize the number of check disks associated with ℓ consecutive triples. We begin by considering small values of ℓ. When $\ell = 2$, define the configuration I_2 to be the unique (5,2)-configuration, which consists of two intersecting triples. An optimal ordering is an I_2-ordering. Horák and Rosa [10] essentially proved the following:

Theorem 3. *Every STS(v) admits an I_2-cyclic ordering.*

Proof. The 1-intersection graph G_1 of the STS has a hamiltonian cycle [10].

Let T be the unique (6,3)-configuration, as depicted in Figure 3.

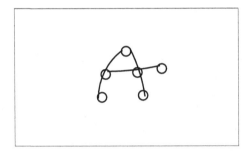

Fig. 3. Optimal Ordering on Three Blocks

An optimal ordering when $\ell = 3$ is a T-ordering. The unique STS(7) has a T-cyclic ordering: 013, 124, 026, 156, 235, 346, 045. The unique STS(9) also has a T-cyclic ordering: 012, 036, 138, 048, 147, 057, 345, 237, 246, 678, 258, 156. We might therefore anticipate that every STS(v) has a T-cyclic ordering, but this does not happen. To establish this, we require a few definitions. A *proper subsystem* of a Steiner triple system (S, T) is a pair (S', T') with $S' \subset S$ and $T' \subset T$, $|S'| > 3$, and (S', T') itself a Steiner triple system. A *Steiner space* is a Steiner triple system with the property that every three elements which do not appear together in a triple are contained in a proper subsystem.

Theorem 4. *No Steiner space admits a T-ordering. Hence, whenever we have $v \equiv 1, 3 \pmod 6$, $v \geq 15$, and $v \notin \{19, 21, 25, 33, 37, 43, 51, 67, 69, 145\}$, there is a Steiner triple system admitting no T-ordering.*

Proof. Suppose that (S, T) is a Steiner space which has a T-ordering. Then consider two consecutive triples under this ordering. These are contained within a proper subsystem. Any triple preceding or following two consecutive triples of a subsystem must also lie in the subsystem. But this forces all triples of T to lie in the subsystem, which is a contradiction.

The conditions on v reflect the current knowledge about the existence of Steiner spaces (see [5]).

A much weaker condition suffices to establish that there is no T-ordering. A T-ordering cannot have any two consecutive triples which appear together in a proper subsystem. By the same token, a T-ordering cannot have any two triples which appear together in a proper subsystem and are separated by only one triple in the ordering. Hence the strong condition on subsystems enforced in Steiner spaces can be relaxed. Of course, our interest is in producing Steiner triple systems that do admit T-orderings. Both STS(13)s admit T-orderings but not cyclic T-orderings. Of the 80 STS(15)s, only fourteen admit cyclic T-orderings; they are numbers 20, 22, 38, 39, 44, 48, 50, 51, 52, 53, 65, 67, 75, and 76. However, 73 of the systems (those numbered 8–80) admit a T-ordering. System 1 is the projective triple system and hence is a Steiner space (see [5]). However, the six systems numbered 2–7 also do not appear to admit a T-ordering. These results have all been obtained with a simple backtracking algorithm.

General constructions for larger orders appear to be difficult to establish. However, we expect that for every order $v \geq 15$ there exists a system having a T-cyclic ordering. For example, let $T_{i0} = \{i, 5+i, 11+i\}$, $T_{i1} = \{i, 2+i, 9+i\}$, and $T_{i2} = \{1+i, 2+i, 5+i\}$, with arithmetic modulo 19. Then an STS(19) with a T-cyclic ordering exists with the triple T_{ij} in position $27i + j$ mod 57 for $0 \leq i < 19$ and $0 \leq j < 3$. A similar solution for $v = 25$ is obtained by setting $T_{i0} = \{i, 1+i, 6+i\}$, $T_{i1} = \{6+i, 8+i, 16+i\}$, $T_{i2} = \{1+i, 8+i, 22+i\}$, and $T_{i3} = \{3+i, 6+i, 22+i\}$, arithmetic modulo 100. Place triple T_{ij} in position $32i + j$ modulo 100. While these small designs indicate that specific systems admitting an ordering can be easily found, we have not found a general pattern for larger orders.

When $\ell = 4$, four triples must involve at least six distinct elements. Indeed, the only (6,4)-configuration is the Pasch configuration. It therefore appears that the best systems from an ordering standpoint (when $\ell = 4$) are precisely those which are poor from the standpoint of erasure correction. However, in our performance experiments, ordering plays a larger role than does the erasure correction capability [4,3]. Hence it is sensible to examine STSs which admit orderings with Pasch configurations placed consecutively. Unfortunately, this does not work in general:

Lemma 1. *No STS(v) for $v > 3$ is Pasch-orderable.*

Proof. Any three triples of a Pasch configuration lie in a unique Pasch configuration. Hence four consecutive triples forming a Pasch configuration for some triple ordering can neither be preceded nor followed by a triple which forms a second Pasch configuration.

It therefore appears that an optimal ordering has exactly $\lceil (v(v-1)/6) - 3 \rceil$ of the sets of four consecutive triples inducing a Pasch configuration; these alternate with sets of four consecutive triples forming a (7,4)-configuration. We have not explored this possibility.

A general theory for all values of ℓ would be worthwhile, but appears to be substantially more difficult than for pessimal orderings.

4   Conclusions

It is natural to ask whether the orderings found here have a real impact on disk array performance. Figures 4 - 6 show the results of performance experiments using various orderings. The desired orderings will provide the lowest response times. The 'good' ordering is a T-ordering when one exists, and otherwise is an ordering found in an effort to maximize the number of consecutive T configurations; it is labeled A in these figures. The 'bad' ordering is a D_3-ordering and is labeled B. The ordering labeled C is one obtained from a random triple ordering. The most significant difference in performance arises in a workload of straight writes. This is as expected because this is where decreasing the actual update penalty has the greatest impact. Although the read workload shows no apparent differences during fault-free mode, it does start to differentiate when multiple failures occur.

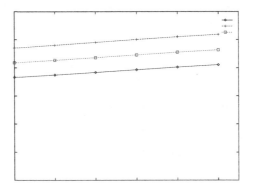

Fig. 4. Ordering Results - Straight Write Workload

The structure of optimal orderings for $\ell \geq 4$ is an open and interesting question. Minimizing disk access through ordering means that the update penalty is only an upper bound on the number of accesses in any write. By keeping the number of check disk accesses consistently lower, performance gains can be achieved. An interesting question is the generalization for reads and writes of different sizes: Should an array be configured specifically for a particular size when optimization is desired? One more issue in optimization of writes in triple erasure codes is that of the large or stripe write [9]. At some point, if we have a large write in an array, all of the check disks are accessed. There is a threshold beyond which it is less expensive to read all of the information disks, compute

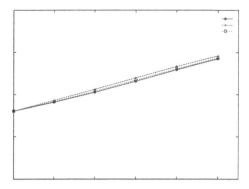

Fig. 5. Ordering Results - Straight Read Workload

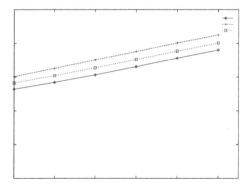

Fig. 6. Ordering Results - Mixed Workload

the new parity and then write out all of the new information disks and all of the check disks. When using an STS(15), a threshold occurs beyond the halfway point. An STS(15) has 35 information and 15 check disks. If 23 disks are to be written they must use at least 14 check disks. In the method of writing described above, 46 information accesses and 28 check disk accesses yield a total of 74 physical disk accesses. A large write instead has 35 reads, followed by $15 + 23 = 28$ writes, for a total of 73 physical accesses. This threshold for all STSs determines to some extent how to optimize disk writes.

Steiner triple systems provide an interesting option for redundancy in large disk arrays. They have the unexpected property of lowering the expected update penalty when ordered optimally.

Acknowledgments

Research of the authors is supported by the Army Research Office (U.S.A.) under grant number DAAG55-98-1-0272 (Colbourn). Thanks to Sanjoy Baruah, Ron Gould, Alan Ling, and Alex Rosa for helpful comments.

References

1. Yeow Meng Chee, Charles J. Colbourn and Alan C. H. Ling. Asymptotically optimal erasure-resilient codes for large disk arrays. *Discrete Applied Mathematics*, to appear.
2. Peter M. Chen, Edward K. Lee, Garth A. Gibson, Randy H. Katz and David A. Patterson. RAID: High-performance, reliable secondary storage. *ACM Computing Surveys* 26 (1994) 145–185.
3. M.B. Cohen and C.J. Colbourn. Steiner triple systems as multiple erasure codes in large disk arrays. submitted.
4. Myra B. Cohen. Performance analysis of triple erasure codes in large disk arrays. Master's thesis, University of Vermont, 1999.
5. Charles J. Colbourn and Alexander Rosa. *Triple Systems*. Oxford University Press, Oxford, 1999.
6. Garth A. Gibson. *Redundant Disk Arrays, Reliable Parallel Secondary Storage*. MIT Press, 1992.
7. M.J. Grannell, T.S. Griggs, and C.A. Whitehead. The resolution of the anti-Pasch conjecture. *Journal of Combinatorial Designs*, to appear.
8. Lisa Hellerstein, Garth A. Gibson, Richard M. Karp, Randy H. Katz and David A. Patterson. Coding techniques for handling failures in large disk arrays. *Algorithmica* 12 (1994) 182–208.
9. Mark C. Holland. *On-Line Data Reconstruction In Redundant Disk Arrays*. PhD thesis, Carnegie Mellon University, 1994.
10. Peter Horák and Alexander Rosa. Decomposing Steiner triple systems into small configurations. *Ars Combinatoria* 26 (1988) 91–105.
11. János Komlós, Gábor Sárközy and Endre Szemerédi. On the Pósa-Seymour conjecture. *Journal of Graph Theory* 29 (1998) 167–176.
12. Edward K. Lee. Software and performance issues in the implementation of a RAID prototype. Technical Report CSD-90-573, University of California at Berkeley, 1990.
13. Edward K. Lee. *Performance Modeling and Analysis of Disk Arrays*. PhD thesis, University of California at Berkeley, 1993.
14. A.C.H. Ling, C.J. Colbourn, M.J. Grannell and T.S. Griggs. Construction techniques for anti-Pasch Steiner triple systems. *Journal of the London Mathematical Society*, to appear.
15. Paul Massiglia. *The RAID Book, A Storage System Technology Handbook, 6th Edition*. The RAID Advisory Board, 1997.
16. Scott A. Vanstone and Paul C. van Oorschot. *An Introduction to Error Correcting Codes with Applications*. Kluwer Academic Publishers, 1989.

Rank Inequalities for Packing Designs and Sparse Triple Systems

Lucia Moura [*]

Department of Computer Science, University of Toronto,
and The Fields Institute for Research in Mathematical Sciences
lucia@cs.toronto.edu

Abstract. Combinatorial designs find numerous applications in computer science, and are closely related to problems in coding theory. Packing designs correspond to codes with constant weight; 4-sparse partial Steiner triple systems (4-sparse PSTSs) correspond to erasure-resilient codes able to correct all (except for "bad ones") 4-erasures, which are useful in handling failures in large disk arrays [4,10]. The study of polytopes associated with combinatorial problems has proven to be important for both algorithms and theory. However, research on polytopes for problems in combinatorial design and coding theories have been pursued only recently [14,15,17,20,21]. In this article, polytopes associated with t-(v, k, λ) packing designs and sparse PSTSs are studied. The subpacking and sparseness inequalities are introduced. These can be regarded as rank inequalities for the independence systems associated with these designs. Conditions under which subpacking inequalities define facets are studied. Sparseness inequalities are proven to induce facets for the sparse PSTS polytope; some extremal families of PSTS known as Erdös configurations play a central role in this proof. The incorporation of these inequalities in polyhedral algorithms and their use for deriving upper bounds on the packing numbers are suggested. A sample of 4-sparse $PSTS(v)$, $v \leq 16$, obtained by such an algorithm is shown; an upper bound on the size of m-sparse PSTSs is presented.

1 Introduction

In this article, polytopes associated with problems in combinatorial design and coding theories are investigated. We start by defining the problems in which we are interested, and then describe their polytopes and motivations for this research. Throughout the paper, we denote by $\binom{V}{k}$ the family of sets $\{B \subseteq V : |B| = k\}$. Let $v \geq k \geq t$. A t-(v, k, λ) *design* is a pair $(V, \mathcal{B})$ where V is a v-set and $\mathcal{B}$ is a collection of k-subsets of V called *blocks* such that every t-subset of V is contained in exactly λ blocks of $\mathcal{B}$. Design theorists are concerned with the existence of these designs. A t-(v, k, λ) *packing design* is defined by replacing the condition "in exactly λ blocks" in the above definition by "in at most λ blocks". The objective is to determine the *packing number*, denoted by $D_\lambda(v, k, t)$, which

---

[*] Supported by Natural Sciences and Engineering Research Council of Canada PDF

G. Gonnet, D. Panario, and A. Viola (Eds.): LATIN 2000, LNCS 1776, pp. 105–114, 2000.
© Springer-Verlag Berlin Heidelberg 2000

is the *maximum* number of blocks in a t–(v, k, λ) packing design. The existence of a t–(v, k, λ) design can be decided by checking whether the packing number $D_\lambda(v, k, t)$ is equal to $\lambda \binom{v}{t} / \binom{k}{t}$. Thus, the determination of the packing number is the most general problem and we will concentrate on it. Designs place a central role in the theory of error-correcting codes, and, in particular, t–$(v, k, 1)$ packing designs correspond to constant weight codes of weight k, length v and minimum distance $2(k - t + 1)$. For surveys on packing designs see [18,19].

Determining the packing number is a hard problem in general, although the problem has been solved for specific sets of parameters. For instance, the existence of Steiner Triple Systems (i.e. 2–$(v, 3, 1)$ designs), and the packing number for Partial Steiner Triple Systems (PSTS) (i.e. 2–$(v, 3, 1)$ packing designs) have been settled. On the other hand, the study of triple systems is an active area of research with plenty of open problems. Interesting problems arise in the study of STSs and PSTSs avoiding prescribed sub-configurations (see the survey [8]). Let us denote by $STS(v)$ the Steiner triple system ($PSTS(v)$ for a partial one) on v points. A (p, l)-configuration in a (partial) Steiner triple system is a set of l blocks (of the (partial) Steiner triple system) spanning p elements. Let $m \geq 4$. An $STS(v)$ is said to be m-sparse if it avoids every $(l + 2, l)$-configuration for $4 \leq l \leq m$. Erdös (see [12]) conjectured that for all $m \geq 4$ there exists an integer v_m such that for every admissible $v \geq v_m$ there exist an m-sparse $STS(v)$. Again the objective is to determine the sparse packing number, denoted by $D(m, v)$, which is the *maximum* number of blocks in an m-sparse $PSTS(v)$. The 4-sparse PSTSs are the same as anti-Pasch ones, since Pasches are the only $(6, 4)$-configurations. A 4-sparse (or anti-Pasch) $STS(v)$ is known to exist for all $v \equiv 3 \pmod 6$ [2]. For the remaining case, i.e. the case $v \equiv 1 \pmod 6$, there are constructions and partial results. Anti-mitre Steiner triple systems were first studied in [6]. The 5-*sparse* Steiner triple systems are the systems that are both anti-Pasch and anti-mitre. Although there are some results on 5-sparse STSs [6,13], the problem is far from settled. In spite of Erdös conjecture, no m-sparse Steiner triple system is known for $m \geq 6$. The study of m-sparse PSTSs gives rise to interesting extremal problems in hypergraph theory; in addition, these designs have applications in computer science. For instance, the 4-sparse (or anti-Pasch) PSTSs correspond to erasure-resilient codes that tolerates all (except bad) 4-erasures, which are useful in applications for handling failures in large disk arrays [4,10].

Let $\mathcal{D}$ be the set of all packing designs of the same kind and with the same parameters (for instance, the set of all 2-$(10, 3, 1)$ packing designs or the set of all 5-sparse $PSTS(10)$). Let $P(\mathcal{D})$ be the polytope in $\mathbb{R}^{\binom{v}{k}}$ given by the convex hull of the incidence vectors of the packing designs in $\mathcal{D}$. Thus, determining the packing number associated with $\mathcal{D}$, amounts to solving the following optimization problem

$$\begin{cases} \text{maximize} & \sum_{B \in \binom{V}{k}} x_B \\ \text{Subject to} & x \in P(\mathcal{D}). \end{cases}$$

If we had a description of $P(\mathcal{D})$ in terms of linear inequalities, this problem could be solved via linear programming. Unfortunately, it is unlikely for us to find complete descriptions of polytopes for hard combinatorial problems. On the other hand, some very effective computational methods use partial descriptions of a problem's polytope [3]. Therefore, it is of great interest to find classes of facets for these polytopes. It is also important to design efficient separation algorithms for a class of facets. Given a point outside a polytope and a class of valid inequalities for the polytope, a separation algorithm determines an inequality that is violated by the point or decides one does not exist. This is fundamental in branch-and-cut or other polyhedral algorithms that work with partial descriptions of polytopes.

Polytopes for general t–(v, k, λ) packing designs were first discussed in [14]; their clique facets have been determined for all packings with $\lambda = 1$ and $k - t \in \{1, 2\}$ for all t and v [16]. A polyhedral algorithm for t–$(v, k, 1)$ packings and designs was proposed and tested in [17]. A related work that employs incidence matrix formulations for 2-(v, k, λ) design polytopes can be found in [20].

In this paper, we present two new classes of inequalities: the subpacking and the sparseness inequalities. They are types of *rank inequalities* when one regards the packing designs as independence systems, as discussed in Section 2. In Section 3, we focus on the subpacking inequalities, which are valid inequalities for both t–(v, k, λ) packing designs and sparse PSTSs. We study conditions under which these inequalities induce facets for the packing design polytope. In Section 4, we discuss sparseness inequalities. Given $m \geq 4$, the l-sparseness inequalities, $2 \leq l \leq m$, are valid for the m-sparse PSTS polytope, and proven to always be facet inducing. In Section 5, we show the results of our branch-and-cut algorithm for determining the sparse packing number for 4-sparse $PSTS(v)$ with $v \leq 16$. The algorithm follows the lines of the one described in [17], but employs sparse facets. With these 4-sparse packing numbers in hand, we develop a simple bound that uses the previous packing number and Chvátal-Gomory type of cuts to give an upper bound on the next packing numbers. Further research is discussed in Section 6.

2 Independence Systems, Packing Designs and their Polytopes

In this section, we define some terminology about independence systems and collect some results we use from the independence system literature. We closely follow the notation in [11]. Along the section, we translate the concepts to the context of combinatorial designs.

Let $N = \{v_1, v_2, \ldots, v_n\}$ be a finite set. An *independence system* on N is a family $\mathcal{I}$ of subsets of N closed under inclusion, i.e. satisfying the property: $J \in \mathcal{I}$ and $I \subseteq J$ implies $I \in \mathcal{I}$, for all $J \in \mathcal{I}$. Any set in $\mathcal{I}$ is called *independent* and any set outside $\mathcal{I}$ is called *dependent*. Any minimal (with respect to set inclusion) dependent set is called a circuit, and an independent system is characterized by its family of circuits, which we denote by $\mathcal{C}$. The *independence number* of $\mathcal{I}$, denoted by $\alpha(\mathcal{I})$, is the maximum size of an independent set in $\mathcal{I}$. Given a subset S of N, the *rank* of S is defined by $r(S) = \max\{|I| : I \in \mathcal{I} \text{ and } I \subseteq S\}$. Note that $\alpha(\mathcal{I}) = r(N)$.

If the circuits in $\mathcal{C}$ have size 2, $G = (N, \mathcal{C})$ forms a graph with N as the nodeset, $\mathcal{C}$ as the edgeset and $\mathcal{I}$ forms the set of independent or stable sets of G.

Remark 1. (*Packing Designs*) Given t, v, k, λ, let $\mathcal{I}$ be the family of all t-(v, k, λ) packing designs on the same v-set V. Let, $N = \binom{V}{k}$, then $\mathcal{I}$ is clearly an independence system on N. The packing number is the independence number. Each circuit in $\mathcal{C}$ corresponds to a subset of $\binom{V}{k}$ of cardinality $\lambda + 1$ such that its k-sets contain a common t-subset of V. For $\lambda = 1$, $\mathcal{C}$ is simply formed by the pairs of k-sets of V which intersect in at least t points, and the underlying graph is obvious.

Following the definition in [9], an *Erdös configuration of order n, $n \geq 1$*, in a (partial) Steiner triple system is any $(n + 2, n)$-configuration, which contains no $(l + 2, l)$-configuration, $1 < l < n$. In fact, this is equivalent to requiring that $4 \leq l < n$, since there cannot be any $(4, 2)$- or $(5, 3)$-configurations in a PSTS.

Remark 2. (*Sparse PSTSs.*) Let $\mathcal{I}$ be the independence system of the $2 - (v, 3, 1)$ packing designs on the same v-set V. Let $\mathcal{C}$ be its collection of circuits, namely, the family of all pairs of triples of V whose intersection has cardinality 2. Adding m-sparseness requirements to $\mathcal{I}$ amounts to removing from $\mathcal{I}$ the packing designs that are not m-sparse, and adding extra circuits to $\mathcal{C}$. The circuits to be added to $\mathcal{C}$ are precisely the Erdös configurations of order l, for all $4 \leq l \leq m$.

Before we discuss valid inequalities for the independence system polytope, we recall some definitions. A *polyhedron* $P \subseteq \mathbb{R}^n$ is the set of points satisfying a finite set of linear inequalities. A *polytope* is a bounded polyhedron. A polyhedron $P \subseteq \mathbb{R}^n$ is of *dimension k*, denoted by $dim P = k$, if the maximum number of affinely independent points in P is $k + 1$. We say that P is *full dimensional* if $dim P = n$. Let $d \in \mathbb{R}^n$ and $d_0 \in \mathbb{R}$. An inequality $d^T x \leq d_0$ is said to be *valid* for P if it is satisfied by all points of P. A subset $F \subseteq P$ is called a *face* of P if there exists a valid inequality $d^T x \leq d_0$ such that $F = P \cap \{x \in \mathbb{R}^n : d^T x = d_0\}$; the inequality is said to *represent* or to *induce* the face F. A *facet* is a face of P with dimension $(dim P) - 1$. If P is full dimensional (which can be assumed w.l.o.g. for independence systems), then each facet is determined by a unique (up to multiplication by a positive number) valid inequality. Moreover, the minimal system of inequalities representing P is given by the inequalities inducing its facets.

Consider again an independence system $\mathcal{I}$ on N. The *rank inequality* associated with a subset S of N is defined by

$$\sum_{i \in S} x_i \leq r(S), \tag{1}$$

and is obviously a valid inequality for the independence system polytope $P(\mathcal{I})$. Necessary or sufficient conditions for a rank inequality to induce a facet have been discussed [11]. We recall some definitions. A subset S of N is said to be *closed* if $r(S \cup \{i\}) \geq r(S) + 1$ for all $i \in N \setminus S$. S is said to be *nonseparable* if $r(S) < r(T) + r(S \setminus T)$ for all nonempty proper subset T of S.

A necessary condition for (1) to induce a facet is that S be closed and non-separable. This was observed by Laurent [11], and was stated by Balas and Zemel [1] for independent sets in graphs. A sufficient condition for (1) to induce a facet is given in the next theorem. Let $\mathcal{I}$ be an independence system on N and let S be a subset of N. Let $\mathcal{C}$ be the family of circuits of $\mathcal{I}$ and let $\mathcal{C}_S$ denote its restriction to S. The *critical graph* of $\mathcal{I}$ on S, denoted by $G_S(\mathcal{I})$, is defined as having S as its nodeset and with edges defined as follows: $i_1, i_2 \in S$ are adjacent if and only if the removal of all circuits of $\mathcal{C}_S$ containing $\{i_1, i_2\}$ increases the rank of S.

Theorem 1. *(Laurent [11], Chvátal [5] for graphs) Let $S \subseteq N$. If S is closed and the critical graph $G_S(\mathcal{I})$ is connected, then the rank inequality (1) associated with S induces a facet of the polytope $P(\mathcal{I})$.*

Proposition 1. *(Laurent [11], Cornuejols and Sassano [7]) The following are equivalent*

1. *The rank inequality (1) induces a facet of $P(\mathcal{I})$.*
2. *S is closed and the rank inequality (1) induces a facet of $P(\mathcal{I}_S)$.*

3   Subpacking Inequalities for t–(v, k, λ) Packings

Let us denote by $P_{t,v,k,\lambda}$ the polytope associated with the t–(v, k, λ) packing designs on the same v-set V, and by $\mathcal{I}_{t,v,k,\lambda}$ the corresponding independence system on $N = \binom{V}{k}$. Let $S \subseteq V$. Then, it is clear that $r(\binom{S}{k}) = D_\lambda(|S|, k, t)$ and the rank inequality associated with $\binom{S}{k}$ is given by

$$\sum_{B \in \binom{S}{k}} x_B \le D_\lambda(|S|, k, t). \tag{2}$$

We call this the *subpacking inequality* associated with S, which is clearly valid for $P_{t,v,k,\lambda}$. In this section, we investigate conditions for this inequality to be facet inducing. The next proposition gives a sufficient condition for a subpacking inequality not to induce a facet.

Proposition 2. *If there exists a t-(v, k, λ) design, then*

$$\sum_{B \in \binom{V}{k}} x_B \le D_\lambda(v, k, t) \tag{3}$$

does not induce a facet of $P_{t,v,k,\lambda}$.

Proof. Since there exists a t-(v, k, λ) design, it follows $D_\lambda(v, k, t) = \lambda\binom{v}{t}/\binom{k}{t}$. Then, equation (3) can be obtained by adding the clique facets: $\sum_{B \supseteq T} x_B \le \lambda$, for all $T \subseteq V$, $|T| = t$. Thus, (3) cannot induce a facet. $\square$

The next proposition addresses the extendibility of facet inducing subpacking inequalities from $P_{t,|S|,k,\lambda}$ to $P_{t,v,k,\lambda}$, $v \ge |S|$.

Proposition 3. *Let $S \subseteq V$. Then, the following are equivalent:*

1. *the subpacking inequality (2) induces a facet of $P_{t,v,k,\lambda}$.*
2. *the subpacking inequality (2) induces a facet of $P_{t,|S|,k,\lambda}$; and for all $B' \in \binom{V}{k} \setminus \binom{S}{k}$ there exists t–$(|S|, k, \lambda)$ packing design $(S, \mathcal{B})$ with $|\mathcal{B}| = D_\lambda(|S|, k, t)$ such that $(S, \mathcal{B} \cup \{B'\})$ is a t–(v, k, λ) packing design.*

Proof. The last condition in 2 is equivalent to $\binom{S}{k}$ being closed for the independence system $\mathcal{I}_{t,v,k,\lambda}$; thus, the equivalence comes directly from Theorem 1. □

For the particular case of $k = t + 1$ and $\lambda = 1$, facet inducing subpacking inequalities are always extendible.

Proposition 4. *(Guaranteed extendibility of a class of subpacking facets) Let $k = t + 1$ and $\lambda = 1$. Then, the subpacking inequality*

$$\sum_{B \in \binom{S}{t+1}} x_B \leq D_1(|S|, t + 1, t) \tag{4}$$

associated with $S \subseteq V$ induces a facet for $P_{t,v,t+1,1}$ if and only if it induces a facet for $P_{t,|S|,t+1,1}$.

The proof of Proposition 4 involves showing that the second condition in item 2 of Proposition 3 holds for $k = t + 1$ and $\lambda = 1$.

Theorem 2. *(Facet defining subpacking inequalities for PSPSs) Let $S \subseteq V$, $|S| \leq 10$. The subpacking inequality associated with S induces a facet of $P_{2,v,3,1}$, $v \geq |S|$, if and only if $|S| \in \{4, 5, 10\}$.*

Sketch of the proof. Since there exist $STS(v)$ for all $v \equiv 1, 3 \pmod 6$, Proposition 2 covers cases $|S| \in \{7, 9\}$. It remains to deal with the cases $|S| \in \{4, 5, 6, 8, 10\}$ (see Table 1). Subpacking inequalities with $|S| = 4$ are facet-inducing clique inequalities. For the case $|S| \in \{5, 10\}$, we show that the corresponding critical graphs are connected, which (by Theorem 1) is sufficient to show the inequalities induce facets of $P_{2,|S|,3,1}$. Proposition 4 guarantees they also define facets of $P_{2,v,3,1}$. For the case $|S| \in \{6, 8\}$, we show that the corresponding subpacking inequalities can be written as (non-trivial) linear combinations of other valid inequalities, which implies that they do not induce facets. □

Remark 3. (Separation of subpacking inequalities) For a constant C, subpacking inequalities with $|S| \leq C$ can be separated in polynomial time. This is the case, since there are exactly $\sum_{s=4}^{C} \binom{v}{s} \in O(v^C)$ inequalities to check, which is a polynomial in the number of variables of the problem, which is $\binom{v}{k}$.

4   Sparseness Facets for Sparse PSTSs

Let us denote by $P_{m,v}$ the polytope associated with m-sparse $PSTS(v)$ on the same v-set V, and by $\mathcal{I}_{m,v}$ the corresponding independence system. The main contribution of this section is a class of facet inducing inequalities for $P_{m,v}$, which we call sparseness inequalities, given by Theorem 3.

Table 1. Summary of facet inducing subpacking inequalities of $P_{2,v,3,1}$ for $|S| \leq 10$.

$|S|$	$D_1(|S|,3,2)$ or $r(\binom{S}{3})$	$\sum_{B \cdot \binom{S}{3}} x_B \leq D_1(|S|,3,2)$ is facet inducing, $v \geq |S|$	Reference
4	1	Yes	maximal clique [17]
5	2	Yes	Theorem 2
6	4	No	Theorem 2
7	7	No	$\exists\, STS(7)$ + Proposition 2
8	8	No	Theorem 2
9	12	No	$\exists\, STS(9)$ + Proposition 2
10	13	Yes	Theorem 2
1,3 (mod 6)	$(|S|^2 - |S|)/6$	No	$\exists\, STS(|S|)$ + Proposition 2

Lemma 1. *(Lefmann et al.[12, Lemma 2.3])*
Let l, r be positive integers, $l \geq 1$. Then any $(l+2, l+r)$-configuration in a Steiner triple system contains a $(l+2, l)$-configuration.

The proofs of the next two lemmas are omitted in this extended abstract.

Lemma 2. *(Construction of an Erdös configuration, for all $n \geq 4$)*
Consider the following recursive definition:

$$\mathcal{E}_4 = \{E_1 = \{1,2,5\}, E_2 = \{3,4,5\}, E_3 = \{1,3,6\}, E_4 = \{2,4,6\}\},$$
$$\mathcal{E}_5 = \mathcal{E}_4 \setminus \{E_4\} \cup \{\{2,4,7\}, E_5 = \{5,6,7\}\}$$
$$\mathcal{E}_{n+1} = \mathcal{E}_n \setminus \{E_n\} \cup \{E_n \setminus \{n+2\} \cup \{n+3\}\} \cup$$
$$\cup \{E_{n+1} = \{n+2, n+3, 1+((n-1) \bmod 4)\}\}, n \geq 5.$$

Then, for all $n \geq 4$, $\mathcal{E}_n$ is an Erdös configuration of order n.

Lemma 3. *Let $v-2 \geq l \geq 4$ and let T be an $(l+2)$-subset of V. Let $R \in \binom{V}{3} \setminus \binom{S}{3}$. Then, there exists an Erdös configuration $\mathcal{S}$ of order l on the points of T and a triple $S \in \mathcal{S}$, such that $\mathcal{S} \setminus \{S\} \cup \{R\}$ is an l-sparse $PSTS(v)$.*

Theorem 3. *(m-sparseness facets) Let $m \geq 4$. Then, for any $2 \leq l \leq m$ and any $(l+2)$-subset T of V, the inequality*

$$s(T): \quad \sum_{B \in \binom{T}{3}} x_B \leq l-1$$

defines a facet for $P_{m,v}$.

Proof. Inequalities $s(T)$ with $l \in \{1,2\}$ are facet inducing for $P_{2,v,3,1}$ (see Table 1), but even though the inclusion $P_{m,v} \subseteq P_{2,v,3,1}$ is in general proper, it is easy to show they remain facet inducing for $P_{m,v}$. Thus, we concentrate on $l \geq 4$.

The validity of $s(T)$ comes from the definition of l-sparse PSTSs, i.e. the fact that $r(\binom{T}{3})) = l - 1$ for $\mathcal{I}_{m,v}$. Lemma 3 implies that $\mathcal{I}_{m,v}$ is closed. Thus, by Theorem 1, it is sufficient to show that the critical graph $G_{\binom{T}{3}}(\mathcal{I}_{m,v})$ is connected.

Let $\mathcal{E}$ be an Erdös configuration of order l on the points of T. There must be two triples in $\mathcal{E}$ whose intersection is a single point, call those triples B_1 and B_2. We claim $\mathcal{E} \setminus \{B_1\}$ and $\mathcal{E} \setminus \{B_2\}$ are m-sparse 2-$(v, 3, 1)$ packings. Indeed, $|\mathcal{E} \setminus \{B_i\}| = |\mathcal{E}| - 1 = l - 1$, and since $\mathcal{E}$ was $(l-1)$-sparse, so is $\mathcal{E} \setminus \{B_i\}$, $i = 1, 2$. Thus, there exists an edge in the critical graph $G_{\binom{T}{3}}(\mathcal{I}_{m,v})$ connecting triples B_1 and B_2. By permuting T, we can show this is true for any pair of triples which intersects in one point. That is, there exists an edge in $G_{\binom{T}{3}}(\mathcal{I}_{m,v})$ connecting C_1 and C_2, for any $C_1, C_2 \in \binom{T}{3}$ with $|C_1 \cap C_2| = 1$. It is easy to check that this graph is connected. □

Remark 4. The following is an integer programming formulation for the optimization problem associated with $P_{m,v}$, in which all the inequalities are facet inducing (see Theorem 3). Note that the second type of inequalities can be omitted from the integer programming formulation, since for integral points they are implied by the first type of inequalities (the first type guarantees that x is a PSTS).

$$
\begin{cases}
\text{maximize} & \sum_{B \in \binom{V}{3}} x_B \\
\text{Subject to} & \sum_{B \in \binom{T}{3}} x_B \leq 1, & \text{for all } T \subseteq V, |T| = 4, \\
& \sum_{B \in \binom{T}{3}} x_B \leq 2, & \text{for all } T \subseteq V, |T| = 5, \\
& \sum_{B \in \binom{T}{3}} x_B \leq 3, & \text{for all } T \subseteq V, |T| = 6, \\
& \qquad \vdots & \vdots \\
& \sum_{B \in \binom{T}{3}} x_B \leq m - 1, & \text{for all } T \subseteq V, |T| = m + 2, \\
& x \in \{0, 1\}^{\binom{V}{3}}
\end{cases}
$$

Remark 5. (Separation of m-sparse facets) For constant $m \geq 4$, l-sparse facets, $l \leq m$, can be separated in polynomial time. This is the case, since there are exactly $\sum_{l=4}^{m+2} \binom{v}{l} \in O(v^{m+2})$ inequalities to check, which is a polynomial in the number of variables of the problem, which is $\binom{v}{3}$.

5   Using Facets for Lower and Upper Bounds

In this section, we illustrate some interesting uses of valid inequalities for packing design problems. Recall that $D(m, v)$ denotes the maximum size of an m-sparse $PSTS(v)$. We show an upper bound on $D(m, v)$ based on valid subpacking inequalities for m-sparse PSTSs. We also display the results of an algorithm that uses 4-sparse facets to determine $D(4, v)$.

Proposition 5. *(Upper bound for m-sparse number) Let $m \geq 4$. Then,*

$$
D(m, v) \leq U(m, v) := \left\lfloor \frac{D(m, v - 1) \cdot v}{v - 3} \right\rfloor.
$$

Table 2. The anti-Pasch (4-sparse) PSTS number for small v

	exact*	upper bounds **	
v	$D(4,v)$	$D_1(v,3,2)$	$U(4,v)$
6	3	4	4
7	5	7	5
8	8	8	8
9	12	12	12
10	12	13	17
11	15	17	16
12	19	20	20
13	≥ 24	26	24
14	28	28	30
15	35	35	35
16	37	37	43

* results from branch-and-cut algorithm

** upper bounds from known packing numbers and from Proposition 5

To the best of our knowledge the determination of $D(4,v)$ for $v \in [10,13]$ are new results.

Proof. There are v rank inequalities of the form $\sum_{B \in \binom{T}{3}} x_B \leq D(m, v-1)$, for $T \in \binom{V}{v-1}$. Each triple appears in $v-3$ of these inequalities. Thus, adding these inequalities yields $\sum_{B \in \binom{V}{3}} x_B \leq \frac{D(m,v-1) \cdot v}{v-3}$. Since the left-hand side is integral, we take the floor function on the right-hand side. The inequality is valid in particular for x being the incidence vector of a maximal m-sparse $STS(v)$, in which case the left-hand side is equal to $D(m,v)$.   □

In Table 2, we show values for $D(4,v)$ obtained by our algorithm. To the general algorithm in [17], we added 4-sparse inequalities. Due to their large number, the 4-sparse inequalities are not included in the original integer programming formulation, but are added whenever violated. For $v = 13$, it was not possible to solve the problem to optimality but a solution of size 24 was obtained; since this matches one of the upper bounds, we conclude $D(4,13) = 24$. All other cases were solved to optimality.

6 Conclusions and Further Work

In this article, we derive and study new classes of valid and facet inducing inequalities for the packing designs and m-sparse PSTS polytopes. We also exemplify how this knowledge can be used in algorithms to construct designs as well as for deriving upper bounds on packing numbers. A number of extensions of this work could be pursued. For instance, we are currently investigating how to generalize results from Table 1 in order to determine the facet inducing subpacking inequalities of PSTSs for all $|S|$. We are also working on the design of separation algorithms for m-sparse facets that would be more efficient than the naive one which checks all inequalities (see complexity in Remark 5). Other directions for further research are the study of other rank inequalities and the investigation of new upper bounds on the lines suggested in Section 5. In an expanded version of this article, we intend to include the proofs that were omitted in this extended abstract as well as some of the extensions mentioned above.

References

1. E. Balas and E. Zemel. Critical cutsets of graphs and canonical facets of set-packing polytopes. *Math. Oper. Res.*, 2:15–19, 1977.
2. A.E. Brouwer. Steiner triple systems without forbidden configurations. Technical Report ZW104/77, Mathematisch Centrum Amsterdam, 1977.
3. A. Caprara and M. Fischetti. Branch-and-cut algorithms. In Dell'Amico et al, eds., *Annotated Bibliographies in Combinatorial Optimization*, John Wiley & Sons, 1997.
4. Y.M. Chee, C.J. Colbourn, and A.C.H. Ling. Asymptotically optimal erasure-resilient codes for large disk arrays. *Discrete Appl. Math.*, to appear.
5. V. Chvátal. On certain polytopes associated with graphs. *J. Combin. Theory. Ser. B*, 18:138–154, 1975.
6. C.J. Colbourn, E. Mendelsohn, A. Rosa, and J. Širáň. Anti-mitre Steiner triple systems. *Graphs Combin.*, 10:215–224, 1994.
7. G. Cornuéjols and A. Sassano. On the 0,1 facets of the set covering polytope. *Math. Programming*, 43:45–55, 1989.
8. M.J. Grannell and T.S. Griggs. Configurations in Steiner triple systems. *Combinatorial Designs and their applications*, 103–126, Chapman & Hall/CRC Res. Notes Math. 403, Chapman & Hall/CRC, 1999.
9. M.J. Grannell, T.S. Griggs, and E. Mendelsohn. A small basis for four-line configurations in Steiner triple systems. *J. Combin. Des.*, 3(1):51–59, 1995.
10. L. Hellerstein, G.A. Gibson, R.M. Karp, R.H. Katz, D.A. Paterson. Coding techniques for handling failures in large disk arrays. *Algorithmica*, 12:182–208, 1994.
11. M. Laurent. A generalization of antiwebs to independence systems and their canonical facets. *Math. Programming*, 45:97–108, 1989.
12. H. Lefmann, K.T. Phelps, and V. Rödl. Extremal problems for triple systems. *J. Combin. Des.*, 1:379–394, 1993.
13. A.C.H. Ling. A direct product construction for 5-sparse triple systems. *J. Combin. Des.*, 5:444–447, 1997.
14. L. Moura. Polyhedral methods in design theory. In Wallis, ed., *Computational and Constructive Design Theory*, Math. Appl. 368:227–254, 1996.
15. L. Moura. Polyhedral Aspects of Combinatorial Designs. PhD thesis, University of Toronto, 1999.
16. L. Moura. Maximal s-wise t-intersecting families of sets: kernels, generating sets, and enumeration. *J. Combin. Theory. Ser. A*, 87:52–73, 1999.
17. L. Moura. A polyhedral algorithm for packings and designs. *Algorithms–ESA'99. Proceedings of the 7th Annual European Symposium held in Prague, 1999*, Lecture Notes in Computer Science 1643, Springer-Verlag, Berlim, 1999.
18. W.H. Mills and R.C. Mullin. Coverings and packings. In Dinitz and Stinson, eds., *Contemporary Design Theory: a collection of surveys*, 371–399. Wiley, 1992.
19. D.R. Stinson. Packings. In Colbourn and Dinitz, eds., *The CRC handbook of combinatorial designs*, 409–413, CRC Press, 1996.
20. D. Wengrzik. Schnittebenenverfahren für Blockdesign-Probleme. Master's thesis, Universität Berlin, 1995.
21. E. Zehendner. Methoden der Polyedertheorie zur Herleitung von oberen Schranken für die Mächtigkeit von Blockcodes. Doctoral thesis, Universität Augsburg, 1986.

The Anti-Oberwolfach Solution: Pancyclic 2-Factorizations of Complete Graphs

Brett Stevens

Department of Mathematics and Statistics
Simon Fraser University
Burnaby BC V5A 1S6
brett@math.sfu.ca

Abstract. We pose and completely solve the existence of pancyclic 2-factorizations of complete graphs and complete bipartite graphs. Such 2-factorizations exist for all such graphs, except a few small cases which we have proved are impossible. The solution method is simple but powerful. The pancyclic problem is intended to showcase the power this method offers to solve a wide range of 2-factorization problems. Indeed, these methods go a long way towards being able to produce arbitrary 2-factorizations with one or two cycles per factor.

1 Introduction

Suppose that there is a conference being held at Punta del Este, Uruguay. There are $2n+1$ people attending the conference and it is to be held over n days. Each evening there is a dinner which everyone attends. To accommodate the many different sizes of conferences, the Las Dunas Hotel has many different sizes of tables. In fact, they have every table size from a small triangular table to large round tables seating $2n+1$ people. When this was noticed, the organizers, being knowledgeable in combinatorics, asked themselves if a seating arrangement could be made for each evening such that every person sat next to every other person exactly once over the course of the conference and each size table was used at least once.

Such a schedule, really a decomposition of K_{2n+1} into spanning graphs all with degree 2 (collections of cycles), would be an example of a 2-factorization of K_{2n+1}. Due to their usefulness in solving scheduling problems, 2-factorizations have been well studied. The Oberwolfach problem asks for a 2-factorization in which each subgraph in the decomposition has the same pattern of cycles and much work has been done toward its solution [2,7]. This corresponds to the hotel using the exact same set of tables each night. Often other graphs besides odd complete graphs are investigated. Complete graphs of even order with a perfect matching removed so the graph has even degree have received much attention [1]. In such solutions each person would miss sitting next to exactly one other during the conference. Oberwolfach questions have also been posed and solved for complete bipartite graphs [7]. The problem posed in the introductory

G. Gonnet, D. Panario, and A. Viola (Eds.): LATIN 2000, LNCS 1776, pp. 115–122, 2000.

paragraph asks that every size cycle appear and so is called the pancyclic 2-factorization problem, or, since it forces such different cycle sizes, the title of 'anti-Oberwolfach problem' emphasizes this contrast. There are analogous formulations for an even number of people with a complete matching removed (co-author avoiding to prevent conflict) and for bipartite graphs as well (the seating arrangements alternating computer scientist and mathematician to foster cross disciplinary communication).

The Conference Organizers soon noted that tables of size $n - 1$ and $n - 2$, although available, were forbidden since the remaining people would be forced to sit at tables of size 1 or 2 which did not exist and would preclude every pair being neighbors exactly once. After realizing this and doing a preliminary count, the organizers then asked themselves for a schedule that would include the first evening with everyone seated around one large table of size $2n + 1$, an evening with a size three table paired with a size $2n - 2$ table, an evening with a size four table paired with a size $2n - 3$ table and so forth up to an evening with size n table paired with a size $n + 1$ table. There was one evening remaining and the organizers thought it would be nice to have everyone seated again at one table for the final dinner together.

If the solution methods from the Oberwolfach problem can be paired with methods for the anti-Oberwolfach problem, then it is conceivable that that general 2-factorization problems can be tackled with great power. This would enable us to answer many different and new scheduling and tournament problems. Indeed, the pancyclic question is recreational in nature but we use it as a convenient context in which to present powerful and very serious construction methods that can contribute to a broader class of 2-factorizations.

Another primary motivation for this problem is recent papers investigating the possible numbers of cycles in cycle decompositions of complete graphs [3] and in 2-factorizations[4,5]. For each n, the number of cycles that appear in an anti-Oberwolfach solution are admissible so the question was asked if this specific structure was possible. We show that the answer to all versions of the problem, complete odd graphs, complete even graphs minus a complete matching, and complete bipartite graphs, is affirmative, except for small cases which we have proved impossible. The solution method is very similar to Piotrowski's approach to 2-factorization problems: we modify pairs of Hamiltonian cycles into pairs of 2-factors with the desired cycle structures.

In this paper we offer first some definitions and discussion of 2-factorizations, formalizing the notions discussed above. Then we solve the standard and bipartite formulations of the anti-Oberwolfach problem. We end with a discussion of the solution method, possible extensions of the problem, and the power these methods provide for constructing very general classes of 2-factorizations.

2   Definitions and Discussion

Definition 1 *A k-factorization of a graph G, is a decomposition of G into spanning subgraphs all regular of degree k. Each such subgraph is called a k-factor.*

We are interested in a special class of 2-factorizations, but also use 1-factors (perfect matchings) on occasion.

Definition 2 *A* pancyclic *2-factorization of a graph, G, of order n, is a 2-factorization of G where a cycle of each admissible size, $3, 4, \ldots, n-4, n-3, n$, appears at least once in some 2-factor.*

There is a similar definition for the bipartite graphs in which no odd cycles can appear:

Definition 3 *A* pancyclic *2-factorization of a bipartite graph, G, of even order n, is a 2-factorization of G where a cycle of each admissible size, $4, 6, \ldots, n-4, n$, appears at least once in some 2-factor.*

We ask whether such 2-factorizations exist for complete odd graphs K_{2n+1}, complete even graphs, with a 1-factor removed to make the degree even, $K_{2n} - nK_2$, and complete bipartite graphs, some with a 1-factor removed, $K_{2n,2n}$ and $K_{2n+1,2n+1} - (2n+1)K_2$.

In every case, counting shows that the all the 2-factors that are not Hamiltonian (an n-cycle) must be of the form: an i-cycle and a $n-i$ cycle. We define here a notation to refer to the different structure of 2-factors:

Definition 4 *An $i, (n-i)$-factor is a 2-factor of an order n graph, G, that is the disjoint union of an i-cycle and a $(n-i)$-cycle.*

In each case the solution is similar. For each graph in question, G, we present a 2-factorization, $\{F_0, F_1, \ldots, F_k\}$, and a cyclic permutation σ of a subset of the vertices of G so that $F_i = \sigma^i(F_0)$ and σ^{k+1} is the identity. We decompose the union of consecutive pairs of 2-factors, $F_i \cup F_{i+1}$, into two other 2-factors with desired cycle structures by swapping pairs of edges. The cyclic automorphism group guarantees that any two unions of any two consecutive 2-factors are isomorphic. Thus we can formulate general statements about decomposition of the complete graphs into these unions and the possible decompositions of these unions. A few cases require more sophisticated manipulation. In certain cases we swap only one pair of edges; in others, we use up to four such swaps. These methods demonstrate the power of Piotrowski's approach of decomposing pairs of Hamiltonian cycles from a Walecki decomposition into the desired 2-factors

3   Main Results

We demonstrate the solution method in more detail for the case K_{2n+1}. In the other cases the solution method is essentially the same with minor adjustments and a slight loss of economy.

3.1   The Solution for K_{2n+1}

Walecki's decomposition of K_{2n+1} into Hamiltonian 2-factors give us the starting point of our construction.

Lemma 1. *There exists a decomposition of K_{2n+1} into Hamiltonian 2-factors that are cyclic developments of each other.*

The first of the Hamiltonian 2-factors is shown in Figure 1. Each of the remaining $n - 1$ 2-factors is a cyclic rotation of the first.

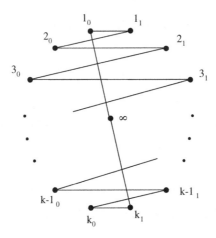

Fig. 1. A Walecki 2-factor of K_{2n+1}.

The union of two consecutive Hamilton cycles from the Walecki decomposition is isomorphic to the graph given in Figure 2. This graph can be decomposed

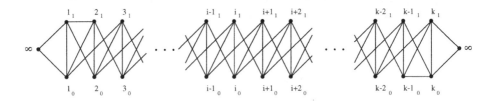

Fig. 2. The union of two consecutive Walecki 2-factors.

into two other Hamiltonian 2-factors that are not identical to the original Walecki 2-factors. These are shown in Figure 3. It is these two Hamiltonian 2-factors that can be decomposed into 2-factors with various cycle structures.

Lemma 2. *The graph in Figure 2 can be decomposed into two 2-factors such that the first is an $2i+1, 2(n-i)$-factor and the second is a $2j+1, 2(n-j)$-factor for any $1 \le i \ne j \le n - 2$. Alternatively the second can remain a Hamiltonian 2-factor, with no restriction on i.*

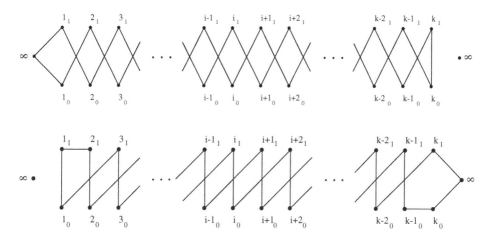

Fig. 3. Decomposition of the union of two consecutive Walecki 2-factors into two other Hamiltonian 2-factors.

Proof. The first decomposition is achieved by swapping four edges between the two graphs of Figure 3 and is shown in Figure 4.

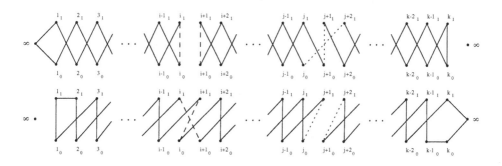

Fig. 4. Decomposition into a $2i + 1, 2(n - i)$-factor and a $2j + 1, 2(n - j)$-factor.

The second decomposition is achieved by swapping only two edges between the two graphs of Figure 3 and is shown in Figure 5.

In both figures the sets of swapped edges are shown as dashed or dotted lines.

The flexibility of the parameters i and j together with the decomposition of K_{2n+1} into n cyclically derived Hamiltonian 2-factors gives the main result.

Theorem 1 *There exists a pancyclic 2-factorization of K_{2n+1} for all $n \geq 1$.*

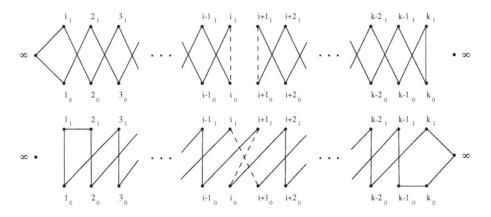

Fig. 5. Decomposition into a $2i + 1, 2(n - i)$-factor and a Hamiltonian 2-factor.

3.2   The Remaining Cases: $K_{2n} - nK_2$, $K_{2n,2n}$ and $K_{2n+1,2n+1} - (2n + 1)K_2$

Decomposing graphs with odd degree into 2-factors is impossible since each 2-factor accounts for two edges from each vertex. In these cases it is customary to remove a 1-factor and attempt to decompose the resulting graph which has even degree. The existence of pancyclic 2-factorizations of $K_{2n} - nK_2$, $K_{2n,2n}$ and $K_{2n+1,2n+1} - (2n + 1)K_2$ is achieved in the same manner as that of K_{2n+1}. We decompose the graph into 2-factors (usually Hamiltonian) that are cyclic developments of each other, so that all unions of pairs of consecutive 2-factors are isomorphic. The union of two consecutive 2-factors has a structure very similar to that in the case K_{2n+1} and they can be broken into smaller cycles in almost exactly the same manner. There are two minor, though notable, differences. When decomposing $K_{2n} - nK_2$, it is necessary in one fourth of the cases to reinsert the removed 1-factor and remove another to be able to construct odd numbers of 2-factors with different parities. In the bipartite case it is sometimes necessary to apply the edge swapping additional times since the original decomposition into 2-factors may not have produced Hamiltonian 2-factors. Again, in each case the complete solution is achievable.

Theorem 2   *There exists a pancyclic 2-factorization of $K_{2n} - nK_2$ for all $n \geq 1$.*

Theorem 3   *There exists a pancyclic 2-factorization of $K_{n,n}$ for all even $n > 4$ and $K_{n,n} - nK_2$ for all odd $n > 1$. The cases $n = 1, 2, 4$ are all impossible.*

In all cases the union of the edges from each set of four swapped edges form an induced 4-cycle, and the remainder of the graphs are paths connecting pairs of vertices from the 4-cycle. This induced 4-cycle and connected paths are the underlying structure of the construction. Consideration of this structure allows the swapping to be formalized and made rigorous, so that the proofs can rest on a foundation other than nice figures. Unfortunately the statements of these swapping lemmas are lengthy and technical and space does not permit their inclusion.

4   Conclusion

As a demonstration of a powerful method for a wide range of 2-factorization problems, of similar type to Piotrowski's Oberwolfach constructions, we have solved the pancyclic 2-factorization for four infinite families of complete or nearly complete graphs, K_{2n+1}, $K_{2n}-nK_2$, $K_{2n,2n}$ and $K_{2n+1,2n+1}-(2n+1)K_2$. In each case, pancyclic 2-factorizations exist for all n except for a very few small n where the solution is shown not to exist. Moreover, in each case the solution method is similar. We start with a 2-factorization of the graph in question with a cyclic automorphism group. The union of consecutive pairs of the 2-factors is shown to be decomposable into two 2-factors with a wide range of cycle structures by judicious swapping of the two pairs of opposite edges of induced 4-cycles. This flexibility of decomposition and the automorphism group allow the desired solution to be constructed.

The plethora of induced 4-cycles in the union of consecutive 2-factors from the various 2-factorizations allow us not only to construct the various solutions in many different ways, but to go far beyond the problem solved here. In K_{2n+1} it seems that the swapping lemmas can only produce one odd cycle per factor and at most two in $K_{2n} - nK_2$. Beyond this restriction there is a great deal of flexibility in the application of the swapping lemmas. The use of these methods to solve the pancyclic 2-factorization problem indicates the strength and range of the swapping lemmas. We propose that the methods outlined in this article might be powerful for constructing Oberwolfach solutions, and other 2-factorization and scheduling problems. One very interesting problem is the construction of 2-factorizations with prescribed lists of cycle types for each 2-factor. If the list can only contain 2-factors with one or two cycles, then the methods presented here nearly complete the problem. The only obstacle towards the solution of these problems is the construction of pairs of 2-factors with the same cycle type, the Oberwolfach aspect of the question. P. Gvozdjak is currently working on such constructions.

There are other pancyclic decomposition questions that can be asked. The Author and H. Verrall are currently working on the directed analogue of the anti-Oberwolfach problem. Other obvious pancyclic problems can be formulated for higher λ for both directed and undirected graphs; 2-path covering pancyclic decompositions, both resolvable and non-resolvable . In each of these cases we gain the flexibility to ask for different numbers within each size class of cycles, possibly admitting digons and losing other restrictions enforced by the tightness of the case solved here.

5   Acknowledgments

I would like to thank Profs. E. Mendelsohn and A. Rosa for suggesting this problem and their support and encouragement. I was supported by NSF Graduate Fellowship GER9452870, the Department of Mathematics at the University of Toronto, a PIMS postdoctoral fellowship and the Department of Mathematics and Statistics at Simon Fraser University.

References

1. B. Alspach and S. Marshall. Even cycle decompositions of complete graphs minus a 1-factor. *Journal of Combinatorial Designs*, 2(6):441–458, 1994.
2. B. Alspach, P. Schellenberg, D. Stinson, and D. Wagner. The Oberwolfach problem and factors of uniform odd length cycles. *J. Combin. Theory. Ser. A*, 52:20–43, 1989.
3. E. J. Billington and D. E. Bryant. The possible number of cycles in cycle systems. *Ars Combin.*, 52:65–70, 1999.
4. I. J. Dejter, F. Franek, E. Mendelsohn, and A. Rosa. Triangles in 2-factorizations. *J. Graph Theory*, 26(2):83–94, 1997.
5. M. J. Grannell and A. Rosa. Cycles in 2-factorizations. *J. Combin. Math. Combin. Comput.*, 29:41–64, 1999.
6. E. Lucas. *Récréations Mathématiques*, volume 2. Gauthier-Villars, Paris, 1883.
7. W. Piotrowski. The solution of the bipartite analogue of the Oberwolfach problem. *Discrete Math.*, 97:339–356, 1991.

Graph Structure of the Web: A Survey

Prabhakar Raghavan[1]

IBM Almaden Research Center K53/B1, 650 Harry Road, San Jose CA 95120.
pragh@almaden.ibm.com

1 Summary

The subject of this survey is the directed graph induced by the hyperlinks between Web pages; we refer to this as the *Web graph*. Nodes represent static html pages and hyperlinks represent directed edges between them. Recent estimates [5] suggest that there are several hundred million nodes in the Web graph; this quantity is growing by several percent each month. The average node has roughly seven hyperlinks (directed edges) to other pages, making for a total of several billion hyperlinks in all.

There are several reasons for studying the Web graph. The structure of this graph has already led to improved Web search [7,10,11,12,25,34], more accurate topic-classification algorithms [13] and has inspired algorithms for enumerating emergent cyber-communities [28]. Beyond the intrinsic interest of the structure of the Web graph, measurements of the graph and of the behavior of users as they traverse the graph, are of growing commercial interest. These in turn raise a number of intriguing problems in graph theory and the segmentation of Markov chains. For instance, Charikar *et al.* [14] suggest that analysis of surfing patterns in the Web graph could be exploited for targeted advertising and recommendations. Fagin *et al.* [18] consider the limiting distributions of Markov chains (modeling users browsing the Web) that occasionally undo their last step.

In this lecture we will cover the following themes from our recent work:

- How the structure of the Web graph has been exploited to improve the quality of Web search.
- How the Web harbors an unusually large number of certain clique-like subgraphs, and the efficient enumeration of these subgraphs for the purpose of discovering communities of interest groups in the Web.
- A number of measurements of degree sequences, connectivity, component sizes and diameter on the Web. The salient observations include:
  1. In-degrees on the Web follow an inverse power-law distribution.
  2. About one quarter the nodes of the Web graph lie in a giant strongly connected component; the remaining nodes lie in components that give some insights into the evolution of the Web graph.
  3. The Web is not well-modeled by traditional random graph models such as $G_{n,p}$.

G. Gonnet, D. Panario, and A. Viola (Eds.): LATIN 2000, LNCS 1776, pp. 123–125, 2000.

– A new class of random graph models for evolving graphs. In particular, some characteristics observed in the Web graph are modeled by random graphs in which the destinations of some edges are created by probabilistically *copying* from other edges at random. This raises the prospect of the study of a new class of random graphs, one that also arises in other contexts such as the graph of telephone calls [3].

Pointers to these algorithms and observations, as well as related work, may be found in the bibliography below.

Acknowledgements

The work covered in this lecture is the result of several joint pieces of work with a number of co-authors. I thank the following colleagues for collaborating on these pieces of work: Andrei Broder, Soumen Chakrabarti, Byron Dom, David Gibson, Jon Kleinberg, S. Ravi Kumar, Farzin Maghoul, Sridhar Rajagaopalan, Raymie Stata, Andrew Tomkins and Janet Wiener.

References

1. S. Abiteboul, D. Quass, J. McHugh, J. Widom, and J. Wiener. The Lorel Query language for semistructured data. *Intl. J. on Digital Libraries,* 1(1):68-88, 1997.
2. R. Agrawal and R. Srikanth. Fast algorithms for mining association rules. *Proc. VLDB,* 1994.
3. W. Aiello, F. Chung and L. Lu. A random graph model for massive graphs. To appear in the *Proceedings of the ACM Symposium on Theory of Computing,* 2000.
4. G. O. Arocena, A. O. Mendelzon, G. A. Mihaila. Applications of a Web query language. *Proc. 6th WWW Conf.,* 1997.
5. K. Bharat and A. Broder. A technique for measuring the relative size and overlap of public Web search engines. *Proc. 7th WWW Conf.,* 1998.
6. K. Bharat and M. R. Henzinger. Improved algorithms for topic distillation in a hyperlinked environment. *Proc. ACM SIGIR,* 1998.
7. S. Brin and L. Page. The anatomy of a large-scale hypertextual Web search engine. *Proc. 7th WWW Conf.,* 1998. See also `http://www.google.com`.
8. A.Z. Broder, S.R. Kumar, F. Maghoul, P. Raghavan, S. Rajagopalan, R. Stata, A. Tomkins and J. Wiener. Graph structure in the web: experiments and models. *Submitted for publication.*
9. B. Bollobás. *Random Graphs,* Academic Press, 1985.
10. J. Carrière and R. Kazman. WebQuery: Searching and visualizing the Web through connectivity. *Proc. 6th WWW Conf.,* 1997.
11. S. Chakrabarti, B. Dom, D. Gibson, J. Kleinberg, P. Raghavan and S. Rajagopalan. Automatic resource compilation by analyzing hyperlink structure and associated text. *Proc. 7th WWW Conf.,* 1998.
12. S. Chakrabarti, B. Dom, S. R. Kumar, P. Raghavan, S. Rajagopalan, and A. Tomkins. Experiments in topic distillation. *SIGIR workshop on hypertext IR,* 1998.
13. S. Chakrabarti and B. Dom and P. Indyk. Enhanced hypertext classification using hyperlinks. *Proc. ACM SIGMOD,* 1998.

14. M. Charikar, S. R. Kumar, P. Raghavan, S. Rajagopalan and A. Tomkins. On targeting Markov segments. Proc. *ACM Symposium on Theory of Computing*, 1999.
15. H. T. Davis. *The Analysis of Economic Time Series*. Principia press, 1941.
16. R. Downey, M. Fellows. Parametrized Computational Feasibility. In *Feasible Mathematics II*, P. Clote and J. Remmel, eds., Birkhauser, 1994.
17. L. Egghe, R. Rousseau, *Introduction to Informetrics*, Elsevier, 1990.
18. R. Fagin, A. Karlin, J. Kleinberg, P. Raghavan, S. Rajagopalan, R. Rubinfeld, M. Sudan, A. Tomkins. Random walks with "back buttons". To appear in the *Proceedings of the ACM Symposium on Theory of Computing*, 2000.
19. D. Florescu, A. Levy and A. Mendelzon. Database techniques for the World Wide Web: A survey. *SIGMOD Record*, 27(3): 59-74, 1998.
20. E. Garfield. Citation analysis as a tool in journal evaluation. *Science*, 178:471–479, 1972.
21. N. Gilbert. A simulation of the structure of academic science. *Sociological Research Online*, 2(2), 1997.
22. G. Golub, C. F. Van Loan. *Matrix Computations*, Johns Hopkins University Press, 1989.
23. M. R. Henzinger, P. Raghavan, and S. Rajagopalan. Computing on data streams. *AMS-DIMACS series,* special issue on computing on very large datasets, 1998.
24. M. M. Kessler. Bibliographic coupling between scientific papers. *American Documentation*, 14:10–25, 1963.
25. J. Kleinberg. Authoritative sources in a hyperlinked environment, *J. of the ACM*, 1999, to appear. Also appears as IBM Research Report RJ 10076(91892) May 1997.
26. J. Kleinberg, S. Ravi Kumar, P. Raghavan, S. Rajagopalan and A. Tomkins. The Web as a graph: measurements, models and methods. Invited paper in Proceedings of the *International Conference on Combinatorics and Computing, COCOON*, 1999. Springer-Verlag Lecture Notes in Computer Science.
27. D. Konopnicki and O. Shmueli. Information gathering on the World Wide Web: the W3QL query language and the W3QS system. *Trans. on Database Systems,* 1998.
28. S. R. Kumar, P. Raghavan, S. Rajagopalan and A. Tomkins. Trawling emerging cyber-communities automatically. *Proc. 8th WWW Conf.*, 1999.
29. L. V. S. Lakshmanan, F. Sadri, and I. N. Subramanian. A declarative approach to querying and restructuring the World Wide Web. *Post-ICDE Workshop on RIDE*, 1996.
30. R. Larson. Bibliometrics of the World Wide Web: An exploratory analysis of the intellectual structure of cyberspace. *Ann. Meeting of the American Soc. Info. Sci.*, 1996.
31. A. J. Lotka. The frequency distribution of scientific productivity. *J. of the Washington Acad. of Sci.*, 16:317, 1926.
32. A. Mendelzon, G. Mihaila, and T. Milo. Querying the World Wide Web, *J. of Digital Libraries* 1(1):68–88, 1997.
33. A. Mendelzon and P. Wood. Finding regular simple paths in graph databases. *SIAM J. Comp.*, 24(6):1235-1258, 1995.
34. E. Spertus. ParaSite: Mining structural information on the Web. *Proc. 6th WWW Conf.*, 1997.
35. G. K. Zipf. Human behavior and the principle of least effort. *New York: Hafner*, 1949.

Polynomial Time Recognition of Clique-Width ≤ 3 Graphs (Extended Abstract)

Derek G. Corneil[1], Michel Habib[2], Jean-Marc Lanlignel[2], Bruce Reed[3], and Udi Rotics[1]

[1] Department of Computer Science, University of Toronto, Toronto, Canada *
[2] LIRMM, CNRS and University Montpellier II, Montpellier, France *
[3] Equipe de Combinatoire, CNRS, University Paris VI, Paris, France *

Abstract. The Clique-width of a graph is an invariant which measures the complexity of the graph structures. A graph of bounded tree-width is also of bounded Clique-width (but not the converse). For graphs G of bounded Clique-width, given the bounded width decomposition of G, every optimization, enumeration or evaluation problem that can be defined by a Monadic Second Order Logic formula using quantifiers on vertices but not on edges, can be solved in polynomial time.

This is reminiscent of the situation for graphs of bounded tree-width, where the same statement holds even if quantifiers are also allowed on edges. Thus, graphs of bounded Clique-width are a larger class than graphs of bounded tree-width, on which we can resolve fewer, but still many, optimization problems efficiently.

In this paper we present the first polynomial time algorithm ($O(n^2 m)$) to recognize graphs of Clique-width at most 3.

1 Introduction

The notion of the Clique-width of graphs was first introduced by Courcelle, Engelfriet and Rozenberg in [CER93]. The clique-width of a graph G, denoted by $cwd(G)$, is defined as the minimum number of labels needed to construct G, using the four graph operations: creation of a new vertex v with label i (denoted $i(v)$), disjoint union ($\oplus$), connecting vertices with specified labels (η) and renaming labels (ρ). Note that $\oplus$ is the disjoint union of two labeled graphs, each vertex of the new graph retains the label it had previously. $\eta_{i,j}$ ($i \neq j$), called the "join" operation, causes all edges (that are not already present) to be created between every vertex of label i and every vertex of label j. $\rho_{i \to j}$ causes all vertices of label i to assume label j. As an example of these notions see the graph in Fig. 4 together with its 3-expression in Fig. 4 and the parse tree associated with the expression in Fig. 4. A detailed study of clique-width is presented in [CO99].

---

* email: dgc,rotics @cs.toronto.edu
* email: habib,lanligne @lirmm.fr
* email: reed@moka.ccr.jussieu.fr

G. Gonnet, D. Panario, and A. Viola (Eds.): LATIN 2000, LNCS 1776, pp. 126–134, 2000.

Also a study of the clique-width of graphs with few P_4s (i.e., a path on four vertices) and on perfect graph classes is presented in [MR99,GR99]. For example, distance hereditary graphs and P_4-sparse graphs have bounded clique-width (3 and 4 respectively) whereas unit interval graphs, split graphs and permutation graphs all have unbounded clique-width.

The motivation for studying clique-width is analogous to that of tree-width. In particular, given a parse tree which shows how to construct a graph G using k labels and the operations above, every decision, optimization, enumeration or evaluation problem on G which can be defined by a Monadic Second Order Logic formula ψ, using quantifiers on vertices but not on edges, can be solved in time $c_k \cdot O(n + m)$ where c_k is a constant which depends only on ψ and k, where n and m denote the number of vertices and edges of the input graph, respectively. For details, see [CMRa,CMRb].

Furthermore clique-width is "more powerful" than tree-width in the sense that if a class of graphs is of bounded tree-width then it is also of bounded clique-width [CO99]. (In particular for every graph G, $cwd(G) \leq 2^{twd(G)+1} + 1$, where $twd(G)$ denotes the tree-width of G).

One of the central open questions concerning clique-width is determining the complexity of recognizing graphs of clique-width at most k, for fixed k. It is easy to see that graphs of clique-width 1 are graphs with no edges. The graphs of clique-width at most 2 are precisely the cographs (i.e., graphs without P_4) [CO99]. In this paper we present a polynomial time algorithm ($O(n^2 m)$) to determine if a graph has clique-width at most 3. For graphs of Clique-width ≤3 the algorithm also constructs the 3-expression which defines the graph.

An implementation that achieves the $O(n^2 m)$ bound would be quite complicated, because a linear modular decomposition algorithm is needed. However the other steps of the algorithm present no major difficulty: we use only standard data structures, and the Ma-Spinrad split decomposition algorithm. So if we fall back on an easy modular decomposition algorithm (see Sec. 6), there is a slightly slower ($O(n^2 m \log n)$), easily implementable version of the algorithm.

Unfortunately, there does not seem to be a succinct forbidden subgraph characterization of graphs with clique-width at most 3, similar to the P_4-free characterization of graphs with clique-width at most 2. In fact every cycle C_n with $n \geq 7$ has clique-width 4, thereby showing an infinite set of minimal forbidden induced subgraphs for Clique-width ≤3 [MR99].

2   Background

We first introduce some notation and terminology. The graphs we consider in this paper are undirected and loop-free. For a graph G we denote by $V(G)$ (resp. $E(G)$) the set of vertices (resp. edges) of G. For $X \subseteq V(G)$, we denote by $G[X]$ the subgraph of G induced by X. We denote by $G \setminus X$ the subgraph of G induced by $V(G) \setminus X$. We say that vertex v is *universal to* X if v is adjacent to all vertices in $X \setminus \{v\}$ and that v is *universal in* G if v is universal to $V(G)$. On the other hand v *misses* X if v misses (i.e., is not adjacent to) all vertices in $X \setminus \{v\}$. We denote by $N(v)$ the neighborhood of v in G, i.e., the set of vertices in G adjacent to v. We denote by $N[v]$ the closed neighborhood of v, i.e., $N[v] = N(v) \cup \{v\}$.

A labeled graph is a graph with integer labels associated with its vertices, such that each vertex has exactly one label. We denote by $\langle G : V_1, \ldots, V_p \rangle$ the labeled graph G with labels in $\{1, \ldots, p\}$ where V_i denotes the set of vertices of G having label i (some of these sets may be empty).

The definition of the Clique-width of graphs (see the introduction) extends naturally to labeled graphs. The *Clique-width* of a labeled graph $\langle G : V_1, \ldots, V_p \rangle$ denoted by $cwd(G : V_1, \ldots, V_p)$ is the minimum number of labels needed to construct G such that all vertices of V_i have label i (at the end of the construction process), using the four operations $i(v)$, η, ρ and $\oplus$ (see the introduction for the definition of these operations). Note that, for instance, the cycle with 4 vertices (the C_4) is of Clique-width ≤ 3, but the C_4 labeled $1-1-2-2$ consecutively around the circle is not.

We say that a graph is *2-labeled* if exactly two of the label sets are non-empty, and *3-labeled* if all three of them are non-empty.

Without loss of generality, we may assume that our given graphs are *prime*, in the sense that they have no modules. (A *module* of G is an induced subgraph H, $1 < |H| < |G|$, such that each vertex in $G \setminus H$ is either universal to H or misses H). This assumption follows from the easily verifiable observation (see [CMRa]) that for every graph G which is not a cograph (i.e., is of clique-width > 2), and has a module H, $cwd(G) = max\{cwd(H), cwd(G \setminus (H - x))\}$, where x is any vertex of H.

Given a connected 3-labeled graph, the last operation in a parse tree which constructs it must have been a join. In principle this yields three possibilities. However, if two different join operations are possible, the graph has a module: for example, if both $\eta_{1,2}$ and $\eta_{1,3}$ are possible, the vertices of label 2 and 3 form a module. So since we are only considering prime graphs we can determine the last operation of the parse tree.

Unfortunately we cannot continue this process on the subproblems as deleting the join edges may disconnect the graph and leave us with 2-labeled subproblems. In fact it turns out that solving Clique-width restricted to 3-labeled graphs is no easier than solving it for 2-labeled graphs.

In contrast, if we attempt to find in this top-down way the parse tree for a 2-labeled prime graph, then we never produce a subproblem with only 1 non-empty label set, because its vertices would form a module (as the reader may verify). This fact motivates the following definition: for partition $A \cup B$ of $V(H)$, let $2-LAB(H : A, B)$ denote the problem of determining whether $cwd(H : A, B) \leq 3$ (and finding a corresponding decomposition tree). Since A and B form a disjoint partition of $V(H)$ we will also denote it as $2-LAB(H : A, -)$. If we can find a polynomial time algorithm to solve this problem, then our problem reduces to finding a small set S of possible 2-labelings such that at least one of them is of Clique-width ≤ 3 iff G is of Clique-width ≤ 3. We first discuss how we solve $2-LAB$, then discuss how to use it to solve the general Clique-width ≤ 3 problem.

3   Labeled Graphs

The 2–LAB problem is easier to solve than the general Clique-width ≤3 problem, because there are fewer possibilities. The last operation must have been a relabeling, and before that edges were created. With our top-down approach, we have to introduce a third label set, that is to split one of the two sets A or B in such a way that all edges are present between two of the three new sets (Fig. 1 shows the two possibilities when the set of vertices labeled with 1 is split); and there are four ways to do this in general. Each of these four ways corresponds to one of the ways of introducing the third label set, namely: consider the vertices of A that are universal to B; consider the vertices of B that are universal to A; consider the co-connected components of both A and B (these are the connected components of the complement of the set).

Fig. 1. 2–LAB procedure main idea

If there is only one possible way of relabeling the vertices and undoing a join, then we have a unique way of splitting our problem into smaller problems; we do so and continue, restricting our attention to these simpler subproblems.

The difficulty arises when there is more than one possible join that could be undone. As mentioned in the previous section, if all three label sets are nonempty, then this possibility will not arise because it would imply the existence of a module. In the general case this situation may arise, but again it implies very strong structural properties of the graph, which allow us to restrict our attention to just one of the final possible joins. The proof that we need consider just one final join even when there is more than one possibility will be described in the journal version of the paper (or see [WWW]).

We then remove the edges (adding a join node to the decomposition tree), which disconnects the graph, and we can apply again the above algorithm, until either we have a full decomposition tree, or we know that the input graph with the initial labeling of A and B is not of Clique-width ≤3.

4   Algorithm Outline

We now know how to determine if the Clique-width of $\langle G : A, B \rangle$ is ≤ 3 for any partition $A \cup B$ of $V(G)$. Our problem thus reduces to finding a small set S of possible 2-labelings such that at least one of them is of Clique-width ≤3 iff G is of Clique-width ≤3.

We are interested in the last join operation in a parse tree corresponding to a 3-expression which defines G (if such an expression exists); without loss of generality we can assume that this is a $\eta_{1,2}$. The first case is when there is only one vertex x of label 1. In this case the parse tree is a solution of 2–$LAB(G : \{x\}, -)$. More generally, if the graph obtained by deleting the edges from the last join has a 3-labeled connected component, then it turns out that there is a simple procedure for finding the corresponding 2–LAB problems.

Do all graphs with clique-width at most 3 have a parse tree which ends in this way? Unfortunately the answer is no. The graph in Fig. 4 has clique-width 3 but it is easy to show that there is no parse tree formed in the manner described above.

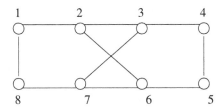

Fig. 2. An example graph.

$$t = \eta_{1,2}(l \oplus r)$$

$$l = \rho_{1 \bullet \; 3} \circ \eta_{1,3}\Big(\eta_{1,2}\big(1_{(8)} \oplus 2_{(7)}\big) \oplus \eta_{2,3}\big(3_{(1)} \oplus 2_{(2)}\big)\Big)$$

$$r = \rho_{2 \bullet \; 3} \circ \eta_{2,3}\Big(\eta_{1,2}\big(1_{(6)} \oplus 2_{(5)}\big) \oplus \eta_{1,3}\big(1_{(3)} \oplus 3_{(4)}\big)\Big)$$

Fig. 3. A 3-expression t for the graph of Fig. 4.

Thus, we need to consider other final joins, when the graph obtained by deleting the edges from the last join has no 3-labeled connected component. This leads to the notion of *cuts* (often called *joins* in the literature) first studied by Cunningham [Cun82]. A *cut* is a disjoint partition of $V(G)$ into $(X : Y)$ where $|X|, |Y| > 1$ together with the identification of subsets $\tilde{X} \subseteq X, \tilde{Y} \subseteq Y$, called the *boundary sets*, where $E(G) \cap (X \times Y) = \tilde{X} \times \tilde{Y}$. Note that since we assume our graphs are module free, $\tilde{X} \subset X$ and $\tilde{Y} \subset Y$. For the graph in Fig. 4 $X = \{1, 2, 7, 8\}, \tilde{X} = \{2, 7\}, Y = \{3, 4, 5, 6\}$ and $\tilde{Y} = \{3, 6\}$.

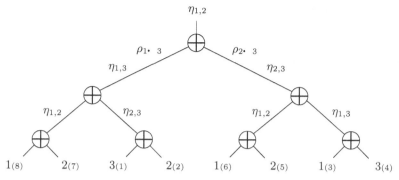

Fig. 4. The parse tree corresponding to the 3-expression of Fig. 4.

Note how the partition of $V(G)$ by a cut $(X : Y)$ is reflected in a parse tree of G, where $G_1 = G[X]$ and $G_2 = G[Y]$. This suggest the possibility of an algorithm to examine every cut of G and to try to find a parse tree that reflects that cut. There are a number of problems with this approach. First, the number of cuts may grow exponentially with n. (In particular, consider the graph consisting of $(n-1)/2$ P_3s that all share a common endpoint.) Fortunately, as we will see later, we only need consider at most $O(n)$ cuts. Secondly, we would need a polynomial time algorithm to find such a set of $O(n)$ cuts. In fact the algorithm by Ma and Spinrad [MS94] does this for us. (Another approach for finding a polynomial size set of cuts is described in [WWW].) For any of these cuts (say cut $(X : Y)$) we can see if it corresponds to an appropriate decomposition by solving $2\text{--}LAB(G : \tilde{X} \cup \tilde{Y}, -)$.

We now present the formal specifications of our algorithm.

5   Formal Description

Our algorithm has the following outline: (Note that although the algorithm is described as a recognition algorithm, it can easily be modified to produce a 3-expression which defines the input graph, if such an expression exists.)

Given graph J use Modular Decomposition to find the prime graphs $J_1, \ldots, J_k$ associated with J.

<u>for</u> $i := 1$ to k

        <u>if</u> $\neg CWD3(J_i)$ <u>then</u>

            STOP $(cwd(J) > 3)$

STOP $(cwd(J) \leq 3)$

<u>function</u> $CWD3(G)$

{*This function is true iff prime graph G has $cwd(G) \leq 3$.*}

{*First see if there is a parse tree with a final join with a 3-labeled connected component.* }

<u>for</u> each $x \in V(G)$
   <u>if</u> $2\text{-}LAB(G : \{x\}, -)$ or $2\text{-}LAB(G : \{z \in N(x)|N[z] \not\subseteq N[x]\}, -)$
   <u>then</u>
      <u>return</u> <u>true</u>

{*Since there is no such parse tree, determine if there is a parse tree for which the final join corresponds to a cut.*}

Produce a set of cuts $\{(X_1 : Y_1), (X_2 : Y_2), \ldots, (X_l : Y_l)\}$ so that if there is a parse tree whose final join corresponds to a cut, there is one whose final join corresponds to a cut in this set (using e.g. Ma-Spinrad).

<u>for</u> $i := 1$ to l
   <u>if</u> $2\text{-}LAB(G : \tilde{X}_i \cup \tilde{Y}_i, -)$ <u>then</u>
      <u>return</u> <u>true</u>

<u>return</u> <u>false</u>

6   Correctness and Complexity Issues

In this section we give more details regarding the sufficiency, for our purposes, of the set of cuts determined by the Ma-Spinrad algorithm and then briefly discuss the complexity of our algorithm.

As mentioned in Sect. 4, the number of cuts in a graph may grow exponentially with the size of the graph. We prove however, that if none of the cuts identified by the Ma-Spinrad algorithm show that $cwd(G) \leq 3$ then no cut can establish $cwd(G) \leq 3$. In order to prove this we first introduce some notation.

We say that cut $(X : Y)$ is *connected* if both $G[X]$ and $G[Y]$ are connected, *1-disconnected* if exactly one of $G[X]$ and $G[Y]$ is disconnected and *2-disconnected* if both $G[X]$ and $G[Y]$ are disconnected. We say that two cuts $(X : Y)$ and $(W : Z)$ *cross* iff $X \cap W \neq \emptyset$, $X \cap Z \neq \emptyset$, $Y \cap W \neq \emptyset$ and $Y \cap Z \neq \emptyset$. We denote by $\mathcal{C}_T$ the set of cuts produced by the Ma-Spinrad algorithm. Recall that for every cut $(X : Y)$ our algorithm calls $2\text{-}LAB(G : \tilde{X} \cup \tilde{Y}, -)$ to check whether this cut can establish $cwd(G) \leq 3$. We denote it as the *call to $2\text{-}LAB$ on behalf of cut $(X : Y)$*. Suppose all the calls to $2\text{-}LAB$ on behalf of the cuts in $\mathcal{C}_T$ failed and there is a cut $(X : Y)$ not in $\mathcal{C}_T$ such that the call to $2\text{-}LAB$ on behalf of $(X : Y)$ succeeds. We show that there is a cut in $\mathcal{C}_T$ $(W : Z)$ which crosses $(X : Y)$. Furthermore we show that if $(X : Y)$ is connected then $\tilde{X} \cup \tilde{Y} = \tilde{W} \cup \tilde{Z}$. Thus the call to $2\text{-}LAB$ on behalf of cut $(X : Y)$ is the same as the call to $2\text{-}LAB$ on behalf of cut $(W : Z)$, a contradiction. Thus $(X : Y)$ is not connected. We show that if $(X : Y)$ is 2-disconnected then $(X : Y)$ must be in $\mathcal{C}_T$, again a contradiction. Thus (X, Y) must be 1-disconnected. In this case we also reach a contradiction, as described in the full version of the paper.

We now turn briefly to complexity issues. As shown in [CH94] modular decomposition can be performed in linear time. The Ma-Spinrad algorithm can be implemented in $O(n^2)$ time. Function 2-LAB is invoked $O(n)$ times. As shown in the journal version of the paper, the complexity of 2-LAB is $O(mn)$; thus the overall complexity of our algorithm is $O(n^2 m)$.

There is one case in the 2–LAB procedure where we use a modular decomposition tree. Thus for achieving best complexity, a linear modular decomposition algorithm is needed there. Up to now, no such algorithm is known that is also easy to implement. However, if a practical algorithm is sought, one can use an $O(n + m \log n)$ algorithm [HPV99]. The complexity of 2–LAB is then $O(mn \log n)$, and the overall complexity would be $O(mn^2 \log n)$.

7 Concluding Remarks

Having shown that the clique-width at most 3 problem is in P, the key open problem is to determine whether the fixed clique-width problem is in P for constants larger than 3. Even extending our algorithm to the 4 case is a nontrivial and open problem. Although, to the best of our knowledge, it has not been established yet, one fully expects the general clique-width decision problem to be NP-complete.

Acknowledgments

D.G. Corneil and U. Rotics wish to thank the Natural Science and Engineering Research Council of Canada for financial assistance.

References

[CER93]   B. Courcelle, J. Engelfriet, and G. Rozenberg. Handle-rewriting hypergraph grammars. *J. Comput. System Sci.*, 46:218-270, 1993.

[CH94]   A. Cournier and M. Habib. A new linear algorithm for modular decomposition. *Lecture Notes in Computer Science*, 787:68-84, 1994.

[CMRa]   B. Courcelle, J.A. Makowsky, and U. Rotics. Linear time solvable optimization problems on certain structured graph families, extended abstract. Graph Theoretic Concepts in Computer Science, 24th International Workshop, WG'98, volume 1517 of Lecture Notes in Computer Science, pages 1-16. Springer Verlang, 1998. Full paper to appear in Theory of Computing Systems.

[CMRb]   B. Courcelle, J.A. Makowsky, and U. Rotics. On the fixed parameter complexity of graph enumeration problems definable in monadic second order logic. To appear in Disc. Appl. Math.

[CO99]   B. Courcelle and S. Olariu. Upper bounds to the clique-width of graphs. To appear in Disc. Appl. Math.
          (http://dept-info.labri.u-bordeaux.fr/~courcell/ActSci.html), 1999.

[Cun82]   W.H. Cunningham. Decomposition of directed graphs. *SIAM J. Algebraic Discrete Methods*, 3:214-228, 1982.

[GR99]   M. C. Golumbic and U. Rotics. On the clique-width of perfect graph classes (extended abstract). To appear in WG99, 1999.

[HPV99]   M. Habib, C. Paul and L. Viennot. Partition refinement techniques: an interesting algorithmic tool kit. *International Journal of Foundations of Computer Science*, volume 10, 1999, 2:147–170.

[MR99]   J.A. Makowsky and U. Rotics. On the clique-width of graphs with few P_4's. To appear in the International Journal of Foundations of Computer Science (IJFCS), 1999.

[MS94]   T. Ma and J. Spinrad. An $O(n^2)$ algorithm for undirected split decomposition. *Journal of Algorithms*, 16:145-160, 1994.

[WWW]   A complete description of the 2–*LAB* procedure, and also a complete description in French of a different approach to the entire algorithm. http://www.lirmm.fr/~lanligne

On Dart-Free Perfectly Contractile Graphs[*]
Extended Abstract

Cláudia Linhares Sales[1] and Frédéric Maffray[2]

[1] DC-LIA, Bloco 910, CEP 60455-760, Campus do Pici, UFC, Fortaleza-CE, Brazil
[2] CNRS, Laboratoire Leibniz, 46 avenue Félix Viallet, 38031 Grenoble Cedex, France

Abstract. The dart is the five-vertex graph with degrees 4, 3, 2, 2, 1. An even pair is pair of vertices such that every chordless path between them has even length. A graph is perfectly contractile if every induced subgraph has a sequence of even-pair contractions that leads to a clique. We show that a recent conjecture on the forbidden structures for perfectly contractile graphs is satisfied in the case of dart-free graphs. Our proof yields a polynomial-time algorithm to recognize dart-free perfectly contractile graphs.

Keywords: Perfect graphs, even pairs, dart-free graphs, claw-free graphs

1 Introduction

A graph G is *perfect* [1] if every induced subgraph H of G has its chromatic number $\chi(H)$ equal to the maximum size $\omega(H)$ of the cliques of H. One of the most attractive properties of perfect graphs is that some problems that are hard in general, such as optimal vertex-coloring and maximum clique number, can be solved in polynomial time in perfect graphs, thanks to the algorithm of Grötschel, Lovász and Schrijver [7]. However, that algorithm, based on the ellipsoid method, is quite impractical. So, an interesting open problem is to find a combinatorially "simple" polynomial-time algorithm to color perfect graphs. In such an algorithm, one may reasonably expect that some special structures of perfect graphs will play an important role. An *even pair* in a graph G is a pair of non-adjacent vertices such that every chordless path of G between them has an even number of edges. The *contraction* of a pair of vertices x, y in a graph G is the process of removing x and y and introducing a new vertex adjacent to every neighbor of x or y in G. Fonlupt and Uhry [6] proved that *contracting an even pair in a perfect graph yields a new perfect graph with the same maximum clique number.* In consequence, a natural idea for coloring a perfect graph G is, whenever it is possible, to find an even pair in G, to contract it, and to repeat this procedure until a graph G' that is easy to color is obtained. By the result of Fonlupt and Uhry, that final graph G' has the same maximum clique size as G and (since it is perfect) the same chromatic number. Each

[*] This research was partially supported by the cooperation between CAPES (Brazil) and COFECUB (France), project number 213/97. The first author is partially supported by CNPq-Brazil grant number 301330/97.

G. Gonnet, D. Panario, and A. Viola (Eds.): LATIN 2000, LNCS 1776, pp. 135–144, 2000.

vertex of G' represents a stable set of G, so one can easily obtain an optimal coloring of G from any optimal coloring of G'. For many classical perfect graphs one may expect the final graph to be a clique. Thus one may wonder whether every perfect graph admit a sequence of even-pair contractions that leads to a clique. Unfortunately, the answer to this question is negative (the smallest counterexample is the complement of a 6-cycle).

Bertschi [2] proposes to call a graph G *even contractile* if it admits a sequence of even-pair contractions leading to a clique, and *perfectly contractile* if every induced subgraph of G is even contractile. The class of perfectly contractile graphs contains many known classes of perfect graphs, such as Meyniel graphs, weakly triangulated graphs, and perfectly orderable graphs, see [4].

Everett and Reed [5] have proposed a conjecture characterizing perfectly contractile graphs. In order to present it, we need some technical definitions. A *hole* is a chordless cycle of length at least five, and an *antihole* is the complement a hole. A hole or antihole is even (resp. odd) if it has an even (odd) number of vertices. We denote by $\bar{C}_6$ the complement of a hole on six vertices.

Definition 1 (Stretcher). *A* stretcher *is any graph that can be obtained by subdividing the three edges of $\bar{C}_6$ that do not lie in a triangle in such a way that the three chordless paths between the two triangles have the same parity. A stretcher is odd (resp. even) if the three paths are odd (even) (see figure 1).*

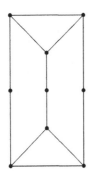

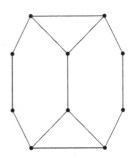

An even stretcher   An odd stretcher

Conjecture 1 (Perfectly Contractile Graph Conjecture [5]). A graph is perfectly contractile if and only if it contains no odd hole, no antihole, and no odd stretcher.

Note that there is no even pair in an odd hole or in an antihole, but odd stretchers may have even pairs. So, the 'only if' part of the conjecture is established if we can check that every sequence of even-pair contractions in an odd stretcher leads to a graph that is not a clique; this is less obvious but was done formally in [9].

The above conjecture has already been proved for planar graphs [9], for claw-free graphs [8] and for bull-free graphs [3].

Here, we are interested in the *dart-free* perfectly contractile graphs. Recall that the *dart* is the graph on five vertices with degree sequence $(4, 3, 2, 2, 1)$; in other words, a dart is obtained from a 4-clique by removing one edge and adding a new vertex adjacent to exactly one of the remaining vertices of degree three. We will call *tips* of the dart its two vertices of degree two.

A dart

A graph is *dart-free* if it does not contain a dart as an induced subgraph. Dart-free graphs form a large class of interest in the realm of perfect graphs as it contains all diamond-free graphs and all claw-free graphs. Dart-free graphs were introduced by Chvátal, and Sun [11] proved that the Strong Perfect Graph Conjecture is true for this class, that is, a dart-free graph is perfect if only if it contain no odd hole and no odd antihole. Chvátal, Fonlupt, Sun and Zemirline [12] devised a polynomial-time algorithm to recognize dart-free graphs. On the other hand, the problem of coloring the vertices of a dart-free perfect graph in polynomial time using only simple combinatorial arguments remains open.

We will prove that Everett and Reed's conjecture on perfectly contractile graphs is also true for dart-free graphs, that is:

Theorem 1 (Main Theorem). *A dart-free graph is perfectly contractile if and only if it contains no odd hole, no antihole, and no odd stretcher.*

Moreover, we will present a polynomial-time combinatorial algorithm to color optimally the perfectly contractile dart-free graphs. In order to prove our main theorem, we will use the decomposition structure found by Chvátal, Fonlupt, Sun and Zemirline [12]. It is presented in the next section.

We finish this section with some terminology and notation. We denote by $N(x)$ the subset of vertices of G to which x is adjacent. The complement of a graph G is denoted by $\bar{G}$. If $\{x, y\}$ is an even pair of a graph G, the graph obtained by the contraction of x and y is denoted by G/xy. It will be convenient here to call two vertices x, y of a graph *twins* when they are adjacent and $N(x) \cup \{x\} = N(y) \cup \{y\}$ (the usual definition of twins does not necessarily require them to be adjacent). A *claw* is a graph isomorphic to the complete bipartite graph $K_{1,3}$. A *double-claw* is a graph with vertices u_1, u_2, u_3, v_1, v_2 and edges v_1v_2 and u_iv_j

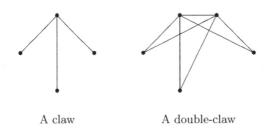

A claw   A double-claw

$(1 \leq i \leq 3, 1 \leq j \leq 2)$. Two twins x, y are called *double-claw twins* if they are the vertices v_1, v_2 in a double-claw as above. The *join* of two vertex-disjoint graphs $G_1 = (V_1, E_1)$ and $G_2 = (V_2, E_2)$ is the graph G with vertex-set $V_1 \cup V_2$ and edge-set $E_1 \cup E_2 \cup F$, where F is set of all pairs made of one vertex of G_1 and one vertex of G_2.

2   Decomposition of Dart-Free Perfect Graphs

We present here the main results from [12] and adopt the same terminology. We call DART-FREE the algorithm from [12] to recognize dart-free perfect graphs.

When a graph G has a pair of twins x, y, Lovász's famous Replication Lemma [10] ensures that G is perfect if and only if $G - x$ (or $G - y$) is perfect. So, the initial step of algorithm DART-FREE is to remove one vertex among every pair of twins in the graph. Dart-free graphs without twins have some special properties.

Definition 2 (Friendly graph [12]). *A graph G is* friendly *if the neighborhood $N(x)$ of every vertex x of G that is the center of a claw induces vertex-disjoint cliques.*

Theorem A ([12]) *Let G be a dart-free graph without twins. If G and $\bar{G}$ are connected, then G is friendly.*

Theorem B ([12]) *A graph G is friendly if and only if it contains no dart and no pair of double-claw twins.*

Let G be a dart-free graph. Let W be the subset of all vertices of G that have at least one twin, and let T be a subset of W such that every pair of twins of G has at least one of them in T. Using Theorem A, one can find in polynomial time a family $\mathcal{F}$ of pairwise vertex-disjoint friendly graphs such that: (a) the elements of $\mathcal{F}$ are induced subgraphs of $G - T$, and (b) G is perfect if and only if every element of $\mathcal{F}$ is perfect. This family can be constructed as follows: first, put $G - T$ in $\mathcal{F}$; then, as long as there exists an element H of $\mathcal{F}$ such that either H or $\bar{H}$ is disconnected, replace in $\mathcal{F}$ the graph H by its connected components (if H is disconnected) or by the complements of the connected components of $\bar{H}$

(if $\bar{H}$ is disconnected). In consequence, the problem of deciding whether a dart-free graph is perfect is reduced to deciding whether a friendly graph is perfect or not. For this purpose, friendly graphs are decomposed further.

Definition 3 (Bat [12]). *A bat is any graph that can be formed by a chordless path $a_1a_2 \cdots a_m$ ($m \geq 6$) and an additional vertex z adjacent to a_1, a_i, a_{i+1} and a_m for some i with $3 \leq i \leq m-3$ and to no other vertex of the path. A graph G is* bat-free *if it does not contain any bat as an induced subgraph.*

Given a graph G and a vertex z, a z-edge is any edge whose endpoints are both adjacent to a given vertex z. The graph obtained from G by removing vertex z and all z-edges is denoted by $G * z$.

Definition 4 (Rosette [12]). *A graph G is said to have a* rosette *centered at a vertex z of G if $G * z$ is disconnected and the neighborhood of z consists of vertex-disjoint cliques.*

Theorem C ([12]) *Every friendly graph G containing no odd hole either is bat-free, or has a clique-cutset, or has a rosette.*

Definition 5 (Separator [12]). *A* separator *S is a cutset with at most two vertices such that, if S has two non-adjacent vertices, each component of $G - S$ has at least two vertices.*

Theorem D ([12]) *Every bat-free friendly graph G containing no odd hole either is bipartite, or is claw-free, or has a separator.*

A decomposition of G along special cutsets can be defined as follows:

- *Clique-cutset decomposition:* Let C be a clique-cutset of G and let B_1, ..., B_k be the connected components of $G - C$. The graph G is decomposed into the *pieces* of G with respect to C, which are the induced subgraphs $G_i = G[B_i \cup C]$ ($i = 1, \ldots, k$).
- *Rosette decomposition:* Consider a rosette centered at a vertex z of G, and let $B_1, \ldots B_k$ ($k \geq 2$) be the connected components of $G * z$. The graph G is decomposed into $k + 1$ graphs $G_1, \ldots, G_k, H$ defined as follows. For $i = 1, \ldots, k$, the graph G_i is $G[B_i \cup \{z\}]$. The graph H is formed from $G[N(z)]$ by adding vertices $w_1, \ldots, w_k$ and edges from w_i to all of $N(z) \cap B_i$ ($i = 1, \ldots, k$).
- *Separator decomposition:* When S is a separator of size one or two with its two vertices adjacent, S is a clique-cutset and the decomposition is as above. When $S = \{u, v\}$ is a separator of G with u, v non-adjacent, let $B_1, \ldots, B_k$ be the components of $G - S$, and let P be a chordless path between u and v in G. The graph G is decomposed into k graphs $G_1, \ldots, G_k$ defined as follows. If P is even, G_i is obtained from $G[B_i \cup S]$ by adding one vertex w_i with edges to u and v. If P is odd, set $G_i = G[B_i \cup S] + uv$.

Algorithm BAT-FREE builds a decomposition tree T of a friendly graph G. At the initial step, G is the root and the only node of the tree. At the general step, let G' be any node of T. If G' can be decomposed by one of the special cutsets (clique or rosette), then add in T, as children of G', the graphs into which it is decomposed. More precisely, the clique-cutset decomposition is applied first, if possible; the rosette decomposition is applied only if the clique-cutset decomposition cannot be applied. Since each leaf H of T is friendly and has no clique-cutset and no rosettes, Theorem C ensures that either H is bat-free or G was not perfect. So, the second phase of the algorithm examines the leaves of T: each leaf H of T must either be bipartite, or be claw-free, or contain a separator, or else G is not perfect, by Theorem D. If H contains a separator, a separator decomposition is applied. When no separator decomposition is possible, G is perfect if and only if all the remaining leaves of T are either bipartite or claw-free.

3   Dart-Free Perfectly Contractile Graphs

This section is dedicated to the proof of Theorem 1. In this proof, we will use the decomposition of dart-free perfect graphs obtained by the BAT-FREE algorithm. We organize this section following the steps of the decomposition. First, we examine the friendly graphs.

Theorem 2. *A friendly graph G is perfectly contractile if and only if it contains no odd stretcher, no antihole and no odd hole.*

Proof. As observe at the beginning, no perfectly contractile graph can contain an odd hole, an antihole, or an odd stretcher. Conversely, suppose that G has no odd stretcher, no antihole and no odd hole as induced subgraph, and let us prove by induction on the number of vertices of G that G is perfectly contractile. The fact is trivially true when G has at most six vertices. In the general case, by Theorem C, G either is bat-free or has a clique-cutset or a rosette. We are going to check each of these possibilities. The following lemmas prove Theorem 2 respectively when G has a clique-cutset and when G has a rosette. Their proofs are omitted and will appear in the full version of the paper.

Lemma 1. *Let G be a friendly graph, with no odd hole, no odd stretcher and no antihole. If G has a clique cutset, then G is perfectly contractile.*

Lemma 2. *Let G be a friendly graph with no odd hole, no antihole and no odd stretcher. If G has a rosette centered at some vertex z, then:*
   (i) Every piece of G with respect to z is perfectly contractile; and
   (ii) G is perfectly contractile if every piece of G has a sequence of even-pair contractions leading to a clique such that each graph g in this sequence either is dart-free or contains a dart whose tips form an even pair of g.

Lemma 3. *Let G be a friendly graph with no odd hole, no antihole and no odd stretcher. If G is bat-free, then G is perfectly contractile.*

For this lemma, we prove that: (a) If G has a separator S, then G is perfectly contractile; (b) If G is bipartite or claw-free, G is perfectly contractile. These facts imply Lemma 3.

The following lemmas will ensure the existence of a sequence of even-pair contractions whose intermediate graphs are dart-free.

Lemma 4. *Every bipartite graph admits a sequence of even-pair contractions that lead to a clique and whose intermediate graphs are dart-free.*

Lemma 5. *Every claw-free graph admits a sequence of even-pair contractions that lead to a clique and whose intermediate graphs either are dart-free or have a dart whose tips form an even pair.*

Lemma 4 is trivial; the proof of Lemma 5 is based on the study of even pairs in claw-free graphs that was done in [8].

Lemmas 4 and 5 imply that every friendly graph that contains no odd hole, no antihole and no odd stretcher admits a sequence of even-pair contractions whose intermediate graphs are dart-free. Therefore, Lemmas 1, 2 and 3 together imply Theorem 2.

3.1   An Algorithm

Now we give the outline of an even-pair contraction algorithm for a friendly graph G without odd holes, antiholes or odd stretchers. The algorithm has two main steps: constructing the decomposition tree, then contracting even pairs in a bottom-up way along the tree.

In the first step, the algorithm uses a queue Q that initially contains only G. While Q is not empty, a graph G' of Q is dequeued and the following sequence of steps are executed at the same time that a decomposition tree T is being built:

1. If G' has a clique-cutset C, put the pieces of G' with respect to C in Q; repeat the first step.
2. If G' has a rosette centered at z, put in Q all the pieces of G' with respect to the rosette; except H; repeat the first step.
3. If G' has a separator $\{a, b\}$ and $\{a, b\}$ forms an even pair, contract a and b, put G'/ab in Q; repeat the first step. If $\{a, b\}$ forms an odd pair, put $G' + ab$ in Q; repeat the first step.

The second step examines the tree decomposition T in a bottom-up way. For each leaf G' of T, we have:

1. If G' is a bipartite graph, then a sequence of even-pair contractions that turns G' into a K_2 is easily obtained.
2. If G' is a claw-free graph, then a sequence of even-pair contractions that turns G' into a clique can be obtained by applying the algorithm described in [8].

Now, since every leaf is a clique, we can glue the leaves by the cutsets that produced them, following the tree decomposition. Three cases appear:

1. Suppose that $G_1, \ldots, G_k$ are the pieces produced by a separator $\{a, b\}$. Since $G_1, \ldots, G_k$ are cliques, glueing $G_1, \ldots, G_k$ by $\{a, b\}$ (more exactly, by vertices that correspond to a and b) will produce a graph with a clique-cutset and such that every piece is a clique. This is a special kind of triangulated graph, in which a sequence of even-pair contractions leading to a clique can easily be obtained.

2. Suppose that $G_1, \ldots, G_k$ are the pieces produced by a rosette and let G' be the graph obtained by glueing all these pieces (which are cliques) according to the rosette. Let G'' be the graph obtained from G' by removing (i) every vertex that sees all the other vertices, and (ii) every vertex whose neighbours form a clique; it is easy to check that any sequence of even-pair contractions for G'' yields a sequence of even-pair contractions for G'. Moreover, we can prove that G'' is friendly and bat-free; so it must either have a separator or be a bipartite graph or a claw-free graph. Thus, an additional step of decomposition and contraction will give us the desired sequence of even-pair contractions for G''.

3. Suppose that $G_1, \ldots, G_k$ are the pieces produced by a clique-cutset. Again, the graph G' obtained by glueing these pieces along the clique-cutset is a triangulated graph for which a sequence of even-pair contractions that turns it into a clique can be easily obtained.

Finally, we can obtain a sequence of even-pair contractions for G by concatenating all the sequences mentioned above.

Proof of Theorem 1

Let G be a dart-free graph with no odd hole, no antihole and no odd stretcher. If G has no twins and G and $\bar{G}$ are connected, then G is friendly and so, by Theorem 2, perfectly contractile. If $\bar{G}$ is disconnected then we can show that G has a very special structure which is easy to treat separately. If G is disconnected, it is sufficient to argue for each component of G. Hence we are reduced to the case where G is connected and is obtained from a friendly graph by replication (making twins). We conduct this proof by induction and along three steps. First, we modify slightly the construction of family $\mathcal{F}$ described in section 2. As we have seen, $\mathcal{F}$ was obtained from a dart-free graph without twins. Unfortunately, twins cannot be bypassed easily in the question of perfect contractibility (Note: it would follow from Everett and Reed's conjecture that replication preserves perfect contractibility; but no proof of this fact is known). However, by Theorem B, we need only remove double-claw twins from a dart-free graph G to be able to construct a family of friendly graphs from G. It is not hard to see that if a dart-free graph G such that G and $\bar{G}$ are connected contains a double claw, then it must contain double-claw twins (see the proof of Theorem A in [12]). So Theorem A can be reformulated as follows:

Theorem E ([12]) *Let G be a dart-free graph without double-claw twins. If G and $\bar{G}$ are connected, then G is friendly.*

Therefore, instead of removing all the twins of a dart-free graph G, we can afford to remove only the double-claw twins, as follows: initialize $G' = G$ and $T = \emptyset$; as long as G' has a pair of double-claw twins x, y, set $G' = G' - x$ and $T = T + x$. Observe that if G contains no odd stretcher, no antihole and no odd hole, then so does G'. Now let the family $\mathcal{F}'$ be obtained from G' just like $\mathcal{F}$ was obtained in the previous section.

The second step is to examine each friendly graph F of $\mathcal{F}'$, to add back its twins and to prove that it is perfectly contractile. Since G' contains no odd stretcher, no antihole and no odd hole, so does F; hence, by Theorem 2, F is perfectly contractile. Denote by T_F the set of double-claw twins of F, and by $F + T_F$ the graph obtained by adding back the twins in F. Since F is friendly, we can consider the tree decomposition of F. Five cases appear:

(1) F contains a clique-cutset. Then $F + T_F$ contains a clique-cutset C. Every piece of $F + T_F$ with respect to C is an induced subgraph of G, and so it is perfectly contractile. Moreover, clearly, each even pair of $F + T_F$ is an even pair of the whole graph.

(2) F contains a rosette (centered at a vertex z). Suppose first that z has no twins in T_F. Then $F + T_F$ also contains a rosette centered at z, and the proof works as usual. Now suppose that z has a set of twins $T(z) \in T_F$. We can generalize the rosette in following way: remove the vertices $z + T(z)$ and all the z edges, and construct the pieces $G_1, \ldots, G_k$ as before, except that $z + T(z)$ (instead of z alone) lies in every piece. Each piece is an induced subgraph of G, so it is perfectly contractile. Moreover we can prove that each even pair in a piece is an even pair of the whole graph. The desired result can then be obtained as in the twin-free case above.

(3) F has a separator $\{a, b\}$. Let $A = \{a, a_1, \ldots, a_l\}$ and $B = \{b, b_1, \ldots, b_r\}$ ($r \geq l \geq 0$) be the sets of twins of a and b respectively. If $\{a, b\}$ is an even pair, then we do the following sequence of contractions: $\{a, b\}, \{a_1, b_1\}, \ldots, \{a_l, b_l\}$. A lemma (whose proof is omitted here) ensures that this is a valid sequence of even-pair contractions. The result is a graph with a clique-cutset $C \cup R$, where C consists of the l contracted vertices and R is made of the $r - l$ remaining vertices of B. For each piece G_1 of this graph with respect to C, the graph $G_1 - R$ is isomorphic to a piece of F/ab with vertex ab replicated l times. This fact, together with the fact that all the pieces of F with respect to $\{a, b\}$ are perfectly contractile, and with the induction hypothesis, implies that $F + T_F$ is perfectly contractile. If $\{a, b\}$ is an odd pair, a different type of modification of the construction of the pieces is introduced; we skip this subcase for the sake of shortness.

(4) F is bipartite. Then the vertices of $F + T_F$ can be divided into two sides such that every connected component of each side is a clique and every clique from one side sees all or none of the vertices of any clique on the other side. It is easy to check directly that such a graph is perfectly contractile.

(5) F is claw-free. Then $F + T_F$ is claw-free. So $F + T_F$, as F, is perfectly contractile.

Lemma 6. *If F is a perfectly contractile friendly graph, then $F + T_F$ is perfectly contractile.*

The third and last step of the proof of Theorem 1 is the following lemma:

Lemma 7. *A dart-free graph G' without double-claw twins is perfectly contractile if and only if every friendly graph H of $\mathcal{F}'$ is perfectly contractile.*

Finally, given a dart-free graph G that contains no odd hole, no antihole and no odd stretcher, we obtain a graph G' that contains no double-twins and is decomposable into friendly graphs. By Theorem 2 these graphs are perfectly contractile, and by Lemma 6, adding the twins back to these graphs preserves their perfectly contractability. So, the modified family $\mathcal{F}'$ is a set of perfectly contractile graphs. By Lemma 7, G is perfectly contractile, and the proof of Theorem 1 is now complete.

4   Conclusion

The many positive results gathered in the past few years about Conjecture 1 (see [4]) motivate us to believe strongly in its validity and to continue our study of this conjecture.

References

1. C. Berge. Les problèmes de coloration en théorie des graphes. *Publ. Inst. Stat. Univ. Paris*, 9:123–160, 1960.
2. M. Bertschi. Perfectly contractile graphs. *J. Comb. Theory, Series B*, 50:222–230, 1990.
3. C.M.H. de Figueiredo, F. Maffray, and O. Porto. On the structure of bull-free perfect graphs. *Graphs and Combinatorics*, 13:31–55, 1997.
4. H. Everett, C.M.H. de Figueiredo, C. Linhares Sales, F. Maffray, O. Porto, and B. Reed. Path parity and perfection. *Disc. Math.*, 165/166:233–252, 1997.
5. H. Everett and B.A. Reed. Problem session on path parity. *DIMACS Workshop on Perfect Graphs, Princeton, NJ*, June 1993.
6. J. Fonlupt and J.P. Uhry. Transformations which preserve perfectness and h-perfectness of graphs. *"Bonn Workshop on Combinatorial Optimization 1981", Ann. Disc. Math.*, 16:83–95, 1982.
7. M. Grötschel, L. Lovász, and A. Schrijver. Polynomial algorithms for perfect graphs. *"Topics on Perfect Graphs", Ann. Disc. Math.*, 21:325–356, 1984.
8. C. Linhares-Sales and F. Maffray. Even pairs in claw-free perfect graphs. *J. Comb. Theory, Series B*, 74:169–191, 1998.
9. C. Linhares-Sales, F. Maffray, and B.A. Reed. On planar perfectly contractile graphs. *Graphs and Combinatorics*, 13:167–187, 1997.
10. L. Lovász. Normal hypergraphs and perfect graphs. *Disc. Math.*, 2:253–267, 1972.
11. L. Sun. Two classes of perfect graphs. *J. Comb. Theory, Series B*, 53:273–292, 1991.
12. L. Sun V. Chvátal, J. Fonlupt and A. Zemirline. Recognizing dart-free perfect graphs. Tech. Report 92778, Institut für Ökonometrie und Operations Research, Rheinische Friedrich Wilhelms Universität, Bonn, Germany, 1992.

Edge Colouring Reduced Indifference Graphs

Celina M.H. de Figueiredo[1], Célia Picinin de Mello[2], and Carmen Ortiz[3]

[1] Instituto de Matemática, Universidade Federal do Rio de Janeiro, Brazil
`celina@cos.ufrj.br`
[2] Instituto de Computação, Universidade Estadual de Campinas, Brazil
`celia@dcc.unicamp.br`
[3] Escuela de Ingeniería Industrial, Universidad Adolfo Ibañez, Chile
`cortiz@uai.cl`

Abstract. The chromatic index problem – finding the minimum number of colours required for colouring the edges of a graph – is still unsolved for indifference graphs, whose vertices can be linearly ordered so that the vertices contained in the same maximal clique are consecutive in this order. Two adjacent vertices are twins if they belong to the same maximal cliques. A graph is reduced if it contains no pair of twin vertices. A graph is overfull if the total number of edges is greater than the product of the maximum degree by $\lfloor n/2 \rfloor$, where n is the number of vertices. We give a structural characterization for neighbourhood-overfull indifference graphs proving that a reduced indifference graph cannot be neighbourhood-overfull. We show that the chromatic index for all reduced indifference graphs is the maximum degree.

1   Introduction

In this paper, G denotes a simple, undirected, finite, connected graph. The sets $V(G)$ and $E(G)$ are the vertex and edge sets of G. Denote $|V(G)|$ by n and $|E(G)|$ by m. A graph with just one vertex is called *trivial*. A *clique* is a set of vertices pairwise adjacent in G. A *maximal clique* of G is a clique not properly contained in any other clique. A *subgraph* of G is a graph H with $V(H) \subseteq V(G)$ and $E(H) \subseteq E(G)$. For $X \subseteq V(G)$, denote by $G[X]$ the *subgraph induced by* X, that is, $V(G[X]) = X$ and $E(G[X])$ consists of those edges of $E(G)$ having both ends in X. For $Y \subseteq E(G)$, the *subgraph induced by* Y is the subgraph of G whose vertex set is the set of endpoints of edges in Y and whose edge set is Y; this subgraph is denoted by $G[Y]$. The notation $G \setminus Y$ denotes the subgraph of G with $V(G \setminus Y) = V(G)$ and $E(G \setminus Y) = E(G) \setminus Y$. A graph G is H-*free* if G does not contain an isomorphic copy of H as an induced subgraph. Denote by C_n the chordless cycle on n vertices and by $2K_2$ the complement of the chordless cycle C_4. A *matching* M of G is a set of pairwise non adjacent edges of G. A matching M of G *covers* a set of vertices X of G when each vertex of X is incident to some edge of M. The graph $G[M]$ is also called a *matching*.

For each vertex v of a graph G, the *adjacency* $\mathrm{Adj}_G(v)$ of v is the set of vertices that are adjacent to v. The *degree* of a vertex v is $\deg(v) = |\mathrm{Adj}_G(v)|$.

G. Gonnet, D. Panario, and A. Viola (Eds.): LATIN 2000, LNCS 1776, pp. 145–153, 2000.

The *maximum degree* of a graph G is then $\Delta(G) = \max_{v \in V(G)} \deg(v)$. We use the simplified notation Δ when there is no ambiguity. We call Δ-*vertex* a vertex with maximum degree. The set $N[v]$ denotes the *neighbourhood* of v, that is, $N[v] = \text{Adj}_G(v) \cup \{v\}$. A subgraph induced by the neighbourhood of a vertex is simply called a *neighbourhood*. We call Δ-*neighbourhood* the neighbourhood of a Δ-vertex. Two vertices v e w are *twins* when $N[v] = N[w]$. Equivalently, two vertices are twins when they belong to the same set of maximal cliques. A graph is *reduced* if it contains no pair of twin vertices. The *reduced graph* G' of a graph G is the graph obtained from G by collapsing each set of twins into a single vertex and removing possible resulting parallel edges and loops.

The *chromatic index* $\chi'(G)$ of a graph G is the minimum number of colours needed to colour the edges of G such that no adjacent edges get the same colour. A celebrated theorem by Vizing [12,10] states that $\chi'(G)$ is always Δ or $\Delta + 1$. Graphs with $\chi'(G) = \Delta$ are said to be in *Class* 1; graphs with $\chi'(G) = \Delta + 1$ are said to be in *Class* 2. A graph G satisfying the inequality $m > \Delta(G)\lfloor n/2 \rfloor$, is said to be an *overfull* graph [8]. A graph G is *subgraph-overfull* [8] when it has an overfull subgraph H with $\Delta(H) = \Delta(G)$. When the overfull subgraph H can be chosen to be a *neighbourhood*, we say that G is *neighbourhood-overfull* [4]. Overfull, subgraph-overfull, and neighbourhood-overfull graphs are in Class 2.

It is well known that the recognition problem for the set of graphs in Class 1 is NP-complete [9]. The problem remains NP-complete for several classes, including comparability graphs [1]. On the other hand, the problem remains unsolved for *indifference graphs*: graphs whose vertices can be linearly ordered so that the vertices contained in the same maximal clique are consecutive in this order [11]. We call such an order an *indifference order*. Given an indifference graph, for each maximal clique A, we call *maximal edge* an edge whose endpoints are the first and the last vertices of A with respect to an indifference order. Indifference graphs form an important subclass of interval graphs: they are also called unitary interval graphs or proper interval graphs. The reduced graph of an indifference graph is an indifference graph with a unique indifference order (except for its reverse). This uniqueness property was used to describe solutions for the recognition problem and for the isomorphism problem for the class of indifference graphs [2].

It has been shown that every odd maximum degree indifference graph is in Class 1 [4] and that every subgraph-overfull indifference graph is in fact neighbourhood-overfull [3]. It has been conjectured that every Class 2 indifference graph is neighbourhood-overfull [4,3]. Note that the validity of this conjecture implies that the edge-colouring problem for indifference graphs is in P.

The goal of this paper is to investigate this conjecture by giving another positive evidence for its validity. We describe a structural characterization for neighbourhood-overfull indifference graphs. This structural characterization implies that no reduced indifference graph is neighbourhood-overfull. We prove that all reduced indifference graphs are in Class 1 by exhibiting an edge colouring with Δ colours for every indifference graph with no twin Δ-vertices. In order to construct such an edge colouring with Δ colours, we decompose an arbitrary indifference graph with no twin Δ-vertices into two indifference graphs: a matching covering all Δ-vertices and an odd maximum degree indifference graph.

The characterization for neighbourhood-overfull indifference graphs is described in Section 2. The decomposition and the edge colouring of indifference graphs with no twin Δ-vertices is in Section 3. Our conclusions are in Section 4.

2   Neighbourhood-Overfull Indifference Graphs

In this section we study the overfull Δ-neighbourhoods of an indifference graph. Since it is known that every odd maximum degree indifference graph is in Class 1 [4], an odd maximum degree indifference graph contains no overfull Δ-neighbourhoods. We consider the case of even maximum degree indifference graphs. A nontrivial complete graph with even maximum degree Δ is always an overfull Δ-neighbourhood. We characterize the structure of an overfull Δ-neighbourhood obtained from a complete graph by removal of a set of edges.

Theorem 1. *Let $K_{\Delta+1}$ be a complete graph with even maximum degree Δ. Let $F = K_{\Delta+1} \setminus R$, where R is a nonempty subset of edges of $K_{\Delta+1}$. Then, the graph F is an overfull indifference graph with maximum degree Δ if and only if $H = G[R]$ is a $2K_2$-free bipartite graph with at most $\Delta/2 - 1$ edges.*

The proof of Theorem 1 is divided into two lemmas.

Lemma 1. *Let $K_{\Delta+1}$ be a complete graph with even maximum degree Δ. Let R be a nonempty subset of edges of $K_{\Delta+1}$. If $F = K_{\Delta+1} \setminus R$ is an overfull indifference graph with maximum degree Δ, then $H = G[R]$ is a $2K_2$-free bipartite graph and $|R| < \Delta/2$.*

Proof. Let R be a nonempty subset of edges of $K_{\Delta+1}$, a complete graph with even maximum degree Δ. Let $F = K_{\Delta+1} \setminus R$ be an overfull indifference graph with maximum degree Δ. Note that $\Delta > 2$.

Because F is an overfull graph, $|V(F)|$ is odd and there are at most $\Delta/2 - 1$ missing edges joining vertices of F. Hence $|R| < \Delta/2$.

Suppose, by contradiction, that the graph $H = G[R]$ contains a $2K_2$ as an induced subgraph. Then F contains a chordless cycle C_4 as an induced subgraph, a contradiction to F being an indifference graph. Since H is a graph free of $2K_2$, we conclude that the graph H does not contain the chordless cycle C_k, $k \geq 6$. We show that the graph H does not contain neither a C_5 nor a C_3 as an induced subgraph. Assume the contrary. If H contains a C_5 as an induced subgraph, then F contains a C_5 as an induced subgraph, since C_5 is a self-complementary graph, a contradiction to F being an indifference graph. If H contains a C_3 as an induced subgraph, then F contains a $K_{1,3}$ as an induced subgraph, since, by hypothesis, F has at least one vertex of degree Δ, a contradiction to F being an indifference graph. Therefore, H is a bipartite graph, without $2K_2$, and $|R| < \Delta/2$.   □

Lemma 2. *Let $K_{\Delta+1}$ be a complete graph with even maximum degree Δ. If $H = G[R]$ is a $2K_2$-free bipartite graph induced by a nonempty set R of edges of size $|R| < \Delta/2$, then $F = K_{\Delta+1} \setminus R$ is an overfull indifference graph with maximum degree Δ.*

Proof. By definition of F and because $|R| < \Delta/2$, we have that F is an overfull graph and it has vertices of degree Δ. We shall prove that F is an indifference graph by exhibiting an indifference order on the vertex set $V(F)$ of F.

Since $H = G[R]$ is a $2K_2$-free bipartite graph, H is connected with unique bipartition of its vertex set into sets X and Y. Now, label the vertices $x_1, x_2, \ldots, x_k$ of X and label the vertices $y_1, y_2, \ldots, y_\ell$ of Y according to *degree ordering*: labels correspond to vertices in no increasing vertex degree order, i.e., $\deg(x_1) \geq \deg(x_2) \geq \cdots \geq \deg(x_k)$ and $\deg(y_1) \geq \deg(y_2) \geq \cdots \geq \deg(y_\ell)$, respectively.

This degree ordering induces the following properties on the vertices of the adjacency of each vertex of X and Y:

- The adjacency of a vertex of H defines an interval on the degree order, i.e., $\mathrm{Adj}_H(x_i) = \{y_j : 1 \leq p \leq j \leq p + q \leq \ell\}$ and $\mathrm{Adj}_H(y_j) = \{x_i : 1 \leq r \leq i \leq r + s \leq k\}$.
  Indeed, let a be a vertex of H such that $\mathrm{Adj}_H(a)$ is not an interval. Then $\mathrm{Adj}_H(a)$ has at least two vertices b and d, and there is a vertex c such that $ac \notin R$ between b and d. Without loss of generality, suppose that $\deg(b) \geq \deg(c) \geq \deg(d)$. Since $\deg(c) \geq \deg(d)$, there is a vertex e such that e is adjacent to c but is not adjacent to d. It follows that, when either $\deg(e) \leq \deg(a)$ or $\deg(a) \leq \deg(e)$, H has an induced $2K_2$ (ec and ad), a contradiction.
- The adjacency-sets of the vertices of H are ordered with respect to set inclusion according to the following *containment property*: $\mathrm{Adj}_H(x_1) \supseteq \mathrm{Adj}_H(x_2) \supseteq \cdots \supseteq \mathrm{Adj}_H(x_k)$ and $\mathrm{Adj}_H(y_1) \supseteq \mathrm{Adj}_H(y_2) \supseteq \cdots \supseteq \mathrm{Adj}_H(y_\ell)$.
  For, suppose there are a and b in X with $\deg(a) \geq \deg(b)$ and $\mathrm{Adj}_H(a) \not\supseteq \mathrm{Adj}_H(b)$. Hence, there are vertices c and d such that c is adjacent to a but not to b, and d is adjacent to b but not to a. The edges ac and bd induce a $2K_2$ in H, a contradiction.
- $x_1 y_1$ is a *dominating edge* of H, i.e., every vertex of H is adjacent to x_1 or to y_1.
  This is a direct consequence of the two properties above.

When H is not a complete bipartite graph, let i and j be the smallest indices of vertices of X and Y, respectively, such that $x_i y_j$ is not an edge of H. Note that, because $x_1 y_1$ is a dominating edge of H, we have i and j greater than 1. Define the following partition of $V(H)$:

$$A := \{x_1, x_2, \ldots, x_{i-1}\}; \qquad S := \{x_i, x_{i+1}, \ldots, x_k\};$$

$$B := \{y_1, y_2, \ldots, y_{j-1}\}; \qquad T := \{y_j, y_{j+1}, \ldots, y_\ell\}.$$

Note that S and T can be empty sets and that the graph induced by $A \cup B$ is a complete bipartite subgraph of H.

Now we describe a total ordering on $V(F)$ as follows. We shall prove that this total ordering gives the desired indifference order.

- First, list the vertices of $X = A \cup S$ as $x_1, x_2, \ldots, x_k$;
- Next, list all the Δ-vertices of F, $v_1, v_2, \ldots, v_s$;
- Finally, list the vertices of $Y = T \cup B$ as $y_l, y_{l-1}, \ldots, y_1$.

The ordering within the sets A, S, D, T, B, where D denotes the set of the Δ-vertices of F, is induced by the ordering of $V(F)$.

By the containment property of the adjacency-sets of the vertices of H and because each adjacency defines an interval, the consecutive vertices of the same degree are twins in F. Hence, it is enough to show that the ordering induced on the reduced graph F' of F is an indifference order.

For simplicity, we use the same notation for vertices of F and F', i.e., we call vertices in F' corresponding to vertices of X by $x_1', x_2', \ldots, x_{k'}'$, and we call vertices corresponding to vertices of Y by $y_1', y_2', \ldots, y_{\ell'}'$. Note that the set D contains only one representative vertex in F', and we denote this unique representative vertex by v_1'.

By definition of F', $x_1'v_1', x_2'y_{\ell'}', \ldots, x_{i-1}'y_{j+1}', x_i'y_j', \ldots, x_{k'}'y_2', v_1'y_1'$ are edges of F'. Since vertex v_1' is a representative vertex of a Δ-vertex of F, it is also a Δ-vertex of F'. Thus v_1' is adjacent to each vertex of F'. Each edge listed above, distinct from $x_1'v_1'$ and $v_1'y_1'$, has form $x_p'y_q'$. We want to show that x_p' is adjacent to all vertices from x_{p+1}' up to y_q' with respect to the order. For suppose, without loss of generality, that x_p' is not adjacent to some vertex z between x_p' and y_q' with respect to the order. Now by the definition of the graphs F and H, every edge of $K_{\Delta+1}$ not in F belongs to graph H. Since H is a bipartite graph, with bipartition of its vertex set into sets X and Y, we have in F all edges linking vertices in X, and so we have $z \neq x_s$, $p \leq s \leq k$. Vertex z is also distinct from y_s, $q \leq s \leq l$, by the properties of the adjacency in H. Hence, x_p' is adjacent to all vertices from x_{p+1}' up to y_q' with respect to the order. It follows that each edge listed above defines a maximal clique of F'. Hence, this ordering satisfies the property that vertices belonging to the same maximal clique are consecutive and we conclude that this ordering on $V(F')$ is the desired indifference order. This conclusion completes the proofs of both Lemma 2 and Theorem 1.   □

Corollary 1. *Let G be an indifference graph. A Δ-neighbourhood of G with at most $\Delta/2$ vertices of maximum degree is not neighbourhood-overfull.*

Proof. Let F be a Δ-neighbourhood of G with at most $\Delta/2$ vertices of degree Δ. If Δ is odd, then F is not neighbourhood-overfull. If Δ is even, then we use the notation of Lemma 1 and Lemma 2. The hypothesis implies $|X|+|Y| \geq (\Delta/2)+1$. Since vertex x_1 misses every vertex of Y and since vertex y_1 misses every vertex of X, there are at least $|X|+|Y|-1$ missing edges having as endpoints x_1 or y_1. Hence, there are at least $|X|+|Y|-1 \geq \Delta/2$ missing edges in F, and F cannot be neighbourhood-overfull.   □

Corollary 2. *An indifference graph with no twin Δ-vertices is not neighbour-hood-overfull.*

Proof. Let G be an indifference graph with no twin Δ-vertices. The hypothesis implies that every Δ-neighbourhood F of G contains precisely one vertex of degree Δ. Now Corollary 1 says F is not neighbourhood-overfull and therefore G itself cannot be neighbourhood-overfull.   □

Corollary 3. *A reduced indifference graph is not neighbourhood-overfull.*   □

3   Reduced Indifference Graphs

We have established in Corollary 2 of Section 2 that an indifference graph with no twin Δ-vertices is not neighbourhood-overfull, a necessary condition for an indifference graph with no twin Δ-vertices to be in Class 1. In this section, we prove that every indifference graph with no twin Δ-vertices is in Class 1. We exhibit a Δ-edge colouring for an even maximum degree indifference graph with no twin Δ-vertices. Since every odd maximum degree indifference graph is in Class 1, this result implies that all indifference graphs with no twin Δ-vertices, and in particular that all reduced indifference graphs are in Class 1.

Let $E_1, \ldots, E_k$ be a partition of the edge set of a graph G. It is clear that if the subgraphs $G[E_i]$, $1 \le i \le k$, satisfy $\Delta(G) = \sum_i \Delta(G[E_i])$ and, if for each i, $G[E_i]$ is in Class 1, then G is also in Class 1. We apply this decomposition technique to our given indifference graph with even maximum degree and no twin Δ-vertices.

We partition the edge set of an indifference graph G with even maximum degree Δ and with no twin Δ-vertices into two sets E_1 and E_2, such that $G_1 = G[E_1]$ is an odd maximum degree indifference graph and $G_2 = G[E_2]$ is a matching.

Let G be an indifference graph and $v_1, v_2, \ldots, v_n$ an indifference order for G. By definition, an edge $v_i v_j$ is maximal if there does not exist another edge $v_k v_\ell$ with $k \le i$ and $j \le \ell$. Note that an edge $v_i v_j$ is maximal if and only if the edges $v_{i-1} v_j$ and $v_i v_{j+1}$ do not exist. In addition, every maximal edge $v_i v_j$ defines a maximal clique having v_i as its first vertex and v_j as its last vertex. Thus, every vertex is incident to zero, one, or two maximal edges. Moreover, given an indifference graph with an indifference order and an edge that is maximal with respect to this order, the removal of this edge gives a smaller indifference graph: the original indifference order is an indifference order for the smaller indifference graph.

Based on Lemma 3 below, we shall formulate an algorithm for choosing a matching of an indifference graph with no twin Δ-vertices that covers every Δ-vertex of G.

Lemma 3. *Let G be a non trivial graph. If G is an indifference graph with no twin Δ-vertices, then every Δ-vertex of G is incident to precisely two maximal edges.*

Proof. Let G be a non trivial indifference graph without twin Δ-vertices and let v be a Δ-vertex of G. Consider $v_1, v_2, \ldots, v_n$ an indifference order for G. Because v is a Δ-vertex of G and G is not a clique, we have $v = v_j$, with $j \neq 1, n$. Let v_i and v_k be the leftmost and the rightmost vertices with respect to the indifference order that are adjacent to v_j, respectively. Suppose that $v_i v_j$ is not a maximal edge. Then $v_{i-1}v_j$ or $v_i v_{j+1}$ is an edge in G. The existence of $v_{i-1}v_j$ contradicts v_i being the leftmost neighbour of v_j. Because v_j and v_{j+1} are not twins, the existence of $v_i v_{j+1}$ implies $\deg(v_{j+1}) \geq \Delta + 1$, a contradiction. Analogously, we have that $v_j v_k$ is also a maximal edge.   □

We now describe an algorithm for choosing a set of maximal edges that covers all Δ-vertices of G.

Input: an indifference graph G with no twin Δ-vertices with an indifference order $v_1, \ldots, v_n$ of G.
Output: a set of edges M that covers all Δ-vertices of G.

1. For each Δ-vertex of G, say v_j, in the indifference order, put in a set $\mathcal{E}$ the edge $v_i v_j$, where v_i is its leftmost neighbour with respect to the indifference order. Each component of the graph $G[\mathcal{E}]$ is a path. (Each component H of $G[\mathcal{E}]$ has $\Delta(H) \leq 2$ and none of the components is a cycle, by the maximality of the chosen edges.)
2. For each path component P of $G[\mathcal{E}]$, number each edge with consecutive integers starting from 1. If a path component P_i contains an odd number of edges, then form a matching M_i of $G[\mathcal{E}]$ choosing the edges numbered by odd integers. If a path component P_j contains an even number of edges, then form a matching M_j choosing the edges numbered by even integers.
3. The desired set of edges M is the union $\bigcup_k M_k$.

We claim that the matching M above defined covers all Δ-vertices of G. For, if a path component of $G[\mathcal{E}]$ contains an odd number of edges, then M covers all of its vertices. If a path component of $G[\mathcal{E}]$ contains an even number of edges, then the only vertex not covered by M is the first vertex of this path component. However, by definition of $G[\mathcal{E}]$, this vertex is not a Δ-vertex of G.

Theorem 2. *If G is an indifference graph with no twin Δ-vertices, then G is in Class 1.*

Proof. Let G be an indifference graph with no twin Δ-vertices. If G has odd maximum degree, then G is in Class 1 [4].

Suppose that G is an even maximum degree graph. Let $v_1, \ldots, v_n$ be an indifference order of G. Use the algorithm described above to find a matching M for G that covers all Δ-vertices of G. The graph $G \setminus M$ is an indifference graph with odd maximum degree because the vertex sets of G and $G \setminus M$ are the same and the indifference order of G is also an indifference order for $G \setminus M$. Moreover, since M is a matching that covers all Δ-vertices of G, we have that

$\Delta(G \setminus M) = \Delta - 1$ is odd. Hence, the edges of $G \setminus M$ can be coloured with $\Delta - 1$ colours and one additional colour is needed to colour the edges in the matching M. This implies that G is in Class 1.   □

Corollary 4. *All reduced indifference graphs are in Class* 1.   □

4   Conclusions

We believe our work makes a contribution to the problem of edge-colouring indifference graphs in three respects.

First, our results on the colouring of indifference graphs show that, in all cases we have studied, neighbourhood-overfullness is equivalent to being Class 2, which gives positive evidence to the conjecture that for any indifference graph neighbourhood-overfullness is equivalent to being Class 2. It would be interesting to extend these results to larger classes. We established recently [5] that every odd maximum degree dually chordal graph is Class 1. This result shows that our techniques are extendable to other classes of graphs.

Second, our results apply to a subclass of indifference graphs defined recently in the context of clique graphs. A graph G is a *minimum indifference graph* if G is a reduced indifference graph and, for some indifference order of G, every vertex of G is the first or the last element of a maximal clique of G [6]. Given two distinct minimum indifference graphs, their clique graphs are also distinct [6]. This property is not true for general indifference graphs [7]. Note that our results apply to minimum indifference graphs: no minimum indifference graph is neighbourhood-overfull, every minimum indifference graph is in Class 1, and we can edge-colour any minimum indifference graph with Δ colours.

Third, and perhaps more important, the decomposition techniques we use to show these results are new and proved to be simple but powerful tools.

Acknowledgments. We are grateful to João Meidanis for many insightful and inspiring discussions on edge colouring. We thank Marisa Gutierrez for introducing us to the class of minimum indifference graphs. This work was partially supported by Pronex/FINEP, CNPq, CAPES, FAPERJ and FAPESP, Brazilian research agencies. This work was done while the first author was visiting IASI, the Istituto di Analisi dei Sistemi ed Informatica, with financial support from CAPES grant AEX0147/99-0. The second author is visiting IASI on leave from IC/Unicamp with financial support from FAPESP grant 98/13454-8.

References

1. L. Cai and J. A. Ellis. NP-completeness of edge-colouring some restricted graphs. *Discrete Appl. Math.*, 30:15–27, 1991.
2. C. M. H. de Figueiredo, J. Meidanis, and C. P. de Mello. A linear-time algorithm for proper interval graph recognition. *Inform. Process. Lett.*, 56:179–184, 1995.

3. C. M. H. de Figueiredo, J. Meidanis, and C. P. de Mello. Local conditions for edge-coloring. Technical report, DCC 17/95, UNICAMP, 1995. To appear in *J. Combin. Mathematics and Combin. Computing* 31, (1999).

4. C. M. H. de Figueiredo, J. Meidanis, and C. P. de Mello. On edge-colouring indifference graphs. *Theoret. Comput. Sci.*, 181:91–106, 1997.

5. C. M. H. de Figueiredo, J. Meidanis, and C. P. de Mello. Total-chromatic number and chromatic index of dually chordal graphs. *Inform. Process. Lett.*, 70:147–152, 1999.

6. M. Gutierrez and L. Oubiña. Minimum proper interval graphs. *Discrete Math.*, 142:77–85, 1995.

7. B. Hedman. Clique graphs of time graphs. *J. Combin. Theory Ser. B*, 37:270–278, 1984.

8. A. J. W. Hilton. Two conjectures on edge-colouring. *Discrete Math.*, 74:61–64, 1989.

9. I. Holyer. The NP-completeness of edge-coloring. *SIAM J. Comput.*, 10:718–720, 1981.

10. J. Misra and D. Gries. A constructive proof of Vizing's theorem. *Inform. Process. Lett.*, 41:131–133, 1992.

11. F. S. Roberts. On the compatibility between a graph and a simple order. *J. Combin. Theory Ser. B*, 11:28–38, 1971.

12. V. G. Vizing. On an estimate of the chromatic class of a p-graph. *Diskrete Analiz.*, 3:25–30, 1964. In Russian.

Two Conjectures on the Chromatic Polynomial

David Avis[1], Caterina De Simone[2], and Paolo Nobili[3]

[1] School of Computer Science, McGill University, 3480 University Street, Montreal,
Canada, H3A2A7. `avis@cs.mcgill.ca`*
[2] Istituto di Analisi dei Sistemi ed Informatica (IASI), CNR, Viale Manzoni 30,
00185 Rome, Italy. `desimone@iasi.rm.cnr.it`
[3] Dipartimento di Matematica, Università di Lecce, Via Arnesano, 73100 Lecce,
Italy; and IASI-CNR. `nobili@iasi.rm.cnr.it`

Abstract. We propose two conjectures on the chromatic polynomial
of a graph and show their validity for several classes of graphs. Our
conjectures are stronger than an older conjecture of Bartels and Welsh
[1].

Keywords: Vertex colorings, chromatic polynomials of graphs

The goal of this paper is to propose two conjectures on the chromatic polynomial
of a graph and prove them for several classes of graphs. Our conjectures are
stronger than a conjecture of Bartel and Welsh [1] that was recently proved by
Dong [2].

Let G be a graph. The chromatic polynomial of G is related to the colourings
(of the vertices) of G so that no two adjacent vertices get the same colour. If we
denote by $c_k(G)$ the number of ways to colour the vertices of G with exactly k
colours, then the *chromatic polynomial* of G is:

$$P(G, \lambda) = \sum_{k=1}^{n} c_k(G)(\lambda)_k,$$

where $(\lambda)_k = \binom{\lambda}{k} k!$.

Let $\omega(G)$ denote the *clique number* of G (maximum number of pairways
adjacent vertices) and let $\chi(G)$ denote the chromatic number of G (minimum
number of colours used in a colouring).

We propose the following two conjectures on $P(G, \lambda)$:

Conjecture 1

$$\frac{P(G, \lambda - 1)}{P(G, \lambda)} \leq \frac{\lambda - \chi(G)}{\lambda} \left(\frac{\lambda - 1}{\lambda} \right)^{n - \chi(G)} \qquad \forall \lambda \geq n; \tag{1}$$

Conjecture 2

$$\frac{P(G, \lambda - 1)}{P(G, \lambda)} \leq \frac{\lambda - \omega(G)}{\lambda} \left(\frac{\lambda - 1}{\lambda} \right)^{n - \omega(G)} \qquad \forall \lambda \geq n. \tag{2}$$

---

* This research was performed while the author was visiting IASI.

G. Gonnet, D. Panario, and A. Viola (Eds.): LATIN 2000, LNCS 1776, pp. 154–162, 2000.
© Springer-Verlag Berlin Heidelberg 2000

Conjectures 1 and 2 are related to a conjecture of Bartels and Welsh [1], known as the *Shameful Conjecture*:

The Shamefule Conjecture: *For every graph G with n vertices,*

$$\frac{P(G, n-1)}{P(G, n)} \leq \left(\frac{n-1}{n}\right)^n.$$

The Shameful Conjecture was recently proved by Dong [2], who showed that for every connected graph G with n vertices,

$$\frac{P(G, \lambda - 1)}{P(G, \lambda)} \leq \frac{\lambda - 2}{\lambda}\left(\frac{\lambda - 2}{\lambda - 1}\right)^{n-2} \qquad \forall \lambda \geq n. \tag{3}$$

What relates Conjectures 1 and 2 to Bartel's and Welsh's conjecture is that both are stronger than their conjecture. To show that, let first prove the following easy inequality:

$$\frac{m - k}{m} \leq \left(\frac{m-1}{m}\right)^k \qquad \text{for every two integers } m, k \text{ with } m \geq k \geq 0, \tag{4}.$$

The validity of (4) immediately comes from the following inequality, which is strict if $k > 1$:

$$\frac{m-k}{m} = \prod_{i=1}^{k} \frac{m-i}{m-i+1} \leq \prod_{i=1}^{k} \frac{m-1}{m}.$$

Now, (4) immediately implies that Conjecture 1 is stronger than Conjecture 2 (write (4) with $m = \lambda - \omega(G)$ and $k = \chi(G) - \omega(G)$). To see that Conjecture 2 implies the Shameful Conjecture, write (2) with $\lambda = n$, that is:

$$\frac{P(G, n-1)}{P(G, n)} \leq \frac{n - \omega(G)}{n}\left(\frac{n-1}{n}\right)^{n - \omega(G)},$$

and apply inequality (4) with $m = n$ and $k = \omega(G)$.

Moreover, inequality (3) is not stronger than inequalities (1) and (2): in fact, it is easy to show that for every graph G with n vertices and $2\omega(G) \geq n+2$, the right hand side of inequality (2) is smaller than the right hand side of inequality (3).

The two upper bounds given in our conjectures can be considered as interpolations between the respective ratios for the empty graphs O_n (graph with n vertices and no edges) and the complete graphs K_n (graphs with n vertices and all edges), for which the conjectured bounds are clearly tight. Their strength allowed us to define operations on graphs that maintain the validity of the conjectured bounds. In particular, we prove the validity of our conjectures for several classes of graphs and then use these classes of graphs as building blocks to enlarge the class of graphs for which our conjectures are true.

In [1], it was introduced the concept of the *mean colour number* of a graph G with n vertices, as

$$\mu(G) = n \left(1 - \frac{P(G, n-1)}{P(G, n)} \right).$$

Since the Shameful Conjecture is true, it immediately yields a bound for $\mu(G)$, that is

$$\mu(G) \geq n \left(1 - \left(\frac{n-1}{n} \right)^n \right),$$

which is tight only for the graph $G = O_n$. If Conjecture 2 were true, then we could get a better bound for $\mu(G)$, that is

$$\mu(G) \geq n \left(1 - \frac{n - \omega(G)}{n} \left(\frac{n-1}{n} \right)^{n-\omega(G)} \right),$$

which is tight when G is the disjoint union of a clique and a stable set.

The next four theorems will give operations for building families of graphs which satisfy Conjecture 1 (or Conjecture 2) from the basic graph O_1.

For this purpose, we first need to give some notations and definitions. Let G be a graph with n vertices, if G is a tree then we shall denote it by T_n, and if G is a cycle then we shall denote it by C_n.

A *universal* vertex in a graph is a vertex which is adjacent to all other vertices. A *clique-cutset* in a graph G is a clique whose removal from G disconnects the graph. If G has induced subgraphs G_1 and G_2 such that $G = G_1 \cup G_2$ and $G_1 \cap G_2 = K_t$ (for some t), then we say that G arises from G_1 and G_2 by *clique identification* (see Figure 1). Clearly, if G arises by clique identification from two other graphs, then G has a clique-cutset (namely, the clique K_t).

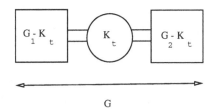

Fig. 1. Clique cutset K_t

A graph is *chordal* (or *triangulated*) if it contains no induced cycles other than triangles. It is well known that a graph is chordal if it can be constructed recursively by clique identifications, starting from complete graphs.

Let uv be an edge of a graph G. By $G_{|uv}$ we denote the graph obtained from G by *contracting* the edge uv into a new vertex which becomes adjacent to all the former neighbours of u and v. We say that G is *contractable* to a graph F if G contains a subgraph that becomes F after a series of edge contractions and edge deletions. A graph is *series-parallel* if it is not contractable to K_4.

Finally, let uv be an edge of a graph G. *Subdividing* the edge uv means to delete uv and add a new vertex x which is adjacent to only u and v. It is well known that a series-parallel multigraph can be constructed recursively from a K_2 by the operations of subdiving and of doubling edges.

Now we are ready to prove our results.

Theorem 1 *(Disjoint union) Let H be a graph obtained from two graphs G_1 and G_2 by disjoint union. If both G_1 and G_2 satisfy Conjecture 1 then H also satisfies Conjecture 1.*

Proof. : Assume that H has n vertices. Let n_i denote the number of vertices of G_i $(i = 1, 2)$ and let $\lambda \geq n(= n_1 + n_2)$. Assume that $\chi(H) = \chi(G_2) \geq \chi(G_1)$. Since

$$P(H, \lambda) = P(G_1, \lambda)P(G_2, \lambda),$$

we have

$$\frac{P(H, \lambda - 1)}{P(H, \lambda)} = \frac{P(G_1, \lambda - 1)}{P(G_1, \lambda)} \frac{P(G_2, \lambda - 1)}{P(G_2, \lambda)}.$$

Since both G_1 and G_2 satisfy Conjecture 1, we have

$$\frac{P(H, \lambda - 1)}{P(H, \lambda)} \leq \frac{\lambda - \chi(G_1)}{\lambda} \frac{\lambda - \chi(G_2)}{\lambda} \left(\frac{\lambda - 1}{\lambda}\right)^{n - \chi(G_1) - \chi(G_2)}.$$

But (3) implies that

$$\frac{\lambda - \chi(G_1)}{\lambda} \left(\frac{\lambda - 1}{\lambda}\right)^{-\chi(G_1)} \leq 1,$$

and so we are done.

Theorem 2 *(Add a universal vertex) Let H be a graph obtained from some graph G by adding a universal vertex. If G satisfies Conjecture 1 then H also satisfies Conjecture 1.*

Proof. : Assume that H has n vertices. Write $\chi = \chi(H) = \chi(G) + 1$ and let $\lambda \geq n$. Since $P(H, \lambda) = \lambda P(G, \lambda - 1)$, we have:

$$\frac{P(H, \lambda - 1)}{P(H, \lambda)} = \frac{\lambda - 1}{\lambda} \frac{P(G, \lambda - 2)}{P(G, \lambda - 1)}.$$

But then, since G satisfies Conjecture 1,

$$\frac{P(H, \lambda - 1)}{P(H, \lambda)} \leq \frac{\lambda - 1}{\lambda} \frac{\lambda - \chi}{\lambda - 1} \left(\frac{\lambda - 2}{\lambda - 1}\right)^{n - \chi},$$

and so we are done because $\frac{\lambda - 2}{\lambda - 1} < \frac{\lambda - 1}{\lambda}$.

Theorem 3 *(Clique identification) Let H be a graph obtained from two graphs G_1 and G_2 by clique identification. If both G_1 and G_2 satisfy Conjecture 1 then H also satisfies Conjecture 1.*

Proof. : Set $\chi = \chi(H)$. Without loss of generality, we can assume that $\chi(G_2) \geq \chi(G_1)$, and so $\chi = \chi(G_2)$. Let n_i denote the number of vertices of G_i ($i = 1, 2$) and let $G_1 \cap G_2 = K_t$. Clearly, H has $n = n_1 + n_2 - t$ vertices. Now, let $\lambda \geq n$. Since

$$P(H, \lambda) = \frac{P(G_1, \lambda) P(G_2, \lambda)}{P(K_t, \lambda)} = \frac{P(G_1, \lambda) P(G_2, \lambda)}{(\lambda)_t},$$

we have

$$\frac{P(H, \lambda - 1)}{P(H, \lambda)} = \frac{\lambda}{\lambda - t} \frac{P(G_1, \lambda - 1)}{P(G_1, \lambda)} \frac{P(G_2, \lambda - 1)}{P(G_2, \lambda)}.$$

Since both G_1 and G_2 satisfy (1), we have

$$\frac{P(H, \lambda - 1)}{P(H, \lambda)} \leq \frac{\lambda}{\lambda - t} \frac{\lambda - \chi(G_1)}{\lambda} \frac{\lambda - \chi(G_2)}{\lambda} \left(\frac{\lambda - 1}{\lambda}\right)^{n_1 + n_2 - \chi(G_1) - \chi(G_2)},$$

that is

$$\frac{P(H, \lambda - 1)}{P(H, \lambda)} \leq \frac{\lambda - \chi(G_1)}{\lambda - t} \left(\frac{\lambda - 1}{\lambda}\right)^{t - \chi(G_1)} \frac{\lambda - \chi}{\lambda} \left(\frac{\lambda - 1}{\lambda}\right)^{n - \chi}.$$

Hence, to prove the theorem, we only need show that

$$\frac{\lambda - \chi(G_1)}{\lambda - t} \left(\frac{\lambda - 1}{\lambda}\right)^{t - \chi(G_1)} \leq 1,$$

that is

$$\left(\frac{\lambda - 1}{\lambda}\right)^{\chi(G_1) - t} \geq \frac{\lambda - \chi(G_1)}{\lambda - t}.$$

Now, since $\chi(G_1) \geq t$, (3) (with $m = \lambda$ and $k = \chi(G_1) - t$) implies that

$$\left(\frac{\lambda - 1}{\lambda}\right)^{\chi(G_1) - t} \geq \frac{\lambda - \chi(G_1) + t}{\lambda}.$$

But

$$\frac{\lambda - \chi(G_1) + t}{\lambda} \geq \frac{\lambda - \chi(G_1)}{\lambda - t},$$

and so we are done.

Theorem 4 *(Edge subdivision) Let G be a graph with n vertices, let uv be an edge of G, let r be a positive integer, and let H be the graph obtained from G by deleting edge uv and by adding the new vertices $x_1, \cdots, x_r$ and connecting each of them to both u and v (see Figure 2). If the following two properties hold*

(a) $\min\{d_G(u), d_G(v)\} \leq \frac{n + r + 1}{2}$,

(b) both G and $G_{|uv}$ satisfy Conjecture 2,

then the graph H also satisfies Conjecture 2.

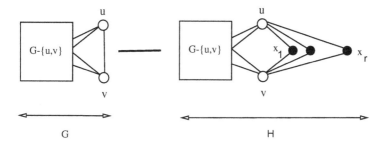

Fig. 2. Subdivide edge uv

Before proving Theorem 4, we need the following technical lemma.

Lemma 1 *Let x and r be two integers with $x > r \geq 1$. Then*

$$\frac{(x-1)^r(x+1)^{r+1} - x^{2r+1}}{x[(x-1)^r x^r - (x+1)^r(x-2)^r]} < \frac{x}{2}.$$

Proof. **of Theorem 4.**

Write $G'' = G_{|uv}$, $\omega = \omega(H)$, $\omega' = \omega(G)$, and $\omega'' = \omega(G'')$. Since

$$P(H,\lambda) = P(H+uv,\lambda) + P(H_{|uv},\lambda) = (\lambda-2)^r P(G,\lambda) + (\lambda-1)^r P(G'',\lambda),$$

we have

$$\frac{P(H,\lambda-1)}{P(H,\lambda)} = \left(\frac{\lambda-3}{\lambda-2}\right)^r \frac{P(G,\lambda-1)}{P(G,\lambda)}\alpha + \left(\frac{\lambda-2}{\lambda-1}\right)^r \frac{P(G'',\lambda-1)}{P(G'',\lambda)}(1-\alpha),$$

where

$$\alpha = \frac{(\lambda-2)^r P(G,\lambda)}{(\lambda-2)^r P(G,\lambda) + (\lambda-1)^r P(G'',\lambda)}.$$

To prove the theorem we have to show that, for every $\lambda \geq n+r$,

$$\frac{P(H,\lambda-1)}{P(H,\lambda)} \leq \frac{\lambda-\omega}{\lambda}\left(\frac{\lambda-1}{\lambda}\right)^{n+r-\omega}.$$

For this purpose, write

$$R = \left(\frac{\lambda-3}{\lambda-2}\right)^r, \quad S = \left(\frac{\lambda-2}{\lambda-1}\right)^r.$$

Since by assumption both G and G'' satisfy Conjecture 2, we have

$$\frac{P(H,\lambda-1)}{P(H,\lambda)} \leq R\alpha\,\frac{\lambda-\omega'}{\lambda}\left(\frac{\lambda-1}{\lambda}\right)^{n-\omega'} + S(1-\alpha)\,\frac{\lambda-\omega''}{\lambda}\left(\frac{\lambda-1}{\lambda}\right)^{n-1-\omega''}.$$

Hence, to prove the theorem we only need show that

$$R\alpha \frac{\lambda - \omega'}{\lambda - \omega}\left(\frac{\lambda}{\lambda - 1}\right)^{s'} + S(1 - \alpha)\frac{\lambda - \omega''}{\lambda - \omega}\left(\frac{\lambda}{\lambda - 1}\right)^{s''} \leq 1, \qquad (4)$$

where $s' = r + \omega' - \omega$ and $s'' = r + \omega'' - \omega + 1$.

For this purpose, first note that either $\omega' = \omega$ or $\omega' = \omega + 1$ and that either $\omega'' = \omega$ or $\omega'' = \omega + 1$. But, since

$$\frac{\lambda - \omega^*}{\lambda - \omega}\left(\frac{\lambda}{\lambda - 1}\right)^{\omega^* - \omega} \leq 1$$

where $\omega^* = \omega'$ or $\omega^* = \omega''$, it follows that we only need show the validity of (4) in the case $\omega' = \omega'' = \omega$, that is

$$\left(\frac{\lambda - 3}{\lambda - 2}\right)^r \left(\frac{\lambda}{\lambda - 1}\right)^r \alpha + \left(\frac{\lambda - 2}{\lambda - 1}\right)^r \left(\frac{\lambda}{\lambda - 1}\right)^{r+1} (1 - \alpha) \leq 1. \qquad (5)$$

Now, inequality (5) is equivalent to the following

$$\alpha\left[\left(\frac{\lambda - 3}{\lambda - 2}\right)^r - \left(\frac{\lambda - 2}{\lambda - 1}\right)^r \frac{\lambda}{\lambda - 1}\right] \leq \left(\frac{\lambda - 1}{\lambda}\right)^r - \left(\frac{\lambda - 2}{\lambda - 1}\right)^r \frac{\lambda}{\lambda - 1}.$$

Since the coefficient of α in this inequality is strictly negative, we can divide both sides by this term and simplify terms to get the equivalent inequality:

$$\alpha \geq \left(\frac{\lambda - 2}{\lambda}\right)^r \frac{(\lambda - 2)^r \lambda^{r+1} - (\lambda - 1)^{2r+1}}{(\lambda - 2)^{2r}\lambda - (\lambda - 3)^r(\lambda - 1)^{r+1}}.$$

Replacing the expression for α, we have

$$\frac{P(G, \lambda)}{P(G'', \lambda)} \geq \frac{(\lambda - 2)^r \lambda^{r+1} - (\lambda - 1)^{2r+1}}{(\lambda - 1)[(\lambda - 2)^r(\lambda - 1)^r - \lambda^r(\lambda - 3)^r]}. \qquad (6)$$

Hence, in order to prove the theorem, we only need show that (6) holds. Now, Lemma 1 implies that

$$\frac{(\lambda - 2)^r \lambda^{r+1} - (\lambda - 1)^{2r+1}}{(\lambda - 1)[(\lambda - 2)^r(\lambda - 1)^r - \lambda^r(\lambda - 3)^r]} \leq \frac{\lambda - 1}{2}.$$

Hence it is sufficient to show that for every $\lambda \geq n + r$,

$$\frac{P(G, \lambda)}{P(G'', \lambda)} \geq \frac{\lambda - 1}{2}.$$

For this purpose, consider any λ colouring of the graph G''. Since G'' has less than λ vertices, this colouring can be extended to a λ colouring of the graph G as follows: give to vertex u (respectively, v) the same colour as that given to the vertex in G'' arising from contracting uv, and give to vertex v (respectively, u)

any of the $\lambda - d_G(u)$ (respectively, $\lambda - d_G(v)$) colours not used by the neighbours of vertex u (respectively, v). In other words,

$$P(G, \lambda) \geq P(G'', \lambda)(\lambda - \min\{d_G(u), d_G(v)\}).$$

Now, by assumption, $\min\{d_G(u), d_G(v)\} \leq \frac{n+r+1}{2}$, and so, since $\lambda \geq n + r$,

$$\lambda - \min\{d_G(u), d_G(v)\} \geq \lambda - \frac{n+r+1}{2} \geq \frac{\lambda - 1}{2},$$

and we are done. The theorem follows.

The previous four theorems give operations for building families of graphs which satisfy Conjecture 1 (or Conjecture 2) from the basic graph O_1.

The following corollary follows immediately from Theorems 2 and 3:

Corollary 1 *Every chordal graph satisfies Conjecture 1.*

In particular, the empty graphs O_n and the trees T_n satisfy Conjecture 1. Moreover, Theorems 3 and 4 can be used to prove the following result:

Theorem 5 *Every series-parallel graph satisfies Conjecture 2.*

Proof. : Let H be a series-parallel graph with m vertices. If m is small then the theorem is obviously true. Hence, we can assume that every series-parallel graph with fewer vertices than H verifies Conjecture 2. Moreover, we can assume that H has no clique-cutset: for otherwise H would arise as clique-identification from two other series-parallel graphs and so we could apply Theorem 3.

Now, by definition, H comes from some other series-parallel graph H' by either duplicating some edge of H' or by subdividing some edge of H'. Since in the first case H will still verify Conjecture 2, we only need show the validity of the theorem when H is constructed from H' by subdividing some edge uv of H'.

Let x be the unique vertex of H that is not a vertex of H'. Set

$$T = \{y \in V(H') : d_{H'}(y) = 2, yu \in E(H'), yv \in E(H')\}.$$

Write $T = \{x_1, \cdots, x_{r-1}\}$, with $r \geq 1$. Let G denote the graph obtained from H' by removing all vertices in T. It follows that H can be built from G by subdividing edge uv with the r vertices $x_1, \cdots, x_{r-1}, x_r$ with $x_r = x$, as shown in Figure 2. Clearly, G is also series-parallel, and so it verifies Conjecture 2. Let n denote the number of vertices of G. Note that H has $n + r$ vertices. Since the graph $G_{|uv}$ is also series-parallel, we can apply Theorem 4. Hence, to prove the theorem, we only need show that

$$\min\{d_G(u), d_G(v)\} \leq \frac{n+r+1}{2}.$$

For this purpose, set

$$A = \{y \in V(G) : yu \notin E(G), yv \notin E(G)\}$$
$$B = \{y \in V(G) : yu \in E(G), yv \in E(G)\}$$
$$C = V(G) - (A \cup B \cup \{u, v\}).$$

Now, if B contains at most one vertex, then $d_G(u) + d_{G'}(v) \leq n + 1$ and we are done. Hence we can assume that B contains at least two vertices. Clearly, B is a stable set in G' (for otherwise, G would contain a K_4).

First, note that:

- no vertex in C is adjacent to some vertex in B.

To see this, assume the contrary: there exists some vertex $z \in C$ which is adjacent to some vertex $y \in B$. Without loss of generality, we can assume that $zu \in E(G)$. Since H has no clique-cutset, it follows that $\{u, y\}$ is not a clique-cutset in G, and so there must exists a path P in $G - \{u, y\}$ joining z to v. But then contracting all edges of $P - \{v\}$, we get a K_4, contradicting the assumption that G is series-parallel.

Next, note that

- every vertex in A is adjacent to at most one vertex in B.

This is obviously true because G is not contractable to K_4.

Since by assumption H and hence G has no clique cutset, every vertex in B is adjacent to some vertex in $A \cup C$, it follows that $|A| \geq |B|$ (recall that no vertex in B is adjacent to some vertex in C), and so $n = 2 + |A| + |B| + |C| \geq 2 + 2|B| + C$. But then $d_G(u) + d_{G'}(v) = |C| + 2|B| + 2 \leq n$, and so $\min\{d_G(u), d_G(v)\} \leq \frac{n}{2}$, and we are done.

References

1. J.E. Bartels, D. Welsh, The Markov Chain of Colourings, In: Lecture Notes of Computer Science 920, Proceedings of the 4th International IPCO Conference (Copenhagen, Denmark) 373–387 (1995).
2. F.M. Dong, Proof of a Chromatic Polynomial Conjecture, to appear on J. of Combin. Theory Ser. B.

Finding Skew Partitions Efficiently[*]

Celina M. H. de Figueiredo[1], Sulamita Klein[1], Yoshiharu Kohayakawa[2], and
Bruce A. Reed[3]

[1] Instituto de Matemática and COPPE, Universidade Federal do Rio de Janeiro,
Brazil. {celina,sula}@cos.ufrj.br
[2] Instituto de Matemática e Estatística, Universidade de São Paulo, Brazil.
yoshi@ime.usp.br
[3] CNRS, Université Pierre et Marie Curie, Institut Blaise Pascal, France.
reed@ecp6.jussieu.fr

Abstract. A skew partition as defined by Chvátal is a partition of the
vertex set of a graph into four nonempty parts A, B, C, D such that
there are all possible edges between A and B, and no edges between C
and D. We present a polynomial-time algorithm for testing whether a
graph admits a skew partition. Our algorithm solves the more general
list skew partition problem, where the input contains, for each vertex,
a list containing some of the labels A, B, C, D of the four parts. Our
polynomial-time algorithm settles the complexity of the original partition
problem proposed by Chvátal, and answers a recent question of Feder,
Hell, Klein and Motwani.

1 Introduction

A *skew partition* is a partition of the vertex set of a graph into four nonempty
parts A, B, C, D such that there are all possible edges between A and B, and
no edges between C and D. We present a polynomial-time algorithm for testing
whether a graph admits a skew partition, as well as for the more general list skew
partition problem, where the input contains, for each vertex, a list containing
some of the four parts.

Many combinatorial problems can be described as finding a partition of the
vertices of a given graph into subsets satisfying certain properties *internally*
(some parts may be required to be independent, or sparse in some other sense,
others may conversely be required to be complete or dense), and *externally* (some
pairs of parts may be required to be completely nonadjacent, others completely
adjacent). In [10], Feder et al. defined a parameterized family of graph problems
of this type.

The basic family of problems they considered is as follows: partition the
vertex set of a graph into k parts $A_1, A_2, \ldots, A_k$ with a fixed "pattern" of requirements as to which A_i are independent or complete and which pairs A_i, A_j

[*] Research partially supported by CNPq, MCT/FINEP PRONEX Project 107/97,
CAPES(Brazil)/COFECUB(France) Project 213/97, FAPERJ, and by FAPESP
Proc. 96/04505-2.

G. Gonnet, D. Panario, and A. Viola (Eds.): LATIN 2000, LNCS 1776, pp. 163–172, 2000.

are completely nonadjacent or completely adjacent. These requirements may be conveniently encoded by a symmetric $k \times k$ matrix M in which the diagonal entry $M_{i,i}$ is 0 if A_i is required to be independent, 2 if A_i is required to be a clique, and 1 otherwise (no restriction). Similarly, the off-diagonal entry $M_{i,j}$ is 0, 1, or 2, if A_i and A_j are required to be completely nonadjacent, have arbitrary connections, or are required to be completely adjacent, respectively. Following [10], we call such a partition an M-partition.

Many combinatorial problems just ask for an M-partition. For instance a k-coloring is an M-partition where M is the adjacency matrix of the complete k-graph, and, more generally, H-coloring (homomorphism to a fixed graph H [13]) is an M-partition where M is the adjacency matrix of H. It is known that H-coloring is polynomial-time solvable when H is bipartite and NP-complete otherwise [13]. When M is the adjacency matrix of H plus twice the identity matrix (all diagonal elements are 2), then M-partitions reduce to the so-called (H, K)-partitions which were studied by MacGillivray and Yu [15]. When H is triangle-free then (H, K)-partition is polynomial-time solvable, otherwise it is NP-complete.

Other well-known problems ask for M-partitions in which all parts are restricted to be nonempty (e.g., skew partitions, clique cutsets, stable cutsets). In yet other problems there are additional constraints, such as those in the definition of a homogeneous set (requiring one of the parts to have at least 2 and at most $n - 1$ vertices). For instance, Winkler asked for the complexity of deciding the existence of an M-partition, where M has the rows $1101, 1110, 0111$, and 1011, such that all parts are nonempty and there is at least one edge between parts A and B, B and C, C and D, and D and A. This has recently been shown NP-complete by Vikas [17].

The most convenient way to express these additional constraints turns out to be to allow specifying for each vertex (as part of the input) a "list" of parts in which the vertex is allowed to be. Specifically, the *list-M-partition problem* asks for an M-partition of the input graph in which each vertex is placed in a part which is in its list. Both the basic M-partition problem ("Does the input graph admit an M-partition?"), and the problem of existence of an M-partition with all parts nonempty, admit polynomial-time reductions to the list-M-partition problem, as do all of the above problems with the "additional" constraints. List partitions generalize list-colorings, which have proved very fruitful in the study of graph colorings [1, 12]. They also generalize list-homomorphisms which were studied earlier [7, 8, 9].

Feder et al. [10] were the first to introduce and investigate the list version of these problems. It turned out to be a useful generalization, since list problems recurse more conveniently. This enabled them to classify the complexity (as polynomial-time solvable or NP-complete) of list-M-partition problems for all 3×3 matrices M and some 4×4 matrices M. For other 4×4 matrices M they were able to produce sub-exponential algorithms - including one for the skew partition problem described below. This was the first sub-exponential algorithm for the problem, and an indication that the problem is not likely to be NP-

complete. We were motivated by their approach, and show that in fact one can use the mechanism of list partitions to give a *polynomial-time* algorithm for the problem.

A *skew partition* is an M-partition, where M has the rows $1211, 2111, 1110$, and 1101, such that all parts are nonempty. List Skew Partition (LSP) is simply the list-M-partition problem for this M. We can solve skew partition by solving at most n^4 LSP problems such that $v_i \in A_i$ for $1 \leq i \leq 4$, for all possible quadruples $\{v_1, v_2, v_3, v_4\}$ of vertices of the input graph.

The skew partition problem has interesting links to perfect graphs, and is one of the main problems in the area. Before presenting our two algorithms we discuss perfect graphs and their link to skew partition.

2   Skew Cutsets and the Strong Perfect Graph Conjecture

A graph is *perfect* if each induced subgraph admits a vertex colouring and a clique of the same size. A graph is *minimal imperfect* if it is not perfect but all of its proper induced subgraphs are perfect. Perfect graphs were first defined by Berge [2] who was interested in finding a good characterization of such graphs. He proposed the *strong perfect graph conjecture*: the only minimal imperfect graphs are the odd chordless cycles of length at least five and their complements. Since then researchers have enumerated a list of properties of minimal imperfect graphs. The strong perfect graph conjecture remains open and is considered a central problem in computational complexity, combinatorial optimization, and graph theory.

Chvátal [4] proved that no minimal imperfect graph contains a structure that he called a *star cutset*: a vertex cutset consisting of a vertex and some of its neighbours. Chvátal exhibited a polynomial-time recognition algorithm for graphs with a star cutset. He also conjectured that no minimal imperfect graph contains a *skew partition*. Recalling our earlier definition, a skew partition is a partition of the vertex set of a graph into four nonempty parts A, B, C, D such that there are all possible edges between A and B, and no edges between C and D. We call each of the four nonempty parts A, B, C, D a *skew partition set*. We say that $A \cup B$ is a *skew cutset*. The complexity of testing for the existence of a skew cutset has motivated many publications [5, 10, 14, 16]. Recently, Feder et al. [10] described a quasi-polynomial algorithm for testing whether a graph admits a skew partition, which strongly suggested that this problem was not NP-complete. In this paper, we present a polynomial-time recognition algorithm for testing whether a graph admits a skew partition.

Cornuéjols and Reed [5] proved that no minimal imperfect graph contains a skew partition in which A and B are both stable sets. Actually, they proved the following more general result. Let a *complete multi-partite* graph be one whose vertex set can be partitioned into stable sets $S_1, \ldots, S_k$, such that there are all possible edges between S_i and S_j, for $i \neq j$. They proved that no minimal imperfect graph contains a skew cutset that induces a complete multi-partite graph. Their work raised questions about the complexity of testing for the existence either of a complete bipartite cutset or of a complete multi-partite cutset in a graph.

Subsequently, Klein and de Figueiredo [16] showed how to use a result of Chvátal [3] on matching cutsets in order to establish the NP-completeness of recognizing graphs with a stable cutset. In addition, they established the NP-completeness of recognizing graphs with a complete multi-partite cutset. In particular, their proof showed that it is NP-complete to test for the existence of a complete bipartite cutset, even if the cutset induces a $K_{1,p}$.

As shown by Chvátal [4], to test for the existence of a star cutset is in P, whereas to test for the existence of the special star cutset $K_{1,p}$ is NP-complete, as shown in [16]. The polynomial-time algorithm described in this paper offers an analogous complexity situation: to test for the existence of a skew cutset is in P, whereas to test for the existence of a complete bipartite cutset is NP-complete [16].

3   Overview

The goal of this paper is to present a polynomial-time algorithm for the following decision problem:

Skew Partition Problem
Input: a graph $G = (V, E)$.
Question: Is there a skew partition A, B, C, D of G?

We actually consider list skew partition (LSP) problems, stated as decision problems as follows:

List Skew Partition Problem
Input: a graph $G = (V, E)$ and for each vertex $v \in V$, a subset L_v of $\{A, B, C, D\}$.
Question: Is there a skew partition A, B, C, D of G such that each v is contained in some element of the corresponding L_v?

Throughout the algorithm we have a partition of V into at most 15 sets indexed by the nonempty subsets of $\{A, B, C, D\}$, i.e., $\{S_L | L \subseteq \{A, B, C, D\}\}$, such that Property 1 is always satisfied. For convenience, we denote $S_{\{A\}}$ by S_A. Note that the relevant inputs for LSP have S_A, S_B, S_C, and S_D nonempty.

Property 1. If the algorithm returns a skew partition, then if v is in S_L, then the returned skew partition set containing v is in L.

Initially, we set $S_L = \{v | L_v = L\}$.

We also restrict our attention to LSP instances which satisfy the following property:

Property 2. If $v \in S_L$, for some L with $A \in L$, then it sees every vertex of S_B. If $v \in S_L$, for some L with $B \in L$, then it sees every vertex of S_A. If $v \in S_L$, for some L with $C \in L$, then it is non-adjacent to every vertex of S_D. If $v \in S_L$, for some L with $D \in L$, then it is non-adjacent to every vertex of S_C.

Both Properties 1 and 2 hold throughout the algorithm.

Remark 1. Since S_B must be contained in B, we know that if v is to be in A for some solution to the problem, then v must see all of S_B. Thus if some $v \in S_A$ misses a vertex of S_B, then there is no solution to the problem and we need not continue. If there is some L with A properly contained in L and a vertex v in S_L which misses a vertex of S_B, then we know in any solution to the problem v must be contained in some element of $L \setminus A$. So we can reduce to a new problem where we replace S_L by $S_L \setminus v$, we replace $S_{L \setminus A}$ by $S_{L \setminus A} + v$ and all other S_L are as before. Such a reduction reduces $\sum_L |S_L||L|$ by 1. Since this sum is at most $4n$, after $O(n)$ similar reductions we must obtain an LSP problem satisfying Property 2 (or halt because the original problem has no solution).

In our discussion we often create new LSP instances and whenever we do so, we always perform this procedure to reduce to an LSP problem satisfying Property 2.

For an instance I of LSP we have $\{S_L(I)|L \subseteq \{A, B, C, D\}$, but we drop the (I) when it is not needed for clarity.

We will consider a number of restricted versions of the LSP problems:

- MAX-3-LSP: an LSP problem satisfying Property 2 such that $S_{ABCD} = \emptyset$;
- MAX-2-LSP: an LSP problem satisfying Property 2 such that if $|L| > 2$, then $S_L = \emptyset$;
- AC-TRIV LSP: an LSP problem satisfying Property 2 such that $S_{AC} = \emptyset$;

  Remark 2. It is easy to obtain a solution to an instance of AC-TRIV-LSP as follows: $A = S_A$, $B = \bigcup_{B \in L} S_L$, $C = S_C$, and $D = \bigcup_{D \in L, B \notin L} S_L$. By Property 2 this is indeed a skew partition.

- BD-TRIV LSP, AD-TRIV LSP, BC-TRIV-LSP. These problems are defined and solved similarly as AC-TRIV LSP.

Our algorithm for solving LSP requires four subalgorithms which replace an instance of LSP by a polynomial number of instances of more restricted versions of LSP.

Algorithm 1 *Takes an instance of LSP and returns in polynomial time a list $\mathcal{L}$ of a polynomial number of instances of MAX-3-LSP such that*

(i) a solution to any problem in $\mathcal{L}$ is a solution of the original problem, and
(ii) if none of the problems in $\mathcal{L}$ have a solution, then the original problem has no solution.

Algorithm 2 *Takes an instance of MAX-3-LSP and returns in polynomial time a list $\mathcal{L}$ of a polynomial number of instances of MAX-3-LSP such that:*
(i) and (ii) of Algorithm 1 hold, and

(iii) for each problem in $\mathcal{L}$, either $S_{ABC} = \emptyset$ or $S_{ABD} = \emptyset$.

Algorithm 3 *Takes an instance of MAX-3-LSP and returns in polynomial time a list $\mathcal{L}$ of a polynomial number of instances of MAX-3-LSP such that:*
(i) and (ii) of Algorithm 1 hold, and

(iii) for each problem in $\mathcal{L}$, either $S_{BCD} = \emptyset$ or $S_{ACD} = \emptyset$.

Algorithm 4 *Takes an instance of MAX-3-LSP such that*

(a) either S_{ABC} or S_{ABD} is empty, and
(b) either S_{BCD} or S_{ACD} is empty

and returns a list $\mathcal{L}$ of a polynomial number of problems each of which is an instance of one of MAX-2-LSP, AC-TRIV LSP, AD-TRIV LSP, BC-TRIV LSP or BD-TRIV LSP such that (i) and (ii) of Algorithm 1 hold.

We also need two more algorithms for dealing with the most basic instances of LSP.

Algorithm 5 *Takes an instance of MAX-2-LSP and returns either*

(i) a solution to this instance of MAX-2-LSP, or
(ii) the information that this problem instance has no solution.

Remark 3. Algorithm 5 simply applies 2-SAT as discussed in [6]; we omit the details.

Algorithm 6 *Takes an instance of AC-TRIV LSP or AD-TRIV LSP or BC-TRIV LSP or BD-TRIV LSP and returns a solution using the partitions discussed in the Remark 2.*

To solve an instance of LSP we first apply Algorithm 1 to obtain a list L_1 of instances of MAX-3-LSP. For each problem instance I on L_1, we apply Algorithm 2 and let L_I be the output list of problem I. We let L_2 be the concatenation of the lists $\{L_I | I \in L_1\}$. For each I in L_2, we apply Algorithm 3. Let L_3 be the concatenation of the lists $\{L_I | I \in L_2\}$. For each problem instance I on L_3, we apply Algorithm 4. Let L_4 be the concatenation of the lists $\{L_I | I \in L_3\}$. Each element of L_4 can be solved using either Algorithm 5 or Algorithm 6 in polynomial time. If any of these problems has a solution S, then by the specifications of the algorithms, S is a solution to the original problem. Otherwise, by the specifications of the algorithms, there is no solution to the original problem. Clearly, the whole algorithm runs in polynomial time.

4   Algorithm 1

We now present Algorithm 1. The other algorithms are similar and their details are left for a longer version of this paper. The full details and proofs are in the technical report [11].

Algorithm 1 recursively applies Procedure 1 which runs in polynomial time.

Procedure 1 Input: *An instance I of LSP.*
Output: *Four instances I_1, I_2, I_3, I_4 of LSP such that, for $1 \le j \le 4$, we have*
$|S_{ABCD}(I_j)| \le \frac{9}{10}|S_{ABCD}(I)|$.

It is easy to prove inductively that recursively applying Procedure 1 yields a polynomial time implementation of Algorithm 1 which when applied to an input graph with n vertices creates as output a list $\mathcal{L}$ of instances of LSP such that
$|\mathcal{L}| \le 4^{\log_{\frac{10}{9}} n} \le n^{14}$.

Let $n = |S_{ABCD}(I)|$. For any skew partition $\{A, B, C, D\}$, let $A' = A \cap S_{ABCD}(I)$, $B' = B \cap S_{ABCD}(I)$, $C' = C \cap S_{ABCD}(I)$, and $D' = D \cap S_{ABCD}(I)$.

Case 1: There exists a vertex v in S_{ABCD} such that $\frac{n}{10} \le |S_{ABCD} \cap N(v)| \le \frac{9n}{10}$.
Branch according to whether $v \in A$, $v \in B$, $v \in C$, or $v \in D$ with instances I_A, I_B, I_C, I_D respectively. We define I_A by initially setting $S_A(I_A) = v + S_A(I)$ and reducing so that Property 2 holds. We define I_B, I_C, I_D similarly. We note that by Property 2, if $v \in C$, then $D \cap N(v) = \emptyset$. So, $S_{ABCD}(I_C) \subset S_{ABCD}(I) \setminus N(v)$.

Because there are at least $\frac{n}{10}$ vertices in $S_{ABCD} \cap N(v)$, this means that $|S_{ABCD}(I_C)| \le \frac{9n}{10}$.

Symmetrically, $|S_{ABCD}(I_D)| \le \frac{9n}{10}$.

Similarly, by Property 2, $S_{ABCD}(I_A) \subset S_{ABCD}(I) \cap N(v)$, so $|S_{ABCD}(I_A)| \le \frac{9n}{10}$. Symmetrically $|S_{ABCD}(I_B)| \le \frac{9n}{10}$.   □

Case 2: There are at least $\frac{n}{10}$ vertices v in S_{ABCD} such that $|S_{ABCD} \cap N(v)| < \frac{n}{10}$ and there are at least $\frac{n}{10}$ vertices v in S_{ABCD} such that $|S_{ABCD} \cap N(v)| > \frac{9n}{10}$.
Let $W = \{v \in S_{ABCD} : |S_{ABCD} \cap N(v)| > \frac{9n}{10}\}$ and $X = \{v \in S_{ABCD} : |S_{ABCD} \cap N(v)| < \frac{n}{10}\}$. Branch according to $|A'| \ge \frac{n}{10}$, or $|B'| \ge \frac{n}{10}$, or $|C'| \ge \frac{n}{10}$, or $|D'| \ge \frac{n}{10}$ with corresponding instances $I_{A'}, I_{B'}, I_{C'}$ and $I_{D'}$. Each of these choices forces either all the vertices in W or all the vertices in X to have smaller label sets, as follows. If $|A'| \ge \frac{n}{10}$, then every vertex in B has $\frac{n}{10}$ neighbours in $S_{ABCD}(I)$, so $B \cap X = \emptyset$. Thus, $S_{ABCD}(I_{A'}) = S_{ABCD}(I) \setminus X$, and $|S_{ABCD}(I_{A'})| \le \frac{9n}{10}$. If $|B'| \ge \frac{n}{10}$, then a symmetrical argument shows that $X \cap A = \emptyset$. Thus, $S_{ABCD}(I_{B'}) = S_{ABCD}(I) \setminus X$, and $|S_{ABCD}(I_{B'})| \le \frac{9n}{10}$. If $|C'| \ge \frac{n}{10}$, then every vertex in D has at least $\frac{n}{10}$ non-neighbours in $S_{ABCD}(I)$. Hence $W \cap D = \emptyset$, $S_{ABCD}(I_{C'}) = S_{ABCD}(I) \setminus W$, and so $|S_{ABCD}(I_{C'})| \le \frac{9n}{10}$. If $|D'| \ge \frac{n}{10}$, then a symmetrical argument shows that $|S_{ABCD}(I_{D'})| \le \frac{9n}{10}$.   □

Case 3: There are over $\frac{9n}{10}$ vertices in W.

We will repeatedly apply the following procedure to various $W' \subseteq W$ with $|W'| \geq \frac{8n}{10}$. We recursively define a partition of W' into three sets O, T, and NT such that:

- There are all edges between O and T;
- For every w in NT, there exists v in O such that w misses v;
- The complement of O is connected.

Start by choosing v_1 in W' and setting: $O = \{v_1\}$, $T = N(v_1) \cap W'$, and $NT = W' \setminus (N(v_1) \cup \{v_1\})$. Note that for each vertex v of W', since v misses at most $\frac{n}{10}$ vertices of S_{ABCD}, $|N(v) \cap W'| > |W'| - \frac{n}{10}$. So $|NT| = |W' \setminus (N(v_1) \cup \{v_1\})| < \frac{n}{10}$. Grow O by moving an arbitrary vertex v from NT to O, and by moving $T \setminus N(v)$ from T to NT until:

(i) $|O| + |NT| \geq \frac{n}{10}$; or
(ii) $NT = \emptyset$.

If the growing process stops with condition (i), i.e., $|O| + |NT| \geq \frac{n}{10}$, and v_i was the last vertex added to O, then adding v_i to O increased $|NT|$ by at most $|W' \setminus (N(v_i) \cup \{v_i\})| < \frac{n}{10}$. Thus, $|O| + |NT| < \frac{n}{10} + \frac{n}{10} = \frac{n}{5}$. So, $|T| \geq \frac{8n}{10} - \frac{n}{5} = \frac{6n}{10} \geq \frac{n}{10}$.

Our first application of the procedure is with $W' = W$. If we stop because (i) holds, then we define four new instances of LSP according to the intersection of the skew partition sets A, B, C or D with O, as follows:

(a) $I_1 : C \cap O \neq \emptyset$,
(b) $I_2 : C \cap O = \emptyset$, $D \cap O \neq \emptyset$,
(c) $I_3 : O \subseteq A$,
(d) $I_4 : O \subseteq B$.

Recall that the complement of O is connected, which implies that if $O \cap (C \cup D) = \emptyset$, then either $O \subseteq A$ or $O \subseteq B$. If $O \subseteq A$, then $NT \cap B = \emptyset$ since $(\forall w \in NT)(\exists v \in O)$ such that $vw \notin E$. Thus, $(O \cup NT) \cap S_{ABCD}(I_3) = \emptyset$. Hence $|S_{ABCD}(I_3)| \leq \frac{9n}{10}$. A symmetrical argument shows that $|S_{ABCD}(I_4)| \leq \frac{9n}{10}$.

If $C \cap O \neq \emptyset$, then $D \cap T = \emptyset$. Thus, $T \cap S_{ABCD}(I_1) = \emptyset$, which implies $|S_{ABCD}(I_1)| \leq \frac{9n}{10}$. A symmetrical argument shows that $|S_{ABCD}(I_2)| \leq \frac{9n}{10}$. Thus, if our application of the above procedure halts with output (i), then we have found the four desired output instances of LSP.

Otherwise, the growing process stops with condition (ii), i.e., $NT = \emptyset$ and $|O| < \frac{n}{10}$. Set $O_1 = O$ and reapply the algorithm to $W' = W \setminus O_1$ to obtain O_2. More generally, having constructed disjoint sets $O_1, \ldots, O_i$ with $|\cup_{j=1}^i O_j| < n/10$, we construct O_{i+1} by applying the algorithm to $W_i = W \setminus \cup_{j=1}^i O_j$. Note $|W_i| > \frac{8n}{10}$.

We continue until $|\cup_{j=1}^i O_j| \geq \frac{n}{10}$ or condition (i) occurs. If condition (i) ever occurs, then we proceed as above. Otherwise, we stop after some iteration i^* such that $|\cup_{i < i^*} O_i| < \frac{n}{10}$ and $|\cup_{i \leq i^*} O_i| \geq \frac{n}{10}$. Since $|O_{i^*}| < \frac{n}{10}$, we have that

$|\cup_{i \leq i^*} O_i| \leq \frac{n}{5}$. Also, all the edges between sets $Z = \cup_{j=1}^{i} O_j$ and $Y = W \backslash \cup_{j=1}^{i} O_j$ exist, which implies that $C \cap Z = \emptyset$ or $D \cap Y = \emptyset$.

We now define two new instances of LSP according to the intersection of skew partition sets C or D with Z, as follows:

(a) I_1: $C \cap Z = \emptyset$,
(b) I_2: $D \cap Y = \emptyset$.

In either output instance I_i, $|S_{ABCD}(I_i)| \leq \frac{9n}{10}$.   □

Note that the case $|X| > \frac{9n}{10}$ is symmetric to Case 3 (consider $\overline{G}$) and is omitted.

5   Concluding Remarks

It is evident to the authors that the techniques we have developed will apply to large classes of list-M-partition problems. We intend to address this in future work.

References

1. N. Alon, and M. Tarsi. Colorings and orientations of graphs. *Combinatorica* **12** (1992) 125–134.
2. C. Berge. Les problèmes de coloration en théorie des graphes. *Publ. Inst. Stat. Univ. Paris* **9** (1960) 123–160.
3. V. Chvátal. Recognizing decomposable graphs. *J. Graph Theory* **8** (1984) 51–53.
4. V. Chvátal. Star-Cutsets and perfect graphs. *J. Combin. Theory Ser. B* **39** (1985) 189–199.
5. G. Cornuéjols, and B. Reed. Complete multi-partite cutsets in minimal imperfect graphs. *J. Combin. Theory Ser. B* **59** (1993) 191–198.
6. H. Everett, S. Klein, and B. Reed. An optimal algorithm for finding clique-cross partitions. *Congr. Numer.* **135** (1998) 171–177.
7. T. Feder, and P. Hell. List homomorphisms to reflexive graphs. *J. Combin. Theory Ser. B* **72** (1998) 236–250.
8. T. Feder, P. Hell, and J. Huang. List homomorphisms to bipartite graphs. Manuscript.
9. T. Feder, P. Hell, and J. Huang. List homomorphisms to general graphs. Manuscript.
10. T. Feder, P. Hell, S. Klein, and R. Motwani. Complexity of graph partition problems. *Proceedings of the thirty-first ACM Symposium on Theory of Computing, STOC'99* (1999) 464–472.
11. C. M. H. de Figueiredo, S. Klein, Y. Kohayakawa, and B. Reed. Finding Skew Partitions Efficiently. Technical Report **ES-503/99**, COPPE/UFRJ, Brazil. Available at ftp://ftp.cos.ufrj.br/pub/tech_reps/es50399.ps.gz.
12. H. Fleischner, and M. Stiebitz. A solution of a coloring problem of P. Erdös. *Discrete Math.* **101** (1992) 39–48.
13. P. Hell, and J. Nešetřil. On the complexity of H-coloring. *J. Combin. Theory Ser. B* **48** (1990) 92–110.

14. C .T. Hoàng. *Perfect Graphs*. Ph.D. Thesis, School of Computer Science, McGill University, Montreal, (1985).

15. G. MacGillivray, and M.L. Yu. Generalized partitions of graphs. *Discrete Appl. Math.* **91** (1999) 143–153.

16. S. Klein, and C. M. H. de Figueiredo. The NP-completeness of multi-partite cutset testing. *Congr. Numer.* **119** (1996) 216–222.

17. N. Vikas. Computational complexity of graph compaction. *Proceedings of the tenth ACM-SIAM Symposium on Discrete Algorithms* (1999) 977–978.

On the Competitive Theory and Practice of Portfolio Selection

(Extended Abstract)

Allan Borodin[1], Ran El-Yaniv[2], and Vincent Gogan[3]

[1] Department of Computer Science, University of Toronto
bor@cs.toronto.edu
[2] Department of Computer Science, Technion
rani@cs.technion.ac.il
[3] Department of Computer Science, University of Toronto
vincent@cs.toronto.edu

Abstract. Given a set of say m stocks (one of which may be "cash"), the online portfolio selection problem is to determine a portfolio for the ith trading period based on the sequence of prices for the preceding $i - 1$ trading periods. Competitive analysis is based on a worst case perspective and such a perspective is inconsistent with the more widely accepted analyses and theories based on distributional assumptions. The competitive framework does (perhaps surprisingly) permit non trivial upper bounds on relative performance against CBAL-OPT, an optimal offline constant rebalancing portfolio. Perhaps more impressive are some preliminary experimental results showing that certain algorithms that enjoy "respectable" competitive (i.e. worst case) performance also seem to perform quite well on historical sequences of data. These algorithms and the emerging competitive theory are directly related to studies in information theory and computational learning theory and indeed some of these algorithms have been pioneered within the information theory and computational learning communities. We present a mixture of both theoretical and experimental results, including a more detailed study of the performance of existing and new algorithms with respect to a standard sequence of historical data cited in many studies. We also present experiments from two other historical data sequences.

1 Introduction

This paper is concerned with the *portfolio selection (PS)* problem, defined as follows. Assume a market with m securities. The securities can be stocks, bonds, currencies, commodities, etc. For each trading day $i \geq 0$, let $\mathbf{v}_i = (v_{i,1}, v_{i,1}, \ldots, v_{i,m})$ be the price vector for the ith period, where $v_{i,j}$, the price or value of the jth security, is given in the "local" currency, called here *cash* or *dollars*. For analysis it is often more convenient to work with *relative prices* rather than prices. Define $x_{i,j} = v_{i,j}/v_{i-1,j}$ to be the relative price of the jth security corresponding to the ith period.[1] Denote by

---

[1] Here we are greatly simplifying the nature of the market and assuming that $x_{i+1,j}$ is the ratio of opening price on the $i + 1^{st}$ day to the opening price on the ith day. That is, we are assuming that a trader can buy or sell at the opening price. Later we try to compensate for this by incorporating bid-ask spreads into transaction costs.

G. Gonnet, D. Panario, and A. Viola (Eds.): LATIN 2000, LNCS 1776, pp. 173–196, 2000.

$\mathbf{x}_i = (x_{i,1}, \ldots, x_{i,m})$ the *market vector* of relative prices corresponding to the ith day. A *portfolio* $\mathbf{b}$ is specified by the proportions of current dollar wealth invested in each of the securities

$$\mathbf{b} = (b_1, \ldots, b_m), \qquad b_j \leq 1, \qquad \sum b_j = 1 .$$

The return of a portfolio $\mathbf{b}$ w.r.t. a market vector $\mathbf{x}$ is $\mathbf{b} \cdot \mathbf{x} = \sum_j b_j x_j$. The (compound) *return* of a sequence of portfolios $B = \mathbf{b}_1, \ldots, \mathbf{b}_n$ w.r.t. a market sequence $X = \mathbf{x}_1, \ldots, \mathbf{x}_n$ is

$$R(B, X) = \prod_{i=1}^{n} \mathbf{b}_i \cdot \mathbf{x}_i .$$

A PS algorithm is any deterministic or randomized rule for specifying a sequence of portfolios. If ALG is a deterministic (respectively, randomized) PS algorithm then its (expected) return with respect to a market sequence X is denoted by ALG(X).

The basic PS problem described here ignores several important factors such as transaction commissions, buy-sell spreads and risk tolerance and control.

A simple strategy, advocated by many financial advisors, is to simply divide up the amount of cash available and to *buy and hold* a portfolio of the securities. This has the advantage of minimizing transaction costs and takes advantage of the natural tendency for the market to grow. In addition, there is a classical algorithm, due to Markowitz [Mar59], for choosing the weightings of the portfolio so as to minimize the variance for any target expected return.

An alternative approach to portfolio management is to attempt to take advantage of volatility (exhibited in price fluctuations) and to actively trade on a "day by day" basis. Such trading can sometimes lead to returns that dramatically outperform the performance of the best security.

For example, consider the class of *constant rebalanced* algorithms. An algorithm in this class, denoted CBAL$_\mathbf{b}$ is specified by a fixed portfolio $\mathbf{b}$ and maintains a constant weighting (by value) amongst the securities. Thus, at the beginning of each trading period CBAL$_\mathbf{b}$ rebalances its portfolio so that it is $\mathbf{b}$-balanced. The constant rebalanced algorithms are motivated by several factors. In particular, it can be shown that the optimal offline algorithm in this class, CBAL-OPT, can lead to exponential returns that dramatically outperform the best stock (see e.g. [Cov91]).

One objective in studying the portfolio selection problem is to arrive at online trading strategies that are *guaranteed, in some sense*, to perform well. What is the choice of performance measure? We focus on a *competitive analysis* framework whereby the performance of an online portfolio selection strategy is compared to that of a benchmark algorithm on every input. One reasonable benchmark is the return provided by the best stock. For more active strategies, an optimal algorithm (that has complete knowledge of the future) could provide returns so extreme that any reasonable approach is doomed when viewed in comparison. Specifically, any online algorithm competing against the optimal offline algorithm, called OPT is at best m^n-competitive where n is the number of trading days and m is the number of stocks.

In the more classical (and perhaps most influential) approach, the PS problem is broken down into two stages. First, one uses statistical assumptions and historical data to create a model of stock prices. After this the model is used to predict future price movements. The technical difficulties encountered in the more traditional approach (i.e. formulating a realistic yet tractable statistical model) motivates competitive analysis. The competitive analysis approach starts with minimal assumptions, derives algorithms within this worst case perspective and then perhaps adds statistical or distributional assumptions as necessary (to obtain analytical results and/or to suggest heuristic improvements to the initial algorithms). It may seem unlikely that this approach would be fruitful but some interesting results have been proven. In particular, Cover's *Universal Portfolio* algorithm [Cov91] was proven to possess important theoretical properties. We are mainly concerned with competitive analyses against CBAL-OPT and say that a portfolio selection algorithm ALG is c-competitive (w.r.t. CBAL-OPT) if the supremum, over all market sequences X, of the ratio CBAL-OPT$(X)/$ALG(X) is less than or equal c.

Instead of looking at ALG's competitive ratio we can equivalently measure the degree of "universality" of ALG. Following Cover [Cov91], we say that ALG is *universal* if for all X,

$$\frac{\log \text{CBAL-OPT}(X)}{n} - \frac{\log \text{ALG}(X)}{n} \to 0 \ .$$

In this sense, the "regret" experienced by an investor that uses a universal online algorithm approaches zero as time goes on. Clearly the rate at which this regret approaches zero corresponds to the competitive ratio and ALG is universal if and only if its competitive ratio is $2^{o(n)}$. One motivation for measuring performance by universality is that it corresponds to the minimization of the regret, using a logarithmic utility function (see [BE98]). On the other hand, it obscures the convergence rate and therefore we prefer to use the competitive ratio. When the competitive ratio of a PS algorithm (against CBAL-OPT) can be bounded by a polynomial in n (for a fixed number of stocks), we shall say that the algorithm is *competitive*.

2 Some Classes of PS Algorithms

2.1 Buy-And-Hold (BAH) Algorithms

The simplest portfolio selection policy is buy-and-hold (BAH): Invest in a particular portfolio and let it sit for the entire duration of the investment. Then, in the end, cash the portfolio out of the market. The optimal offline algorithm, BAH-OPT, invests in the best performing stock for the relevant period. Most investors would probably consider themselves to be very successful if they were able to achieve the return of BAH-OPT.

2.2 Constant-Rebalanced (CBAL) Algorithms

The constant-rebalanced (CBAL) algorithm CBAL$_\mathbf{b}$ has an associated fixed portfolio $\mathbf{b} = (b_1, \ldots, b_m)$ and operates as follows: at the beginning of each trading period it makes trades so as to rebalance its portfolio to $\mathbf{b}$ (that is, a fraction b_i is invested in the ith stock, $i = 1, \ldots, m$). It is easy to see that the return of CBAL-OPT is bounded from below by the return of BAH-OPT since every BAH strategy can be thought of as an extremal CBAL algorithm. It has been empirically shown that in real market sequences, the return of CBAL-OPT can dramatically outperform the best stock (see e.g. Table 3).

Example 1 (Cover and Gluss [CG86], Cover [Cov91]). Consider the case $m = 2$ with one stock and cash. Consider the market sequence

$$X = \begin{pmatrix} 1 \\ 1/2 \end{pmatrix}, \begin{pmatrix} 1 \\ 2 \end{pmatrix}, \begin{pmatrix} 1 \\ 1/2 \end{pmatrix}, \begin{pmatrix} 1 \\ 2 \end{pmatrix}, \ldots$$

$$\text{CBAL}_{(\frac{1}{2},\frac{1}{2})}(X) = \left[\left(\frac{1}{2} \cdot \left(1 + \frac{1}{2}\right)\right) \left(\frac{1}{2} \cdot (1 + 2)\right) \right]^{n/2}$$

$$= \left[\frac{3}{4} \cdot \frac{3}{2} \right]^{n/2} = \left(\frac{9}{8} \right)^{n/2} .$$

Thus, for this market, the return of $\text{CBAL}_{(\frac{1}{2},\frac{1}{2})}$ is exponential in n while the best stock is moving nowhere.

Under the assumption that the daily market vectors are identically and independently and identically distributed (i.i.d), there is yet another motivating reason for considering CBAL algorithms.

Theorem 1 (Cover and Thomas [CT91]). *Let $X = \mathbf{x}_1, \ldots, \mathbf{x}_n$ be i.i.d. according to some distribution $F(\mathbf{x})$. Then, for some $\mathbf{b}$, $\text{CBAL}_\mathbf{b}$ performs at least as good (in the sense of expected return) as the best online PS algorithm.*

Theorem 1 tells us that it is sufficient to look for our best online algorithm in the set of CBAL algorithms, provided that the market is generated by an i.i.d. source. One should keep in mind, however, that the i.i.d. assumption is hard to justify. (See [Gre72,BCK92] for alternative theories.)

2.3 Switching Sequences (Extremal Algorithms)

Consider any sequence of stock indices,

$$J^{(n)} = j_1, j_2, \ldots, j_n, \qquad j_i \in \{1, \ldots, m\} .$$

This sequence prescribes a portfolio management policy that switches its entire wealth from stock j_i to stock j_{i+1}. An algorithm for generating such sequences can be deterministic or randomized. Ordentlich and Cover [OC96] introduce switching sequence algorithms (called extremal strategies). As investment strategies, switching sequences may seem to be very speculative but from a theoretical perspective they are well motivated. In particular, for any market sequence, the true optimal algorithm (called OPT) is a switching sequence[2] .

Example 2. Consider the following market sequence X:

$$\begin{pmatrix} 0 \\ 2 \end{pmatrix}, \begin{pmatrix} 0 \\ 2 \end{pmatrix}, \begin{pmatrix} 0 \\ 2 \end{pmatrix}, \begin{pmatrix} 0 \\ 2 \end{pmatrix}, \begin{pmatrix} 2 \\ 0 \end{pmatrix}, \begin{pmatrix} 2 \\ 0 \end{pmatrix}, \begin{pmatrix} 2 \\ 0 \end{pmatrix}, \begin{pmatrix} 2 \\ 0 \end{pmatrix}$$

Starting with $1, the switching sequence $2, 2, 2, 2, 1, 1, 1, 1$ returns $256. In contrast, any (mixture of) buy and holds will go bankrupt on X returning $0 and for all $\mathbf{b}$, $\text{CBAL}_\mathbf{b}(X) \leq \text{CBAL}_{(\frac{1}{2},\frac{1}{2})}(X) = \text{CBAL-OPT}(X) = \$1$.

---

[2] Here we are assuming either no transaction costs or a simple transaction cost model such as a fixed percentage commission.

3   Some Basic Properties

In this section we derive some basic properties concerning PS algorithms.

3.1   Kelly Sequences

A *Kelly* (or a "horse race") market vector for m stocks is a vector of the form

$$\underbrace{(0,0,1,0,0,0,0,0,0,0,0,0)}_{m} \ .$$

That is, except for one stock that retains its value, all stocks crash.

Let $\mathcal{K}_m^n$ be the set of all length n Kelly market sequences over m stocks. (When either m and/or n is clear from the context it will be omitted.) There are m^n Kelly market sequences of length n. Kelly sequences were introduced by Kelly [Kel56] to model horse race gambling. We can use them to derive lower bounds for online PS algorithms (see e.g. [CO96] and Lemma 6 below).

The following simple but very useful lemma is due to Ordentlich and Cover [OC96].

Lemma 1. *Let* ALG *be any online algorithm. Then*

$$\sum_{K \in \mathcal{K}_m^n} \text{ALG}(K) = 1 \ .$$

Proof. We sketch the proof for the case $m = 2$. The proof is by induction on n. For the base case, $n = 1$, notice that ALG must specify its first portfolio, $(b, 1 - b)$, before the (Kelly) market sequence (in this case, of length one) is presented. The two possible Kelly sequences are $K_1 = \binom{0}{1}$ and $K_2 = \binom{1}{0}$. Therefore, $\text{ALG}(K_1) + \text{ALG}(K_2) = (1-b) + b = 1$. The induction hypothesis states that the lemma holds for $n - 1$ days. The proof is complete when we add up the two returns corresponding to the two possible Kelly vectors for the first day.

$\square$

Lemma 1 permits us to relate CBALS to online switching sequences as shown in the next two lemmas.

Lemma 2. *Let* ALG *be any PS algorithm. There exists an algorithm* ALG$'$, *which is a mixture of switching sequences, that is equivalent (in terms of expected return) to* ALG, *over Kelly market sequences.*

Proof. Fix n. By Lemma 1 we have $\sum_{K_\ell \in \mathcal{K}^n} \text{ALG}(K_\ell) = 1$ with $\text{ALG}(K_\ell) \geq 0$ for all $K_\ell \in \mathcal{K}^n$. Therefore, $\{\text{ALG}(K_\ell)\}_\ell$ is a probability distribution. For a sequence of Kelly market vectors $K = k_1, k_2, \ldots, k_n$, denote by $S_K = s(k_1), s(k_2), \ldots, s(k_n)$ the switching sequence where $s(k_i)$ is the index of the stock with relative price one in k_i. Let ALG$' = \sum_{K_\ell \in \mathcal{K}^n} \text{ALG}(K_\ell) \cdot S_{K_\ell}$ be the mixture of switching sequences that assigns a weight $\text{ALG}(K_\ell)$ to the sequence S_{K_ℓ}. Clearly, for each Kelly market sequence K we have $\text{ALG}'(K) = \text{ALG}(K)$.

Lemma 2 can be extended in the following way.

Lemma 3. *If* ALG $=$ CBAL$_\mathbf{b}$ *is a constant-rebalanced algorithm then (for known n) we can achieve the same return as* CBAL$_\mathbf{b}$ *on any market sequence using the same method. That is, use the mixture* ALG$' = \sum_{K_\ell \in \mathcal{K}^n}$ CBAL$_\mathbf{b}(K_\ell) \cdot S_{K_\ell}$.

Proof. To prove this, consider any $\mathbf{b} = (b_1, \ldots, b_m)$. The return of CBAL$_\mathbf{b}$ over a Kelly market sequence $K_\ell = k_1^\ell, \ldots, k_n^\ell$ is CBAL$_\mathbf{b}(K_\ell) = \prod_i b_{s(k_i^\ell)}$ which gives the weight of S_{K_ℓ} in ALG$'$. Therefore, for an arbitrary market sequence we have

$$\text{ALG}'(X) = \sum_{K_\ell \in \mathcal{K}^n} \prod_i b_{s(k_i^\ell)} S_{K_\ell}(X)$$

$$= \prod_{i=1}^{n} \sum_{j=1}^{m} b_i x_{ij}$$

$$= \prod_i \mathbf{b} \cdot \mathbf{x}_i = \text{CBAL}_\mathbf{b}(X).$$

Theorem 2. *(i) A (Mixture of)* CBAL *algorithms can emulate a (mixture) of* BAH *algorithms. (ii) A (mixture of)* SS *algorithms can emulate a* CBAL *algorithm and hence any mixture of* CBAL*S.*

Lemma 4. *The competitive ratio is invariant under scaling of the relative prices. In particular, it is invariant under scaling of each day independently Thus we can assume without loss of generality that all relative prices are in $[0, 1]$*

Lemma 5. *In the game against* OPT *it is sufficient to prove lower bounds using only Kelly market sequences.*

Proof. Let $X = \mathbf{x}_1, \ldots, \mathbf{x}_n$ be an arbitrary market sequence. Using Lemma 4 we can scale each day independently so that in each market vector $\mathbf{x}_i = x_{i1} \ldots, x_{im}$, the maximum relative price of each day (say it is $x_{i,j_{\max}}$) equals 1. Now we consider the "Kelly projection" X' of the market X; that is, in the market vector $\mathbf{x}_i'$, $x_{i,j_{\max}}' = 1$ and $x_{i.\ell} = 0$ for $\ell \neq j_{\max}$. For any algorithm ALG, we have ALG$(X) \geq$ ALG(X'), but OPT always satisfies (when there are no commissions) OPT$(X) =$ OPT(X').   □

3.2 Lower Bound Proof Technique

Using Lemma 1 we can now describe a general lower bound proof technique due to Cover and Ordentlich.

Lemma 6 (Portfolio Selection Lower Bound). *Let* OPT$_\mathcal{C}$ *be an optimal offline algorithm from a class $\mathcal{C}$ of offline algorithms (e.g.* CBAL-OPT *when $\mathcal{C}$ is the class of constant rebalanced portfolios). Then $r_m^n = \sum_{K \in \mathcal{K}_m^n}$ OPT$_\mathcal{C}(K)$ is a lower bound on the competitive ratio of any online algorithm relative to* OPT$_\mathcal{C}$.

Proof. We use Lemma 1. Clearly, the maximum element in any set is not smaller than any weighted average of all the elements in the set. Let $Q_K = \frac{\text{ALG}(K)}{\sum_K \text{ALG}(K)} = \text{ALG}(K)$. We then have

$$\max_{K \in \mathcal{K}_m^n} \frac{\text{OPT}_C(K)}{\text{ALG}(K)} \geq \sum_{K \in \mathcal{K}_m^n} Q_K \cdot \frac{\text{OPT}_C(K)}{\text{ALG}(K)}$$

$$= \sum_{K \in \mathcal{K}_m^n} \text{ALG}(K) \cdot \frac{\text{OPT}_C(K)}{\text{ALG}(K)}$$

$$= \sum_{K \in \mathcal{K}_m^n} \text{OPT}_C(K) .$$

$\square$

4 Optimal Bounds against BAH-OPT and OPT

In some cases, Lemma 6 provides the means for easily proving lower bounds. For example, consider the PS game against BAH-OPT For any market sequence, BAH-OPT invests its entire wealth in the best stock. Therefore, $\sum_{K \in \mathcal{K}_m^n} \text{BAH-OPT}(K) = m$, and m is a lower bound on the competitive ratio of any investment algorithm for this problem. Moreover, m is a tight bound since the algorithm that invests $1/m$ of its initial wealth in each of the m stocks achieves a competitive ratio of m.

Similarly, for any Kelly market sequence K we have $\text{OPT}(K) = 1$. Therefore, as there are m^n Kelly sequences of length n, we have, $\sum_{K \in \mathcal{K}_m} \text{OPT}(K) = m^n$ and thus m^n is a lower bound on the competitive ratio of any online algorithm against OPT. In this case, $\text{CBAL}_{(1/m,\ldots,1/m)}$ achieves the optimal bound!

5 How Well Can We Perform against CBAL-OPT

Comparison against CBAL-OPT has received considerable attention in the literature. The best known results are summarized in Table 1. Using the lower bound technique from Section 3.2, we first establish the Ordentlich and Cover [OC96] lower bound.

The following lemma is immediate but very useful.

Lemma 7. *Consider a Kelly market sequence $X^n = \mathbf{x}_1, \ldots, \mathbf{x}_n$ over m stocks. We can represent X^n as a sequence $X^n = x_1, \ldots, x_n \in \{1, 2 \ldots, m\}^n$. Let CBAL_b be a constant-rebalanced algorithm. Suppose that in the sequence X^n there are n_j occurrences of (stock) j with $\sum_j n_j = n$. We say that such a Kelly sequence has type $(n_1, n_2, \ldots, n_m)$. For a sequence X^n of type $(n_1, n_2, \ldots, n_m)$, the return $R(\mathbf{b}, X^n)$ of CBAL_b on the sequence X_n is*

$$R(\mathbf{b}, X^n) = \left(\prod_{j=1}^{n_1} b_1\right) \left(\prod_{j=1}^{n_2} (b_2)\right) \cdots \left(\prod_{j=1}^{n_m} (b_m)\right) = \prod b_j^{n_j} .$$

That is, the return of any CBAL depends only only the type.

The development in Section 6.2 gives an alternative and more informative proof in that it determines the optimal portfolio $\mathbf{b}^*$ for any Kelly sequence $\mathbf{x}^n$ of a given type; namely $\mathbf{b}^* = (n_1/n, \ldots, n_m/n)$.

We then can apply Lemma 6 to the class $\mathcal{C}$ of CBAL algorithms to obtain:

Lemma 8. *Let* ALG *be any online PS algorithm. Then a lower bound for the competitive ratio of* ALG *against* CBAL-OPT *is*

$$r_m^n = \sum_{(n_1,\ldots,n_m):\Sigma n_j = n} \binom{n}{n_1 n_2 \ldots n_m} \prod_j \left(\frac{n}{n_j}\right)^{n_j}.$$

Proof. Consider the case $m = 2$. There are clearly $\binom{n}{n_1}$ length n Kelly sequences X having type $(n_1, n_2) = (n_1, n - n_1)$ and for each such sequence CBAL-OPT$(X) = \prod_j \left(\frac{n}{n_j}\right)^{n_j}$.

Using a mixture of switching sequences and a min-max analysis, Ordentlich and Cover provide a matching upper bound (for *any* market sequence of length n) showing that r_m^n is the optimal bound for the competitive ratio in the context of a known horizon n.

Table 1. The best known results w.r.t. CBAL

	$m = 2$	$m \geq 2$	Source
Lower[3]	$r_2^n \simeq \sqrt{\pi n/2}$	$r_m^n \simeq \frac{\pi}{\Gamma(\frac{m}{2})}\left(\frac{n}{2}\right)^{\frac{m-1}{2}}$	[OC96]
Upper (known n)	r_2^m	r_m^n	[OC96]
Upper	$2\sqrt{n+1}$	$2(n+1)^{\frac{m-1}{2}}$	[CO96]

From the discussion above it follows immediately that any one CBAL (or any finite mixture of CBALS) cannot be competitive relative to CBAL-OPT; indeed for any $m \geq 2$, the competitive ratio of any CBAL against CBAL-OPT will grow exponentially in n.

5.1 Cover's Universal Portfolio Selection Algorithm

The *Universal Portfolio* algorithms presented by Cover [Cov91] are special cases of the class of "μ-weighted" algorithms which we denote by W_μ. A rather intuitive understanding of the μ-weighted algorithms was given by Cover and Ordentlich. These algorithms are parameterized by a distribution μ, over the set of all portfolios $\mathcal{B}$. Cover and Ordentlich show the following result:

$$\text{wealth of } W_\mu = E_{\mu(\mathbf{b})} [\text{wealth of CBAL}_\mathbf{b}] .$$

---

[3] The Gamma function is defined as $\Gamma(x) = \int_0^\infty e^{-t} t^{x-1} dt$. It can be shown that $\Gamma(1) = 1$ and that $\Gamma(x + 1) = x\Gamma(x)$. Thus if $n \geq 1$ is an integer, $\Gamma(n + 1) = n!$. Note also that $\Gamma(1/2) = \sqrt{\pi}$.

This observation is interesting because the definition of w_μ (see [CO96]) is in terms of a sequence of adaptive portfolios (depending on the market sequence) that progressively give more weight to the better-performing constant rebalanced portfolios. But the above observation shows that the return of these μ-weighted algorithms is equivalent to a "non-learning algorithm". That is, a mixture of CBALS specifies a randomized trading strategy that is in some sense independent of the stock market data. Of course, the composite portfolio determined by a mixture of CBALS does depend on the market sequence.

Cover and Ordentlich analyze two instances of w_μ. One (called UNI) that uses the uniform distribution (equivalently, the Dirichlet$(1, 1, \ldots, 1)$ distribution) and another (here simply called DIR) that uses the Dirichlet$(\frac{1}{2}, \frac{1}{2}, \ldots, \frac{1}{2})$ distribution. They prove that the uniform algorithm UNI has competitive ratio

$$\binom{n + m - 1}{m - 1} \le (n + 1)^{m-1}, \tag{1}$$

and that this bound is tight. Somewhat surprisingly (in contrast, see the discussion concerning the algorithms in Sections 6.3– 6.5.) this bound can be extracted from UNI by an adversary using only Kelly market sequences; in fact, by using a Kelly sequence $\tilde{X}$ in which one fixed stock "wins" every day. For the case of $m = 2$, this can be easily seen since the return CBAL$_{(b,1-b)}(\tilde{X})$ is b^n for n days and $\int_0^1 b^n db = \frac{1}{n+1}$.

Cover and Ordentlich show that the DIR algorithm has competitive ratio

$$\frac{\Gamma(1/2)\Gamma(n + m/2)}{\Gamma(m/2)\Gamma(n + 1/2)} \le 2(n + 1)^{\frac{m-1}{2}} \ . \tag{2}$$

and here again the bound is tight and achieved using Kelly sequences. Hence for any fixed m, there is a constant gap between the optimal lower bound for fixed horizon and the upper bound provided by DIR for unknown horizon.

It is instructive to consider an elegant proof of the universality of UNI (with a slightly inferior bound) due to Blum and Kalai [BK97].[4] Let $\mathcal{B}$ be the $(m - 1)$-dimensional simplex of portfolio vectors and let μ be any distribution over $\mathcal{B}$. Recall that the return of the μ-weighted alg is a μ-weighted average of the returns of all CBAL$_b$ algs. Let X be any market sequence of length n and let CBAL$_{b^*}$ = CBAL-OPT. Say that b is "near" b^* if $b = \frac{n}{n+1}b^* + \frac{1}{n+1}z$ for some $z \in \mathcal{B}$. Therefore, for each day i we have

$$\text{CBAL}_b(x_i) \ge \frac{n}{n + 1} \cdot \text{CBAL}_{b^*}(x_i)$$

So, for n days,

$$\frac{\text{CBAL}_{b^*}(X)}{\text{CBAL}_b(X)} \le \left(1 + \frac{1}{n}\right)^n \le e.$$

Let $\mathbf{Vol}_m(\cdot)$ denote the m-dimensional volume.

---

[4] See also the web page http://www.cs.cmu.edu/˜akalai/coltfinal/slides.

Under the uniform distribution over $\mathcal{B}$, the probability that b is near $\mathbf{b}^*$ is

$$
\begin{aligned}
\mathbf{Pr}[\text{b near } \mathbf{b}^*] &= \frac{\mathbf{Vol}_{m-1}\left(\frac{n}{n+1}\mathbf{b}^* + \frac{1}{n}z\right)}{\mathbf{Vol}_{m-1}(\mathcal{B})} \\
&= \frac{\mathbf{Vol}_{m-1}\left(\frac{1}{n}z\right)}{\mathbf{Vol}_{m-1}(\mathcal{B})} \\
&= \left(\frac{1}{n+1}\right)^{m-1}.
\end{aligned}
$$

Thus $\left(\frac{1}{n+1}\right)^{m-1}$ fraction of the initial wealth is invested in CBAL's which are "near" CBAL-OPT, each of which is attaining a ratio e. Therefore, the competitive ratio achieved is $e \cdot (n+1)^{m-1}$.

5.2 Expert Advice and the EG Algorithm

The EG algorithm proposed by Helmbold it et al [HSSW98] takes a different approach. It tries to move towards the CBAL-OPT portfolio by using an update function that minimizes an objective function of the form

$$
F^t(\mathbf{b}_{t+1}) = \eta \log(\mathbf{b}_{t+1} \cdot \mathbf{x}_t) - d(\mathbf{b}_{t+1}, \mathbf{b}_t)
$$

where $d(\mathbf{b}, \mathbf{b}')$ is some distance or dissimilarity measure over distributions (portfolios) and η is a learning rate parameter.

The competitive bound proven by Helmbold *et al* for EG is weaker than the bound obtained for UNI. However, EG is computationally much simpler than UNI and experimentally it outperforms UNI on the New York Stock Exchange data (see [HSSW98] and Table 3). The EG algorithm developed from a framework for online regression and a successful body of work devoted to predicting based on expert advice. When trying to select the *best expert* (or a weighting of experts), the EG algorithm is well motivated. It is trying to minimize a loss function based on the weighting of various expert opinions and in this regard it is similar to UNI. However, it is apparent that CBAL-OPT does not make its money (over buy and hold) by seeking out the best stock. If one is maintaining a *constant portfolio*, one is selling rising stocks and buying falling ones. This strategy is advantageous when the falling stock reverses its trend and starts rising. We also note that in order to prove the universality of EG, the value of the learning rate η decreases to zero as the horizon n increases. When $\eta = 0$, EG degenerates to the uniform CBAL$_{(1/m,1/m,...,1/m)}$ which is not universal whereas the small learning rate (as given by their proof) of EG is sufficient to make it universal. It is also the case that if each day the price relatives for all stocks were identical, then (as one would expect) EG will again be identical to the uniform CBAL. Hence when one combines a small learning rate with a "reasonably stable" market (i.e. the price ratives are not too erratic), we might expect the performance of EG to be similar to that of the uniform CBAL and this seems to be confirmed by our experiments.

6 On Some Portfolio Selection Algorithms

In this section we present and discuss several online portfolio selection algorithms. The DELTA algorithm is a new algorithm suggested by the goal of exploiting the rationale of constant rebalancing algorithms. The other algorithms are adaptations of known prediction algorithms.

6.1 The DELTA Algorithm

In this section we define what we call the DELTA(r, w, t_{corr}) algorithm. Informally, this online algorithm operates as follows. There is a risk parameter r between 0 and 1 controlling the fraction of a stocks value that the algorithm is willing to trade away on any given day. Each stock will risk that proportion of its weight if it is climbing in value and is sufficiently anti-correlated with other stock(s). If it is anti-correlated, the at-risk amount is spread amongst the falling stocks proportional to the correlation coefficient [5]. The algorithm takes two other parameters. The "window" length, w, specifies the length of history used in calculating the new portfolio. To take advantage of short term movements in the price relatives, a small window is used. Finally, $t_{corr} < 0$ is a correlation threshold which determines if a stock is sufficiently anti-correlated with another stock (in which case the weighting of the stock is changed).

A theoretical analysis of the DELTA algorithm seems to be beyond reach, at this stage. In Sections 7 and 8 we present experimental studies of the performance of the DELTA algorithm.

6.2 The Relation between Discrete Sequence Prediction and Portfolio Selection

We briefly explore the well established relation between the portfolio selection problem and prediction of discrete sequences. We then discuss the use of some known prediction algorithms for the PS problem.

Simply put, the standard worst case prediction game under the log-loss measure is a special case of the PS game where the adversary is limited to generating only Kelly market vectors. As mentioned in Section 5.1, Cover and Ordentlich showed that the PS algorithms UNI and DIR obtain their worst-case behavior over Kelly market sequences. However, this does not imply that the PS problem is reducible to the prediction problem, and we will see here several examples of prediction algorithms that are not competitive (against CBAL-OPT) in the PS context but are competitive in the prediction game.

Here is a brief description of the prediction problem. (For a recent comprehensive survey of online prediction see [MF98].) In the online prediction problem the online player receives a sequence of observations $x_1, x_2, \ldots, x_{t-1}$ where the x_i are symbols in some alphabet of size m. At each time instance t the player must generate a prediction b_t for the next, yet unseen symbol x_t. The prediction is in general a probability distribution

---

[5] The *correlation coefficient* is a normalized covariance with the covariance divided by the product of the standard deviations; that is,
Cor(X,Y) = Cov(X,Y)/(std(X)*std(Y)) where
Cov(X,Y) = E[X-mean(X)) * (Y-mean(Y))].

$\mathbf{b}_t(x_t) = \mathbf{b}_t(x_t|x_1, \ldots, x_{t-1})$ over the alphabet. Thus, $\mathbf{b}_t$ gives the confidence that the player has on each of the m possible future outcomes. After receiving x_t the player incurs a loss of $l(\mathbf{b}_t, x_t)$ where l is some loss function. Here we concentrate on the log-loss function where $l(\mathbf{b}_t, x_t) = -\log \mathbf{b}_t(x_t)$. The total loss of a prediction algorithm, $B = \mathbf{b}_1, \mathbf{b}_2, \ldots, \mathbf{b}_n$ with respect to a sequence $X = x_1, x_2, \ldots, x_n$ is $L(B, X) = \sum_{i=1}^{n} l(\mathbf{b}_i, x_i)$.

As in the competitive analysis of online PS algorithms, it is customary to measure the worst case competitive performance of a prediction algorithm with respect to a comparison class of offline prediction strategies. Here we only consider the comparison class of *constant predictors* (which correspond to the class of constant rebalanced algorithms in the PS problem). The "competitive ratio"[6] of a strategy B with respect to the comparison class $\mathcal{B}$ is $\max_X [L(B, X) - \inf_{b \in \mathcal{B}} L(b, X)]$.

There is a complete equivalence between (competitive analysis of) online prediction algorithms under the log loss measure with respect to (offline) constant predictors and the (competitive analysis of) online PS algorithms with respect to (offline) constant rebalanced algorithms whenever the only allowable market vectors are Kelly sequences. To see that, consider the binary case $m = 2$ and consider the Kelly market sequence $X = x_1, \ldots, x_n$ where $x_i \in \{0, 1\}$ represents a Kelly market vector ($x_i = 0$ corresponds to the Kelly market vector $\binom{1}{0}$ and $x_i = 1$ corresponds to $\binom{0}{1}$). For ease of exposition we now consider (in this binary case) stock indices in $\{0, 1\}$ (rather than $\{1, 2\}$). Let CBAL$_b$ be the constant-rebalanced algorithm with portfolio $(b, 1 - b)$. The return of CBAL$_b$ is $R(b, X) = \prod_{i=1}^{n} (x_i b + (1 - x_i)(1 - b))$. Let (n_0, n_1) be the type of X (i.e. in X there are n_0 zeros and n_1 ones, $n_0 + n_1 = n$.) Taking the base 2 logarithm of the return and dividing by n we get

$$
\begin{aligned}
\frac{1}{n} \log R(b, X) &= \frac{n_0}{n} \log b + \frac{n_1}{n} \log(1 - b) \\
&= -\frac{n_0}{n} \log \frac{1}{b} - \frac{n_1}{n} \log \frac{1}{1-b} \\
&= -\left(\frac{n_0}{n} \log \frac{n_0/n}{b} + \frac{n_1}{n} \log \frac{n_1/n}{(1-b)} \right) + \left(\frac{n_0}{n} \log \frac{n_0}{n} + \frac{n_1}{n} \log \frac{n_1}{n} \right) \\
&= -D_{KL}\left[\left(\frac{n_0}{n}, \frac{n_1}{n} \right) \| (b, 1 - b) \right] - H\left(\frac{n_0}{n}, \frac{n_1}{n} \right) .
\end{aligned}
\tag{3}
$$

As the KL-divergence $D_{KL}(\cdot \| \cdot)$ is always non-negative, the optimal offline choice for the constant-rebalanced portfolio (CBAL-OPT) , that maximizes $\frac{1}{n} \log R(b, X)$, is $b^* = n_0/n$, in which case the KL-divergence vanishes.

We now consider the competitive ratio obtained by an algorithm ALG against CBAL-OPT. Using the above expression for the log-return of CBAL-OPT we get

$$
\begin{aligned}
\log \frac{\text{CBAL-OPT}}{\text{ALG}} &= \log \text{CBAL-OPT} - \log \text{ALG} \\
&= -nH\left(\frac{n_0}{n}, \frac{n_1}{n} \right) - \log \text{ALG}.
\end{aligned}
$$

---

[6] The more common term in the literature is "regret".

Using Jensen's inequality we have

$$-\log \text{ALG}(X) = -\log \prod_{i=1}^{n} (x_i b_i + (1 - x_i)(1 - b_i))$$

$$= -\sum_{i=1}^{n} \log (x_i b_i + (1 - x_i)(1 - b_i))$$

$$\leq -\sum_{i=1}^{n} (x_i \log b_i + (1 - x_i) \log(1 - b_i))$$

$$= \sum_{i=1}^{n} \left(x_i \log \frac{x_i}{b_i x_i} + (1 - x_i) \log \frac{1 - x_i}{(1 - b_i)(1 - x_i)} \right)$$

$$= \sum_{i=1}^{n} D_{KL}\left[(x_i, 1 - x_i) \| (b_i, 1 - b_i)\right] + \sum_{i=1}^{n} H(x_i, 1 - x_i) \ .$$

Since each of the entropies $H(x_i, 1 - x_i) = 0$ ($x_i = 0$ or $x_i = 1$), we have

$$-\log \text{ALG}(X) \leq \sum_{i=1}^{n} D_{KL}\left[(x_i, 1 - x_i) \| (b_i, 1 - b_i)\right] \ .$$

Putting all this together,

$$\log \frac{\text{CBAL-OPT}}{\text{ALG}} \leq \sum_{i=1}^{n} D_{KL}\left[(x_i, 1 - x_i) \| (b_i, 1 - b_i)\right] - nH\left(\frac{n_0}{n}, \frac{n_1}{n}\right) \ .$$

In the prediction game the online player must generate a prediction b_i for the ith bit of a binary sequence that is revealed online. The first, KL-divergence, term of the bound measures in bits the total redundancy or inefficiency of the prediction. The second, entropy term, measures how predictable the sequence X is. In order to prove a competitive ratio of C, against CBAL-OPT, it is sufficient to prove that

$$\sum_{i=1}^{n} D_{KL}\left[(x_i, 1 - x_i) \| (b_i, 1 - b_i)\right] \leq \log C + nH\left(\frac{n_0}{n}, \frac{n_1}{n}\right) \ . \tag{4}$$

If $x_i = 1$ this expression reduces to $\log \frac{1}{b_i}$, and if $x_i = 0$, it reduces to $\log \frac{1}{1 - b_i}$. Therefore, the summation over all these KL-divergence terms can be expressed as the logarithm of $\prod_i \frac{1}{z_i}$ where $z_i = b_i$ iff $x_i = 1$ and $z_i = 1 - b_i$ if $x_i = 0$. We thus define, for an online prediction algorithm, its "probability product" P corresponding to an input sequence to be $\prod_i \frac{1}{z_i}$, and to prove a competitive ratio of C it is sufficient to prove that $\log(P) = \log \prod_i \frac{1}{z_i} \leq \log C + nH\left(\frac{n_0}{n}, \frac{n_1}{n}\right)$. A similar development to the above can be established for any $m > 2$.

6.3   The Add-beta Prediction Rule

Consider the following method for predicting the $i + 1^{st}$ bit of a binary sequence, based on the frequency of zeros and ones that appeared until (and including) the ith round.

We maintain counts, C_j^t, $j = 0, 1$. Such that C_0^t records the frequency of the zeros and C_1 records the frequency of ones until and including round t. Based on these counts the (online) algorithm predicts that the $t + 1^{st}$ bit will be 0 with probability

$$p_0^{t+1} = \frac{C_0^t + \beta}{2\beta + C_0^t + C_1^t} ,$$

where the parameter beta is a non-negative real. This rule is sometimes called the *add-beta* prediction rule. The instance $\beta = 1$ is called *Laplace law of succession* (see [MF98,CT91,Krich98]), and the case $\beta = 1/2$ is known to be optimal in both distributional and distribution free (worst case) settings [MF98]. In the case where there are m possible outcomes, $1, \ldots, m$, the add-beta rule becomes

$$b_i^{t+1} = \frac{C_i^t + \beta}{m\beta + \sum_{1 \leq j \leq m} C_j^t} .$$

For the PS problem, one can use any prediction algorithm, such as the add-beta rule in the following straightforward manner. Assume that the price relatives are normalized everyday so that the largest price relative equals one. For each market vector $X = x_1, \ldots, x_m$ consider its *Kelly projection* $K(X)$, in which all components are zero except for component $\arg\max x_i$ which is normalized to equal one[7]. At each round such an algorithm yields a prediction for each of the m symbols.

Algorithm M0: The first prediction-based PS online algorithm we consider is called M0 (for "Markov of order zero"). This algorithm simply uses the add-beta prediction rule on the Kelly projections of the market vectors. For the prediction game (equivalently, PS with Kelly markets) one can show that algorithm M0 with $\beta = 1$ is $(n + 1)^{m-1}$-competitive; that is, it achieves an identical competitive ratio to UNI, the uniform-weighted PS algorithm of Cover and Ordentlich (see Section 5.1). Here is a sketch of the analysis for the case $m = 2$. The first observation is that the return of M0 is the same for all Kelly sequences of the same type. This result can be shown by induction on the length of the market sequence. It is now straightforward to calculate the "probability product" $P(J)$ of M0 with respect to a market sequence X of type $J = (n_0, n_1)$ (by explicitly calculating it for a sequence which contains n_0 zeros followed by n_1 ones), which equals to $\frac{(n+1)!}{n_0! n_1!}$. Since $\log\binom{n}{n_1} \leq nH\left(\frac{n_0}{n}, \frac{n_1}{n}\right)$ (see [CT91]), we have

$$\log P(J) = \log\left((n+1)\binom{n}{n_1}\right)$$
$$= \log(n+1) + \log\binom{n}{n_1}$$
$$\leq \log(n+1) + nH\left(\frac{n_0}{n}, \frac{n_1}{n}\right) .$$

---

[7] To simplify the discussion, randomly choose amongst the stocks that achieve the maximum price relative if there is more than one such stock.

Using inequality (4) and the development thereafter it follows that (with $\beta = 1$) algorithm MO is $(n+1)$-competitive in the prediction game (or restricted PS with Kelly sequences) and it is not hard to prove a tight lower bound of $n + 1$ on its competitive ratio using the Kelly sequence of all ones. One can quite easily generalize the above arguments for $m > 2$, and it is possible to prove using a similar (but more involved) analysis that MO based on the add-$\frac{1}{2}$ rule achieves a competitive ratio of $2(n + 1)^{(m-1)/2}$, the same as the universal DIR algorithm for the general PS problem. (See Merhav and Feder [MF98] and the references therein.)

Despite the fact that algorithm MO is competitive in the online prediction game it is not competitive (nor even universal) in the unrestricted PS game against offline constant rebalanced portfolios. For the case $m = 2$ this can be shown using market sequences of the form

$$\left(\begin{pmatrix} 1 - \epsilon \\ 1 \end{pmatrix} \right)^n \left(\begin{pmatrix} 1 \\ 0 \end{pmatrix} \right)^n .$$

It is easy to see that the competitive ratio is $2^{\Omega(n)}$. Cover and Gluss [CG86] show how a less naive learning algorithm is universal under the assumption that the set of possible market vectors is bounded. In doing so, they illustrate how their algorithm (based on the Blackwell [Bl56] approachability-excludability theorem) avoids the potential pitfalls of a naive counting scheme such as MO.

Algorithm T0: One might surmise that the deficiency of algorithm MO (in the PS game) can attributed to the fact that it ignores useful information in market vectors. Like MO, algorithm T0 uses the add-beta rule but now maintains its counters as follows:

$$C_j^{t+1} = C_j^t + \log_2(x_{t+1,j} + 1) .$$

Here we again assume that price relatives have been normalized so that on any given day the maximum is one. Clearly, T0 reduces to MO in the prediction game. Algorithm T0 is also not competitive. This can be shown using sequences of the form

$$\left(\begin{pmatrix} 1 \\ 1 \end{pmatrix} \right)^{tn} \left(\begin{pmatrix} 1 \\ 0 \end{pmatrix} \right)^n .$$

For sufficiently large t, it is clear that the behavior of T0 on the sequence will be similar to that of CBAL$_{(\frac{1}{2}, \frac{1}{2})}$, which is not competitive.

6.4   Prediction-Based Algorithms: Lempel-Ziv Trading

One can then suspect that the non-competitiveness of the add-beta variants (MO, T0) is due to the fact that they ignore dependencies among market vectors (they record only zero-order statistics). In an attempt to examine this possibility we also consider the following trading algorithm based on the Lempel-Ziv compression algorithm [ZL78]. The LZ algorithm was also considered in the context of prediction (see Langdon [Lan83] and Rissanen [Ris83]). Feder [Fed91] and Feder and Gutman [FG92] consider the worst case competitive performance of the algorithm in the context of gambling. Feder shows that the LZ algorithm is universal with respect to the (offline) class of all finite state prediction machines.

Like MO, the PS algorithm based on Lempel-Ziv is the LZ prediction algorithm applied to the Kelly projections of the market vectors. Using the same nemesis sequence as for MO, namely

$$\left(\begin{pmatrix}1-\epsilon\\1\end{pmatrix}\right)^n \left(\begin{pmatrix}1\\0\end{pmatrix}\right)^n ,$$

it is easy to see that a lower bound for the competitive ratio of LZ is $2^{\Omega(\sqrt{(n)})}$ and hence LZ is not competitive by our definition (although it might still be universal).

6.5   Portfolio Selection Work Function Algorithm

In his PhD thesis Ordentlich suggests the following algorithm (which is also a generalization of the MO algorithm with $\beta = 1/2$). For the case $m = 2$ this algorithm chooses the next portfolio $\mathbf{b}_{t+1}$ to be

$$\mathbf{b}_{t+1} = \frac{t-1}{t}\mathbf{b}_t^* + \frac{1}{t}\begin{pmatrix}1/2\\1/2\end{pmatrix},$$

where b_t^* is the optimal constant rebalanced portfolio until time t. With its two components, one that tracks the optimal offline algorithm so far and a second which may be viewed as greedy, this algorithm can be viewed as kind of a work function algorithm (see [BE98]).

Ordentlich shows that the sequence

$$\left(\begin{pmatrix}1\\1-\epsilon\end{pmatrix}\right)^n \begin{pmatrix}0\\1\end{pmatrix}\begin{pmatrix}1\\0\end{pmatrix},$$

produces a competitive ratio of $\Omega(n^2)$ from this algorithm.

We can slightly improve Ordentlich's lower bound to $\Omega(n^{5/2})$. We concatenate $\left(\begin{pmatrix}0\\1\end{pmatrix}\right)^i \left(\begin{pmatrix}1\\0\end{pmatrix}\right)^i$ for $i = 2 \ldots k$ to the end of Ordentlich's sequence, with $k = \Theta(\sqrt{(n)})$ so that the entire input sequence remains of length $O(n)$.

It remains an open question as to whether or not Ordentlich's algorithm is competitive (or at least universal). The $\Omega(n^{(5/2)})$ lower bound shows that this algorithm is not as competitive as UNI and DIR.

7   Experimental Results

We consider three data sets as test suites for most of the algorithms considered in the relevant literature. [8] The first data set is the stock market data as first used by Cover

---

[8] We do not present experiments for the DIR algorithm nor for Ordentlich's "work function algorithm". Even though DIR's worst case competitive bound is better than that of UNI, it has been found in practice (see [HSSW98]) that DIR's performance is worse than UNI. It is computationally time consuming to even approximate DIR and the work function algorithm. In the full version of this paper we plan to present experimental results for these algorithms.

[Cov91] and then Cover and Ordentlich [CO96], Helmbold *et al* [HSSW98], Blum and Kalai [BK97] and Singer [Sin98].[9] This data set contains 5651 daily prices for 36 stocks in the New York Stock Exchange (NYSE) for the twenty two year period July 3rd, 1962 to Dec 31st, 1984. The second data set consists of 88 stocks from the Toronto Stock Exchange (TSE), for the five year period Jan 4th, 1994 to Dec 31st, 1998. The stocks chosen were those that were traded on each of the 1258 trading days in this period. The final data set is for intra day trading in the foreign exchange (FX) market. Specifically, the data covers the bid-ask quotes between USD ($) and Japanese Yen, and between USD and German Marks (DM) for the one year period Oct 1st, 1992 to Sep 30th, 1993. As explained in Section 7.2, we interpret this data as 479081 price relatives for $m = 2$ stocks (i.e. Yen and DM).

7.1   Experiments on NYSE data

All 36 stocks in the sample had a positive return over the entire 22 year sample. The returns ranged from a low of 3.1 to a high (BAH-OPT) of 54. Before running the online algorithms on pairs of stocks, we determined the CBAL-OPT of each stock when traded against cash and a 4% bond as shown in Table 2.

Of the 36 stocks, only three benefited from an active trading strategy against cash. The winning strategy for the remaining 33 stocks was to buy and hold the stock. When cash was replaced by a 4% annualized bond, seven stocks (the ones highlighted in Table 2) benefited from active trading. It is interesting to note that the CBAL-OPT of all 36 stocks is comprised of a portfolio of just 5 stocks. This portfolio has a return of 251.

Stock	Weight
Comm Metals	0.2767
Espey	0.1953
Iroquois	0.0927
Kin Ark	0.2507
Mei Corp	0.1845

Four of these five stocks are also the ones that most benefited from active trading by CBAL-OPT against a bond. The CBAL-OPT of the remaining 31 stocks is still a respectable 69.9, beating BAH-OPT for the entire 36 stocks.

Pairs of Stocks When looking at individual *pairs* of stocks, however, one finds a different story. Instead of just these five or seven stocks benefiting from being paired with one another, one finds that of the possible 630 pairs, almost half (44%) have a CBAL-OPT that is 10% or more than the return of the best stock. It is this fact that has encouraged the consideration of competitive-based online algorithms as this sample does indicate that many stock pairings can benefit from frequent trading. Of course, it can be argued that identifying a "profitable pair" is the real problem.

---

[9] According to Helmbold *et al*, this data set was originally generated by Hal Stern. We do not know what criteria was used in choosing this particular set of 36 stocks.

Stock	BAH	BND(4%)	CBAL-OPT(0%)	CBAL-OPT(4%)
Dupont	3.07	2.41	3.07	3.18
Kin Arc	4.13	2.41	12.54	18.33
Sears	4.25	2.41	4.25	4.25
Lukens	4.31	2.41	4.31	4.93
Alcoa	4.35	2.41	4.35	4.42
Ingersoll	4.81	2.41	4.81	4.81
Texaco	5.39	2.41	5.39	5.39
MMM	5.98	2.41	5.98	5.98
Kodak	6.21	2.41	6.21	6.21
Sher Will	6.54	2.41	6.54	6.54
GM	6.75	2.41	6.75	6.75
Ford	6.85	2.41	6.85	6.85
P and 9	6.98	2.41	6.98	6.98
Pillsbury	7.64	2.41	7.64	7.64
GE	7.86	2.41	7.86	7.86
Dow Chem	8.76	2.41	8.76	8.76
Iroquois	8.92	2.41	9.81	12.08
Kimb Clark	10.36	2.41	10.36	10.36
Fischbach	10.70	2.41	10.70	10.70
IBM	12.21	2.41	12.21	12.21
AHP	13.10	2.41	13.10	13.10
Coke	13.36	2.41	13.36	13.36
Espey	13.71	2.41	14.88	17.89
Exxon	14.16	2.41	14.16	14.16
Merck	14.43	2.41	14.43	14.43
Mobil	15.21	2.41	15.21	15.21
Amer Brands	16.10	2.41	16.10	16.10
Pillsbury	16.20	2.41	16.20	16.20
Arco	16.90	2.41	16.90	16.90
JNJ	17.22	2.41	17.22	17.22
Mei Corp	22.92	2.41	22.92	23.29
HP	30.61	2.41	30.61	30.61
Gulf	32.65	2.41	32.65	32.65
Schlum	43.13	2.41	43.13	43.13
Comm Metals	52.02	2.41	52.02	52.02
Morris	54.14	2.41	54.14	54.14

Table 2. Return of CBAL-OPT when traded against 0 % cash and a 4 % a bond. For example, when Dupont is balanced against cash (respectively, the bond),the return of CBAL-OPT = 3.07 (respectively, 3.18). The seven stocks that profit from active trading against a bond have been highlighted.

We abbreviate this uniform CBAL algorithm as UCBAL$_m$. The uniform buy and hold (denoted UBAH$_m$) and UCBAL$_m$ algorithms give us reasonable (and perhaps more realistic) benchmarks by which to compare online algorithms; that is, while one would certainly expect a good online algorithm to perform well relative to the uniform BAH, it also seems reasonable to expect good performance relative to UCBAL since both algorithms can be considered as naive strategies. We found that for over half (51%) of the stock pairings, CBAL-OPT has a 10% or more advantage over the $(1/2, 1/2)$ CBAL.

The finding that EG (with small learning rate) has no substantial (say, 1%) advantage over UCBAL$_2$ confirms the comments made in Section 5.2. Previous expositions demonstrated impressive returns; that is, where EG outperformed UNI and can significantly outperform the best stock (in a pair of stocks). The same can now be said for UCBAL$_2$. The other interesting result is that DELTA seems to do remarkably better than the UCBAL$_2$ algorithm; it is at least 10% better than UCBAL$_2$ for 344 pairs. In fact, DELTA does at least 10% better than CBAL-OPT a third of the time (204 pairs). In the full version of this paper we will present several other algorithms, some that expand and some that limit the risk.

All Stocks Some of the algorithms were exercised on a portfolio of all 36 stocks. In order to get the most from the data and detect possible biases, the sample was split up into 10 equal time periods. These algorithms were then run on each of the segments. In addition, the algorithms were run on the reverse sequence to see how they would perform in a falling market. These results will be presented in the full paper.

7.2 Experiments on the TSE and FX data

The TSE and FX data are quite different in nature than the NYSE data. In particular, while every stock made money in the NYSE data, 32 of the 88 stocks in the TSE data lost money. The best return was 6.28 (Gentra Inc.) and the worst return was .117 (Pure Gold Minerals). There were 15 stocks having non zero weight in CBAL-OPT, with three stocks (including Gentra Inc.) constituting over 80% of the weight. Unlike its performance on the NYSE data, with respect to the TSE data, UNI does outperform the online algorithms UCBAL, EG, M0 and T0. It does not, however, beat DELTA and it is somewhat worse than UBAH. The FX data was provided to us in a very different form, namely as bid-ask quotes (as they occur) as opposed to (say) closing daily prices. We interpreted each "tick" as a price by taking the average value (i.e. (ask+bid)/2). Since each tick only represents one currency, we merged the ticks into blocks where each block is the shortest number of ticks for which each currency is traded. We then either ignored the bid-ask nature of the prices or we used this information to derive an induced (and seemingly realistic) transaction cost for trading a given currency at any point in time. The Yen decreased with a return of 0.8841 while the DM increased with a return of 1.1568. It should also be noted that the differences in prices for consecutive ticks is usually quite small and thus frequent trading in the context of even small transaction costs (i.e. spreads) can be a very poor strategy. Note that for this particular FX data, CBAL-OPT is simply BAH-OPT.

Table 3 reports on the returns of the various algorithms for all three data sets with and without transaction costs. *Without transaction costs*, we see that simple learning algorithms such as MO can sometimes do quite well, while more aggressive strategies such as DELTA and LZ can have fantastic returns.

8 Portfolio Selection with Commissions and Bid-ask Spreads

Algorithmic portfolio selection with commissions is not so well studied. There can be many commission models. Two simple models are the flat fee model and proportional (or fixed rate) model. In some markets, such as foreign exchange there are no explicit commissions at all (for large volume trades) but a similar (and more complicated) effect is obtained due to buy-sell or bid-ask spreads. In the current reality, with the emerging Internet trading services, the effect of commissions on traders is becoming less significant but bid-ask spreads remain. The data for the FX market contains the bid-ask spreads. When we want to view bid-ask spreads as transaction costs we define the transaction rate as (ask-bid)/(ask+bid). The resulting (induced) transaction costs are quite non uniform (over time) ranging between .00015 and .0094 with a mean transaction rate of .00057.

Table 3 presents the returns for the various algorithms for all data sets using different transaction costs. For the NYSE and TSE data sets, we used fixed transaction cost rates of .1% (i.e. very small) and 2% (more or less "full service"). For the FX data we both artificially introduced fixed rate costs or used the actual bid-ask spreads as discussed above.

For the simple but important model of fixed rate transaction costs, it is not too difficult to extend the competitive analysis results to reflect such costs. In particular, Blum and Kalai [BK97] extend the proof of UNI's competitiveness to this model of transaction costs. Suppose then that there is a transaction rate cost of c ($0 \le c \le 1$); that is, to buy (or sell) \$$d$ of a stock costs \$$(\frac{c}{2})d$ or alternatively, we can say that all transaction costs are payed for by selling at a commission rate of c. Blum and Kalai prove that UNI has a competitive ratio upper bounded by $\binom{(1+c)n+m-1}{m-1}$ generalizing the bound in Equation 1 in Section 5.1. Using their proof in Section 5.1, one can obtain a bound of

$$\frac{\text{CBAL}_{\mathbf{b}^*}(X)}{\text{CBAL}_{\mathbf{b}}(X)} \le \left(1 + \frac{1}{n}\right)^{(1+c)n} \le e^{(1+c)}$$

whenever $\mathbf{b}$ is near $\mathbf{b}^*$ so that the competitive ratio is bounded above by $e^{(1+c)} \cdot (n + 1)^{m-1}$.

Blum and Kalai [BK97], and then Helmbold *et.al* [HSSW98] and Singer [Sin98] [10] present a few experimental results which seem to indicate that although transaction costs are significant, it is still possible to obtain reasonable returns from algorithms such as UNI and EG. Indeed for the NYSE data, EG "beats the market" even in the presence of 2% transaction costs. Our experiments seem to indicate that transaction costs may be much more problematic than some of the previous results and the theoretical competitiveness (say of UNI) suggests. Algorithms such as LZ and our DELTA algorithm can sometimes have exceptionally good returns when there are no transaction costs but disastrous returns with (not unreasonable) costs of 2%.

---

[10] We have been recently informed by Yoram Singer that the experimental results for his adaptive γ algorithm are not correct and, in particular, the results with transaction costs are not as encouraging as reported in [Sin98].

	UBAH	UCBAL	DELTA $(1,10,-.01)$	DELTA $(.1,4,-.01)$	MO$(.5)$
NYSE (0%)	14.4973	27.0752	1.9×10^{08}	326.585	111.849
NYSE (.1%)	14.4901	26.1897	1.7×10^{07}	246.224	105.975
NYSE (2%)	14.3523	13.9218	$1.6 \times 10^{\cdot\,13}$	1.15638	38.0202
TSE (0%)	1.61292	1.59523	4.99295	1.93648	1.27574
TSE (.1%)	1.61211	1.58026	2.91037	1.82271	1.2579
TSE (2%)	1.59679	1.32103	$9.9 \times 10^{\cdot\,05}$	.576564	.962617
FX (0%)	1.02047	1.0225	22094.4	3.88986	1.01852
FX (.1%)	1.01996	.984083	$1.6 \times 10^{\cdot\,26}$	$5.7 \times 10^{\cdot\,05}$	.979137
FX (2%)	1.01026	.475361	$1.1 \times 10^{\cdot\,322}$	$8.2 \times 10^{\cdot\,97}$	.462863
FX (bid-ask)	1.02016	.999421	$1.02 \times 10^{\cdot\,13}$	.00653223	.994662

	TO$(.5)$	EG$(.01)$	LZ	UNI	CBAL-OPT
NYSE (0%)	27.0614	27.0869	79.7863	13.8663	250.592
NYSE (.1%)	26.1773	26.2012	5.49837	13.8176	NC
NYSE (2%)	13.9252	14.6023	$3.5 \times 10^{\cdot\,22}$	9.90825	NC
TSE (0%)	1.59493	1.59164	1.32456	1.60067	6.43390
TSE (.1%)	1.58002	1.58006	.597513	1.58255	NC
TSE (2%)	1.32181	1.34234	$1.5 \times 10^{\cdot\,07}$	1.41695	NC
FX (0%)	1.0225	1.0223	716.781	1.02181	1.15682
FX (.1%)	.984083	.984435	$1.9 \times 10^{\cdot\,32}$	.996199	1.15624
FX (2%)	.475369	.51265	$1.6 \times 10^{\cdot\,322}$	.631336	1.14525
FX (bid-ask)	.999422	.999627	$1.04 \times 10^{\cdot\,17}$	1.00645	1.15661

Table 3. The returns of various algorithms for three different data sets using different transaction costs. Note that UNI and CBAL-OPT have only been approximated. The notation NC indicates an entry which has not yet been calculated.

9   Concluding Remarks and Future Work

From both a theoretical and experimental point of view, it is clear that a competitive based approach to portfolio selection is only just beginning to emerge. In contrast, the related topic of sequence prediction is much better developed and indeed the practice seems to closely follow the theory. There are, of course, at least two significant differences; namely that (first) sequence prediction is a special case of portfolio selection and (second) that transaction costs (or alternatively, bid-ask spreads) are a reality having a significant impact. In the terminology of metrical task systems and competitive analysis (see [BE98] and [BB97]), there is a cost to change states (i.e. portfolios).

On the other hand, PS algorithms can be applied to the area of expert prediction. Specifically, we view each expert as a stock whose log loss for a given prediction can be exponentiated to generate a stock price. Applying a PS algorithm to these prices yields a portfolio which can be interpreted as a mixture of experts. (See Ordentlich [Or96] and Kalai, Chen, Blum and Rosenfeld [KCBR99].)

Any useful online algorithm must at least "beat the market"; that is, the algorithm should be able to consistently equal and many times surpass the performance of the uniform buy and hold. In the absence of transaction costs, all of the online algorithms discussed in this paper were able to beat the market for the NYSE stock data. However, the same was not true for the TSE data, nor for the currency data. Furthermore, when even modest transaction costs (e.g. .1%) were introduced many of the algorithms suffered significant (and sometimes catastrophic) losses. This phenomena is most striking for the DELTA algorithm and for the LZ algorithm which is a very practical algorithm in the prediction setting.

Clearly the most obvious direction for future research is to understand the extent to which a competitive based theory of online algorithms can predict performance with regard to real stock market data. And here we are only talking about a theory that completely disregards feedback on the market of any successful algorithm. We conclude with a few questions of theoretical interest.

1. What is the competitive ratio for Ordentlich's "work function algorithm" and for the Lempel Ziv PS algorithm?
2. Can an online algorithm that only considers the Kelly projection of the market input sequence be competitive (or universal)? What other general classes of online algorithms can analyzed?
3. How can we define a portfolio selection "learning algorithm"? Is there a "true learning" PS algorithm that can attain the worst case competitive bounds of UNI or DIR?
4. To what extent can PS algorithms utilize "side information", as defined in Cover and Ordentlich [CO96]? See the very promising results in Helmbold *et al* [HSSW98].
5. Determine the optimal competitive ratio against CBAL-OPT and against OPT in the context of a fixed commission rate c.
6. Develop competitive bounds within the context of bid-ask spreads.
7. Continue the study of portfolio selection algorithms in the context of "short-selling". (See Vovk and Watkins [VW98].)
8. Consider other benchmark algorithms as the basis of a competitive theory.

Acknowledgments

We thank Rob Schapire for his very constructive comments. We also thank Steve Bellantoni for providing the TSE data.

References

Bl56.   Blackwell, D.: An Analog of the Minimax Theorem for Vector Payoffs. Pacific J. Math., 6, pp 1-8, 1956.

BE98.   Borodin, A., El-Yaniv, R.: Online Computation and Competitive Analysis. Cambridge University Press (1998)

BB97.   Blum, A., Burch, C.: On-line Learning and the Metrical task System Problem. Proceedings of the 10th Annual Conference on Computational Learning Theory (COLT '97), pages 45–53. To appear in Machine Learning.

BK97.   Blum, A., Kalai, A.: Universal portfolios with and without transaction costs. Machine Learning, 30:1, pp 23-30, 1998.

BCK92.   Bollerslev, T., Chou, R.Y., and Kroner, K.F.: ARCH Modeling in Finance: A selective review of the theory and empirical evidence. Journal of Econometrics, 52, 5-59.

BKM93.   Bodie, Z., Kane, A., Marcus, A.J.: Investments. Richard D. Irwin, Inc. (1993)

CG86.   Empirical Bayes Stock Market Portfolios. Advances in Applied Mathematics, 7, pp 170-181, 1986.

CO96.   Cover, T.M., Ordentlich, O.: Universal portfolios with side information. IEEE Transactions on Information Theory, vol. 42 (2), 1996.

Cov91.   Cover, T.M.: Universal portfolios. Mathematical Finance 1(1) (1991) 1–29

CT91.   Cover, T.M., Thomas, J.A.: Elements of Information Theory. John Wiley & Sons, Inc. (1991)

CB99.   Cross J.E. and Barron A.R.: Efficient universal portfolios for past dependent target classes, DIMACS Workshop: On-Line Decision Making, July, 1999.

Fed91.   Feder M.: Gambling using a finite state machine, IEEE Trans. Inform. Theory, vol. 37, pp. 1459–1465, Sept. 1991.

FG92.   Feder M., Gutman M.: Universal Prediction of Individual Sequences IEEE Trans. Inform. Theory, vol. 37, pp. 1459–1465, Sept. 1991.

Gre72.   Green, W.: Econometric Analysis Collier-McMillan, 1972

HSSW98.   Helmbold, D.P., Schapire, R.E., Singer, Y., Warmuth, M.K.: On-line portfolio selection using multiplicative updates. Mathematical Finance, vol. 8 (4), pp.325-347, 1998.

HW98.   Herbster, M., Warmuth, M.K.: Tracking the best expert. Machine Learning, vol. 32 (2), pp.1-29, 1998.

KCBR99.   Kalai, A., Chen S.,, Blum A., and Rosenfeld R.: On-line Algorithms for Combining Language Models. Proceedings of the International Conference on Acoustics, Speech, and Signal Processing (ICASSP), 1999.

Kel56.   Kelly, J.: A new interpretation of information rate. Bell Sys. Tech. Journal **35** (1956) 917–926

Krich98.   R.E. Krichevskiy, Laplace law of succession and universal encoding, *IEEE Trans. on Infor. Theory*. Vol. 44 No. 1, January 1998.

Lan83.   G.G. Langdon, A note on the Lempel-Ziv model for compressing individual sequences. IEEE Trans. Inform. Theory, vol. IT-29, pp. 284–287, 1983

Mar59.   Markowitz, H.: Portfolio Selection: Efficient Diversification of Investments. John Wiley and Sons (1959)

MF98.   Merhav, N., Feder, M.: Universal prediction. IEEE Trans. Inf. Theory **44**(6) (1998) 2124–2147

Or96.   Ordentlich, E.: Universal Investmeny and Universal Data Compression. PhD Thesis, Stanford University, 1996.

OC96.   Ordentlich, E., Cover, T.M.: The cost of achieving the best portfolio in hindsight. Accepted for publication in Mathematics of Operations Research. (Conference version appears in COLT 96 Proceedings under the title of "On-line portfolio selection".)

Ris83.   Rissanen, J.: A universal data compression system. IEEE Trans. Information Theory, vol IT-29, pp. 656-664, 1983.

Sin98.   Singer, Y.: Switching portfolios. International Journal of Neural Systems **84** (1997) 445-455

ZL78.   Ziv, J., Lempel, A.: Compression of individual sequences via variable rate coding. IEEE Trans. Information Theory, vol IT-24, pp. 530-536, 1978.

VW98.   Universal Portfolio Selection Vovk V. and Watkins C.: Universal Portfolio Selection, COLT 1998.

Almost k-Wise Independence and Hard Boolean Functions

Valentine Kabanets

Department of Computer Science
University of Toronto
Toronto, Canada
kabanets@cs.toronto.edu

Abstract. Andreev et al. [3] gave constructions of Boolean functions (computable by polynomial-size circuits) with large lower bounds for read-once branching program (1-b.p.'s): a function in P with the lower bound $2^{n \cdot \text{polylog}(n)}$, a function in quasipolynomial time with the lower bound $2^{n \cdot O(\log n)}$, and a function in LINSPACE with the lower bound $2^{n \cdot \log n \cdot O(1)}$. We point out alternative, much simpler constructions of such Boolean functions by applying the idea of almost k-wise independence more directly, without the use of discrepancy set generators for large affine subspaces; our constructions are obtained by derandomizing the probabilistic proofs of existence of the corresponding combinatorial objects. The simplicity of our new constructions also allows us to observe that there exists a Boolean function in $AC^0[2]$ (computable by a depth 3, polynomial-size circuit over the basis $\{\wedge, \oplus, 1\}$) with the optimal lower bound $2^{n \cdot \log n \cdot O(1)}$ for 1-b.p.'s.

1 Introduction

Branching programs represent a model of computation that measures the space complexity of Turing machines. Recall that a *branching program* is a directed acyclic graph with one source and with each node of out-degree at most 2. Each node of out-degree 2 (a branching node) is labeled by an index of an input bit, with one outgoing edge labeled by 0, and the other by 1; each node of out-degree 0 (a sink) is labeled by 0 or 1. The branching program accepts an input if there is a path from the source to a sink labeled by 1 such that, at each branching node of the path, the path contains the edge labeled by the input bit for the input index associated with that node. Finally, the *size* of a branching program is defined as the number of its nodes.

While there are no nontrivial lower bounds on the size of general branching programs, strong lower bounds were obtained for a number of explicit Boolean functions in restricted models (see, e.g., [12] for a survey). In particular, for *read-once branching programs (1-b.p.'s)* — where, on every path from the source to a sink, no two branching nodes are labeled by the same input index — exponential lower bounds of the form $2^{\Omega(\sqrt{n})}$ were given for explicit n-variable Boolean functions in [17,18,5,7,8,16,10,6,4] among others. Moreover, [7,8,6,4] exhibited Boolean functions in AC^0 that require 1-b.p.'s of size at least $2^{\Omega(\sqrt{n})}$.

G. Gonnet, D. Panario, and A. Viola (Eds.): LATIN 2000, LNCS 1776, pp. 197–206, 2000.
© Springer-Verlag Berlin Heidelberg 2000

After lower bounds of the form $2^{\Omega(\sqrt{n})}$ were obtained for 1-b.p.'s, the natural problem was to find an explicit Boolean function with the truly exponential lower bound $2^{\Omega(n)}$. The first such bound was proved in [1] for the Boolean function computing the parity of the number of triangles in a graph; the constant factor was later improved in [16]. With the objective to improve this lower bound, Savický and Žák [15] constructed a Boolean function in P that requires a 1-b.p. of size at least $2^{n-3\sqrt{n}}$, and gave a probabilistic construction of a Boolean function requiring a 1-b.p. of size at least $2^{n-O(\log n)}$. Finally, Andreev et al. [3] presented a Boolean function in LINSPACE∩P/poly with the optimal lower bound $2^{n-\log n+O(1)}$, and, by derandomizing the probabilistic construction in [15], a Boolean function in QP ∩ P/poly with the lower bound $2^{n-O(\log n)}$, as well as a Boolean function in P with the lower bound $2^{n-\text{polylog}(n)}$; here QP stands for the quasipolynomial time $n^{\text{polylog}(n)}$.

The combinatorics of 1-b.p.'s is quite well understood: a theorem of Simon and Szegedy [16], generalizing the ideas of many papers on the subject, provides a way of obtaining strong lower bounds. A particular case of this theorem states that any 1-b.p. computing an r-mixed Boolean function has size at least $2^r - 1$. Informally, an r-mixed function essentially depends on every set of r variables (see the next section for a precise definition). The reason why this lower-bound criterion works can be summarized as follows. A subprogram of a 1-b.p. G_n starting at a node v does not depend on any variable queried along any path going from the source s of G_n to v, and hence v completely determines a subfunction of the function computed by G_n. If G_n computes an r-mixed Boolean function f_n, then any two paths going from s to v can be shown to query the same variables, whenever v is sufficiently close to s. Hence, such paths must coincide, i.e., assign the same values to the queried variables; otherwise, two different assignments to a set of at most r variables yield the same subfunction of f_n, contradicting the fact that f_n is r-mixed. It follows that, near the source, G_n is a complete binary tree, and so it must have exponentially many nodes.

Andreev et al. [3] construct a Boolean function $f_n(x_1, \ldots, x_n)$ in LINSPACE∩ P/poly that is r-mixed for $r = n - \lceil \log n \rceil - 2$ for almost all n. By the lower-bound criterion mentioned above, this yields the optimal lower bound $\Omega(2^n/n)$ for 1-b.p.'s. A Boolean function in DTIME($2^{\log^2 n}$)∩P/poly that requires a 1-b.p. of size at least $2^{n-O(\log n)}$ is constructed by reducing the amount of randomness used in the probabilistic construction of [15] to $O(\log^2 n)$ advice bits. Since these bits turn out to determine a polynomial-time computable function with the lower bound $2^{n-O(\log n)}$, one gets a function in P with the lower bound $2^{n-O(\log^2 n)}$ by making the advice bits a part of the input.

Both constructions in [3] use the idea of ϵ-biased sample spaces introduced by Naor and Naor [9], who also gave an algorithm for generating small sample spaces; three simpler constructions of such spaces were later given by Alon et al. [2]. Andreev et al. define certain ϵ-discrepancy sets for systems of linear equations over GF(2), and relate these discrepancy sets to the biased sample spaces of Naor and Naor through a reduction lemma. Using a particular construction of a biased sample space (the powering construction from [2]), Andreev et al.

give an algorithm for generating ϵ-discrepancy sets, which is then used to de-randomize both a probabilistic construction of an r-mixed Boolean function for $r = n - \lceil \log n \rceil - 2$ and the construction in [15] mentioned above.

Our results. We will show that the known algorithms for generating small ϵ-biased sample spaces can be applied *directly* to get the r-mixed Boolean function as above, and to derandomize the construction in [15]. The idea of our first construction is very simple: treat the elements (bit strings) of an ϵ-biased sample space as the truth tables of Boolean functions. This will induce a probability distribution on Boolean functions such that, on any subset A of k inputs, the restriction to A of a Boolean function chosen according to this distribution will look almost as if it were a uniformly chosen random function defined on the set A. By an easy probabilistic argument, we will show that such a space of functions will contain the desired r-mixed function, for a suitable choice of parameters ϵ and k.

We indicate several ways of obtaining an r-mixed Boolean function with $r = n - \lceil \log n \rceil - 2$. In particular, using Razborov's construction of ϵ-biased sample spaces that are computable by $AC^0[2]$ formulas [11] (see also [13]), we prove that there are such r-mixed functions that belong to the class of polynomial-size depth 3 formulas over the basis $\{\&, \oplus, 1\}$. This yields the smallest (nonuniform) complexity class known to contain Boolean functions with the optimal lower bounds for 1-b.p.'s. (We remark that, given our lack of strong circuit lower bounds, it is conceivable that the characteristic function of every language in EXP can be computed in nonuniform $AC^0[6]$.)

In our second construction, we derandomize a probabilistic existence proof in [15]. We proceed along the usual path of derandomizing probabilistic algorithms whose analysis depends only on almost k-wise independence rather than full independence of random bits [9]. Observing that the construction in [15] is one such algorithm, we reduce its randomness complexity to $O(\log^3 n)$ bits (again treating strings of an appropriate sample space as truth tables). This gives us a $\text{DTIME}(2^{O(\log^3 n)})$-computable Boolean function of quasilinear circuit-size with the lower bound for 1-b.p.'s slightly better than that for the corresponding quasipolynomial-time computable function in [3], and a Boolean function in quasilinear time, QL, with the lower bound for 1-b.p.'s at least $2^{n-O(\log^3 n)}$, which is only slightly worse than the lower bound for the corresponding polynomial-time function in [3]. In the analysis of our construction, we employ a combinatorial lemma due to Razborov [11], which bounds from above the probability that none of n events occur, given that these events are almost k-wise independent.

The remainder of the paper. In the following section, we state the necessary definitions and some auxiliary lemmas. In Section 3, we show how to construct an r-mixed function that has the same optimal lower bound for 1-b.p. as that in [3], and observe that such a function can be computed in $AC^0[2]$. In Section 4, we give a simple derandomization procedure for a construction in [15], obtaining two more Boolean functions (computable in polynomial time and quasipolynomial time, respectively) that are hard with respect to 1-b.p.'s.

2   Preliminaries

Below we recall the standard definitions of k-wise independence and (ϵ, k)-independence. We consider probability distributions that are uniform over some set $S \subseteq \{0,1\}^n$; such a set is denoted by S_n and called a *sample space*.

Let S_n be a sample space, and let $X = x_1 \ldots x_n$ be a string chosen uniformly from S_n. Then S_n is k-*wise independent* if, for any k indices $i_1 < i_2 < \cdots < i_k$ and any k-bit string α, we have $\mathbf{Pr}[x_{i_1} x_{i_2} \ldots x_{i_k} = \alpha] = 2^{-k}$. Similarly, for S_n and X as above, S_n is (ϵ, k)-*independent* if $|\mathbf{Pr}[x_{i_1} x_{i_2} \ldots x_{i_k} = \alpha] - 2^{-k}| \leqslant \epsilon$ for any k indices $i_1 < i_2 < \cdots < i_k$ and any k-bit string α.

Naor and Naor [9] present an efficient construction of small (ϵ, k)-independent sample spaces; three simpler constructions are given in [2]. Here we recall just one construction from [2], the powering construction, although any of their three constructions could be used for our purposes.

Consider the Galois field $\mathrm{GF}(2^m)$ and the associated m-dimensional vector space over $\mathrm{GF}(2)$. For every element u of $\mathrm{GF}(2^m)$, let $\mathrm{bin}(u)$ denote the corresponding binary vector in the associated vector space. The sample space Pow_N^{2m} is defined as a set of N-bit strings such that each string ω is determined as follows. Two elements $x, y \in \mathrm{GF}(2^m)$ are chosen uniformly at random. For each $1 \leqslant i \leqslant N$, the ith bit ω_i is defined as $\langle \mathrm{bin}(x^i), \mathrm{bin}(y) \rangle$, where $\langle a, b \rangle$ denotes the inner product over $\mathrm{GF}(2)$ of binary vectors a and b.

Lemma 1 ([2]). *The sample space* Pow_N^{2m} *is* $\left(\frac{N}{2^m}, k\right)$-*independent for every* $k \leqslant N$.

As we have mentioned in the introduction, we shall view the strings of the sample space Pow_N^{2m} as the truth tables of Boolean functions of $\log N$ variables. It will be convenient to assume that N is a power of 2, i.e., $N = 2^n$. Thus, the uniform distribution over the sample space $\mathrm{Pow}_{2^n}^{2m}$ induces a distribution $\mathbf{F}_{n,m}$ on Boolean functions of n variables that satisfies the following lemma.

Lemma 2. *Let* A *be any set of* k *strings from* $\{0,1\}^n$, *for any* $k \leqslant 2^n$. *Let* ϕ *be any Boolean function defined on* A. *For a Boolean function* f *chosen according to the distribution* $\mathbf{F}_{n,m}$ *defined above, we have* $|\mathbf{Pr}[f|_A = \phi] - 2^{-k}| \leqslant 2^{-(m-n)}$, *where* $f|_A$ *denotes the restriction of* f *to the set* A.

Proof: The k strings in A determine k indices $i_1, \ldots, i_k$ in the truth table of f. The function ϕ is determined by its truth table, a binary string α of length k. Now the claim follows immediately from Lemma 1 and the definition of (ϵ, k)-independence. ∎

Razborov [11] showed that there exist complex combinatorial structures (such as the Ramsey graphs, rigid graphs, etc.) of exponential size which can be encoded by polynomial-size bounded-depth Boolean formulas over the basis $\{\&, \oplus, 1\}$. In effect, Razborov gave a construction of ϵ-biased sample spaces (using the terminology of [9]), where the elements of such sample spaces are the truth tables of $\mathrm{AC}^0[2]$-computable Boolean functions chosen according to a certain distribution on $\mathrm{AC}^0[2]$-formulas. We describe this distribution next.

For $n, m, l \in \mathbb{N}$, a random formula $\mathbf{F}(n, m, l)$ of depth 3 is defined as

$$\mathbf{F}(n, m, l) = \oplus_{\alpha=1}^{l} \&_{\beta=1}^{m} ((\oplus_{\gamma=1}^{n} \lambda_{\alpha\beta\gamma} x_{\gamma}) \oplus \lambda_{\alpha\beta}), \tag{1}$$

where $\{\lambda_{\alpha\beta}, \lambda_{\alpha\beta\gamma}\}$ is a collection of $(n+1)ml$ independent random variables uniformly distributed on $\{0, 1\}$. The following lemma shows that this distribution determines an ϵ-biased sample space; as observed in [13], a slight modification of the above construction yields somewhat better parameters, but the simpler construction would suffice for us here.

Lemma 3 ([11]). *Let $k, l, m \in \mathbb{N}$ be any numbers such that $k \leqslant 2^{m-1}$, let A be any set of k strings from $\{0, 1\}^n$, and let ϕ be any Boolean function defined on A. For a Boolean function f computed by the random formula $\mathbf{F}(n, m, l)$ defined in (1), we have $|\mathbf{Pr}[f|_A = \phi] - 2^{-k}| \leqslant e^{-l2^{-m}}$, where $f|_A$ denotes the restriction of f to the set A.*

The proof of Lemma 3 is most easily obtained by manipulating certain discrete Fourier transforms. We refer the interested reader to [11] or [13] for details.

Below we give the definitions of some classes of Boolean functions hard for 1-b.p.'s. We say that a Boolean function $f_n(x_1, \ldots, x_n)$ is r-*mixed* for some $r \leqslant n$ if, for every subset X of r input variables $\{x_{i_1}, \ldots, x_{i_r}\}$, no two distinct assignments to X yield the same subfunction of f in the remaining $n - r$ variables. We shall see in the following section that an r-mixed function for $r = n - \lceil \log n \rceil - 2$ has a nonzero probability in a distribution $\mathbf{F}_{n,m}$, where $m \in O(n)$, and in the distribution induced by the random formula $\mathbf{F}(n, m, l)$, where $m \in O(\log n)$ and $l \in \text{poly}(n)$.

It was observed by many researchers that r-mixed Boolean functions are hard for 1-b.p.'s. The following lemma is implicit in [17,5], and is a particular case of results in [7,16].

Lemma 4 ([17,5,7,16]). *Let $f_n(x_1, \ldots, x_n)$ be an r-mixed Boolean function, for some $r \leqslant n$. Then every 1-b.p. computing f_n has size at least $2^r - 1$.*

Following Savický and Žák [15], we call a function $\phi : \{0, 1\}^n \to \{1, 2, \ldots, n\}$ (s, n, q)-*complete*, for some integers s, n, and q, if for every set $I \subseteq \{1, \ldots, n\}$ of size $n - s$ we have

1. for every 0-1 assignment to the variables x_i, $i \in I$, the range of the resulting subfunction of ϕ is equal to $\{1, 2, \ldots, n\}$, and
2. there are at most q different subfunctions of ϕ, as one varies over all 0-1 assignments to x_i, $i \in I$.

Our interest in (s, n, q)-complete functions is justified by the following lemma; its proof is based on a generalization of Lemma 4.

Lemma 5 ([15]). *Let $\phi : \{0, 1\}^n \to \{1, 2, \ldots, n\}$ be an (s, n, q)-complete function. Then the Boolean function $f_n(x_1, \ldots, x_n) = x_{\phi(x_1, \ldots, x_n)}$ requires 1-b.p.'s of size at least $2^{n-s}/q$.*

The following lemma can be used to construct an (s, n, q)-complete function.

Lemma 6 ([15]). *Let A be a $t \times n$ matrix over $\mathrm{GF}(2)$ with every $t \times s$ submatrix of rank at least r. Let $\psi : \{0, 1\}^t \to \{1, 2, \ldots, n\}$ be a mapping such that its restriction to every affine subset of $\{0, 1\}^t$ of dimension at least r has the range $\{1, 2, \ldots, n\}$. Then the function $\phi(\boldsymbol{x}) = \psi(A\boldsymbol{x})$ is $(s, n, 2^t)$-complete.*

A probabilistic argument shows that a $t \times n$ matrix A and a function $\psi : \{0, 1\}^t \to \{1, 2, \ldots, n\}$ exist that satisfy the assumptions of Lemma 6 for the choice of parameters $s, t, r \in O(\log n)$, thereby yielding a Boolean function that requires 1-b.p.'s of size at least $2^{n - O(\log n)}$. Below we will show that the argument uses only limited independence of random bits, and hence it can be derandomized using the known constructions of (ϵ, k)-independent spaces. Our proof will utilize the following lemma of Razborov.

Lemma 7 ([11]). *Let $l > 2k$ be any natural numbers, let $0 < \theta, \epsilon < 1$, and let $\mathcal{E}_1, \ldots, \mathcal{E}_l$ be events such that, for every subset $I \subseteq \{1, \ldots, l\}$ of size at most k, we have $|\mathbf{Pr}[\wedge_{i \in I} \mathcal{E}_i] - \theta^{|I|}| \leqslant \epsilon$. Then $\mathbf{Pr}[\wedge_{i=1}^{l} \bar{\mathcal{E}}_i] \leqslant e^{-\theta l} + \binom{l}{k+1}(\epsilon k + \theta^k)$.*

3   Constructing r-Mixed Boolean Functions

First, we give a simple probabilistic argument showing that r-mixed functions exist for $r = n - \lceil \log n \rceil - 2$. Let f be a Boolean function on n variables that is chosen uniformly at random from the set of all Boolean n-variable functions. For any fixed set of indices $\{i_1, \ldots, i_r\} \subseteq \{1, \ldots, n\}$ and any two fixed binary strings $\alpha = \alpha_1, \ldots, \alpha_r$ and $\beta = \beta_1, \ldots, \beta_r$, the probability that fixing $x_{i_1}, \ldots, x_{i_r}$ to α and then to β will give the same subfunction of f in the remaining $n - r$ variables is 2^{-k}, where $k = 2^{n-r}$. Thus, the probability that f is not r-mixed is at most $\binom{n}{r} 2^{2r} 2^{-k}$, which tends to 0 as n grows.

We observe that the above argument only used the fact that f is random on any set of $2k$ inputs: those obtained after the r variables $x_{i_1}, \ldots, x_{i_r}$ are fixed to α, the set of which will be denoted as A_α, plus those obtained after the same variables are fixed to β, the set of which will be denoted as A_β. This leads us to the following theorem.

Theorem 1. *There is an $m \in O(n)$ for which the probability that a Boolean n-variable function f chosen according to the distribution $\mathbf{F}_{n,m}$ is r-mixed, for $r = n - \lceil \log n \rceil - 2$, tends to 1 as n grows.*

Proof: By Lemma 2, the distribution $\mathbf{F}_{n,m}$ yields a function f which is equal to any fixed Boolean function ϕ defined on a set $A_\alpha \cup B_\beta$ of $2k$ inputs with probability at most $2^{-2k} + 2^{-(m-n)}$. The number of functions ϕ that assume the same values on the corresponding pairs of elements $a \in A_\alpha$ and $b \in A_\beta$ is 2^k. Thus, the probability that f is not r-mixed is at most $\binom{n}{r} 2^{2r} (2^{-k} + 2^{-(m-n-k)})$. If $m = (7 + \delta)n$ for any $\delta > 0$, then this probability tends to 0 as n grows.   ■

By definition, each function from $\mathbf{F}_{n,m}$ can be computed by a Boolean circuit of size $\mathrm{poly}(n, m)$. It must be also clear that checking whether a function from

$\mathbf{F}_{n,m}$, given by a $2m$-bit string, is r-mixed can be done in LINSPACE. It follows from Theorem 1 that we can find an r-mixed function, for $r = n - \lceil \log n \rceil - 2$, in LINSPACE by picking the lexicographically first string of $2m$ bits that determines such a function. By Lemma 4, this function will have the optimal lower bound for 1-b.p.'s, $\Omega(2^n/n)$.

We should point out that any of the three constructions of small (ϵ, k)-independent spaces in [2] could be used in the same manner as described above to obtain an r-mixed Boolean function computable in LINSPACE $\cap$ P/poly, for $r = n - \lceil \log n \rceil - 2$. Applying Lemma 3, we can obtain an r-mixed function with the same value of r.

Theorem 2. *There are $m \in O(\log n)$ and $l \in \mathrm{poly}(n)$ for which the probability that a Boolean n-variable function f computed by the random formula $\mathbf{F}(n, m, l)$ defined in (1) is r-mixed, for $r = n - \lceil \log n \rceil - 2$, tends to 1 as n grows.*

Proof: Proceeding as in the proof of Theorem 1, with Lemma 3 applied instead of Lemma 2, we obtain that the probability that f is not r-mixed is at most $\binom{n}{r}2^{2r}(2^{-k} + 2^{-(l2^{-m}-k)})$. If $m = \lceil \log n \rceil + 3$ and $l = (6 + \delta)n^2$ for any $\delta > 0$, then this probability tends to 0 as n grows. ∎

Corollary 1. *There exists a Boolean function computable by a polynomial-size depth 3 formula over the basis $\{\&, \oplus, 1\}$ that requires a 1-b.p. of size at least $\Omega(2^n/n)$ for all sufficiently large n.*

4   Constructing (s, n, q)-Complete Functions

Let us take a look at the probabilistic proof (as presented in [15]) of the existence of a matrix A and a function ψ with the properties assumed in Lemma 6. Suppose that a $t \times n$ matrix A over GF(2) and a function $\psi : \{0, 1\}^t \to \{1, 2, \dots, n\}$ are chosen uniformly at random. For a fixed $t \times s$ submatrix B of A, if $\mathrm{rank}(B) < r$, then there is a set of at most $r - 1$ columns in B whose linear span contains each of the remaining $s - r + 1$ columns of B. For a fixed set R of such $r - 1$ columns in B, the probability that each of the $s - r + 1$ vectors chosen uniformly at random will be in the linear span of R is at most $(2^{r-1}/2^t)^{s-r+1}$. Thus, the probability that the matrix A is "bad" is at most $\binom{n}{s}\binom{s}{r-1}2^{-(t-r+1)(s-r+1)}$.

For a fixed affine subspace H of $\{0, 1\}^t$ of dimension r and a fixed $1 \leqslant i \leqslant n$, the probability that the range of ψ restricted to H does not contain i is at most $(1 - 1/n)^{2^r}$. The number of different affine subspaces of $\{0, 1\}^t$ of dimension r is at most $2^{(r+1)t}$; the number of different i's is n. Hence the probability that ψ is "bad" is at most $2^{(r+1)t}n(1 - 1/n)^{2^r} \leqslant 2^{(r+1)t}ne^{-2^r/n}$.

An easy calculation shows that setting $s = \lceil (2 + \delta) \log n \rceil$, $t = \lceil (3 + \delta) \log n \rceil$, and $r = \lceil \log n + 2 \log \log n + b \rceil$, for any $\delta > 0$ and sufficiently large b (say, $b = 3$ and $\delta = 0.01$), makes both the probability that A is "bad" and the probability that ψ is "bad" tend to 0 as n grows.

Theorem 3. *There are $d_1, d_2, d_3 \in \mathbb{N}$ such that every $(2^{-d_1 \log^3 n}, d_2 \log^2 n)$-independent sample space over n^{d_3}-bit strings contains both matrix A and function ψ with the properties as in Lemma 6, for $s, r, t \in O(\log n)$.*

Proof: We observe that both probabilistic arguments used only partial independence of random bits. For A, we need a tn-bit string coming from an (ϵ, k)-independent sample space with $k = ts$ and $\epsilon = 2^{-c_1 \log^2 n}$, for a sufficiently large constant c_1. Indeed, for a fixed $t \times s$ submatrix B of A and a fixed set R of $r - 1$ columns in B, the number of "bad" $t \times s$-bit strings α filling B so that the column vectors in R contain in their linear span all the remaining $s - r + 1$ column vectors of B is at most $2^{(r-1)t} 2^{(r-1)(s-r+1)} = 2^{(r-1)(s+t-r+1)}$. If A is chosen from the (ϵ, k)-independent sample space with ϵ and k as above, then the probability that some fixed "bad" string α is chosen is at most $2^{-ts} + \epsilon$. Thus, in this case, the probability that A is "bad" is at most

$$\binom{n}{s}\binom{s}{r-1}\left(2^{-(t-r+1)(s-r+1)} + \epsilon 2^{(r-1)(s+t-r+1)}\right).$$

Choosing the same s, t, and r as in the case of fully independent probability distribution, one can make this probability tend to 0 as n grows, by choosing sufficiently large c_1.

Similarly, for the function ψ, we need a $2^t \lceil \log n \rceil$-bit string from an (ϵ, k)-independent sample space with $k = c_2 \log^2 n$ and $\epsilon = 2^{-c_3 \log^3 n}$, for sufficiently large constants c_2 and c_3. Here we view the truth table of ψ as a concatenation of $2^t \lceil \log n \rceil$-bit strings, where each $\lceil \log n \rceil$-bit string encodes a number from $\{1, \ldots, n\}$. The proof, however, is slightly more involved in this case, and depends on Lemma 7.

Let s, r, and t be the same as before. For a fixed affine subspace $H \subseteq \{0,1\}^t$ of dimension r, such that $H = \{a_1, \ldots, a_l\}$ for $l = 2^r$, and for a fixed $1 \leqslant i \leqslant n$, let $\mathcal{E}_j$, $1 \leqslant j \leqslant l$, be the event that $\psi(a_j) = i$ when ψ is chosen from the (ϵ, k)-independent sample space defined above. Then Lemma 7 applies with $\theta = 2^{-\lceil \log n \rceil}$, yielding that the probability that ψ misses the value i on the subspace H is

$$\mathbf{Pr}[\wedge_{j=1}^l \bar{\mathcal{E}}_j] \leqslant e^{-2^{r-\lceil \log n \rceil}} + \binom{2^r}{k+1}(\epsilon k + 2^{-k\lceil \log n \rceil}). \tag{2}$$

It is easy to see that the first term on the right-hand side of (2) is at most $e^{-4 \log^2 n}$ (when $b = 3$ in r). We need to bound from above the remaining two terms: $\binom{2^r}{k+1} 2^{-k\lceil \log n \rceil}$ and $\binom{2^r}{k+1} \epsilon k$. Using Stirling's formula, one can show that the first of these two terms can be made at most $2^{-4 \log^2 n}$, by choosing c_2 sufficiently large. Having fixed c_2, we can also make the second of the terms at most $2^{-4 \log^2 n}$, by choosing $c_3 > c_2$ sufficiently large. It is then straightforward to verify that the probability that ψ misses at least one value i, $1 \leqslant i \leqslant n$, on at least one affine subspace of dimension r tends to 0 as n grows. ∎

Using any efficient construction of almost independent sample spaces, for example, Pow_N^{2m} with $N = tn \in O(n \log n)$ and $m \in O(\log^2 n)$, we can find a matrix A with the required properties in $\text{DTIME}(2^{O(\log^2 n)})$ by searching through all elements of the sample space and checking whether any of them yields a desired matrix. Analogously, we can find the required function ψ in $\text{DTIME}(2^{O(\log^3 n)})$, by considering, e.g., $\text{Pow}_{N'}^{2m'}$ with $N' = 2^t \lceil \log n \rceil$ and $m' \in O(\log^3 n)$. Thus, constructing both A and ψ can be carried out in quasipolynomial time.

Given the corresponding advice strings of $O(\log^3 n)$ bits, ψ is computable in time polylog(n) and all elements of A can be computed in time $n$polylog(n). So, in this case, the function $\phi(\boldsymbol{x}) = \psi(A\boldsymbol{x})$ is computable in quasilinear time. Hence, by "hard-wiring" good advice strings, we get the function $f_n(\boldsymbol{x}) = x_{\phi(\boldsymbol{x})}$ computable by quasilinear-size circuits, while, by Lemmas 5 and 6, f_n requires 1-b.p.'s of size at least $2^{n-(5+\epsilon)\log n}$, for any $\epsilon > 0$ and sufficiently large n; these parameters appear to be better than those in [3]. By making the advice strings a part of the input, we obtain a function in QL that requires 1-b.p.'s of size at least $2^{n-O(\log^3 n)}$.

We end this section by observing that the method used above to construct an (s, n, q)-complete Boolean function could be also used to construct an r-mixed Boolean function for $r = n - O(\log n)$ by derandomizing Savický's [14] modification of the procedure in [15]. This r-mixed function is also determined by an advice string of length polylog(n), and hence can be constructed in quasipolynomial time.

5   Concluding Remarks

We have shown how the well-known constructions of small ϵ-biased sample spaces [11,9,2] can be directly used to obtain Boolean functions that are exponentially hard for 1-b.p.'s. One might argue, however, that the hard Boolean functions constructed in Sections 3 and 4 are not "explicit" enough, since they are defined as the lexicographically first functions in certain search spaces. It would be interesting to find a Boolean function in P or NP with the optimal lower bound $\Omega(2^n/n)$ for 1-b.p.'s. The problem of constructing a polynomial-time computable r-mixed Boolean function with r as large as possible is of independent interest; at present, the best such function is given in [15] for $r = n - \Omega(\sqrt{n})$. A related open question is to determine whether the minimum number of bits needed to specify a Boolean function with the optimal lower bound for 1-b.p.'s, or an r-mixed Boolean function for $r = n - \lceil \log n \rceil - 2$, can be sublinear.

Acknowledgements. I am indebted to Alexander Razborov for bringing [11] to my attention. I would like to thank Stephen Cook and Petr Savický for their comments on a preliminary version of this paper, and Dieter van Melkebeek for helpful discussions. I also want to express my sincere gratitude to Stephen Cook for his constant encouragement and support. Finally, I am grateful to the anonymous referee for suggestions and favourable comments.

References

1. M. Ajtai, L. Babai, P. Hajnal, J. Komlós, P. Pudlak, V. Rödl, E. Szemerédi, and G. Turán. Two lower bounds for branching programs. In *Proceedings of the Eighteenth Annual ACM Symposium on Theory of Computing*, pages 30–38, 1986.

2. N. Alon, O. Goldreich, J. Håstad, and R. Peralta. Simple constructions of almost k−wise independent random variables. *Random Structures and Algorithms*, 3(3):289–304, 1992. (preliminary version in FOCS'90).

3. A.E. Andreev, J.L. Baskakov, A.E.F. Clementi, and J.D.P. Rolim. Small pseudorandom sets yield hard functions: New tight explicit lower bounds for branching programs. *Electronic Colloquium on Computational Complexity*, TR97-053, 1997.

4. B. Bollig and I. Wegener. A very simple function that requires exponential size read-once branching programs. *Information Processing Letters*, 66:53–57, 1998.

5. P.E. Dunne. Lower bounds on the complexity of one-time-only branching programs. In L. Budach, editor, *Proceedings of the Second International Conference on Fundamentals of Computation Theory*, volume 199 of *Lecture Notes in Computer Science*, pages 90–99, Springer Verlag, Berlin, 1985.

6. A. Gal. A simple function that requires exponential size read-once branching programs. *Information Processing Letters*, 62:13–16, 1997.

7. S. Jukna. Entropy of contact circuits and lower bound on their complexity. *Theoretical Computer Science*, 57:113–129, 1988.

8. M. Krause, C. Meinel, and S. Waack. Separating the eraser Turing machine classes L_e, NL_e, co − NL_e and P_e. *Theoretical Computer Science*, 86:267–275, 1991.

9. J. Naor and M. Naor. Small-bias probability spaces: Efficient constructions and applications. *SIAM Journal on Computing*, 22(4):838–856, 1993. (preliminary version in STOC'90).

10. S. Ponzio. A lower bound for integer multiplication with read-once branching programs. *SIAM Journal on Computing*, 28(3):798–815, 1999. (preliminary version in STOC'95).

11. A.A. Razborov. Bounded-depth formulae over {&, ⊕} and some combinatorial problems. In S. I. Adyan, editor, *Problems of Cybernetics. Complexity Theory and Applied Mathematical Logic*, pages 149–166. VINITI, Moscow, 1988. (in Russian).

12. A.A. Razborov. Lower bounds for deterministic and nondeterministic branching programs. In L. Budach, editor, *Proceedings of the Eighth International Conference on Fundamentals of Computation Theory*, volume 529 of *Lecture Notes in Computer Science*, pages 47–60, Springer Verlag, Berlin, 1991.

13. P. Savický. Improved Boolean formulas for the Ramsey graphs. *Random Structures and Algorithms*, 6(4):407–415, 1995.

14. P. Savický. personal communication, January 1999.

15. P. Savický and S. Zák. A large lower bound for 1-branching programs. *Electronic Colloquium on Computational Complexity*, TR96-036, 1996.

16. J. Simon and M. Szegedy. A new lower bound theorem for read-only-once branching programs and its applications. In J.-Y. Cai, editor, *Advances in Computational Complexity*, pages 183–193. AMS-DIMACS Series, 1993.

17. I. Wegener. On the complexity of branching programs and decision trees for clique function. *Journal of the ACM*, 35:461–471, 1988.

18. S. Zak. An exponential lower bound for one-time-only branching programs. In *Proceedings of the Eleventh International Symposium on Mathematical Foundations of Computer Science*, volume 176 of *Lecture Notes in Computer Science*, pages 562–566, Springer Verlag, Berlin, 1984.

Improved Upper Bounds on the Simultaneous Messages Complexity of the Generalized Addressing Function

Andris Ambainis[1] and Satyanarayana V. Lokam[2]

[1] Department of Computer Science,
University of California at Berkeley,
Berkeley, CA.
ambainis@cs.berkely.edu

[2] Department of Mathematical and Computer Sciences,
Loyola University Chicago,
Chicago, IL 60626.
satya@math.luc.edu

Abstract. We study communication complexity in the model of Simultaneous Messages (SM). The SM model is a restricted version of the well-known multiparty communication complexity model [CFL,KN]. Motivated by connections to circuit complexity, lower and upper bounds on the SM complexity of several explicit functions have been intensively investigated in [PR,PRS,BKL,Am1,BGKL].

A class of functions called the Generalized Addressing Functions (GAF), denoted $GAF_{G,k}$, where G is a finite group and k denotes the number of players, plays an important role in SM complexity. In particular, lower bounds on SM complexity of $GAF_{G,k}$ were used in [PRS] and [BKL] to show that the SM model is exponentially weaker than the general communication model [CFL] for sufficiently small number of players. Moreover, certain unexpected *upper bounds* from [PRS] and [BKL] on SM complexity of $GAF_{G,k}$ have led to refined formulations of certain approaches to circuit lower bounds.

In this paper, we show improved upper bounds on the SM complexity of $GAF_{\mathbb{Z}_2^t,k}$. In particular, when there are three players ($k = 3$), we give an upper bound of $O(n^{0.73})$, where $n = 2^t$. This improves a bound of $O(n^{0.92})$ from [BKL]. The lower bound in this case is $\Omega(\sqrt{n})$ [BKL,PRS]. More generally, for the k player case, we prove an upper bound of $O(n^{H(1/(2k-2))})$ improving a bound of $O(n^{H(1/k)})$ from [BKL], where $H(\cdot)$ denotes the binary entropy function. For large enough k, this is nearly a quadratic improvement. The corresponding lower bound is $\Omega(n^{1/(k-1)}/(k-1))$ [BKL,PRS]. Our proof extends some algebraic techniques from [BKL] and employs a greedy construction of covering codes.

1 Introduction

The Multiparty Communication Model: The model of multiparty communication complexity plays a fundamental role in the study of Boolean function

G. Gonnet, D. Panario, and A. Viola (Eds.): LATIN 2000, LNCS 1776, pp. 207–216, 2000.

complexity. It was introduced by Chandra, Furst, and Lipton [CFL] and has been intensively studied (see the book by Kushilevitz and Nisan [KN] and references therein). In a multiparty communication game, k players wish to collaboratively evaluate a Boolean function $f(x_0, \ldots, x_{k-1})$. The i-th player knows each input argument except x_i; we will refer to x_i as the input *missed* by player i. We can imagine input x_i written on the forehead of player i. The players communicate using a blackboard, visible to all the players. Each player has unlimited computational power. The "algorithm" followed by the players in their exchange of messages is called a *protocol*. The *cost* of a protocol is the total number of bits communicated by the players in evaluating f at a worst-case input. The multiparty communication complexity of f is then defined as the minimum cost of a protocol for f.

The Simultaneous Messages (SM) Model: A restricted model of mulitparty communication complexity, called the Simultaneous Messages (SM) model, recently attracted much attention. It was implicit in a paper by Nisan and Wigderson [NW, Theorem 7] for the case of three players. The first papers investigating the SM model in detail are by Pudlák, Rödl, and Sgall [PRS] (under the name "Oblivious Communication Complexity"), and independently, by Babai, Kimmel, and Lokam [BKL].

In the k-party SM model, we have k players as before with input x_i of $f(x_0, \ldots, x_{k-1})$ written on the forehead of the i-th player. However, in this model, the players are *not* allowed to communicate with each other. Instead, each player simultaneously sends a single message to a *referee* who sees none of the input. The referee announces the value of the function upon receiving the messages from the players. An *SM protocol* specifies how each of the players can determine the message to be sent based on the part of the input that player sees, as well as how the referee would determine the value of the function based on the messages received from the players. All the players and the referee are assumed to have infinite computational power. The cost of an SM protocol is defined to be the *maximum* number of bits sent by a player to the referee, and the *SM-complexity* of f is defined to be the minimum cost of an SM protocol for f on a worst-case input. Note that in the SM model, we use the ℓ_∞-norm of the message lengths as the complexity measure as opposed to the ℓ_1-norm in the general model from [CFL] described above.

The main motivation for studying the SM model comes from the observation that sufficiently strong lower bounds in this restricted model already have some of the same interesting consequences to Boolean circuit complexity as the general multiparty communication model. Moreover, it is proved in [PRS] and [BKL] that the SM model is exponentially weaker than the general communication model when the number of players is at most $(\log n)^{1-\epsilon}$ for any constant $\epsilon > 0$. This exponential gap is proved by comparing the complexities of the Generalized Addressing Function (GAF) in the respective models.

Generalized Addressing Function (GAF): The input to $\mathrm{GAF}_{G,k}$, where G is a group of order n, consists of $n + (k-1) \log n$ bits partitioned among the players as follows: player 0 gets a function $x_0 : G \longrightarrow \{0, 1\}$ (represented as an

n-bit string) on her forehead whereas players 1 through $k-1$ get group elements $x_1, \ldots, x_{k-1}$, respectively, on their foreheads. The output of $\mathrm{GAF}_{G,k}$ on this input is the value of the function x_0 on $x_1 \circ \ldots \circ x_{k-1}$, where $\circ$ represents the group operation in G. Formally,

$$\mathrm{GAF}_{G,k}(x_0, x_1, \ldots, x_{k-1}) := x_0(x_1 \circ \ldots \circ x_{k-1}).$$

In [BKL], a general lower bound is proved on the SM complexity of $\mathrm{GAF}_{G,k}$ for any finite group G. In particular, they prove a lower bound $\Omega\left(n^{1/(k-1)}/(k-1)\right)$ for $\mathrm{GAF}_{\mathbb{Z}_n,k}$ (i.e., G is a cyclic group) and $\mathrm{GAF}_{\mathbb{Z}_2^t,k}$ (i.e., G is a vector space over $GF(2)$). Pudlák, Rödl, and Sgall [PRS] consider the special case of $\mathrm{GAF}_{\mathbb{Z}_2^t,k}$ and prove the same lower bound using essentially the same technique.

Upper Bounds on SM complexity: While the Simultaneous Messages model itself was motivated by lower bound questions, there have been some unexpected developments in the direction of *upper bounds* in this model. We describe some of these upper bounds and their significance below. This paper is concerned with upper bounds on the SM complexity of $\mathrm{GAF}_{\mathbb{Z}_2^t,k}$.

The results of [BKL] and [PRS] include some upper bounds on the SM complexity of $\mathrm{GAF}_{\mathbb{Z}_2^t,k}$ and $\mathrm{GAF}_{\mathbb{Z}_n,k}$, respectively. In [BGKL], upper bounds are also proved on a class of functions defined by certain depth-2 circuits. This class included the "Generalized Inner Product" (GIP) function, which was a prime example in the study and applications of multiparty communication complexity [BNS,G,HG,RW], and the "Majority of Majorities" function.

Babai, Kimmel, and Lokam [BKL] show an $O(n^{0.92})$ upper bound for $\mathrm{GAF}_{\mathbb{Z}_2^t,3}$, i.e., on the 3-party SM complexity of $\mathrm{GAF}_{G,k}$, when $G = \mathbb{Z}_2^t$. More generally, they show an $O(n^{H(1/k)})$ upper bound for $\mathrm{GAF}_{\mathbb{Z}_2^t,k}$. Pudlák, Rödl, and Sgall [PRS] prove upper bounds for $\mathrm{GAF}_{\mathbb{Z}_n,k}$. They show an $O(n \log \log n/ \log n)$ upper bound for $k = 3$, and an $O(n^{6/7})$ for $k \geq c \log n$. (Actually, upper bounds in [PRS] are proved for the so-called "restricted semilinear protocols," but it is easy to see that they imply essentially the same upper bounds on SM complexity.) These upper bounds are significantly improved by Ambainis [Am1] to $O\left(n \log^{1/4} n/2^{\sqrt{\log n}}\right)$ for $k = 3$ and to $O(n^\epsilon)$ for an arbitrary $\epsilon > 0$ for $k = O((\log n)^{c(\epsilon)})$. Note that the upper bounds for $\mathrm{GAF}_{\mathbb{Z}_2^t,k}$ are much better than those for $\mathrm{GAF}_{\mathbb{Z}_n,k}$. It is interesting that the upper bounds, in contrast to the lower bound, appear to depend heavily on the structure of the group G. Specifically, the techniques used in [BKL] and in this paper for $\mathrm{GAF}_{\mathbb{Z}_2^t,k}$ and those used in [PRS] and [Am1] for $\mathrm{GAF}_{\mathbb{Z}_n,k}$ seem to be quite different.

Our upper bounds: In this paper, we give an $O(n^{0.73})$ upper bound on the (3-player) SM complexity of $\mathrm{GAF}_{\mathbb{Z}_2^t,3}$, improving the upper bound of $O(n^{0.92})$ from [BKL] for the same problem. For general k we show an upper bound of $O(n^{H(1/(2k-2))})$ for $\mathrm{GAF}_{\mathbb{Z}_2^t,k}$ improving the upper bound of $O(n^{H(1/k)})$ from [BKL]. For large k, this is nearly a quadratic improvement. The lower bound on the SM complexity of $\mathrm{GAF}_{\mathbb{Z}_2^t,k}$ is $\Omega\left(n^{1/(k-1)}(k-1)\right)$. Our results extend some of the algebraic ideas from [BKL] and employ a greedy construction of *covering codes* of $\{0,1\}^n$.

Significance of Upper Bounds: Upper bounds are obviously useful in assessing the strength of lower bounds. However, upper bounds on SM complexity are interesting for several additional reasons.

First of all, upper bounds on SM complexity of GAF have led to a refined formulation of a communication complexity approach to a circuit lower bound problem. Before the counterintuitive upper bounds proved in [PR,PRS,BKL], it appeared natural to conjecture that the k-party SM complexity of GAF should be $\Omega(n)$ when k is a constant. In fact, proving an $\omega(n/\log\log n)$ lower bound on the *total* amount of communication in a 3-party SM protocol for GAF would have proved superlinear size lower bounds on log-depth Boolean circuits computing an (n-output) explicit function. This communication complexity approach toward superlinear lower bounds for log-depth circuits is due to Nisan and Wigderson [NW] and is based on a graph-theoretic reduction due to Valiant [Va]. However, results from [PR,PRS,BKL] provide $o(n/\log\log n)$ *upper bounds* on the total communication of 3-party SM protocols for $\mathrm{GAF}_{\mathbb{Z}_2^t,3}$ and $\mathrm{GAF}_{\mathbb{Z}_n,3}$ and hence ruled out the possibility of using lower bounds on total SM complexity to prove the circuit lower bound mentioned above. On the other hand, these and similar functions are expected to require superlinear size log-depth circuits. This situation motivated a more careful analysis of Valiant's reduction and a refined formulation of the original communication complexity approach. In the refined approach, proofs of nonexistence of 3-party SM protocols are sought when there are certain constraints on the number of bits sent by *individual players* as opposed to lower bounds on the *total* amount of communication. This new approach is described by Kushilevitz and Nisan in their book [KN, Section 11.3].

Secondly, Pudlák, Rödl, and Sgall [PRS] use their upper bounds on restricted semilinear protocols to disprove a conjecture of Razborov's [Ra] concerning the *contact rank* of tensors.

Finally, the combinatorial and algebraic ideas used in designing stronger upper bounds on SM complexity may find applications in other contexts. For example, Ambainis [Am1] devised a technique to recursively compose SM protocols and used this to improve upper bounds from [PRS]. Essentially similar techniques enabled him in [Am2] to improve upper bounds from [CGKS] on the communication complexity of k-server Private Information Retrieval (PIR) schemes.

1.1   Definitions and Preliminaries

Since we consider the SM complexity of $\mathrm{GAF}_{\mathbb{Z}_2^t,k}$ only, we omit the subscripts for simplicity of notation. We describe the the 3-player protocol in detail. The definitions and results extend naturally to the k-party case for general k.

We recall below the definition of GAF from the Introduction (in the special case when $G = \mathbb{Z}_2^\ell$, and we write A for x_0, the input on forehead of Player 0).

Definition 1. *Assume* $n = 2^\ell$ *for a positive integer* ℓ. *Then, the function* $GAF(A, x_1, \ldots, x_{k-1})$, *where* $A \in \{0,1\}^n$ *and* $x_i \in \{0,1\}^\ell$ *is defined by*

$$GAF(A, x_1, \ldots, x_{k-1}) := A(x_1 \oplus \cdots \oplus x_{k-1}),$$

where A *is viewed as an* ℓ-*input Boolean function* $A : \{0,1\}^\ell \longrightarrow \{0,1\}$.

Note that Player 0 knows only $(k-1) \log n$ bits of information. This information can be sent to the referee if each of players i, for $1 \leq i \leq k-2$, sends x_{i+1} and player $k-1$ sends x_1 to the referee. (This adds a $\log n$ term to the lengths of messages sent by these players and will be insignificant.) Thus, we can assume that player 0 remains silent for the entire protocol and that the referee knows all the inputs $x_1, \ldots, x_{k-1}$. The goal of players 1 through $k-1$ is to send enough information about A to the referee to enable him to compute $A(x_1 \oplus \cdots \oplus x_{k-1})$.

We will use the following notation:

$$\Lambda(\ell, b) := \sum_{j=0}^{b} \binom{\ell}{j}.$$

The following estimates on Λ are well-known (see for instance, [vL, Theorem 1.4.5]):

Fact 1 *For* $0 \leq \alpha \leq 1/2$, $\epsilon > 0$, *and sufficiently large* ℓ,

$$2^{\ell(H(\alpha)-\epsilon)} \leq \Lambda(\ell, \alpha\ell) \leq 2^{\ell H(\alpha)}.$$

∎

The following easily proved estimates on the binary entropy function $H(x) = -x \log x - (1-x) \log(1-x)$ will also be useful:

Fact 2 **(i)** *For* $|\delta| \leq 1/2$,

$$1 - \frac{\pi^2}{3 \ln 2} \delta^2 \leq H\left(\frac{1}{2} - \delta\right) \leq 1 - \frac{2}{\ln 2} \delta^2.$$

(ii) *For* $k \geq 3$, $H(1/k) \leq \log(ek)/k$. ∎

Our results extend algebraic techniques from [BKL]. The main observation in the upper bounds in that paper is that if the function A can be represented as a low-degree polynomial over $GF(2)$, significant, almost quadratic, savings in communication are possible compared to the trivial protocol. We use the following lemma proved in [BKL]:

Lemma 1 (BKL). *Let* f *be an* ℓ-*variate multilinear polynomial of degree at most* d *over* $\mathbb{Z}_2$. *Then* $GAF(f, x, y)$ *has an SM-protocol in which each player sends at most* $\Lambda(\ell, \lfloor d/2 \rfloor)$ *bits.*

2   Simple Protocol

For $z \in \{0,1\}^\ell$, let $|z|$ denote the number of 1's in z.

Lemma 2. *Let $A : \{0,1\}^\ell \longrightarrow \{0,1\}$. Then for each i, $0 \leq i \leq \ell$, there is a multilinear polynomial f_i of degree at most $\ell/2$ such that $f_i(z) = A(z)$ for every z with $|z| = i$.*

Proof: For every $z \in \{0,1\}^\ell$, define the polynomial

$$
\delta_z(x) := \begin{cases} \displaystyle\prod_{z_i=1} x_i, & \text{if } |z| \leq \ell/2, \\ \displaystyle\prod_{z_i=0} (1 - x_i), & \text{if } |z| > \ell/2. \end{cases}
$$

Observe that if $|x| = |z|$, then $\delta_z(x) = 1$ if $x = z$ and $\delta_z(x) = 0$ if $x \neq z$.
     Now, define f_i by

$$
f_i(x) := \sum_{|z|=i} A(z)\delta_z(x).
$$

Clearly, f_i is of degree at most $\ell/2$, since $\delta_z(x)$ is of degree at most $\ell/2$. Furthermore, when $|x| = i$, all terms, except the term $A(x)\delta_x(x)$, vanish in the sum defining f_i, implying $f_i(x) = A(x)$. (Note that when $|x| \neq i$, $f_i(x)$ need not be equal to $A(x)$.)   ∎

Theorem 1. *$GAF(A, x, y)$ can be computed by an SM-protocol in which each player sends at most $\ell\Lambda(\ell, \ell/4) = O(n^{0.82})$ bits.*

Proof: Players 1 and 2 construct the functions f_i for $0 \leq i \leq \ell$ corresponding to A as given by Lemma 2. They execute the protocol given by Lemma 1 for each f_i to enable the Referee to evaluate $f_i(x + y)$. The Referee, knowing x and y, can determine $|x + y|$ and use the information sent by Players 1 and 2 for $f_{|x+y|}$ to evaluate $f_{|x+y|}(x + y)$. By Lemma 2, $f_{|x+y|}(x + y) = A(x + y) = GAF(x + y)$, hence the protocol is correct. By Lemma 1, each of the players send at most $\Lambda(\ell, d_i/2)$ bits for each $0 \leq i \leq \ell$, where $d_i = \deg f_i = \min\{i, \ell - i\} \leq \ell/2$. The claimed bound follows using estimates from Fact 1.   ∎

3   Generalization

We now present a protocol based on the notion of *covering codes*. The protocol in the previous section will follow as a special case.

Definition 2. *An (m, r)-covering code of length ℓ is a set of words $\{c_1, \ldots, c_m\}$ in $\{0,1\}^\ell$ such that for any $x \in \{0,1\}^\ell$, there is a codeword c_i such that $d(x, c_i) \leq r$, where $d(x, y)$ denotes the Hamming distance between x and y.*

Theorem 2. *If there is an (m, r)-covering code of length ℓ, then there is an SM-protocol for $GAF(A, x, y)$ in which each player sends at most $mr\Lambda(\ell, r/2)$ bits.*

Proof: Given an (m,r)-covering code $\{c_1,\ldots,c_m\}$, let us denote by H_{ij}, the Hamming sphere of radius j around c_i:

$$H_{ij} := \{x \in \{0,1\}^\ell : d(x,c_i) = j\}.$$

For $A : \{0,1\}^\ell \longrightarrow \{0,1\}$, it is easy to construct a multilinear polynomial f_{ij} that agrees with A on all inputs from H_{ij}. More specifically, let us write, for each $1 \le i \le m$ and $S \subseteq [\ell]$,

$$\delta_{iS}(x) := \prod_{\{k \in S:c_{ik}=0\}} x_k \cdot \prod_{\{k \in S:c_{ik}=1\}} (1 - x_k),$$

where c_{ik} denotes the k'th coordinate of c_i. Note that $\delta_{iS}(x) = 1$ iff x differs from c_i in the coordinates $k \in S$.

Now, define the polynomial f_{ij} by

$$f_{ij}(x) := \sum_{|S|=j} A(c_i + s)\delta_{iS}(x),$$

where $s \in \{0,1\}^\ell$ denotes the characteristic vector of $S \subseteq [\ell]$. It is easy to verify that $f_{ij}(x) = A(x)$ for all $x \in H_{ij}$. Note also that deg $f_{ij} \le r$.

The players use the protocol given by Lemma 1 on f_{ij} for $1 \le i \le m$ and $1 \le j \le r$. The Referee can determine i and j such that $(x+y) \in H_{ij}$ and evaluate $f_{ij}(x+y) = A(x+y)$ from the information sent by the players. ∎

Remark 1. 1. Trivially, the all-0 vector and the all-1 vector form a $(2, \ell/2)$-covering code of length ℓ. Thus Theorem 1 is a special case of Theorem 2.

2. Assume $\ell = 2^v$, v a positive integer. The *first-order Reed-Muller code* $\mathcal{R}(1,v)$ form a $(2l, (\ell - \sqrt{\ell})/2)$-covering code of length ℓ [vL, Exercise 4.7.10]. Thus we have the following corollary.

Corollary 1. *There is an SM-protocol of complexity $\ell(\ell - \sqrt{\ell})\Lambda(\ell, (\ell - \sqrt{\ell})/4)$.*

4   Limitations

Theorem 3. *Any SM-upper bound obtained via Theorem 2 is at least $n^{0.728}$.*

Proof: For any (m,r)-covering code of length ℓ, we must trivially have

$$m\Lambda(\ell,r) \ge 2^\ell \tag{1}$$

Thus the SM-upper bound from Theorem 2 is at least

$$r2^\ell \frac{\Lambda(\ell,r/2)}{\Lambda(\ell,r)}. \tag{2}$$

Let $r = \alpha\ell$. Using Fact 1 in (2), the upper bound is at least

$$\ell\alpha 2^{\ell(1+H(\alpha/2)-H(\alpha))}.$$

This is minimized at $\alpha = 1 - 1/\sqrt{2}$ giving an upper bound of no less than $0.29289...\ell 2^{0.7284...\ell} \ge n^{0.728}$. ∎

5   Upper Bound for 3 Players

We construct a protocol which gives an upper bound matching the lower bound of Theorem 3. First, we construct a covering code.

Lemma 3. *For any ℓ, r, there is an (m, r)-covering code with $m = O(\ell \frac{2^\ell}{\Lambda(\ell,r)})$.*

Proof: We construct the code greedily. The first codeword c_1 can be arbitrary. Each next codeword c_i is chosen so that it maximizes the number of words x such that $d(x, c_1) > r, d(x, c_2) > r, \ldots, d(x, c_{i-1}) > r$, but $d(x, c_i) \leq r$. We can find c_i by exhaustive search over all words.

Let S_i be the set of words x such that $d(x, c_1) > r, \ldots, d(x, c_i) > r$ and w_i be the cardinality of S_i.

Claim. $w_{i+1} \leq (1 - \frac{\Lambda(\ell,r)}{2^\ell})w_i$.

Proof of Claim: For each $x \in S_i$, there are $\Lambda(\ell, r)$ pairs (x, y) such that $d(x, y) \leq r$. The total number of such pairs is at most $w_i \Lambda(\ell, r)$. By pigeonhole principle, there is a y such that there are at least $w_i \frac{\Lambda(\ell,r)}{2^\ell}$ numbers $x \in S_i$ with $d(x, y) \leq r$.

Recall that c_{i+1} is chosen so that it maximizes the number of such x. Hence, there are at least $w_i \frac{\Lambda(\ell,r)}{2^\ell}$ newly covered words $x \in S_i$ such that $d(x, c_{i+1}) \leq r$. This implies

$$w_{i+1} \leq \left(1 - \frac{\Lambda(\ell, r)}{2^\ell}\right) w_i.$$

This proves the claim.

We have

$$w_0 = 2^\ell,$$

$$w_i \leq \left(1 - \frac{\Lambda(\ell, r)}{2^\ell}\right)^i w_0 = \left(1 - \frac{\Lambda(\ell, r)}{2^\ell}\right)^i 2^\ell.$$

From $1 - x \leq e^{-x}$ it follows that

$$w_i \leq e^{-\frac{\Lambda(\ell,r)}{2^\ell}i} 2^\ell.$$

Let $m = \ln 2 \frac{2^\ell}{\Lambda(\ell,r)}\ell + 1$. Then $w_m < 1$.

Hence, $w_m = 0$, i.e. $\{c_1, \ldots, c_m\}$ is an (m, r)-covering code. ∎

Theorem 4. *There is an SM-protocol for $GAF(A, x, y)$ with communication complexity $O(n^{0.728\ldots+o(1)})$.*

Proof: We apply Theorem 2 to the code of Lemma 3 and get a protocol with communication complexity

$$mr\Lambda(\ell, r/2) = O\left(\ell \frac{2^\ell}{\Lambda(\ell, r)}\right) r\Lambda(\ell, r/2) = O\left(\ell r 2^\ell \frac{\Lambda(\ell, r/2)}{\Lambda(\ell, r)}\right).$$

Let $r = \alpha\ell$, where $\alpha = 1 - 1/\sqrt{2}$ is the constant from Theorem 3. Then, using estimates from Fact 1, the communication complexity is at most

$$O\left(\ell^2 2^\ell \frac{\Lambda(\ell, \alpha\ell/2)}{\Lambda(\ell, \alpha\ell)}\right) = O(n^{0.728\ldots+o(1)}).$$

∎

6   Upper Bounds for k Players

In this section, we generalize the idea from Section 2 to the k-player case. It appears that for large values of k, the simpler ideas from Section 2 already give nearly as efficient a protocol as can be obtained by generalizing Theorem 4. In other words, the covering code (for large k) from Theorem 4 can be replaced by the trivial $(2, \ell/2)$-covering code given by the all-0 and all-1 vectors (cf. Remark 1).

The starting point again is a lemma from [BKL] generalizing Lemma 1:

Lemma 4 (BKL). *Let f be an ℓ-variate multilinear polynomial of degree at most d over $\mathbb{Z}_2$. Then $GAF(f, x_1, \ldots, x_{k-1})$ has a k-player SM protocol in which each player sends at most $\Lambda(\ell, \lfloor d/(k-1) \rfloor)$ bits.*

Theorem 5. *$GAF(A, x_1, \ldots, x_{k-1})$ can be computed by a k-player SM protocol in which each player sends at most $\ell\Lambda(\ell, \ell/(2k-2)) = O(n^{H(1/(2k-2))})$ bits.*

Proof: Similar to Theorem 1. Players 1 thorough k construct the polynomials f_i, $0 \leq i \leq \ell$, corresponding to A as given by Lemma 2. Each f_i is of degree at most $\ell/2$. The players then follow the protocol from Lemma 4 for each of these f_i. ∎

7   Conclusions and Open Problems

We presented improved upper bounds on the SM complexity of of $GAF_{\mathbb{Z}_2^t, k}$. For $k = 3$, we prove an upper bound of $O(n^{0.73})$ improving the previous bound of $O(n^{0.92})$ from [BKL]. For general k, we show an upper bound of $O(n^{H(1/(2k-2))})$ improving the bound of $O(n^{H(1/k)})$ from [BKL]. The first open problem is to improve these bounds to close the gap between upper and lower bounds on the SM complexity of $GAF_{\mathbb{Z}_2^t, k}$. Recall that the lower bound is $\Omega(n^{1/(k-1)}/(k-1))$ [BKL].

The second open problem concerns the SM complexity of $GAF_{\mathbb{Z}_n, k}$, i.e., the generalized addressing function for the cyclic group. Note that the best known upper bounds for the cyclic group [Am1] are significantly weaker than the upper bounds we present here for the vector space $\mathbb{Z}_2^t$. The lower bounds for both $GAF_{\mathbb{Z}_2^t, k}$ and $GAF_{\mathbb{Z}_n, k}$ are $\Omega(n^{1/(k-1)}/(k-1))$ and follow from a general result of [BKL] on SM complexity of $GAF_{G, k}$ for arbitrary finite groups G. In contrast, techniques for upper bounds appear to depend on the specific group G.

References

[Am1]   A. Ambainis. *Upper Bounds on Multiparty Communication Complexity of Shifts.* 13th Annual Symposium on Theoretical Aspects of Computer Science, LNCS, Vol. 1046, pp. 631-642, 1996.

[Am2]   A. Ambainis.: *Upper Bound on the Communication Complexity of Private Information Retrieval.* Proceedings of ICALP'97, Lecture Notes in Computer Science, 1256(1997), pages 401-407.

[B]   B. Bollobás. Random Graphs. Academic Press, 1985, pp. 307-323.

[BE]   L. Babai, P. Erdős. Representation of Group Elements as Short Products. J. Turgeon, A. Rosa, G. Sabidussi, editor. Theory and Practice of Combinatorics, 12, in Ann. Discr. Math., North-Holland, 1982, pp. 21-26.

[BHK]   L. Babai, T. Hayes, P. Kimmel Communication with Help. ACM STOC 1998.

[BGKL]   L. Babai, A. Gál, P. Kimmel, S. V. Lokam. Communictaion Complexity of Simultaneous Messages. Manuscript. A significantly expanded version of [BKL] below.

[BKL]   L. Babai, P. Kimmel, S. V. Lokam. Simultaneous Messages vs. Communication. Proc. of the 12th Symposium on Theoretical Aspects of Computer Science, 1995.

[BNS]   L. Babai, N. Nisan, M. Szegedy. Multiparty Protocols, Pseudorandom Generators for Logspace and Time-Space Trade-offs. Journal of Computer and System Sciences 45, 1992, pp. 204-232.

[BT]   R. Beigel, J. Tarui. On ACC. Proc. of the 32nd IEEE FOCS, 1991, pp. 783-792.

[CFL]   A.K. Chandra, M.L. Furst, R.J. Lipton. Multiparty protocols. Proc. of the 15th ACM STOC, 1983, pp. 94-99.

[CGKS]   B. Chor, O. Goldreich, E. Kushilevitz, M. Sudan. Private Information Retrieval. Proc. of 36th IEEE FOCS, 1995, pp. 41 – 50.

[G]   V. Grolmusz. The BNS Lower Bound for Multi-Party Protocols is Nearly Optimal. Information and Computation, 112, No, 1, 1994, pp. 51-54.

[HG]   J. Håstad, M. Goldmann. On the Power of Small-Depth Threshold Circuits. Computational Complexity, 1, 1991, pp. 113-129.

[HMPST]   A. Hajnal, W. Maass, P. Pudlák, M. Szegedy, G. Turan. Threshold Circuits of Bounded Depth. Proc. of 28th IEEE FOCS, pp. 99 – 110, 1987.

[KN]   E. Kushilevitz, N. Nisan. Communication Complexity, Cambridge University Press, 1997.

[MNT]   Y. Mansour, N. Nisan, P. Tiwari. The Computational Complexity of Universal Hashing. Theoretical Computer Science, 107, 1993, pp. 121-133.

[NW]   N. Nisan, A. Wigderson. Rounds in Communication Complexity Revisited. SIAM Journal on Computing, 22, No. 1, 1993, pp. 211-219.

[PR]   P. Pudlák, V. Rödl. Modified Ranks of Tensors and the Size of Circuits. Proc. 25th ACM STOC, 1993, pp. 523 – 531.

[PRS]   P. Pudlák, V. Rödl, J. Sgall. Boolean circuits, tensor ranks and communication complexity. SIAM J. on Computing 26/3 (1997), pp.605-633.

[Ra]   A. A. Razborov. On rigid matrices. Preprint of Math. Inst. of Acad. of Sciences of USSR (in Russian), 1989.

[RW]   A. A. Razborov, A. Wigderson. $n^{\Omega(\log n)}$ Lower Bounds on the Size of Depth 3 Circuits with AND Gates at the Bottom. Information Processing Letters, Vol. 45, pp. 303–307, 1993.

[Va]   L. Valiant. Graph-Theoretic Arguments in Low-level Complexity. Proc. 6th Math. Foundations of Comp. Sci., Lecture notes in Computer Science, Vol. 53, Springer-Verlag, 1977, pp. 162-176.

[vL]   J. H. van Lint: Introduction to Coding Theory. Springer-Verlag, 1982.

[Y]   A. C-C. Yao. On ACC and Threshold Circuits. Proc. of the 31st IEEE FOCS, 1990, pp. 619-627.

Multi-parameter Minimum Spanning Trees

David Fernández-Baca*

Department of Computer Science, Iowa State University, Ames, IA 50011, USA
fernande@cs.iastate.edu

Abstract. A framework for solving certain multidimensional parametric search problems in randomized linear time is presented, along with its application to optimization on matroids, including parametric minimum spanning trees on planar and dense graphs.

1 Introduction

In the *multi-parameter minimum spanning tree problem*, we are given an edge-weighted graph $G = (V, E)$, where the weight of each edge e is an affine function of a d-dimensional parameter vector $\lambda = (\lambda_1, \lambda_2, \ldots, \lambda_d)$, i.e., $w(e) = a_0(e) + \sum_{i=1}^{d} \lambda_i a_i(e)$. The topology and weight of the minimum spanning tree are therefore functions of λ. Let $z(\lambda)$ be the weight of the minimum spanning tree at λ. The goal is to find

$$z^* = \max_{\lambda} z(\lambda). \tag{1}$$

Problem (1) arises in the context of Lagrangian relaxation. For example, Camerini et al. [5] describe the following problem. Suppose each edge e of G has an installation cost $w(e)$ and d possible maintenance costs $m_i(e)$, one for each of k possible future scenarios, where scenario i has probability p_i. Edge e also has a reliability $q_i(e)$ under each scenario i. Let $\mathcal{T}$ denote the set of all spanning trees of G. Minimizing the total installation and maintenance costs while maintaining an acceptable level of reliability Q under all scenarios is expressible as

$$\min_{T \in \mathcal{T}} \left\{ \sum_{e \in T} \left(w(e) + \sum_{i=1}^{d} p_i m_i(e) \right) : \prod_{e \in T} q_i(e) \geq Q, i = 1, \ldots, d \right\}. \tag{2}$$

A good lower bound on the solution to (2) can be obtained by solving the Lagrangian dual of (2), which has the form (1) with $a_0(e) = w(e) + \sum_{i=1}^{d} p_i m_i(e)$ and $a_i(e) = -\log q_i(e) + (\log Q)/(|V| - 1)^*$.

   In this paper, we give linear-time randomized algorithms for the fixed-dimensional parametric minimum spanning tree problem for planar and dense graphs (i.e., those where $m = \Theta(n^2)$). Our algorithms are based on Megiddo's method

---

* Supported in part by the National Science Foundation under grant CCR-9520946.
* We also need $\lambda \geq 0$; this can easily be handled by our scheme.

G. Gonnet, D. Panario, and A. Viola (Eds.): LATIN 2000, LNCS 1776, pp. 217–226, 2000.
© Springer-Verlag Berlin Heidelberg 2000

of parametric search [24], a technique that turns solutions to fixed-parameter problems (e.g., non-parametric minimum spanning trees) into algorithms for parametric problems. This conversion is often done at the price of incurring a polylogarithmic slowdown in the run time for the fixed-parameter algorithm. Our approach goes beyond the standard application of Megiddo's method to eliminate this slowdown, by applying ideas from the prune-and-search approach to fixed-dimensional linear programming. Indeed, the mixed graph-theoretic/geometric nature of the our problems requires us to use pruning at two levels: geometrically through cuttings and graph-theoretically through sparsification

History and New Results. The parametric minimum spanning tree problem is a special case of the *parametric matroid optimization problem* [14]. The Lagrangian relaxations of several non-parametric matroid optimization problems with side constraints — called *matroidal knapsack problems* by Camerini et al. [5] — are expressible as problems of the form (1). More generally, (1) is among the problems that can be solved by Megiddo's method of parametric search [23,24], originally developed for one-dimensional search, but readily extendible to any fixed dimension [26,10,4].

The power of the parametric search has been widely recognized (see, e.g., [2]). Part of the appeal of the method is its formulation as an easy-to-use "black box." The key requirement is that the underlying *fixed-parameter problem* — i.e., the problem of evaluating $z(\lambda)$ for fixed λ — have an algorithm where all numbers manipulated are affine functions of λ. If this algorithm runs in time $O(T)$, then the parametric problem can be solved in time $O(T^{2d})$ and if there is W-processor, D-step *parallel* algorithm for the fixed parameter problem, the run time can be improved to $O(T(D \log W)^d)$. In some cases, this can be further improved to $O(T(D + \log W)^d)$**. This applies to the parametric minimum spanning tree problem, for which one can obtain $D = O(\log n)$ and $W = O(m)$, where $n = |V|$ and $m = |E|$. The (fixed-parameter) minimum spanning tree problem can be solved in randomized $O(m)$ expected time [21] and $O(m\alpha(m,n) \log \alpha(m,n))$ deterministic time [7]. In any event, by its nature, parametric search introduces a $\log^{O(d)} n$ slowdown with respect to the run time of the fixed-parameter problem. Thus, the algorithms it produces are unlikely to be optimal (for an exception to this, see [11]).

Frederickson [20] was among the first to consider the issue of optimality in parametric search, in the sense that no slowdown is introduced and showed how certain location problems on trees could be solved optimally. Later, certain one-dimensional parametric search problems on graphs of bounded tree-width [18] and the one-dimensional parametric minimum spanning tree problem on planar graphs and on dense graphs [17,19] were shown to be optimally solvable. A key technique in these algorithms is *decimation*: the organization of the search into phases to achieve geometric reduction of problem size. This is closely connected with the *prune-and-search* approach to fixed-dimensional linear programming

---

** Note that the O-notation in all these time bounds hides "constants" that depend on d. The algorithms to be presented here exhibit similar constants.

[25,12,9]. While the geometric nature of linear programming puts fewer obstacles to problem size reduction, a similar effect can be achieved for graph problems through *sparsification* [15,16] a method that has been applied to one-dimensional parametric minimum spanning trees before [1,19].

Here we show that multi-parameter minimum spanning trees can be found in randomized linear expected time on planar and on dense graphs. Our procedures use prune-and-search geometrically through *cuttings* [8], to narrow the search region for the optimal solution, as well as graph-theoretically through sparsification. More generally, we identify *decomposability* conditions that allow parametric problems to be solvable within the same time bound as their underlying fixed-parameter problems.

2   Multidimensional Search

Let h be a hyperplane in $\mathbb{R}^d$, let Λ be a convex subset of $\mathbb{R}^d$, and let $\operatorname{sign}_\Lambda(h)$ be $+1$, 0, or -1, depending, respectively, on whether $h(\lambda) < 0$ for all $\lambda \in \Lambda$, $h(\lambda) = 0$ for some $\lambda \in \Lambda$, or $h(\lambda) > 0$ for all $\lambda \in \Lambda$. Hyperplane h is said to be *resolved* if $\operatorname{sign}_\Lambda(h)$ is known. An *oracle* is a procedure that can resolve any given hyperplane. The following result is known (see also [25]):

Theorem 1 (Agarwal, Sharir, and Toledo [3]). *Given a collection $\mathcal{H}$ of n hyperplanes in $\mathbb{R}^d$ and an oracle $\mathcal{B}$ for Λ, it is possible to find either a hyperplane that intersects Λ or a simplex $\triangle$ that fully contains Λ and intersects at most $n/2$ hyperplanes in $\mathcal{H}$ by making $O(d^3 \log d)$ oracle calls. The time spent in addition to the oracle calls is $n \cdot O(d)^{10d} \log^{2d} d$.*

Corollary 1. *Given a set $\mathcal{H}$ of n hyperplanes and a simplex $\triangle$ containing Λ, a simplex $\triangle'$ intersecting at most n' elements of $\mathcal{H}$ and such that $\Lambda \subseteq \triangle' \subseteq \triangle$ can be found with $O(d^3 \log d \lg(n/n'))$ calls to an oracle.*

If Λ denotes the set of maximizers of function z, we have the following [4]:

Lemma 1. *Locating the position of the maximizers of z relative to a given hyperplane h reduces to carrying out three $(d-1)$-dimensional maximization problems of the same form as (1).*

3   Decomposable Problems

Let m be the problem size. A *decomposable* optimization problem is one whose fixed-parameter version can be solved in two stages:

- A $O(m)$-time *decomposition stage*, independent of λ, which produces a recursive decomposition of the problem represented by a bounded-degree *decomposition tree* D. The nodes of D represent subproblems and its root is the original problem. For each node v in D, m_v is the size of the subproblem associated with v. The children of v are the subproblems into which v is decomposed.

- A $O(m)$-time *optimization stage*, where the decomposition tree is traversed level by level from the bottom up and each node is replaced by a *sparse substitute*. $z(\lambda)$ can be computed in $O(1)$ time from the root's sparse substitute.

Note that after the decomposition stage is done we can evaluate $z(\lambda)$ for multiple values of λ by executing only the optimization stage.

We will make some assumptions about the decomposition stage:

(D1) D is organized into *levels*, with leaves being at level 0, their parents at level 1, grandparents at level 2, and so forth. L_i will denote the set of nodes at level i. The index of the last level is $k = k(n)$.

(D2) There exists a constant $\alpha > 1$, independent of m, such that $|L_i| = O(m/\alpha^i)$ and $m_u = O(\alpha^i)$ for each $u \in L_i$.

We also make assumptions about the optimization stage:

(O1) For each $v \in L_i$, the solution to the subproblem associated with v can be represented by a *sparse substitute* of size $O(\beta(i))$, such that $\beta(i)/\alpha^i < 1/\gamma^i$ for some $\gamma > 1$. For $i = 0$ the sparse substitute can be computed in $O(1)$ time by exhaustive enumeration, while for $i > 0$ the substitute for v depends only on the sparse substitutes for v's children.

(O2) The algorithm for computing the sparse substitute for any node v takes time linear in the total size of the sparse substitutes of v's children. This algorithm is *piecewise affine*; i.e., every number that it manipulates is expressible as an affine combination of the input numbers.

Note that (i) by (O1), the total size of all sparse substitutes for level i is $O(m/\gamma^i)$ and (ii) assumption (O2) holds for many combinatorial algorithms; e.g., most minimum spanning tree algorithms are piecewise affine.

4   The Search Strategy

We now describe our approach in general terms; its applications will be presented in Section 5. We use the following notation. Given a collection of hyperplanes $\mathcal{H}$ in $\mathbb{R}^d$, $\mathcal{A}(\mathcal{H})$ denotes the *arrangement* of $\mathcal{H}$; i.e., the decomposition of $\mathbb{R}^d$ into faces of dimension 0 through d induced by $\mathcal{H}$ [13]. Given $\Gamma \subseteq \mathbb{R}^d$, $\mathcal{A}_\Gamma(\mathcal{H})$ denotes the restriction of $\mathcal{H}$ to Γ.

Overview. The search algorithm simulates the bottom-up, level-by-level execution of the fixed-parameter algorithm for all λ within a simplex $\triangle \subseteq \mathbb{R}^d$ known to contain Λ, the set of maximizers of z. The outcome of this simulation for any node v in the decomposition tree is captured by a *parametric sparse substitute* for v, which consists of (i) a decomposition of $\triangle$ into disjoint regions such that the sparse substitute for each region is unique and (ii) the sparse substitute for each region of the decomposition. Given a parametric sparse substitute for v, obtaining the sparse substitute for v for any fixed $\lambda \in \triangle$ becomes a point location problem in the decomposition.

After simulating the fixed-parameter algorithm, we will have a description of *all* possible outcomes of the computation within $\triangle$, which can be searched to locate some $\lambda^* \in \Lambda$. For efficiency, the simulation of each level is accompanied by the shrinkage of $\triangle$ using Theorem 1. This is the point where we run the risk of incurring the polylogarithmic slowdown mentioned in the Introduction. We get around this with three ideas. First, we shrink $\triangle$ only to the point where the *average* number of regions in the sparse substitute within $\triangle$ for a node at level i is constant-bounded (see also [17]). Second, the oracle used at level i relies on bootstrapping: By Lemma 1, this oracle can be implemented by solving three optimization problems in dimension $d - 1$. If parametric sparse substitutes for all nodes at level $i - 1$ are available, the solution to any such problem will not require reprocessing lower levels.

The final issue is the relationship between node v's parametric sparse substitute $P(v)$ and the substitutes for v's children. An initial approximation to the subdivision for $P(v)$ is obtained by overlapping the subdivisions for the substitutes of the children of v. Within every face F of the resulting subdivision of $\triangle$ there is a unique sparse substitute for each of v's children. However, F may have to be further subdivided because there may still be multiple distinct sparse substitutes for v within F. Instead of producing them all at once, which would be too expensive, we proceed in three stages. First, we get rough subdivisions of the F's through random *cuttings* [8]. Each face of these subdivisions will contain only a relatively small number of regions of $P(v)$. In the second stage, we shrink $\triangle$ so that the total number of regions in the rough subdivisions over all nodes in level i is small. Finally, we generate the actual subdivisions for all v.

Intersection Hyperplanes. Consider a comparison between two values $a(\lambda)$ and $b(\lambda)$ that is carried out when computing the sparse certificate for v or one of its descendants for some $\lambda \in \mathbb{R}^d$. By assumption (O2), $a(\lambda)$ and $b(\lambda)$ are affine functions of λ; their *intersection hyperplane* is $h_{ab} = \{\lambda : a(\lambda) = b(\lambda)\}$. The outcome of the comparison for a given λ-value depends only on which side of h_{ab} contains the value. Let $\mathcal{I}(v)$ consist of all such intersection hyperplanes for v. Then, there is a unique sparse substitute for each face of $\mathcal{A}_\triangle(\mathcal{I}(v))$, since all comparisons are resolved in the same way within it. Thus, our parametric sparse substitutes consist of $\mathcal{A}_\triangle(\mathcal{I}(v))$, together with the substitute for its faces.

For fast retrieval of sparse substitutes, we need a point location data structure for $\mathcal{A}_\triangle(\mathcal{I}(v))$. The complexity of the arrangement and the time needed to build it are $O(n^d)$, where n is the number of elements of $\mathcal{I}(v)$ that intersect the interior of $\triangle$ [13]. A point location data structure with space requirement and preprocessing time $O(n^d)$ can be built which answers point location queries in $O(\log n)$ time [6].

We will make certain assumptions about intersection hyperplanes. Let F be a face of $\mathcal{A}_\triangle(\bigcup\{\mathcal{I}(u) : u \text{ is a child of } v\})$. Then, we have the following.

(H1) The number of elements of $\mathcal{I}(v)$ that intersect the interior of F is $O(m_v^2)$.
(H2) A random sample of size n of the elements of $\mathcal{I}(v)$ that intersect the interior of F can be generated in $O(n)$ time.

Shrinking the Search Region. Let $\kappa_\triangle(v)$ denote the number of faces of $\mathcal{A}_\triangle(\mathcal{I}(v))$ and let $\mathcal{I}_\triangle(v)$ denote the elements of $\mathcal{I}(v)$ that intersect the interior of $\triangle$. The goal of the shrinking algorithm is to reduce the search region $\triangle$ so that, after simulating level r of the fixed-parameter algorithm,

$$\Lambda \subseteq \triangle \qquad \text{and} \qquad \sum_{v \in L_r} |\mathcal{I}_\triangle(v)| \leq m/\alpha^{r(2d+1)}. \tag{3}$$

Lemma 2. *If (3) holds, then $\sum_{v \in L_r} \kappa_\triangle(v) \leq 2m/\alpha^r$.*

Lemma 3. *Suppose we are given a simplex $\triangle$ satisfying (3) for $r = i - 1$ and parametric sparse substitutes within $\triangle$ for all $v \in L_{i-1}$. Then, with high probability, $O(id^4 \log d)$ oracle calls and $O(m/\alpha^{i/4d})$ overhead suffice to find a new simplex $\triangle$ satisfying (3) for $r = i$, together with $\mathcal{I}_\triangle(v)$ for all $v \in L_i$.*

Proof. Let $s = \alpha^i$ and for each $v \in L_i$ let $\mathcal{H}_v$ denote $\bigcup\{\mathcal{I}_\triangle(u) : u \text{ a child of } v\}$.

To shrink the search region, first do the following for each $v \in L_i$ and each face $F \in \mathcal{A}(\mathcal{H}(v))$: ¿From among the elements of $\mathcal{I}(v)$ that arise during the computation of a sparse certificate for some $\lambda \in F$ choose uniformly at random a set C_F of size $s^{1-1/(4d)}$. The total number of elements in all the sets C_F is $(m/s) \cdot s^{1-1/(4d)} = m/s^{1/(4d)}$. By the theory of cuttings [8], with high probability any face in a certain triangulation of $\mathcal{A}(C_F)$, known as the *canonical triangulation*, intersects at most $s/s^{1-1/(4d)}$ elements of $\mathcal{I}_\triangle(v)$.

Next, apply Corollary 1 to the set $\mathcal{H} = \bigcup_{v \in L_i}\{h : h \in \mathcal{H}_v \text{ or } h \in C_F, F \text{ a face of } \mathcal{A}(\mathcal{H}_v)\}$ to find a simplex $\triangle'$ that intersects at most m/s^{d+1} elements of $\mathcal{H}$ and such that $\Lambda \subseteq \triangle' \subseteq \triangle$. Set $\triangle \leftarrow \triangle'$. Since $|\mathcal{H}| = O(m/2^{1/(4d)})$, the total number of oracle calls is $O(d^4 \log d \log s) = O(id^4 \log d)$.

Now, for each $v \in L_i$, we compute $\mathcal{I}_\triangle(v)$ in two steps. First, for each $v \in L_i$ construct the canonical triangulation C_v of the arrangement of $\mathcal{G}_v$, which consists of all hyperplanes in $\mathcal{H}_v \cup \{h \in C_F : F \text{ a face of } \mathcal{A}(\mathcal{H}_v)\}$ intersecting $\triangle$. The total time is $O(m/s)$, since at most m/s^{d+1} v's have non-empty $\mathcal{G}_v$'s and for each such v the number of regions in $\mathcal{A}_v(\mathcal{G}_v)$ is $O(s^d)$. Secondly, enumerate all hyperplanes in $\mathcal{I}_\triangle(v)$ that intersect $\triangle'$. This takes time $O(m/s^{1/4})$. With high probability, at most $(m/s) \cdot s^{1/(4d)} = m/s^{1-1/(4d)}$ hyperplanes will be found.

Finally, we apply Corollary 1 to the set $\mathcal{H} = \bigcup_{v \in L_i} \mathcal{I}_\triangle(v)$ to obtain a simplex $\triangle'$ intersecting at most m/s^{2d+1} of the elements of $\mathcal{H}$ and such that $\Lambda \subseteq \triangle' \subseteq \triangle$. Set $\triangle \leftarrow \triangle'$. The total number of oracle calls is $O(d^4 \log d \log s) = O(id^4 \log d)$. $\square$

A Recursive Solution Scheme. We simulate the execution of the fixed-parameter algorithm level by level, starting at level 0; each step of the simulation produces parametric sparse certificates for all nodes within a given level. We use induction on the level i and the dimension d. For $i = 0$ and every $d \geq 0$, we compute the parametric sparse substitutes for all $v \in L_i$ by exhaustive enumeration. This takes time $O(c_d m)$, for some constant c_d.

Lemma 4. *Let $\triangle$ be a simplex satisfying (3) for $r = i - 1$, and suppose that parametric sparse substitutes within $\triangle$ for all $v \in L_{i-1}$ are known. Then we can,* *with high probability, in time $O(f_d \cdot m/\gamma_d^i)$ find a new simplex whithin which the parametric sparse substitute for the root of D has $O(1)$ regions.*

Proof. To process L_i, $i > 0$, we use recursion on the dimension, d. When $d = 0$, we have the fixed-parameter problem. By (O1), given sparse substitutes for all $v \in L_{i-1}$, we can process L_i in time $O(\beta(i) \cdot m/\alpha^i) = O(f_0 m/\gamma_0^i)$, where $f_0 = 1$, $\gamma_0 = \gamma$. Hence, L_i through L_k can be processed in time $O(f_0 m/\gamma_0^i)$.

Next, consider $d \geq 1$. To process level i, we assume that level $i - 1$ has been processed so that (3) holds for $r = i - 1$. Our goal is to successively maintain (3) for $r = i, i + 1, \ldots, k$. Thus, after simulating level k, $|\mathcal{I}_\triangle(root)| = O(1)$.

The first step is to use Lemma 3 to reduce $\triangle$ to a simplex satisfying (3) for $r = i$ and to obtain $\mathcal{I}_\triangle(v)$ for all $v \in L_i$. The oracle will use the sparse certificates already computed for level $i - 1$: Let h be the hyperplane to be resolved. If $h \cap \triangle = \emptyset$, we resolve h in time $O(d)$ by finding the side of h that contains $\triangle$. Otherwise, by Lemma 1, we must solve three $(d - 1)$-dimensional problems. For each such problem, we do as follows. For every $v \in L_{i-1}$, find the intersection of h with $\mathcal{A}_\triangle(v)$. This defines an arrangement in the $(d - 1)$-dimensional simplex $\triangle' = \triangle \cap h$. The sparse substitute for each region of the arrangement is unique and known; thus, we have parametric sparse substitutes for all $v \in L_{i-1}$. By hypothesis, we can compute z_h^* in time $O(f_{d-1} m/\gamma_{d-1}^i)$.

By Lemma 3, the time for all oracle calls is $O(i \cdot d^4 \log d \cdot f_{d-1} \cdot m/\gamma_{d-1}^i)$. If we discover that h intersects Λ, we return z_h^*. After shrinking $\triangle$, $\mathcal{I}_\triangle(v)$ will be known and we can build $\mathcal{A}_\triangle(\mathcal{I}(v))$. By Lemma 2, this arrangement has $O(m/\alpha^i)$ regions. By assumption (O1) we can find the sparse substitute for each face F of $\mathcal{A}_\triangle(\mathcal{I}(v))$ as follows. First, choose an arbitrary point λ^0 in the interior of F. Next, for each child u of v, find the sparse substitute for u at λ^0. Finally, use these sparse substitutes to compute the sparse substitute for v at λ^0; this will be the sparse substitute for all $\lambda \in F$. Thus, the total time needed to compute the parametric sparse substitutes for $v \in L_i$ is $O(\beta(i) \cdot (m/\alpha^i)) = O(m/\gamma^i)$. The oracle calls dominate the work, taking a total time of $O(i \cdot g_d \cdot f_{d-1} \cdot m/\gamma_{d-1}^i) = O(f_d \cdot m/\gamma_d^i)$, where $f_d = g_d \cdot f_{d-1}$ and γ_d satisfies $1 < \gamma_d < \gamma_{d-1}$. $\square$

Theorem 2. *The optimum solution to a decomposable problem can be found in $O(m)$ time with high probability.*

Proof. Let w be the root of the decomposition tree. After simulating the execution of levels 0 through k, we are left with a simplex $\triangle$ that is crossed by $O(1)$ hyperplanes of $\mathcal{I}(w)$, and $O(1)$ invocations of Theorem 1 suffice to reduce $\triangle$ to a simplex that is crossed by no hyperplane of $\mathcal{I}(w)$. Within this simplex, there is a unique optimum solution, whose cost as a function of λ is, say, $c \cdot \lambda$. The maximizer can now be located through linear programming with objective function $c \cdot \lambda$. The run time is $O(a_d)$, for some value a_d that depends only on d. $\square$

5 Applications

We now show apply Theorem 2 to several matroid optimization problems. We will rely on some basic matroid properties. Let $M = (S, I)$ be a matroid on the set of elements S, where I is the set of independent subsets of S. Let B be a subset of S and let A be a maximum-weight independent subset of B. Then, there exists some maximum-weight independent set of S that does not contain any element of $B \setminus A$. Furthermore the identity of the maximum-weight independent subset of any such B depends on the relative order, but not on the actual values, of the weights of the elements of B. Therefore, the maximum number of distinct comparisons to determine the relative order of elements over all possible choices of λ is $\binom{|B|}{2}$. Thus, (H1) is satisfied. Assumption (H2) is satisfied, since we can get a random sample of size n the intersection hyperplanes for B by picking n pairs of elements from B uniformly at random.

Uniform and Partition Matroids. (This example is for illustration; both problems can be solved by linear programming.) Let S be an m-element set where every $e \in S$ has a weight $w(e) = a_0(e) + \sum_{i=1}^{d} \lambda_i a_i(e)$, let $k \leq m$ be a fixed positive integer, and let $z(\lambda) = \max_{B,|B|=k} \sum_{e \in B} w(e)$. The problem is to find $z^* = \min_\lambda z(\lambda)$. The fixed-parameter problem is *uniform* matroid optimization. D is as follows: If $|S| \leq k$, D consists of a single vertex v containing all the elements of S. Otherwise, split S into two sets of size $m/2$; D consists of a root v connected to the roots of trees for these sets. Thus, D satisfies conditions (D1) and (D2).

The non-parametric sparse substitute for node v consists of the k largest elements among the subset S_v corresponding to v. A sparse substitute for v can be computed from its children in $O(k)$ time. Thus, (O2) is satisfied and, if $k = O(1)$, so is (O1). Hence, z^* can be found in $O(m)$ time.

In *partition matroids*, the set S is partitioned into disjoint subsets $S_1, \ldots, S_r$ and $z(\lambda) = \max\{\sum\{w(e) : e \in B, |B \cap S_i| \leq 1 \text{ for } i = 1, 2, \ldots, r\}$. $z^* = \min_\lambda z(\lambda)$ can be computed in $O(m)$ time by similar techniques.

Minimum Spanning Trees in Planar Graphs. D will represent a recursive separator-based decomposition of the input graph G. Every node v in D corresponds to a subgraph G_v of G with n_v vertices; the root of D represents all of G. The children $u_1, \ldots, u_r$ of v represent a decomposition of G_v into edge-disjoint subgraphs G_{u_i} such that $n_{u_i} \leq n_v/\alpha$ for some $\alpha > 1$, which share a set X_v of *boundary vertices*, such that $|X_v| = O(\sqrt{n_{u_i}})$. D satisfies conditions (D1) and (D2) and can be constructed in time $O(n)$ where n is the number of vertices of G [22].

A sparse substitute for a node x with a set of boundary vertices X is obtained as follows: First, compute a minimum spanning tree T_x of G_x. Next, discard all edges in $E(G_x) \setminus E(T_x)$ and all isolated vertices. An edge e is *contractible* if it has a degree-one endpoint that is not a boundary vertex, or it shares a degree-two non-boundary vertex with another edge f such that $\text{cost}(e) \leq \text{cost}(f)$. Now, repeat the following step while there is a contractible edge in G_v: choose any contractible edge e and contract it. While doing this, keep a running total of the cost of the contracted edges. The size of the resulting graph H_x is $O(|X|) = O(\sqrt{n_x})$. Also, H_x is equivalent to G_x in that if the former is substituted for the latter in the original graph, then the minimum spanning tree of the new graph, together with the contracted edges constitute a minimum spanning tree of the original graph. The sparse substitute computation satisfies (O1) and (O2).

Minimum Spanning Trees in Dense Graphs. D is built in two steps. First, a *vertex partition tree* is constructed by splitting the vertex set into two equal-size parts (to within 1) and then recursively partitioning each half. This results in a complete binary tree of height $\lg n$ where nodes at depth i have $n/2^i$ vertices. ¿From the vertex partition tree we build an *edge partition tree*: For any two nodes x and y of the vertex partition tree at the same depth i containing vertex sets V_x and V_y, create a node E_{xy} in the edge partition tree containing all edges of G in $V_x \times V_y$. The parent of E_{xy} is E_{uw}, where u and w are, respectively, the parents of x and y in the vertex partition tree. An internal node E_{xy} will have three children if $x = y$ and four otherwise. D is built from the edge partition tree by including only non-empty nodes.

Let $u = E_{xy}$ be a node in D. Let G_u be the subgraph of G with vertex set $V_x \cup V_y$ and edge set $E \cap (V_x \times V_y)$. For every j between 0 and the depth of D, (i) there are at most 2^{2j} depth-j nodes, (ii) the edge sets of the graphs associated with the nodes at depth k are disjoint and form a partition of E, and (iii) if u is at depth j, G_u has at most $n/2^j$ vertices and $n^2/2^{2j}$ edges. If G is dense, then $m_u = |E(G_u)| = \Theta(|V(G_u)^2|)$ for all u. Thus, (D1) and (D2) hold. The sparse substitute for G_u is obtained by deleting from G_u all edges not in its minimum spanning forest (which can be computed in $O(|V(G_u)^2|) = O(m_u)$ time). The size of the substitute is $O(\sqrt{m_u})$. Thus, (O1) and (O2) are satisfied.

6   Discussion

Our work shows the extent to which prune-and-search can be used in parametric graph optimization problems. Unfortunately, the heavy algorithmic machinery involved limits the practical use of our ideas. We also suspect that our decomposability framework is too rigid, and that it can be relaxed to solve other problems. Finally, one may ask whether randomization is necessary. Simply substituting randomized cuttings by deterministic ones [6] gives an unacceptable slowdown.

References

1. P. K. Agarwal, D. Eppstein, L. J. Guibas, and M. R. Henzinger. Parametric and kinetic minimum spanning trees. In *Proceedings 39th IEEE Symp. on Foundations of Computer Science*, 1998.
2. P. K. Agarwal and M. Sharir. Algorithmic techniques for geometric optimization. In J. van Leeuwen, editor, *Computer Science Today: Recent Trends and Developments*, volume 1000 of *Lecture Notes in Computer Science*. Springer-Verlag, 1995.
3. P. K. Agarwal, M. Sharir, and S. Toledo. An efficient multidimensional searching technique and its applications. Technical Report CS-1993-20, Computer Science Department, Duke University, July 1993.
4. R. Agarwala and D. Fernández-Baca. Weighted multidimensional search and its application to convex optimization. *SIAM J. Computing*, 25:83–99, 1996.
5. P. M. Camerini, F. Maffioli, and C. Vercellis. Multi-constrained matroidal knapsack problems. *Mathematical Programming*, 45:211–231, 1989.

6. B. Chazelle. Cutting hyperplanes for divide-and-conquer. *Discrete Comput. Geom.*, 9(2):145–158, 1993.

7. B. Chazelle. A faster deterministic algorithm for minimum spanning trees. In *Proceedings 38th IEEE Symp. on Foundations of Computer Science*, pages 22–31, 1997.

8. B. Chazelle and J. Friedman. A deterministic view of random sampling and its use in geometry. *Combinatorica*, 10(3):229–249, 1990.

9. K. L. Clarkson. Linear programming in $O(n \times 3^{d^2})$ time. *Information Processing Letters*, 22:21–24, 1986.

10. E. Cohen and N. Megiddo. Maximizing concave functions in fixed dimension. In P. M. Pardalos, editor, *Complexity in Numerical Optimization*, pages 74–87. World Scientific, Singapore, 1993.

11. R. Cole, J. S. Salowe, W. L. Steiger, and E. Szemerédi. An optimal-time algorithm for slope selection. *SIAM J. Computing*, 18:792–810, 1989.

12. M. E. Dyer. On a multidimensional search technique and its application to the euclidean one-centre problem. *SIAM J. Computing*, 15(3):725–738, 1986.

13. H. Edelsbrunner. *Algorithms in Combinatorial Geometry*. Springer-Verlag, Heidelberg, 1987.

14. D. Eppstein. Geometric lower bounds for parametric matroid optimization. *Discrete Comput. Geom.*, 20:463–476, 1998.

15. D. Eppstein, Z. Galil, G. F. Italiano, and A. Nissenzweig. Sparsification — a technique for speeding up dynamic graph algorithms. *J. Assoc. Comput. Mach.*, 44:669–696, 1997.

16. D. Eppstein, Z. Galil, G. F. Italiano, and T. H. Spencer. Separator-based sparsification I: planarity testing and minimum spanning trees. *J. Computing and Systems Sciences*, 52:3–27, 1996.

17. D. Fernández-Baca and G. Slutzki. Linear-time algorithms for parametric minimum spanning tree problems on planar graphs. *Theoretical Computer Science*, 181:57–74, 1997.

18. D. Fernández-Baca and G. Slutzki. Optimal parametric search on graphs of bounded tree-width. *Journal of Algorithms*, 22:212–240, 1997.

19. D. Fernández-Baca, G. Slutzki, and D. Eppstein. Using sparsification for parametric minimum spanning tree problems. *Nordic Journal of Computing*, 34(4):352–366, 1996.

20. G. N. Frederickson. Optimal algorithms for partitioning trees and locating *p*-centers in trees. Technical Report CSD-TR 1029, Department of Computer Science, Purdue University, Lafayette, IN, October 1990.

21. D. R. Karger, P. N. Klein, and R. E. Tarjan. A randomized linear-time algorithm for finding minimum spanning trees. *J. Assoc. Comput. Mach.*, 42:321–328, 1995.

22. P. N. Klein, S. Rao, M. Rauch, and S. Subramanian. Faster shortest-path algorithms for planar graphs. In *Proceedings of the 26th Annual ACM Symposium on Theory of Computing*, pages 27–37, 1994.

23. N. Megiddo. Combinatorial optimization with rational objective functions. *Math. Oper. Res.*, 4:414–424, 1979.

24. N. Megiddo. Applying parallel computation algorithms in the design of serial algorithms. *J. Assoc. Comput. Mach.*, 30(4):852–865, 1983.

25. N. Megiddo. Linear programming in linear time when the dimension is fixed. *J. Assoc. Comput. Mach.*, 34(1):200–208, 1984.

26. C. H. Norton, S. A. Plotkin, and É. Tardos. Using separation algorithms in fixed dimension. *Journal of Algorithms*, 13:79–98, 1992.

Linear Time Recognition of Optimal L-Restricted Prefix Codes (Extended Abstract)

Ruy Luiz Milidiú[1] and Eduardo Sany Laber[2]

[1] Departamento de Informática, PUC-Rio, Brazil
Rua Marquês de São Vicente 225, RDC, sala 514, FPLF
Gávea, Rio de Janeiro, CEP 22453-900, phone 5521-511-1942
milidiu@inf.puc-rio.br
[2] COPPE/UFRJ,
Caixa Postal 68511 21945-970 Rio de Janeiro, RJ, Brasil
tel: +55 21 590-2552, fax: +55 21 290-6626
laber@inf.puc-rio.br

1 Introduction

Given an alphabet $\Sigma = \{a_1, \ldots, a_n\}$ and a corresponding list of weights $[w_1, \ldots, w_n]$, an optimal prefix code is a prefix code for Σ that minimizes the *weighted length* of a code string, defined to be $\sum_{i=1}^{n} w_i l_i$, where l_i is the length of the codeword assigned to a_i. This problem is equivalent to the following problem: given a list of weights $[w_1, \ldots, w_n]$, find an optimal binary code tree, that is, a binary tree T that minimizes the *weighted path length* $\sum_{i=1}^{n} w_i l_i$, where l_i is the level of the i-th leaf of T from left to right. If the list of weights is sorted, this problem can be solved in $O(n)$ by one of the efficient implementations of Huffman's Algorithm [Huf52]. Any tree constructed by Huffman's Algorithm is called a Huffman tree.

In this paper, we consider optimal L-restricted prefix codes. Given a list of weights $[w_1, \ldots, w_n]$ and an integer L, with $\lceil \log n \rceil \leq L \leq n - 1$, an optimal L-restricted prefix code is a prefix code that minimizes $\sum_{i=1}^{n} w_i l_i$ constrained to $l_i \leq L$ for $i = 1, \ldots, n$. Gilbert [Gil71] recommends the use of these codes when the weights w_i are inaccurately known. Zobel and Moffat [ZM95] describe the use of word-based Huffman codes for compression of large textual databases. Their application allows the maximum of 32 bits for each codeword. For the cases that exceed this limitation, they recommend the use of L-restricted codes.

Some methods can be found in the literature to generate optimal L-restricted prefix codes. Different techniques of algorithm design have been used to solve this problem. The first polynomial algorithm is due to Garey [Gar74]. The algorithm is based on dynamic programming and it has an $O(n^2 L)$ complexity for both time and space. Larmore and Hirschberg [LH90] presented the Package-Merge algorithm. This algorithm uses a greedy approach an runs in $O(nL)$ time, with $O(n)$ space requirement. The authors reduce the original problem to the Coin's Collector Problem, using a nodeset representation of a binary tree. Turpin

G. Gonnet, D. Panario, and A. Viola (Eds.): LATIN 2000, LNCS 1776, pp. 227–236, 2000.

and Moffat [TM96] discuss some practical aspects on the implementation of the Package-Merge algorithm. In [Sch95], Schieber obtains an $O(n2^{O(\sqrt{\log L \log \log n})})$ time algorithm. Currently, this is the fastest strongly polynomial time algorithm for constructing optimal L-restricted prefix codes. Despite, the effort of some researchers, it remains open if there is an $O(n \log n)$ algorithm for this problem.

In this paper we give a linear time algorithm to recognize an optimal L-restricted prefix code. This linear time complexity holds under the assumption that the given list of weights is already sorted. If the list of weights is not sorted, then the algorithm requires an $O(n \log n)$ initial sorting step. This algorithm is based on the nodeset representation of binary trees [LH90].

We assume that we are given an alphabet $\Sigma = \{a_1, \ldots, a_n\}$ with corresponding weights $0 < w_1 \leq \cdots \leq w_n$, an integer L with $\lceil \log n \rceil \leq L < n$ and a list of lengths $l = [l_1, \ldots, l_n]$, where l_i is the length of the codeword assigned to a_i. We say that l is optimal iff $l_1, \ldots, l_n$ are the codewords lengths of an optimal L-restricted prefix code for Σ. The Recognition algorithm that we introduce here determines if l is optimal or not.

The paper is organized as follows. In section 2, we present the nodeset representation for binary trees and we derive some useful properties. In section 3, we present the Recognition algorithm. In section 4, we outline a proof for the algorithm correctness.

2   Trees and Nodesets

For positive integers i and h, let us define a node as an ordered pair (i, h), where i is called the node index and h is the node level. A set of nodes is called a nodeset.

We define the background $R(n, L)$ as the nodeset given by $R(n, L) = \{(i, h) | 1 \leq i \leq n, 1 \leq h \leq L\}$

Let T be a binary tree with n leaves and with corresponding leaf levels $l_1 \geq \ldots \geq l_n$. The treeset $N(T)$ associated to T is defined as the nodeset given by $N(T) = \{(i, h) | 1 \leq h \leq l_i, 1 \leq i \leq n\}$

The background in figure 1 is the nodeset $R(8, 5)$. The nodes inside the polygon are the ones of $N(T)$, where T is the tree with leaves at following levels: $5, 5, 5, 5, 3, 3, 3, 1$.

For any nodeset $A \subset R(n, L)$ we define the complementary nodeset $\overline{A}$ as $\overline{A} = R(n, L) - A$. In figure 1, the nodes outside of the polygon are those of the nodeset $\overline{N(T)}$.

Given a node (i, h), define $width(i, h) = 2^{-h}$ and $weight(i, h) = w_i$. The width and the weight of a nodeset, are defined as the sums of the corresponding widths and weights of their constituent nodes. Let T be a binary tree with n leaves and corresponding leaf levels $l_1 \geq \ldots \geq l_n$. It is not difficult to show [LH90] that $width(N(T)) = n - 1$ and $weight(N(T)) = \sum_{i=1}^{n} w_i l_i$.

In [LH90], Larmore and Hirschberg reduced the problem of finding an optimal code tree with restricted maximal height L, for a given list of weights $w_1, \ldots, w_n$ to the problem of finding the nodeset with width $n - 1$ and minimum weight

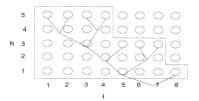

Fig. 1. The background $R(8,5)$ and the treeset $N(T)$ associated to a tree T with leaf levels $5, 5, 5, 5, 3, 3, 3, 1$

included in the background $R(n, L)$. Here, we need a slight variation of the main theorem proved in [LH90].

Theorem 1. *If a tree T is an optimal code tree with restricted maximum height L, then the nodeset associated to T has minimum weight among all nodesets with width $n - 1$ included in $R(n, L)$.*

The proof of the previous theorem is similar to that presented in [LH90]. Therefore, the list $l = [l_1 \geq \cdots \geq l_1]$ is a list of an optimal L-restricted codeword lengths for Σ if and only if the nodeset $N = \{(i, h) | 1 \leq i \leq n, 1 \leq h \leq l_i\}$ has minimum weight among the nodesets in $R(n, L)$ that have width equal to $n - 1$.

In order to find the nodeset with width $n - 1$ and minimum weight, Larmore and Hirschberg used the Package-Merge algorithm. This algorithm was proposed in [LH90] to address the following problem: given a nodeset R and a width d, find a nodeset X included in R with width d and minimum weight. The Package-Merge uses a greedy approach and runs in $O(|R|)$, where $|R|$ denotes the number of nodes in the nodeset R. The Recognition algorithm uses the Package-Merge as an auxiliary procedure.

3   Recognition Algorithm

In this section, we describe a linear time algorithm for recognizing optimal L-restricted prefix codes. The algorithm is divided into two phases.

3.1   First Phase

First, the algorithm scans the list l to check if there is an index i such that $w_i < w_{i+1}$ and $l_i < l_{i+1}$. If that is the case, the algorithm outputs that l is not optimal and stops. In this case, the weighted path length can be reduced interchanging l_i and l_{i+1}. In the negative case, we sort l by non-increasing order of lengths. This can be done in linear time since all the elements of l are integers not greater than n. We claim that l is optimal if and only if the list obtained by sorting l is optimal. In fact, the only case where $l_i < l_{i+1}$ is when $w_i = w_{i+1}$. In this case, we can interchange l_i and l_{i+1} maintaining the same external weighted path length. Hence, we assume that l is sorted with $l_1 \geq \cdots \geq l_n$.

After sorting, the algorithm verifies if $l_1 > L$. In the affirmative case, the algorithm stops and outputs "l is not optimal". In the negative case, l respects the length restriction. Then, the algorithm verifies if $\sum_{i=1}^{n} 2^{-l_i} = 1$. This step can be performed in $O(n)$ since l is sorted. If $\sum_{i=1}^{n} 2^{-l_i} \neq 1$, then the algorithm outputs "l is not optimal". In effect, if $\sum_{i=1}^{n} 2^{-l_i} > 1$, then it follows from McMillan-Kraft inequality [McM56] that there is not a prefix code with codeword lengths given by l. On the other hand, if $\sum_{i=1}^{n} 2^{-l_i} < 1$, then $\sum_{i=1}^{n} w_i l_i$ can be reduced by decreasing one unity the length of the longest codeword length l_1. If $\sum_{i=1}^{n} 2^{-l_i} = 1$, then the algorithm verifies two additional conditions:

1. $l_1, \ldots, l_n$ are the codeword lengths of an unrestricted optimal prefix code for Σ;
2. $l_1 = L$.

Condition 1 can be verified in $O(n)$ by comparing $\sum_{i=1}^{n} w_i l_i$ to the optimal weighted path length obtained by Huffman algorithm. Condition 2 can be checked in $O(1)$. We have three cases that we enumerate below:

Case 1) Condition 1 holds. The algorithm outputs "l is optimal" and stops.

Case 2) Both conditions 1 and 2 do not hold. The algorithm outputs l is not optimal and stops.

Case 3) Only condition 2 holds. The algorithm goes to phase 2.

In the second case, the external weighted path length can be reduced without violating the height restriction by interchanging two nodes that differs by at most one level in the tree with leaf levels given by l.

3.2 Second Phase

If the algorithm does not stop at the first phase, then it examines the treeset N associated to the binary tree with leaf levels $l_1 \geq \ldots \geq l_n$. In this phase, the algorithm determines if N is optimal or not. Recall that it is equivalent to determine if l is optimal or not.

First, we define three disjoint sets. We define the boundary F of a treeset N by

$$F = \{(i, h) \in N | (i, h+1) \notin N\}$$

Let us also define F_2 by

$$F_2 = \{(i, h) \in N - F | h < L - \lceil \log n \rceil - 1 \text{ and } (i+1, h) \notin N - F\}$$

Now, for $h = L - \lceil \log n \rceil - 1, \ldots, L - 1$, let i_h be the largest index i such that $(i, h) \in N - F$. Then, define the nodeset M as

$$M = \{(i, h) | L - \lceil \log n \rceil - 1 \leq h \leq L - 1, i_h - 2^{h + \lceil \log n \rceil - L + 1} < i \leq i_h\}.$$

The way that the nodeset M is defined assures that M contains a nodeset with minimum weight among the nodesets included in $(N - F)$ that have width equal to d, where d is a given diadic [1] number not greater than $2^{\lceil \log n \rceil - L + 1}$.

In figure 2, $R(14, 10)$ is the background. The polygon bounds the nodeset N. The nodes of the boundary F are the nodes inside the polygon that have the letter F written inside. The nodes of the nodeset M are those inside the polygon that have the letter M, while the nodes of the nodeset F_2 are those inside the polygon that have the number 2.

Fig. 2. The nodesets used by the Recognition algorithm

Now, we define three other disjoint nodesets. The upper boundary U of the nodeset N is defined by

$$U = \{(i, h) \in \overline{N} | (i, h - 1) \in N\}$$

Let us also define U_2 by

$$U_2 = \{(i, h) \in \overline{N} - U | h < L - \lceil \log n \rceil - 1 \text{ and } (i - 1, h) \in N \cup U\}$$

Now, for $h = L - \lceil \log n \rceil - 1, \ldots, L - 1$, let i'_h be the smallest index i such that (i, h) belongs to $\overline{U} \cup N$. We define the nodeset P in the following way

$$P = \{(i, h) | L - \lceil \log n \rceil - 1 \leq h \leq L, i'_h \leq i < 2^{h + \lceil \log n \rceil - L + 1} + i'_h\}.$$

In figure 2, the nodes of the upper boundary U are those outside the polygon that have the letter U written inside. The nodes of the nodeset P are those

---

[1] A diadic number is a number that can be written as a sum of integer powers of 2

outside the polygon that have the letter P, while the nodes of the nodeset U_2 are those outside the polygon that have the number 2.

The recognition algorithm performs three steps. The pseudo-code is presented in figure 3.

Recognition Algorithm: Phase 2 ;

1. Evaluate the width and the weight of the nodeset $F \cup M \cup F_2$.

2. Apply the Package-Merge algorithm to obtain the nodeset X with minimum weight among the nodesets included in $F \cup M \cup F_2 \cup U \cup P \cup U_2$ that have width equal to $width(F \cup M \cup F_2)$.

3. If $weight(X) < weight(F \cup M \cup F_2)$, then *outputs N is not optimal*; otherwise *outputs N is optimal*

Fig. 3. The second phase of the recognition algorithm.

As an exemple, let us assume that we are interested to decide if the list of lengths $l = [10, 10, 10, 10, 9, 9, 7, 6, 6, 6, 4, 3, 2, 1]$ is optimal for an alphabet Σ with weights $[1, 1, 2, 3, 5, 8, 13, 21, 34, 61, 89, 144, 233, 377]$ and $L = 10$. The nodeset N associated to l is the one presented in figure 2. The width of the nodeset $F \cup M \cup F_2$ is equal to $2 + 2^{-4} + 2^{-7}$ and its weight is equal to 1646. Let A be the nodeset $\{[(5, 10), (6, 10), (7, 9), (7, 8), (8, 7)]\}$ and let B be the nodeset $\{(10, 6)\}$. The nodeset $N \cup A - B$ has width $2 + 2^{-4} + 2^{-7}$ and weight 1645, and as a consequence, at step 2 the package-merge finds a nodeset with weight smaller than or equal to 1645. Therefore, the algorithms outputs that l is not optimal.

3.3   Algorithm Analysis

The linear time complexity of the algorithm is established below.

Theorem 2. *The Recognition algorithm runs in $O(n)$ time.*

Sketch of the Proof: The phase 1 can be implemented in $O(n)$ time as we showed in section 3.1.

Let us analyze the second phase. Step 1 can be performed in linear time as we argue now. We define $i_s(h)$ and $i_b(h)$, respectively, as the smallest and the biggest index i such that $(i, h) \in F \cup M \cup F_2$. It is easy to verify that $i_b(h)$ is given by the number of elements in l that are greater than or equal to h. On the other hand, $i_s(h)$ can be obtained through $i_b(h)$ and the definitions of F, M and F_2. Hence, the set $\{(i_s(L), i_b(L)), \ldots, (i_s(1), i_b(1))\}$ can be obtained in $O(n)$ time. Since the width of $F \cup M \cup F_2$ is given by $\sum_{h=1}^{L}(i_b(h) - i_s(h) + 1) \times 2^{-h}$, it can be evaluated in linear time.

Now, we consider step 2. We define $i'_s(h)$ and $i'_b(h)$, respectively, as the smallest and the biggest index i such that $(i, h) \in F \cup M \cup F_2 \cup U \cup P \cup U_2$. Observe that

$i_s(h) = i'_s(h)$. Furthermore, $i'_b(h)$ can be obtained through $i_b(h)$ and the definitions of U, P and U_2. Hence, the set $\{(i'_s(L), i'_b(L)), \ldots, i'_s(1), i'_b(1))\}$ is obtained in $O(n)$ time. Now, we show that the number of nodes in $F \cup M \cup F_2 \cup U \cup P \cup U_2$ is $O(n)$. We consider each nodeset separately. It follows from the definition of F and U that each of them has at most n nodes. In addition, both F_2 and U_2 have at most $L - \lceil \log n \rceil - 2$ nodes. Furthermore, from the definitions of M and P one can show that $|M \cup P| < 5n$. Then it follows that $|F \cup M \cup F_2 \cup U \cup P \cup U_2| \leq 7n + 2L$. Therefore, the package-merge runs in $O(n)$ time at step 2.

The step 3 is obviously performed in $O(1)$ ∎

The correctness of the algorithm is a consequence of the theorem stated below.

Theorem 3. *If N is not an optimal nodeset, then it is possible to obtain a new nodeset with width $n - 1$ and weight smaller than N by replacing some nodes in $F \cup M \cup F_2$ by other nodes in $U \cup P \cup U_2$.*

The sketch of the proof of theorem 3 is presented in the next section. The complete proof can be found in the full version of this paper.

Now, we prove that theorem 3 implies on the correctness of the Recognition algorithm.

Theorem 4. *The second phase of the Recognition algorithms is correct.*

Proof: First, we assume that N is not an optimal nodeset. In this case, the theorem 3 assures the existence of nodesets A and B that satisfy the following conditions:

(i) $A \subset U \cup P \cup U_2$ and $B \subset F \cup M \cup F_2$;
(ii) $width(A) = width(B)$;
(iii) $weight(A) < weight(B)$.

These conditions imply that $weight(F \cup M \cup F_2 \cup A - B) < weight(F \cup M \cup F_2)$ and $width(F \cup M \cup F_2 \cup A - B) = widht(F \cup M \cup F_2)$. Let X be the nodeset found by package-merge at step 2 in the second phase. Since $F \cup M \cup F_2 \cup A - B \subset F \cup M \cup F_2 \cup U \cup P \cup U_2$, it follows that $weight(X) \leq weight(F \cup M \cup F_2 \cup A - B)$. Therefore, $weight(X) < weight(F \cup M \cup F_2)$. Hence, the algorithm outputs that N is not optimal.

Now, we assume that N is optimal. In this case, $weight(X) \geq weight(F \cup M \cup F_2)$, otherwise, $N \cup X - (F \cup M \cup F_2)$ would have weight smaller than that of N, what would contradict our assumption. Hence, the algorithm outputs that N is optimal. ∎

4   Correctness

In this section, we outline the proof of theorem 3. We start by defining the concept of a decreasing pair.

Definition 1. *If a pair of nodesets (A, B) satisfy the conditions (i)-(iii) listed below, then we say that (A, B) is a decreasing pair associated to N.*
(i) $A \subset \overline{N}$ and $B \subset N$;
(ii) $width(A) = width(B)$;
(iii) $weight(A) < weight(B)$.

For the sake of simplicity, we use the term DP to denote a decreasing pair associated to N. We can state the following result.

Proposition 1. *The nodeset N is not optimal if and only if there is a DP (A, B) associated to N.*

Proof: If N is not optimal, $(N^* - N, N - N^*)$ is a DP, where N^* is an optimal nodeset. If (A, B) is a DP associated to N, then $N \cup A - B$ has width equal to that of N and has weight smaller than that of N. ∎

Now, we define the concept of good pair (GP).

Definition 2. *A GP is a DP (A, B) that satisfies the following conditions*
(i) For every DP (A', B'), we have that $width(A) \leq width(A')$;
(ii) If $A' \subset \overline{N}$ and $width(A') = width(A)$, then $weight(A) \leq weight(A')$;
(iii) If $B' \subset N$ and $width(B') = width(B)$, then $weight(B) \geq weight(B')$.

Now, we state some properties concerning good pairs.

Proposition 2. *If N is not optimal, then there is a GP associated to N.*

Proof: If N is not optimal, then it follows from proposition 1 that there is at least one DP associated to N. Then, let d be the width of the DP with minimum width associated to N. Furthermore, let A^* be a nodeset with minimum weight among the nodesets included in $\overline{N}$ that have width d, and let B^* be the nodeset with maximum weight among the nodesets included in N that have width d. By definition, (A^*, B^*) is a GP. ∎

Proposition 3. *Let (A^*, B^*) be a GP. If $A' \subset A^*$ and $B' \subset B^*$, then $width(A') \neq width(B')$.*

Proof: Let us assume that $A' \subset A^*$, $B' \subset B^*$ and $width(A') = width(B')$. In this case, let us consider the following partitions $A^* = A' \cup (A^* - A')$ and $B^* = B' \cup (B^* - B')$. Since $weight(A^*) < weight(B^*)$, then either $weight(A') < weight(B')$ or $weight(A^* - A') < weight(B^* - B')$. If $weight(A') < weight(B')$, then (A', B') is a DP and $width(A') < width(A^*)$, that contradicts the fact that (A^*, B^*) is a GP. On the other hand, if $weight(A^* - A') < weight(B^* - B')$, then $(A^* - A', B^* - B')$ is a DP and $width(A^* - A') < width(A^*)$, what also contradicts the fact that (A^*, B^*) is a GP. Hence, $width(A') \neq width(B')$ ∎

Proposition 4. *If (A^*, B^*) is a GP, then the following conditions hold*
(a) $width(A^*) = width(B^*) \leq 2^{-1}$;
(b) $width(A^*) = 2^{-s_1}$, for some integer s_i where $1 \leq s_1 \leq L$;
(c) Either $1 = |A^*| < |B^*|$ or $1 = |B^*| < |A^*|$.

Proof: (a) Let us assume that $width(A^*) = width(B^*) > 2^{-1}$. In this case, one can show, by applying the lemma 1 of [LH90] at most L times, that both A^* and B^* contain a nodeset with width 2^{-1}. However, it contradicts proposition 3. Hence, $width(A^*) = width(B^*) \leq 2^{-1}$.

(b) Now, we assume that $width(A^*) = \sum_{i=1}^{k} 2^{-s_i}$, where $1 \leq s_1 < s_2 \ldots < s_k$ and $k > 1$. In this case, one can show that A^* contains a nodeset A' with width 2^{-s_1} and B^* contains a nodeset B' with width 2^{-s_1}, that contradicts the proposition 3. Hence, $k = 1$ and $width(A^*) = 2^{-s_1}$.

(c) First, we show that either $1 = |A^*|$ or $1 = |B^*|$. Let us assume the opposite, that is, $1 < |A^*|$ and $1 < |B^*|$. Since $width(A^*) = width(B^*) = 2^{-s_i}$ for some $1 \leq s_i \leq L$, then one can show that both A^* and B^* contain a nodeset with width $2^{-(s_i+1)}$, that contradicts the proposition 3. Hence, either $1 = |A^*|$ or $1 = |B^*|$. Furthermore, we cannot have $1 = |A^*| = |B^*|$. In effect, if $1 = |A^*| = |B^*|$, we would have $weight(A^*) \geq weight(B^*)$, and as a consequence, (A^*, B^*) would not be a GP. ■

The previous result allows us to divide our analysis into two cases. In the first case, A^* has only one node. In the second case B^* has only one node. We define two special pairs. The removal good pairs (RGP) and the addition good pairs (AGP).

Definition 3. *If (A^*, B^*) is a GP, $|A^*| = 1$ and for all GP (A, B), with $|A| = 1$, we have $width(B^* - F) \leq width(B - F)$. Then, (A^*, B^*) is a removal good pair (RGP).*

Definition 4. *If (A^*, B^*) is a GP, $|B^*| = 1$ and for all GP (A, B), with $|B| = 1$, we have $width(A^* - U) \leq width(A - U)$. Then, (A^*, B^*) is an addition good pair (AGP).*

We can state the following result.

Proposition 5. *If there is a GP associated to N, then there is either a RGP associated to N or an AGP associated to N.*

Proof: The proof is similar to that of proposition 2. ■

Theorem 5. *If there is a RGP (A, B) associated to N, then there is a RGP (A^*, B^*) associated to N that satisfies the following conditions:*
(a) If $(i, h) \in A^*$, then $(i - 1, h) \in N$;
(b) $|B^* - (F \cup M)| \leq 1$;
(c) If $(i, h) \in B^* - (F \cup M)$, then $(i + 1, h) \notin N - F$.

Proof: We leave this proof for the full version of the paper. The proof requires some additional lemmas and it uses arguments that are similar to that employed in the proof of proposition 4. ■

Theorem 6. *If there is an AGP (A, B) associated to N, then there is an AGP (A^*, B^*) associated to N that satisfies the following conditions:*
(a) If $(i, h) \in B^*$, then $(i + 1, h) \in \overline{N}$;
(b) $|A^* - (U \cup P)| \leq 1$;
(c) If $(i, h) \in |A^* - (U \cup P)|$, then $(i - 1, h) \in N \cup U$.

Proof: We leave this proof to the full paper. ∎

Now, we prove the theorem 3 that implies on the correctness of the Recognition algorithm.

Proof of theorem 3: If N is not an optimal nodeset, then it follows from proposition 2 that there is a GP associated to N. Hence, it follows from proposition 5 that there is either a RGP or an AGP associated to N. We consider two cases:

Case 1) There is a RGP associated to N.

It follows from theorem 5 that there is a RGP (A^*, B^*) associated to N that satisfies the conditions (a), (b) and (c) proposed in that theorem. ¿From the definitions of the nodesets F, M, F_2, U, P, U_2 it is easy to verify that those conditions imply that $B^* \subset F \cup M \cup F_2$ and $A^* \subset U \cup P \cup U_2$.

Case 2) There is an AGP associated to N. The proof is analogous to that of case 1.

Hence, we conclude that if N is not optimal, then there is a DP (A^*, B^*) such that $B^* \subset F \cup M \cup F_2$ e $A^* \subset U \cup P \cup U_2$. Therefore, it is possible to reduce the weight of N by adding some nodes that belong to $U \cup P \cup U_2$ and removing some other nodes that belong to $F \cup M \cup F_2$. ∎

References

[Gar74]   M. R. Garey. Optimal binary search trees with restricted maximal depth. *Siam Journal on Computing*, 3(2):101–110, June 1974.

[Gil71]   E. N. Gilbert. Codes based on innacurate source probabilities. *IEEE Transactions on Information Theory*, 17:304–314, 1971.

[Huf52]   D. A. Huffman. A method for the construction of minimum-redundancy codes. In *Proc. Inst. Radio Eng.*, pages 1098–1101, September 1952. Published as Proc. Inst. Radio Eng., volume 40, number 9.

[LH90]   Lawrence L. Larmore and Daniel S. Hirschberg. A fast algorithm for optimal length-limited Huffman codes. *Journal of the ACM*, 37(3):464–473, July 1990.

[McM56]   B. McMillan. Two inequalities implied by unique decipherability. *IEEE Transaction on Information Theory*, 22:155–156, 1956.

[Sch95]   Baruch Schieber. Computing a minimum-weight k-link path in graphs with the concave Monge property. In *Proceedings of the Sixth Annual ACM-SIAM Symposium on Discrete Algorithms*, pages 405–411, San Francisco, California, 22–24 January 1995.

[TM96]   Andrew Turpin and Alistair Moffat. Efficient implementation of the package-merge paradigm for generating length-limited codes. In Michael E. Houle and Peter Eades, editors, *Proceedings of Conference on Computing: The Australian Theory Symposium*, pages 187–195, Townsville, 29–30 January 1996. Australian Computer Science Communications.

[ZM95]   Justin Zobel and Alistair Moffat. Adding compression to a full-text retrieval system. *Software—Practice and Experience*, 25(8):891–903, August 1995.

Uniform Multi-hop All-to-All Optical Routings in Rings *

Jaroslav Opatrny

Concordia University, Department of Computer Science, Montréal, Canada,
email: opatrny@cs.concordia.ca
WWW home page: http://www.cs.concordia.ca/~faculty/opatrny/

Abstract. We consider all-to-all routing problem in an optical ring network that uses the wavelength-division multiplexing (WDM). Since one-hop, all-to-all optical routing in a WDM optical ring of n nodes needs $\overset{\bullet}{w}(C_n, I_A, 1) = \lceil \frac{1}{2} \lfloor \frac{n^2}{4} \rfloor \rceil$ wavelengths which can be too large even for moderate values of n, we consider in this paper j-hop implementations of all-to-all routing in a WDM optical ring, $j \geq 2$. ¿From among the possible routings we focus our attention on *uniform* routings, in which each node of the ring uses the same communication pattern. We show that there exists a uniform 2-hop, 3-hop, and 4-hop implementation of all-to-all routing that needs at most $\frac{n+3}{3}\sqrt{\frac{n+16}{5}} + \frac{n}{4}$, $\frac{n}{2}\sqrt[3]{\frac{\bullet n/4 \bullet + 4}{5}}$, and $\frac{n+16}{2}\sqrt[4]{\lceil \frac{n}{4} \rceil} + 8$ wavelengths, respectively. These values are within multiplicative constants of lower bounds.

1 Introduction

Optical-fiber transmission systems are expected to provide a mechanism to build high-bandwidth, error-free communication networks, with capacities that are orders of magnitude higher than traditional networks. The high data transmission rate is achieved by transmitting information through optical signals, and maintaining the signal in optical form during switching. *Wavelength-division multiplexing* (or WDM for short) is one of the most commonly used approaches to introduce concurrency into such high-capacity networks [5,6]. In this strategy, the optical spectrum is divided into many different channels, each channel corresponding to a different wavelength. A switched WDM network consists of nodes connected by point-to-point fiber-optic links. Typically, a pair of nodes that is connected by a fiber-optic link is connected by a pair of optic cables. Each cable is used in one direction and can support a fixed number of wavelengths. The switches in nodes are capable of redirecting incoming streams based on wavelengths. We assume that switches cannot change the wavelengths, i.e. there are no wavelength converters.

Thus, a WDM optical network can be represented by a *symmetric digraph*, that is, a directed graph G with vertex set $V(G)$ representing the nodes of the

---

* The work was supported partially by a grant from NSERC, Canada.

G. Gonnet, D. Panario, and A. Viola (Eds.): LATIN 2000, LNCS 1776, pp. 237–246, 2000.

network and edge set $E(G)$ representing optical cables, such that if directed edge $[x, y]$ is in $E(G)$, then directed edge $[y, x]$ is also in $E(G)$. In the sequel, whenever we refer to an edge or a path, we mean a directed edge or a directed path.

Different messages can use the same link (or directed edge) concurrently if they are assigned distinct wavelengths. However, messages assigned the same wavelength must be assigned edge-disjoint paths. In the graph model, each wavelength can be represented by a color.

Given an optical communication network and a pattern of communications among the nodes, one has to design a *routing* i.e. a system of directed paths and an assignment of wavelengths to the paths in the routing so that the given communications can be done simultaneously. We can express the problem more formally as follows.

Given a WDM optical network G, a *communication request* is an ordered pair of nodes (x, y) of G such that a message is to be sent from x to y. A *communication instance* I (or *instance* for short) is a collection of requests.

Let I be an instance in G. A *j-hop* solution [9,10] of I is a routing R in G and an assignment of colors to paths in R such that

1. it is *conflict-free*, i.e., any two paths of R sharing the same directed edge have different colors, and
2. for each request (x, y) in I, a directed path from x to y in R can be obtained by concatenation of at most j paths in R.

Since the cost of an optical switch depends on the number of wavelengths it can handle, it is important to determine paths and a conflict-free color assignment so that the total number of colors is minimized.

In 1-hop, usually called *all optical* solution of I, there is a path from x to y in R for each request (x, y) in I and all communications are done in optical form. In a *j-hop* solution, $j \geq 2$, the signal must be converted into electronic form $j - 1$ times. The conversion into electronic form slows down the transmission, but $j > 1$ can significantly reduce the number of wavelengths needed [10].

For an instance I in a graph G, and a *j-hop* routing R for it, the *j-hop wavelength index* of the routing R, denoted $\vec{w}(G, I, R, j)$, is defined to be the *minimum* number of colors needed for a conflict-free assignment of colors to paths in the routing R. The parameter $\vec{w}(G, I, j)$, the *j-hop optimal wavelength index* for the instance I in G is the minimum value over all possible routings for the given instance I in G. In general, the problem of determining the optimal wavelength index is NP-complete [3].

In this paper, we are interested in *ring* networks. In a ring network there is a link from each node to two other nodes. Thus, the topology of a ring network on n nodes $n \geq 3$, can be represented by a *symmetric directed cycle* C_n (see [4] for any graph terminology not defined here). A symmetric directed cycle C_n, $n \geq 3$, consists of n nodes x_0, x_1, ..., x_{n-1} and there is an arc from x_i to $x_{(i+1) \bmod n}$ and from $x_{(i+1) \bmod n}$ to x_i for $0 \leq i \leq n - 1$, see C_8 in Figure 1. We denote by $p_{i,j}$ a shortest path from x_i to x_j. The *diameter* of C_n, i.e., the maximum among the lengths of the shortest paths among nodes of C_n, denoted by d_n, is equal to $\lfloor \frac{n}{2} \rfloor$.

Fig. 1. Ring C_8

The all-to-all communication instance I_A is the instance that contains all pairs of nodes of a network. I_A has been studied for rings and other types of network topologies [1], [3], [8], [11]. It has been shown for rings in [3] that $\overrightarrow{w}$ $(C_n, I_A, 1) = \lceil \frac{1}{2} \lfloor \frac{n^2}{4} \rfloor \rceil$. The optical switches that are available at present cannot handle hundreds of different wavelengths and thus the value of $\overrightarrow{w}$ $(C_n, I_A, 1)$ can be too large even for moderately large rings. One can reduce the number of wavelength by considering j-hop solutions for the I_A problem for $j \geq 2$.

One possible 2-hop solution of the I_A instance is the routing $\{p_{0,i} : 1 \leq i \leq n-1\} \cup \{(p_{i,0} : 1 \leq i \leq n-1\}$, in which there is a path from x_0 to all other nodes and a path from any node to x_0. Any request (x_i, x_j) in I_A can be obtained by a concatenation of $p_{i,0}$ and $p_{0,j}$, and we can get a conflict-free assignment of colors to all paths in R using $\lceil \frac{n-1}{2} \rceil$ colors. However, this solution has all the drawbacks of having one server x_0 for the network, i.e., a failure of x_0 shuts down the whole network and x_0 is a potential performance bottleneck. This is very obvious if we represent R by the *routing graph* G_R [7], in which there is an edge from x to y if and only if there is a path in R from x to y. In case of the routing above, the routing graph is the star of Figure 2 a).

For better fault-tolerance and better load distribution, we should consider a uniform routing [11] in which each node can communicate directly with the *same number* of nodes as any other node and at the *same distance* along the ring. More specifically, a routing R on a ring of length n is *uniform* if for some integer $m < n$ and some integers $b_1, b_2, \ldots, b_m$ the routing R consists of paths connecting nodes that are at distance b_i, $1 \leq i \leq m$ along the ring, i.e. $R = \{p_{i,i+b_j}, p_{i,i-b_j} : 0 \leq i \leq n-1, 1 \leq j \leq m\}$. In a uniform routing the routing graph is a distributed loop graph [2] of degree $2m$, i.e., for $m = 2$ and $b_1 = 1$ and $b_2 = 3$ we get the routing graph in Figure 2 b). As seen from the figure, this provides much better connectivity and can give a uniform load on the nodes. Furthermore, the routing decisions can be the same in all nodes.

Thus, the problem that we consider in this paper is the following:
Given a ring C_n and integer j, $j \geq 2$, find a routing $R_{n,j}$ such that:

1. $R_{n,j}$ is uniform routing on C_n,
2. $R_{n,j}$ is a j-hop solution of I_A,

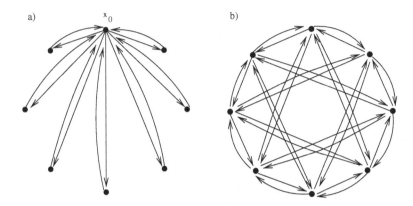

Fig. 2. The routing graph in C_8 of a) a non-uniform, b) a uniform routing

3. the number of colors needed for $R_{n,j}$ is substantially less than $\lceil \frac{1}{2} \lfloor \frac{n^2}{4} \rfloor \rceil$.

In Section 2, we show that there exists a uniform 2-hop routing $R_{n,2}$ for I_A that needs at most $\frac{n+3}{3} \sqrt{\frac{n+16}{5}} + \frac{n}{4}$ colors. We show that this is within a constant factor of a lower bound.

Uniform j-hop solutions of I_A, $j \geq 3$ are considered in Section 3. We show that there exist a uniform 3-hop routing $R_{n,3}$ that needs at most $\frac{n}{2} \sqrt[3]{\frac{\lceil n/4 \rceil + 4}{5}}$ colors, and a 4-hop $R_{n,4}$ with at most $\frac{n+16}{2} \sqrt[4]{\lceil \frac{n}{4} \rceil} + 8$ colors. We present conclusions and open questions in Section 4. Most proofs in this extended abstract are either omitted or only outlined. Complete proofs appear in [12].

2   Uniform, 2-hop All-to-All Routing

Let n be an integer, $n \geq 5$, and C_n be a ring of length n with nodes $x_0, x_1, \ldots, x_{n-1}$. A routing R is a 2-hop solution of all-to-all problem in C_n, if any request x_i, x_j, $i \neq j$ can be obtained as a concatenation of at most two paths in R. Since on a ring any pair of distinct nodes is at distance between 1 and $d_n = \lfloor n/2 \rfloor$, we have that routing R is a uniform 2-hop solution of the all-to-all instance on C_n if there is a set of integers $B = \{b_1, b_2, \ldots, b_m\}$ such that R contains all paths on the ring of length in B, and any integer between 1 and d_n can be obtained using at most one addition or subtraction of two elements in B.

Lemma 1. *Let k and m be positive integers, $k \leq \frac{m}{4}$, and*
$B_{k,m} = B_{k,m}^1 \cup B_{k,m}^2 \cup B_{k,m}^3 \cup B_{k,m}^4$ *where* $B_{k,m}^1 = \{1, 2, \ldots, \lfloor k/2 \rfloor\}$,
$B_{k,m}^2 = \{\lfloor m/2 \rfloor, \lfloor m/2 \rfloor + 1, \lfloor m/2 \rfloor + 2, \ldots, \lfloor m/2 \rfloor + k - 1\}$,
$B_{k,m}^3 = \{m - tk + 1, m - (t-1)k + 1, \ldots, m - 3k + 1, m - 2k + 1\}$ *where*
$t = \lceil (m - \lfloor m/2 \rfloor - 2k + 2)/k \rceil$, *and* $B_{k,m}^4 = \{m - k + 1, m - k + 2, \ldots, m - 1, m\}$.
Then any integer in the set $\{1, 2, \ldots, 2m\}$ can be obtained using at most one addition or subtraction of two elements in $B_{k,m}$.

Proof. It is easy to verify that by at most one addition or subtraction we can generate set $\{2m - 2k + 2, \ldots, 2m\}$ from integers in $B_{k,m}^4$, set $\{2m - (t+1)k + 2, \ldots, 2m - 2k + 1\}$ from $B_{k,m}^4$ and $B_{k,m}^3$, set $\{m + \lfloor m/2 \rfloor - k + 1, \ldots, m + \lfloor m/2 \rfloor + k - 1\}$, from $B_{k,m}^4$ and $B_{k,m}^2$, set $\{m + \lfloor m/2 \rfloor - tk + 1, \ldots, m + \lfloor m/2 \rfloor - k\}$, from $B_{k,m}^2$ and $B_{k,m}^3$, set $\{2\lfloor m/2 \rfloor, \ldots, 2\lfloor m/2 \rfloor + 2k - 2\}$, from $B_{k,m}^2$, set $\{m - k + 1, \ldots, m\}$, from $B_{k,m}^4$, set $\{m - k - \lfloor k/2 \rfloor + 1, \ldots, m - k\}$ from $B_{k,m}^1$ and $B_{k,m}^4$, set $\{m - tk - \lfloor k/2 \rfloor + 1, \ldots, m - 2k + \lfloor k/2 \rfloor + 1\}$, from $B_{k,m}^3$ and $B_{k,m}^1$, set $\{\lfloor m/2 \rfloor, \ldots, \lfloor m/2 \rfloor + k + \lfloor k/2 \rfloor - 1\}$, from $B_{k,m}^2$ and $B_{k,m}^1$, set $\{m - \lfloor m/2 \rfloor - 2k + 2, \ldots, m - \lfloor m/2 \rfloor\}$, from $B_{k,m}^2$ and $B_{k,m}^4$, set $\{k, \ldots, m - \lfloor m/2 \rfloor - 2k + 2\}$, from $B_{k,m}^3$ and $B_{k,m}^4$, and set $\{1, \ldots, k\}$ from $B_{k,m}^4$. Since $m - tk + 1 \leq \lfloor m/2 \rfloor + 2k - 1$, we reach the conclusion of the lemma. □

A lower bound on the number of colors needed for a routing based on set $B_{k,m}$ is equal to the sum of elements in $B_{k,m}$. One way to minimize the sum is to keep the size of the set $B_{k,m}$ as small as possible. The size of the set $B_{k,m}$ is equal to $|B_{k,m}| = \lfloor \frac{k}{2} \rfloor + k + \lceil \frac{(m - \lfloor m/2 \rfloor - 2k + 2)}{k} \rceil - 1 + k \leq \frac{5}{2}k + (m+4)/(2k)$. For an integer m, we obtain the smallest set $B_{k,m}$ that generates $\{1, 2, \ldots, 2m\}$ given in the next lemma by minimizing the value of $|B_{k,m}|$ with respect to k.

Lemma 2. Let m be a positive integer and $s(m) = \lfloor \sqrt{\frac{m+4}{5}} \rfloor$. Then we can generate the set $\{1, 2, \ldots, 2m\}$ by at most one operation of addition or subtraction on integers in the set $B_{s(m),m}$. Set $B_{s(m),m}$ contains at most $\sqrt{5(m+4)}$ integers.

If a node v in C_n can communicate directly with all nodes at distance in $B_{s(m),m}$ from v, where $m = \lceil \frac{n}{4} \rceil$, then by Lemma 2 node v can communicate in 2-hops with every node at distance d_n in C_n. This gives us a way to define a uniform routing on C_n which is a 2-hop solution of I_A.

Lemma 3. Let n be an integer, $n \geq 5$, and let $R_{n,2}$ to be the routing in C_n such that any node in C_n can communicate directly with all nodes at distance in $B_{s(\lceil n/4 \rceil)\lceil n/4 \rceil}$. Then $R_{n,2}$ is a uniform, 2-hop solution of I_A on C_n.

We determine an upper bound on the wavelength index of $R_{n,2}$ by a repeated application of the following lemma.

Lemma 4. Let n be an integer, $n \geq 5$, and $P = \{p_1, p_2, \ldots p_k\}$ be a set of positive integers such that $\sum_{i=1}^{k} p_i < n$. Let I_P be an instance in C_n such that every node in C_n communicates with all nodes at distance in P, and R be a routing of shortest paths.
If $p_1 + p_2 + \cdots + p_k$ divides n then $\overrightarrow{w}(C_n, I_P, R, 1) = p_1 + p_2 + \cdots + p_k$.
If $(p_1 + p_2 + \cdots + p_k) \bmod n = 1$ then $\overrightarrow{w}(C_n, I_P, R, 1) \leq p_1 + p_2 + \cdots + p_k + 1$.
If $(p_1 + p_2 + \cdots + p_k) < m$ then $\overrightarrow{w}(C_n, I_P, R, 1) \leq \overrightarrow{w}(C_n, I_m, R, 1)$, where I_m is an instance in C_n such that every node in C_n communicates with a node at distance m.

Theorem 1. *For any integer $n \geq 5$ there exists a uniform routing $R_{n,2}$ on C_n which is a 2-hop solution of I_A such that*

$$\vec{w}\,(C_n, I_A, R_{n,2}, 2) \leq \frac{n+3}{3}\sqrt{\frac{n+16}{5}} + \frac{n}{4}$$

Proof. Let $d = \lfloor n/2 \rfloor$, $m = \lceil d/2 \rceil$, $s(m) = \lfloor \sqrt{\frac{m+4}{5}} \rfloor$. We define $R_{n,2}$ to consist, for every node v in C_n, of paths from v to nodes at distances in set $B_{s(m),m}$. As specified in Lemma 1, $B_{s(m),m} = B^1_{s(m),m} \cup B^2_{s(m),m} \cup B^3_{s(m),m} \cup B^4_{s(m),m}$ where
$B^1_{s(m),m} = \{1, 2, 3, \ldots, \lfloor s(m)/2 \rfloor\}$, $B^2_{s(m),m} = \{m, m+1, m+2, \ldots, m+s(m)-1\}$,
$B^3_{s(m),m} = \{2m-ts(m)+1, m-(t-1)s(m)+1, \ldots, 2m-3s(m)+1, 2m-2s(m)+1\}$
and $t = \lceil (m - \lfloor m/2 \rfloor - 2s(m) + 2)/s(m) \rceil$ and
$B^4_{s(m),m}\{2m - s(m) + 1, 2m - s(m) + 2, \ldots, 2m - 1, 2m\}$. Since any distance on the ring can be obtained as a combination of two elements in $B_{s(m),m}$, we have that $R_{n,2}$ is a 2-hop solution of I_A .

We now determine an upper bound on the wavelength index of $R_{n,2}$. Assume first that n is divisible by 4. In this case m divides n. Thus, by Lemma 4 the wavelength index for all paths of length m in C_n is equal to m. Similarly, for any integer i, $1 \leq i \leq \lfloor s(m)/2 \rfloor$ the wavelength index for all paths of length i and $m - i$ is equal to m, and for any integer i, $0 \leq i \leq \lfloor s(m)/3 \rfloor$ the wavelength index for all paths of length $\lfloor m/2 \rfloor + i$, $\lfloor m/2 \rfloor + s(m) - 1 - 2i$ and $m - s(m) + i + 1$ is at most $2m$. This deals with all paths whose length is in $B^1_{s(m),m}$, two thirds of paths whose length is in $B^2_{s(m),m}$, and 5/6 of paths whose length is in $B^4_{s(m),m}$. In total, the number of colors needed for these paths is equal to $m(1 + \lfloor s(m)/2 \rfloor) + 2(\lfloor s(m)/3 \rfloor)$.

For any i, $1 \leq i \leq s(m)/6$, all paths of length $m - \lfloor s(m)/2 \rfloor - i$ need at most m colors, so these paths contribute at most $m(\lfloor s(m)/6 \rfloor$ to the wavelength index.

For $0 \leq i \leq \lfloor s(m)/3 \rfloor$ we group together paths of length $\lfloor m/2 \rfloor + \lfloor s(m)/3 \rfloor + 2i + 1$, $\lfloor m/2 \rfloor + s(m) - 2i - 1$ and $m - (i+2)s(m) + 1$. Since $\lfloor m/2 \rfloor + \lfloor s(m)/3 \rfloor + 2i + 1 + \lfloor m/2 + s(m) - 2i - 1 + m - (i+1)s(m) + 1 \leq 2m - (i+1)s(m) + \lfloor s(m)/3 \rfloor$, any of these path-lengths groups needs at most $2m$ colors and they contribute at most $2m(\lfloor s(m)/3 \rfloor$ to the wavelength index.

Any remaining path-lengths in $B^1_{s(m),m}$ need at most m colors and there are at most $s(m) - \lfloor s(m)/3 \rfloor$ of such path-lengths. Thus, $\vec{w}\,(C_n, I_A, R_{n,2}, 2) \leq m(1 + \lfloor s(m)/2 \rfloor + 2(\lfloor s(m)/3 \rfloor) + \lfloor s(m)/6 \rfloor + 2(\lfloor s(m)/3 \rfloor) + s(m) - \lfloor s(m)/3 \rfloor \leq m(1 + 8s(m)/3) \leq \frac{n}{3}\sqrt{\frac{n+16}{5}} + n/4$.

If n is not divisible by 4 then $m = \lceil \lfloor n/2 \rfloor/2 \rceil$ and $4(m - 1) < n$ and $2(2m - 1) \leq n$. Thus, in this case we group the path-lengths together similarly as above, except that in each group we put the path-lengths whose total is either $m - 1$ or less or $2m - 1$ or less. This however may require one more color per each group, which increases the wavelength index at most by $s(m)$ and we get that $\vec{w}\,(C_n, I_A, R_{n,2}, 2) \leq \frac{n+3}{3}\sqrt{\frac{n+16}{5}} + n/4$. $\square$

The theorem shows that for a ring of length n, the ratio of the number colors needed for 1-hop all-to-all routing over the number of wavelengths needed for a uniform 2-hop all-to-all routing is approximately $0.8\sqrt{n}$. This can be very significant in some applications.

Any set B of integers that generates the set $\{1, 2, \ldots, n/2\}$ by at most one addition or subtraction must contain at least $\sqrt{n/4}$ integers. In order to generate all integers between $n/4$ and $n/2$, set B must contain at least $\sqrt{n/8}$ integers between $n/8$ and $n/4$ This gives us the following lower bound on the value of $\overrightarrow{w}(C_n, I_A, 2)$.

Theorem 2. $\overrightarrow{w}(C_n, I_A, 2) \geq \frac{n}{16}\sqrt{\frac{n}{2}}$.

Thus, the value of $\overrightarrow{w}(C_n, I_A, R_{n,2}, 2)$ is within a constant factor of the lower bound.

3   Uniform j-Hop, $j \geq 3$, All-to-All Routing

We can obtain results for uniform j-hop, $j \geq 3$, all-to-all routing using repeatedly the results from the previous section. A routing R is a j-hop all-to-all routing in a C_n, if any request x_i, x_l can be obtained as a concatenation of at most j paths in R. Thus, R is a uniform, j-hop, all-to-all routing on C_n if there is a set of integers $B(j) = \{b_1, b_2, \ldots, b_m\}$ such that R contains all paths on the ring of length in $B(j)$, and any integer between 1 and $\lfloor n/2 \rfloor$ can be obtained using at most $j - 1$ operations of addition or subtraction of j elements in $B(j)$.

Any integer between 1 and $2m$ can be obtained using at most 1 operation of addition or subtraction of 2 elements in set $B_{k,m} = B_{k,m}^1 \cup B_{k,m}^2 \cup B_{k,m}^3 \cup B_{k,m}^4$ from Lemma 1. As seen from the proof of Lemma 1, integers in set $B_{k,m}^3$ are not involved in additions or subtractions with integers from $B_{k,m}^3$ in order to obtain $\{1, 2, \ldots, 2m\}$. Since set $B_{k,m}^3$ is a linear progression containing at most $\frac{m+2}{2k}$ integers between $m/2$ and m, we can obtain all integers in this set, similarly as in Lemma 1, by at most one addition or subtraction from integers in sets $D_{b,k}^1$, $D_{b,k}^2$, $D_{b,k}^3$ where $|D_{b,k}^1 \cup D_{b,k}^2 \cup D_{b,k}^3| \leq \frac{1}{2}\sqrt{\frac{5(m+4)}{k}} + 2$. Thus, any integer between 1 and $2m$ can be obtained by at most 2 additions or subtractions from integers in set $D_{k,m} = B_{k,m}^1 \cup B_{k,m}^2 \cup D_{k,m}^1 \cup D_{k,m}^2 \cup D_{k,m}^3 \cup B_{k,m}^4$ and the total size of $D_{k,m}$ is at most $\lfloor \frac{k}{2} \rfloor + k + \frac{1}{2}\sqrt{\frac{5(m+4)}{k}} + 2 + k$. By minimizing the value of the size with respect to k we obtain the next lemma.

Lemma 5. *Let m be a positive integer and $r(m) = \lfloor \sqrt[3]{\frac{m+4}{20}} \rfloor$. Then we can generate the set $\{1, 2, \ldots, 2m\}$ by at most 2 operations of addition or subtraction from integers in the set $D_{r(m),m}$. Set $D_{r(m),m}$ contains at most $3\sqrt[3]{m+4}$ integers.*

Clearly, if every node in cycle C_n communicates directly with nodes at distance in $C_{s(\lceil n/4 \rceil)), \lceil n/4 \rceil}$, then every node can communicate with any other node in at most 3-hops.

Lemma 6. *Let n be an integer, $n \geq 5$, and let $R_{n,3}$ to be the routing in C_n such that any node in C_n can communicate directly with all nodes at distance in $D_{r(\lceil n/4 \rceil), \lceil n/4 \rceil}$. Then $R_{n,3}$ is uniform, 3-hop solution of I_A on C_n.*

Theorem 3. *For any integer $n \geq 5$ there exists a uniform routing $R_{n,3}$ on C_n which is a 3-hop solution of I_A such that*

$$\vec{w}\,(C_n, I_A, R_{n,3}, 3) \leq \frac{n}{2} \sqrt[3]{\frac{\lceil n/4 \rceil + 4}{5}}$$

Proof. Let $R_{n,3}$ be the 3-hop routing from Lemma 5. We calculate the wavelength index similarly as in Theorem 1. For path-lengths in sets $B^1_{k,m}$, $B^3_{k,m}$ and $B^4_{k,m}$ we use the same methods as in Theorem 1. We thus obtain that the wavelength index of all these paths is at most

$$\lceil n/4 \rceil (1 + r(m)/2 + 2r(m)/3 + 2r(m)/3) = \lceil n/4 \rceil (1 + \tfrac{11}{6} \lfloor \sqrt[3]{\tfrac{\lceil n/4 \rceil + 4}{20}} \rfloor).$$

Since all path-lengths in sets $D^1_{k,m}$, $D^2_{k,m}$, $D^3_{k,m}$ are bounded from above by $\frac{n}{8}$, the wavelength index of all these paths is at most $\frac{5n}{8} \lfloor \sqrt[3]{\frac{\lceil n/4 \rceil + 4}{20}} \rfloor$. Thus the wavelength index of $R_{n,3}$ is at most $\frac{n}{2} \sqrt[3]{\frac{\lceil n/4 \rceil + 4}{5}} \rfloor$.   □

Any integer between 1 and $2m$ can be obtained using at most 1 operation of addition or subtraction of 2 elements in set $B_{k,m} = B^1_{k,m} \cup B^2_{k,m} \cup B^3_{k,m} \cup B^4_{k,m}$ from Lemma 1.

Since integers in any of the sets $B^1_{k,m}$, $B^2_{k,m}$, $B^3_{k,m}$, and $B^4_{k,m}$ form a linear progression, we can obtain all integers in set $B^i_{k,m}$, $1 \leq i \leq 4$, similarly as in Lemma 1, by at most one additions or subtractions from integers in a set of integers that contain at most $\sqrt{5(\frac{|B^i_{k,m}|}{2} + 4)}$ integers. This implies that any integer between 1 and $2m$ can be obtained by at most 3 additions or subtractions on sets of integers that contain at most $\sqrt{5(\frac{k}{4} + 4)} + \sqrt{5(\frac{m+4}{4k} + 4)} + \sqrt{5(\frac{k}{2} + 4)} + \sqrt{5(\frac{k}{2} + 4)}$. By substituting $s(m) = \sqrt{\frac{m+4}{5}}$ for k, we obtain the next lemma.

Lemma 7. *Let m be a positive integer. There exists set E_m such that the set $\{1, 2, \ldots, 2m\}$ can be generated from integers in E_m by at most 3 operations of addition or subtraction. Set E_m contains at most $4(\sqrt[4]{m + 4} + 4)$ integers and any of its elements is less than or equal to $\lceil m/2 \rceil$.*

Clearly, if every node in cycle C_n communicates directly with nodes at distance in $E_{\lceil n/4 \rceil}$, then every node can communicate with any other node in at most 4-hops.

Lemma 8. *Let n be an integer, $n \geq 5$, and let $R_{n,4}$ to be the routing in C_n such that any node in C_n can communicate directly with all nodes at distance in $E_{\lceil n/4 \rceil}$. Then $R_{n,4}$ is a uniform, 4-hop solution of I_A on C_n.*

Theorem 4. *For any integer $n \geq 5$ there exists a uniform routing $R_{n,4}$ on C_n which is a 4-hop solution of I_A such that*

$$\vec{w}\left(C_n, I_A, R_{n,2}, 2\right) \leq \frac{n+16}{2} \sqrt[4]{\left\lceil \frac{n}{4} \right\rceil} + 8$$

Proof. Let $R_{n,4}$ be the routing from Lemma 8. By Lemma 4, all cycles of length k, $k \leq \lceil n/8 \rceil$ need at most $\lceil n/8 \rceil + 1$ colors. Thus,
$\vec{w}\left(C_n, I_A, R_{n,2}, 2\right) \leq (\lceil \frac{n}{8} \rceil + 1)(4\sqrt[4]{\lceil \frac{n}{4} \rceil + 4} + 4) \leq \frac{n+16}{2}\sqrt[4]{\lceil \frac{n}{4} \rceil} + 8$
Note that in this proof the bound on the wavelength index is calculated less precisely than those for 2-hop and 3-hop case. □

We can show, similarly as in the 2-hop case that the result above is within a constant factor of a lower bound.

Clearly, the process that we used for deriving the wavelength index of 3-hop and 4-hop wavelength indices can be extended for higher number of hops.

4 Conclusions

We gave an upper bound on the wavelength index of a uniform j-hop all-to-all communication instance in a ring of length n for $2 \leq j \leq 4$, which is within a multiplicative constant of a lower bound. The results show that there is a large reduction in the value of the wavelength index when going from a 1-hop to a 2-hop routing, since we replace one factor of n by $\sqrt{n}$. However, the rate of reduction diminishes for subsequent number of hops, since we only replace one factor of $\sqrt{n}$ by $\sqrt[3]{n}$ when going from a 2-hop to a 3-hop routing, etc. For example, for a cycle on 100 nodes we get
$\vec{w}\left(C_{100}, I_A, 1\right) = 1250$, $\vec{w}\left(C_{100}, I_A, 2\right) \leq 165$, $\vec{w}\left(C_{100}, I_A, 3\right) \leq 115$.

The value of the upper bounds on the wavelength index that we obtained depends on the size of the set $B_{s(\lceil n/4 \rceil), \lceil n/4 \rceil}$ from which we can generate the set $\{1, 2, \ldots, \lfloor (n-1)/2 \rfloor\}$ using at most one operation of addition or subtraction. Obviously, if we obtain an improvement on the size of $B_{s(\lceil n/4 \rceil), \lceil n/4 \rceil}$, we could improve the upper bounds on value of the wavelength index of a uniform j-hop all-to-all communication instance, $j \geq 2$. This seems to be an interesting combinatorial problem that, to the best of our knowledge, has not been studied previously. It would be equally interesting to get a *better lower bound* on the size of a set that generates set $\{1, 2, \ldots, \lfloor (n-1)/2 \rfloor\}$ using at most one operation of addition or subtraction. Thus we propose the following open problems:

Open Problem 1: *Find a tight lower bound on the size of a set of integers B_m^1 such that any integer in the set $\{1, 2, \ldots, m\}$ can be obtained by at most one operation of addition or subtraction from integers in B_m^1.*

Open Problem 2: *Find a set of integers B_m^1 such that any integer in the set $\{1, 2, \ldots, m\}$ can be obtained by at most one operation of addition or subtraction from integers in B_m^1 and which is smaller in size than the set given in this paper.* Similar open problems can be asked for a higher number of operation.

Answer to open problems 2 does not yet solve the wavelength index of a uniform j-hop all-to-all communication instance in rings, as it is necessary to devise a coloring of the paths in the ring. As of now, there is no general algorithm that can give a good color assignment to paths in case of a uniform instance on a ring. This leads us to propose the following open problem:

Open Problem 3: *Give an algorithm that, given a uniform instance I in C_n, finds a good approximation of $\overrightarrow{w}\,(C_n, I, 1)$.*

References

1. B. Beauquier, S. Pérennes, and T. David. All-to-all routing and coloring in weighted trees of rings. In *Proceedings of Eleven Annual ACM Symposium on Parallel Algorithms and Architectures*, pages 185–190, 1999.
2. J. Bermond, F. Comellas, and D. Hsu. Distributed loop computer networks: A survey. *J. of Parallel and Distributed Computing*, 24:2–10, 1995.
3. J.-C. Bermond, L. Gargano, S. Perennes, and A. A. Rescigno. Efficient collective communication in optical networks. *Lecture Notes in Computer Science*, 1099:574–585, 1996.
4. J. Bondy and U. Murty. *Graph Theory with Applications*. Macmillan Press Ltd, 1976.
5. C. Bracket. Dense wavelength division multiplexing networks: Principles and applications. *IEEE J. Selected Areas in Communications*, 8:948–964, 1990.
6. N. Cheung, K. Nosu, and G. Winzer. An introduction to the special issue on dense WDM networks. *IEEE J. Selected Areas in Communications*, 8:945–947, 1990.
7. D. Dolev, J. Halpern, B. Simons, and R. Strong. A new look at fault tolerant network routing. In *Proceedings of ACM 10th STOC Conference*, pages 526–535, 1984.
8. L. Gargano, P. Hell, and S. Perennes. Colouring paths in directed symmetric trees with applications to WDM routing. In *24th International Colloquium on Automata, Languages and Programming*, volume 1256 of *Lecture Notes in Computer Science*, pages 505–515, Bologna, Italy, 7–11 July 1997. Springer-Verlag.
9. B. Mukherjee. WDM-based local lightwave networks, part I: Single-hop systems. *IEEE Network Magazine*, 6(3):12–27, May 1992.
10. B. Mukherjee. WDM-based local lightwave networks, part II: Multihop systems. *IEEE Network Magazine*, 6(4):20–32, July 1992.
11. L. Narayanan, J. Opatrny, and D. Sotteau. All-to-all optical routing in chordal rings of degree four. In *Proceedings of the Symposium on Discrete Algorithms*, pages 695–703, 1999.
12. J. Opatrny. Low-hop, all-to-all optical routing in rings. Technical report, Concordia University, 1999.

A Fully Dynamic Algorithm
for Distributed Shortest Paths

Serafino Cicerone[1], Gabriele Di Stefano[1], Daniele Frigioni[12], and Umberto Nanni[2]

[1] Dipartimento di Ingegneria Elettrica, Università dell'Aquila, I-67040 Monteluco di Roio - L'Aquila, Italy. {cicerone,gabriele,frigioni}@infolab.ing.univaq.it

[2] Dipartimento di Informatica e Sistemistica, Università di Roma "La Sapienza", via Salaria 113, I-00198 Roma, Italy. nanni@dis.uniroma1.it

Abstract. We propose a fully-dynamic distributed algorithm for the *all-pairs* shortest paths problem on general networks with positive real edge weights. If Δ_σ is the number of pairs of nodes changing the distance after a single edge modification σ (*insert, delete, weight-decrease,* or *weight-increase*) then the message complexity of the proposed algorithm is $O(n\Delta_\sigma)$ in the worst case, where n is the number of nodes of the network. If $\Delta_\sigma = o(n^2)$, this is better than recomputing everything from scratch after each edge modification.

1 Introduction

The importance of finding shortest paths in graphs is motivated by the numerous theoretical and practical applications known in various fields as, for instance, in combinatorial optimization and in communication networks (e.g., see [1,10]). We consider the distributed *all-pairs shortest paths* problem, which is crucial when processors in a network need to route messages with the minimum cost.

The problem of *updating* shortest paths in a dynamic distributed environment arises naturally in practical applications. For instance, the *OSPF* protocol, widely used in Internet (e.g., see [9,13]), updates the routing tables of the nodes after a change to the network, by using a distributed version of Dijkstra's algorithm. In this and many other crucial applications the worst case complexity of the adopted protocols is never better than recomputing the shortest paths from scratch. Therefore, it is important to find distributed algorithms for shortest paths that do not recompute everything from scratch after each change to the network, because this could result very expensive in practice.

If the topology of a network is represented as a graph, where nodes represent processors and edges represent links between processors, then the typical update operations on a dynamic network can be modeled as insertions and deletions of edges and update operations on the weights of edges. When arbitrary sequences of the above operations are allowed, we refer to the *fully dynamic problem*; if only insertions and weight decreases (deletions and weight increases) of edges are allowed, then we refer to the *incremental* (*decremental*) problem.

G. Gonnet, D. Panario, and A. Viola (Eds.): LATIN 2000, LNCS 1776, pp. 247–257, 2000.

Previous works and motivations.   Many solutions have been proposed in the literature to find and update shortest paths in the sequential case on graphs with non-negative real edge weights (e.g., see [1,10] for a wide variety). The state of the art is that no efficient fully dynamic solution is known for general graphs that is faster than recomputing everything from scratch after each update, both for single-source and all-pairs shortest paths. Actually, only *output bounded* fully dynamic solutions are known on general graphs [4,11].

Some attempts have been made also in the distributed case [3,5,7,12]. In this field the efficiency of an algorithm is evaluated in terms of *message, time* and *space* complexity as follows. The *message complexity* of a distributed algorithm is the total number of messages sent over the edges. We assume that each message contains $O(\log n + R)$ bits, where R is the number of bits available to represent a real edge weight, and n is the number of nodes in the network. In practical applications messages of this kind are considered of "constant" size. The *time complexity* is the total (normalized) time elapsed from a change. The *space complexity* is the space usage per node.

In [5], an algorithm is given for computing all-pairs shortest paths requiring $O(n^2)$ messages, each of size n. In [7], an efficient incremental solution has been proposed for the distributed all-pairs shortest paths problem, requiring $O(n \log(nW))$ amortized number of messages over a sequence of edge insertions and edge weight decreases. Here, W is the largest positive *integer* edge weight. In [3], Awerbuch *et al.* propose a general technique that allows to update the all-pairs shortest paths in a distributed network in $\Theta(n)$ amortized number of messages and $O(n)$ time, by using $O(n^2)$ space per node. In [12], Ramarao and Venkatesan give a solution for updating all-pairs shortest paths that requires $O(n^3)$ messages and time and $O(n)$ space. They also show that, in the worst case, the problem of updating shortest paths is as difficult as computing shortest paths.

The results in [12] have a remarkable consequence. They suggest that two possible directions can be investigated in order to devise efficient fully dynamic algorithms for updating all-pairs shortest paths: i) to study the trade-off between the message, time and space complexity for each kind of dynamic change; ii) to devise algorithms that are efficient in different complexity models (with respect to worst case and amortized analyses).

Concerning the first direction, in [7] an efficient incremental solution has been provided, and the difficulty of dealing with edge deletions has been addressed. This difficulty arises also in the sequential case (see for example [2]).

In this paper, the second direction is investigated. We observed that the *output complexity* [4,10] was a good candidate (it is a robust measure of performance for dynamic algorithms in the sequential case [4,10,11]). This notion applies when the algorithms operate within a framework where explicit updates are required on a given data structure. In such a framework, output complexity allows to evaluate the cost of dynamic algorithms in terms of the number of updates to the output information of the problem that are needed after any input update. Here we show the merits of this model also in the field of distributed computation, improving over the results in [12].

Results of the paper. The novelty of this paper is a new efficient and practical solution for the fully dynamic distributed all-pairs shortest paths problem. To the best of our knowledge, the proposed algorithm represents the *first* fully dynamic distributed algorithm whose message complexity compares favorably with respect to recomputing everything from scratch after each edge modification. This result is achieved by explicitly devising an algorithm whose main purpose is to minimize the cost of each output update.

We use the following model. Given an input change σ and a source node s, let $\delta_{\sigma,s}$ be the set of nodes changing either the distance or the parent in the shortest paths tree rooted at s as a consequence of σ. Furthermore, let $\delta_\sigma = \cup_{s \in V} \delta_{\sigma,s}$ and $\Delta_\sigma = \sum_{s \in V} |\delta_{\sigma,s}|$. We evaluate the message and time complexity as a function of Δ_σ. Intuitively, this parameter represents a lower bound to the number of messages of constant size to be sent over the network after the input change σ. In fact, if the distance from u to v changes due to σ, then at least u and v have to be informed about the change.

We design an algorithm that updates only the distances and the shortest paths that actually change after an edge modification. In particular, if *maxdeg* is the maximum degree of the nodes in the network, then we propose a fully dynamic algorithm for the distributed all-pairs shortest paths problem requiring in the worst case: $O(maxdeg \cdot \Delta_\sigma)$ messages and $O(\Delta_\sigma)$ time for *insert* and *weight-decrease* operations; $O(\max\{|\delta_\sigma|, maxdeg\} \cdot \Delta_\sigma)$ messages and time for *delete* and *weight-increase* operations. The space complexity is $O(n)$ per node. The given bounds compare favourably with respect to the results of [12] when $\Delta_\sigma = o(n^2)$, and it is only a factor (bounded by $\max\{|\delta_\sigma|, maxdeg\}$ in the worst case) far from the optimal one, that is the (hypothetical) algorithm that sends over the network a number of messages equal to the number of pairs of nodes affected by an edge modification.

2 Network Model and Notation

We consider *point-to-point communication networks*, where a processor can generate a single message at a time and send it to all its neighbors in one time step. Messages are delivered to their respective destinations within a finite delay, but they might be delivered out of order. The distributed algorithms presented in this paper allow communications only between neighbors. We assume an asynchronous message passing system; that is, a sender of a message does not wait for the receiver to be ready to receive the message. In a *dynamic* network when a modification occurs concerning an edge (u, v), we assume that only nodes u and v are able to detect the change. Furthermore, we do not allow changes to the network that occur while the proposed algorithm is executed.

We represent a computer network, where computers are connected by communication links, by an *undirected weighted graph* $G = (V, E, w)$, where V is a finite set of n *nodes*, one for each computer; E is a finite set of m *edges*, one for each link; and w is a *weight function* from E to positive real numbers. The *weight* of the edge $(u, v) \in E$ is denoted as $w(u, v)$. For each node u, $N(u)$ contains the neighbors of u. A *path* between two nodes u and v is a finite sequence $p = \langle u = v_0, v_1, \ldots, v_k = v \rangle$ of distinct nodes such that, for each $0 \leq i < k$, $(v_i, v_{i+1}) \in E$, and the *weight of the path* is $weight(p) = \sum_{0 \leq i < k} w(v_i, v_{i+1})$. The *distance* $d(u, v)$ between any pair of nodes u and v is the minimum weight of all possible paths connecting u to v. A *shortest path* from u to v is defined as any path p such that $weight(p) = d(u, v)$. If $s \in V$ is an arbitrary source node, we denote as T_s a shortest paths tree of G rooted at s; for any $u \in V$, $T_s(u)$ denotes the subtree of T_s rooted at u. We assume that each node u knows: *i)* the identities of all nodes, $1, 2, \ldots, n$; *ii)* the identity of each node in $N(u)$; *iii)* for each $u_i \in N(u)$, the edge connecting u to u_i, and the weight $w(u, u_i)$.

3   The Fully-Dynamic Algorithm

We describe the algorithms handling *weight-decrease* and *weight-increase* operations, being straightforward the extension to *insert* and *delete* operations, respectively.

We use the following data structures. A *routing table* $RT[\cdot, \cdot]$, needed to store the information on the all-pairs shortest paths. Each node u in G, maintains only the set of records $RT[u, \cdot]$, one record $RT[u, v]$ for each possible destination $v \in V \setminus \{u\}$. Each record has two fields: $RT[u, v].weight$, and $RT[u, v].via$, where *weight* is the distance between u and v, and *via* is the neighbor of u in the path used to determine the *weight*. In the following, each subcomponent of the routing table $RT[u, v].field$ will be also denoted as $field(u, v)$. The space required to store the routing table is clearly $O(n)$ per node.

For each $v \in V$, $d'(s, v)$ denotes the distance from s to v in the graph G' obtained from G after an edge modification (in general, we denote by γ' any parameter γ after an edge modification).

After an edge modification, for each source s, the proposed procedures correctly update $weight(v, s)$ as $d'(v, s)$, and $via(v, s)$ as the neighbor of v in the path used to determine $weight(v, s)$ in G'. Notice that, the procedures implicitly maintain a shortest paths tree T_s for each source s; T_s is the tree induced by the set of edges $(u, via(u, s))$, for each node u reachable from s.

Both for weight-decrease and for weight-increase operations, we describe the behavior of the algorithm with respect to a fixed source s. To obtain the algorithm for updating all-pairs shortest paths, it is sufficient to apply the algorithm with respect to all the possible sources.

3.1   Decreasing the Weight of an Edge

Suppose that a weight decrease operation σ is performed on edge (x, y), that is, $w'(x, y) = w(x, y) - \epsilon$, $\epsilon > 0$. In this case, if $d(s, x) = d(s, y)$, then $\delta_{\sigma, s} = \emptyset$, and no recomputation is needed. Otherwise, without loss of generality, we assume that $d(s, x) < d(s, y)$. In this case, if $d'(s, y) < d(s, y)$ then all nodes that belong to $T_s(y)$ are in $\delta_{\sigma,s}$. On the other hand, there could exist nodes not contained in $T_s(y)$ that belong to $\delta_{\sigma,s}$. In any case, every node in $\delta_{\sigma,s}$ decreases its distance from s as a consequence of σ.

The algorithm shown in Fig. 1 is based on the following property: if $v \in \delta_{\sigma,s}$, then there exists a shortest path connecting v to s in G' that contains the path from v to y in T_y as subpath. This implies that $d'(v,s)$ can be computed as $d'(v,s) = d(v,y) + w'(x,y) + d(x,s)$.

Node v receives "$weight(u,s)$" from u.

```
1.    if via(v,y) = u then
2.       begin
3.          if weight(v,s) > w(v,u) + weight(u,s) then
4.             begin
5.                weight(v,s) := w(v,u) + weight(u,s)
6.                via(v,s) := u
7.                for each vᵢ ∈ N(v) \ {u} do send "weight(v,s)" to vᵢ
8.             end
9.       end
```

Fig. 1. The decreasing algorithm of node v.

Based on this property, the algorithm performs a visit of T_y starting from y. This visit finds all the nodes in $\delta_{\sigma,s}$ and updates their routing tables. Each of the visited nodes v performs the algorithm of Fig. 1. When v figures out that it belongs to $\delta_{\sigma,s}$ (line 3), it sends "$weight(v,s)$" to all its neighbors. This is required because v does not know its children in T_y (since y is arbitrary, maintaining this information would require $O(n^2)$ space per node). Only when a node, that has received the message "$weight(u,s)$" from a neighbor u, performs line 1, it figures out whether it is child of a node in T_y.

Notice that the algorithm of Fig. 1 is performed by every node v distinct from y. The algorithm for y is slightly different: *(i)* y starts the algorithm when it receives the message "$weight(u,s)$" from $u \equiv x$. This message is sent to y as soon as x detects the decrease on edge (x,y); *(ii)* y does not perform the test of line 1; *(iii)* the weight $w(v,u)$ at lines 3 and 5 coincides with $w'(x,y)$.

Theorem 1. *Updating all-pairs shortest paths over a distributed network with n nodes and positive real edge weights, after a weight-decrease or an insert operation, requires $O(maxdeg \cdot \Delta_\sigma)$ messages, $O(\Delta_\sigma)$ time, and $O(n)$ space.*

3.2   Increasing the Weight of an Edge

Suppose that a weight increase operation σ is performed on edge (x,y), that is, $w'(x,y) = w(x,y) + \epsilon$, $\epsilon > 0$. In order to distinguish the set of required updates determined by the operation, we borrow from [4] the idea of coloring the nodes with respect to s, as follows:

- $color(q, s) = white$ if q changes neither the distance from s nor the parent in T_s (i.e., $weight'(q, s) = weight(q, s)$ and $via'(q, s) = via(q, s)$);

- $color(q, s) = pink$ if q preserves its distance from s, but it must replace the old parent in T_s (i.e., $weight'(q, s) = weight(q, s)$ and $via'(q, s) \neq via(q, s)$);

- $color(q, s) = red$ if q increases the distance from s (i.e., $weight'(q, s) > weight(q, s)$).

According to this coloring, the nodes in $\delta_{\sigma, s}$ are exactly the red and pink nodes. Without loss of generality, let us assume that $d(s, x) < d(s, y)$. In this case it is easy to see that if $v \notin T_s(y)$, then $v \notin \delta_{\sigma, s}$. In other words, all the red and pink nodes belong to $T_s(y)$.

Initially all nodes are white. If v is pink or red, then either v is child of a red node in $T_s(y)$, or $v \equiv y$. If v is red then the children of v in $T_s(y)$ will be either pink or red. If v is pink or white then the other nodes in $T_s(v)$ are white.

By the above discussion, if we want to bound the number of messages delivered over the network, to update the shortest paths from s as a function of the number of output updates, then we cannot search the whole $T_s(y)$. In fact, if $T_s(y)$ contains a pink node v, then the nodes in $T_s(v)$ remain white and do not require any update. For each red or pink node v, we use the following notation:

- $AP_s(v)$ denotes the set of *alternative parents* of v with respect to s, that is, a neighbor q of v belongs to $AP_s(v)$ when $d(s, q) + w(q, v) = d(s, v)$.

- $BNR_s(v)$ denotes the *best non-red neighbor* of v with respect to s, that is, a non-red neighbor q of v, such that the quantity $d(s, q) + w(q, v)$ is minimum.

If $AP_s(v)$ is empty and $BNR_s(v)$ exists, then $BNR_s(v)$ represents the best way for v to reach s in G' by means of a path that does not contain red nodes.

The algorithm that we propose for handling weight-increase operations consists of three phases, namely the *Coloring*, the *Boundarization*, the *Recomputing* phase. In the following we describe in detail these three phases. We just state in advance that the coloring phase does not perform any update to $RT[\cdot, s]$. A pink node v updates $via(v, s)$ during the boundarization phase, whereas a red node v updates both $weight(v, s)$ and $via(v, s)$ during the recomputing phase.

Coloring phase. During this phase each node in $T_s(y)$ decides its color. At the beginning all these nodes are white. The pink and red nodes are found starting from y and performing a pruned search of $T_s(y)$. The coloring phase of a generic node v is given in Fig. 2. Before describing the algorithm in detail, we remark that it works under the following assumptions.

A1. If a node v receives a request for $weight(v, s)$ and $color(v, s)$ from a neighbor (line 7), then it answers immediately.
A2. If a red node v receives the message "$color(z, s) = red$" from $z \in N(v)$, then it immediately sends "end-coloring" to z (see line 1 for red nodes).

When v receives "$color(z, s) = red$" from z, it understands that has to decide its color. The behavior of v depends on its current color. Three cases may arise:

The *red* node v receives the message "*color*$(z, s) = red$" from $z \in N(v)$.

1.   send to z the message "end-coloring"; HALT

The *non-red* node v receives the message "*color*$(z, s) = red$" from $z \in N(v)$.

1.   **if** *color*$(v, s) = white$ **then**
2.   **begin**
3.   **if** $z \neq via(v, s)$ **then** send to z the message "end-coloring"; HALT
4. $AP_s(v) := \emptyset$
5.   **for each** $v_i \in N(v) \setminus \{z\}$ **do**
6.   **begin**
7.   ask v_i for *weight*(v_i, s) and *color*(v_i, s)
8.   **if** *color*$(v_i, s) \neq red$ **and** *weight*$(v, s) = w(v, v_i) + weight(v_i, s)$
9.   **then** $AP_s(v) := AP_s(v) \cup \{v_i\}$
10.   **end**
11. **end**
12. **if** $z \in AP_s(v)$ **then** $AP_s(v) := AP_s(v) \setminus \{z\}$
13. **if** $AP_s(v) \neq \emptyset$
14. **then** *color*$(v, s) := pink$
15. **else begin**
16.   *color*$(v, s) := red$
17.   **for each** $v_i \in N(v) \setminus \{z\}$ send to v_i the message "*color*$(v, s) = red$"
18.   **for each** $v_i \in N(v) \setminus \{z\}$ wait from v_i the message "end-coloring"
19.   **end**
20. send to z the message "end-coloring";

Fig. 2. The coloring phase of node v

1. *v is white:* In this case, v tests whether z is its parent in $T_s(y)$ or not. If $z \neq via(v, s)$ (line 3), then the color of v remains white and v communicates to z the end of its coloring. If $z = via(v, s)$, then v finds all the alternative parents with respect to s, and records them into $AP_s(v)$ (lines 4–10). If $AP_s(v) \neq \emptyset$ (line 13), then v sets its color to pink (line 14) and communicates to z the end of its coloring (line 20). If $AP_s(v) = \emptyset$ (line 15), then v does the following: *i*) sets its color to red (line 16); *ii*) propagates the message "*color*$(v, s) = red$" to each neighbor but z (line 17); *iii*) waits for the message "end-coloring" from each of these neighbors (line 18); *iv*) communicates the end of its coloring phase to z (line 20).

2. *v is pink:* In this case, the test at line 12 is the first action performed by v. If z is an alternative parent of v, then z is removed from $AP_s(v)$ (since z is now red). After this removing, v performs the test at line 13: if there are still elements in $AP_s(v)$, then v remains pink and sends to z the message concerning the end of its coloring phase (lines 14 and 20); otherwise, v becomes red and propagates the coloring phase to its neighbors (lines 15–19), as already described in case 1 above.

3. *v is red:* In this case, *v* performs a different procedure: it simply communicates to *z* the end of its coloring phase (see line 1 for red nodes). This is done to guarantee that Assumption A2 holds.

According to this strategy, at the end of the coloring phase node *y* is aware that each node in $T_s(y)$ has been correctly colored. The algorithm of Fig. 2 is performed by every node distinct from *y*. The algorithm for *y* is slightly different. In particular, at line 20, *y* does not send "end-coloring" to $z \equiv x$. Instead, *y* starts the *boundarization phase* described below and shown in Fig. 3.

Node *v* receives the message "start boundarization(ϵ)" from $z = via(v, s)$.

1.   **if** $color(v, s) = pink$ **then**
2.   **begin**
3. $via(v, s) := q$, where *q* is an arbitrary node in $\text{AP}_s(v)$
4. $color(v, s) := white;$ **HALT**
5.   **end**
6.   **if** $color(v, s) = red$ **then**
7.   **begin**
8. $\ell_v := weight(v, s) + \epsilon$
9. $\text{BNR}_s(v) := \textbf{nil}$
10. $\text{PINK-CHILDREN}_s(v) := \emptyset; \text{RED-CHILDREN}_s(v) := \emptyset$
11.   **for each** $v_i \in N(v) \setminus \{z\}$ **do**
12.   **begin**
13.   *v* asks v_i for $weight(v_i, s)$, $via(v_i, s)$, and $color(v_i, s)$
14.   **if** $color(v_i, s) \neq red$ **and** $\ell_v > w(v, v_i) + weight(v_i, s)$ **then**
15.   **begin**
16. $\ell_v := w(v, v_i) + weight(v_i, s)$
17. $\text{BNR}_s(v) := v_i$
18.   **end**
19.   **if** $color(v_i, s) = pink$ **and** $via(v_i, y) = v$
20.   **then** $\text{PINK-CHILDREN}_s(v) := \text{PINK-CHILDREN}_s(v) \cup \{v_i\}$
21.   **if** $color(v_i, s) = red$ **and** $via(v_i, y) = v$
22.   **then** $\text{RED-CHILDREN}_s(v) := \text{RED-CHILDREN}_s(v) \cup \{v_i\}$
23.   **end**
24.   **if** $\text{BNR}_s(v) = \textbf{nil}$
25.   **then** $B_s(v) := \emptyset$ {*v* is not boundary for *s*}
26.   **else** $B_s(v) := \{\langle v; \ell_v \rangle\}$ {*v* is boundary for *s*}
27.   **for each** $v_i \in \text{PINK-CHILDREN}_s(v) \cup \text{RED-CHILDREN}_s(v)$
28.   **do** send "start boundarization(ϵ)" to v_i
29.   **for each** $v_i \in \text{RED-CHILDREN}_s(v)$ **do**
30.   **begin**
31.   wait the message "$B_s(v_i)$" from v_i
32. $B_s(v) := B_s(v) \cup B_s(v_i)$
33.   **end**
34.   send "$B_s(v)$" to $via(v, s)$
35.   **end**

Fig. 3. The boundarization phase of node *v*

Boundarization phase. During this phase, for each red and pink node v, a path (not necessarily the shortest one) from v to s is found. Note that, as in Assumption A1 of Coloring, if a node v receives a request for $weight(v,s)$, $via(v,s)$ and $color(v,s)$ from a neighbor (line 13), then it answers immediately.

When a pink node v receives the message "start boundarization(ϵ)" ((ϵ) is the increment of $w(x,y)$) from $via(v,s)$, it understands that the coloring phase is terminated; at this point v needs only to choose arbitrarily $via(v,s)$ among the nodes in $\text{AP}_s(v)$, and to set its color to white (lines 2–5).

When a red node v receives the message "start boundarization(ϵ)" from $via(v,s)$, it has no alternative parent with respect to s. At this point, v computes the shortest between the old path from v to s (whose weight is now increased by ϵ (line 8)), and the path from v to s via $\text{BNR}_s(v)$ (if any). If $\text{BNR}_s(v)$ exists, then v can reach s through a path containing no red nodes. In order to find $\text{BNR}_s(v)$, v asks every neighbor v_i for $weight(v_i,s)$, $via(v_i,s)$ and $color(v_i,s)$ (see lines 12–23). At the same time, using $color(v_i,s)$ and $via(v_i,s)$, v finds its pink and red children in $T_s(y)$ and records them into $\text{PINK-CHILDREN}_s(v)$ and $\text{RED-CHILDREN}_s(v)$ (see lines 20 and 22).

If $\text{BNR}_s(v)$ exists and the path from v to s via $\text{BNR}_s(v)$ is shorter than $weight(v,s) + \epsilon$, then v is called *boundary for s*. In this case, v initializes $B_s(v)$ as $\{\langle v; \ell_v \rangle\}$ (line 26), where ℓ_v is the weight of the path from v to s via $\text{BNR}_s(v)$.

When v terminates the boundarization phase, the set $B_s(v)$ contains all the pairs $\langle z; \ell_z \rangle$ such that $z \in T_s(v)$ is a boundary node. In fact, at line 28 v sends to each node $v_i \in \text{PINK-CHILDREN}_s(v) \cup \text{RED-CHILDREN}_s(v)$ the value ϵ (to propagate the boundarization), and waits to receive $B_s(v_i)$ from $v_i \in \text{RED-CHILDREN}_s(v)$ (lines 30–33). Notice that v does not wait for any message from a pink children $v_i \in \text{PINK-CHILDREN}_s(v)$, because $B_s(v_i)$ is empty. Whenever v receives $B_s(v_i)$ from a child v_i, it updates $B_s(v)$ as $B_s(v) \cup B_s(v_i)$ (line 32). Finally, at line 34, v sends $B_s(v)$ to y via $via(v,s)$.

At the end of the boundarization phase, the set $B_s(y)$, containing all the boundary nodes for s, has been computed and stored in y. Notice that, the algorithm of Fig. 3 is performed by every node distinct from y. The algorithm for y is slightly different. In particular, at line 34, y does not send "$B_s(y)$" to $via(y,s)$. Instead, y uses this information to start *recomputing phase*. In the recomputing phase, y broadcasts through $T_s(y)$ the set $B_s(y)$ to each red node.

Recomputing phase. In this phase, each red node v computes $weight'(v,s)$ and $via'(v,s)$. The recomputing phase of a red node v is shown in Fig. 4, and described in what follows. Let us suppose that the red node v has received the message "$B_s(y)$" from $via(v,y)$.

Concerning the shortest path from v to s in G' two cases may arise: a) it coincides with the shortest path from v to s in G; b) it passes through a boundary node. In case b) two subcases are possible: $b1$) the shortest path from v to s in G' passes through $\text{BNR}_s(v)$; $b2$) the shortest path from v to s in G' contains a boundary node different from v.

Node v performs the following local computation: for each $b \in B_s(y)$, it computes w_{min} as $\min_b\{weight(v,b) + \ell_b\}$ (line 1), and b_{min} as the boundary node such that $w_{min} = weight(v, b_{min}) + \ell_{b_{min}}$ (line 2). After v has updated $weight(v, s)$ as $weight(v, s) + \epsilon$ (line 3), v compares $weight(v, s)$ (the new weight of the old path from v to s) with w_{min} (line 4) and correctly computes $weight'(v, s)$ (line 6), according to cases a) and b) above. At lines 7–9, $via(v, s)$ is computed according to cases $b1$) and $b2$) above. Finally, by using the information contained in RED-CHILDREN$_s(v)$, v propagates $B_s(y)$ to the red nodes in $T_s(v)$.

The node v receives "$B_s(y)$" from $via(v, y)$.

1. $w_{min} := \min\{weight(v, b) + \ell_b \mid \langle b; \ell_b \rangle \in B_s(y)\}$
2.   let b_{min} be a node such that $w_{min} = weight(v, b_{min}) + \ell_{b_{min}}$
3. $weight(v, s) := weight(v, s) + \epsilon$
4.   **if** $weight(v, s) > w_{min}$ **then**
5.   **begin**
6. $weight(v, s) := w_{min}$
7.   **if** $b_{min} = v$
8.   **then** $via(v, s) := \text{BNR}_s(v)$
9.   **else** $via(v, s) := via(v, b_{min})$
10.   **end**
11. $color(v, s) := white$
12.   **for each** $v_i \in$ RED-CHILDREN$_s(v)$ **do** send "$B_s(y)$" to v_i

Fig. 4. The recomputing phase of node v

It is easy to show that each phase is deadlock free.

Theorem 2. *Updating all-pairs shortest paths over a distributed network with n nodes and positive real edge weights, after a weight-increase or a delete operation, requires $O(\max\{|\delta_\sigma|, maxdeg\} \cdot \Delta_\sigma)$ messages, $O(\max\{|\delta_\sigma|, maxdeg\} \cdot \Delta_\sigma)$ time, and $O(n)$ space.*

References

1. R. K. Ahuia, T. L. Magnanti and J. B. Orlin. *Network Flows: Theory, Algorithms and Applications.* Prentice Hall, Englewood Cliffs, NJ (1993).
2. G. Ausiello, G. F. Italiano, A. Marchetti-Spaccamela and U. Nanni. Incremental algorithms for minimal length paths. *Journal of Algorithms*, **12**, 4 (1991), 615–638.
3. B. Awerbuch, I. Cidon and S. Kutten. Communication-optimal maintenance of replicated information. *Proc. IEEE Symp. on Found. of Comp. Sc.*, 492–502, 1990.
4. D. Frigioni, A. Marchetti-Spaccamela and U. Nanni. Fully dynamic output bounded single source shortest paths problem. *Proc. ACM-SIAM Symp. on Discrete Algorithms*, 212–221, 1996. Full version, *Journal of Algorithms*, to appear.
5. S. Haldar. An all pair shortest paths distributed algorithm using $2n^2$ messages. *Journal of Algorithms*, **24** (1997), 20–36.
6. P. Humblet. Another adaptive distributed shortest path algorithm. *IEEE transactions on communications*, **39**, n. 6 (1991), 995–1003.
7. G. F. Italiano. Distributed algorithms for updating shortest paths. *Proc. Int. Workshop on Distributed Algorithms.* LNCS 579, 200–211, 1991.

8. J. McQuillan. Adaptive routing algorithms for distributed computer networks. BBN Rep. 2831, Bolt, Beranek and Newman, Inc., Cambridge, MA 1974.
9. J. T. Moy. *OSPF - Anatomy of an Internet Routing Protocol.* Addison, 1998.
10. G. Ramalingam. Bounded incremental computation. LNCS 1089, 1996.
11. G. Ramalingam and T. Reps. On the computational complexity of dynamic graph problems. *Theoretical Computer Science*, **158**, (1996), 233–277.
12. K. V. S. Ramarao and S. Venkatesan. On finding and updating shortest paths distributively. *Journal of Algorithms*, **13** (1992), 235–257.
13. A.S. Tanenbaum. Computer networks. Prentice Hall, Englewood Cliffs, NJ (1996).

Integer Factorization and Discrete Logarithms

Andrew Odlyzko

AT&T Labs, Florham Park, NJ 07932, USA
amo@research.att.com
http://www.research.att.com/~amo

Abstract. Integer factorization and discrete logarithms have been known for a long time as fundamental problems of computational number theory. The invention of public key cryptography in the 1970s then led to a dramatic increase in their perceived importance. Currently the only widely used and trusted public key cryptosystems rely for their presumed security on the difficulty of these two problems. This makes the complexity of these problems of interest to the wide public, and not just to specialists.

This lecture will present a survey of the state of the art in integer factorization and discrete logarithms. Special attention will be devoted to the rate of progress in both hardware and algorithms. Over the last quarter century, these two factors have contributed about equally to the progress that has been made, and each has stimulated the other. Some projections for the future will also be made.

Most of the material covered in the lecture is available in the survey papers [1,2] and the references listed there.

References

1. A.M. Odlyzko, The future of integer factorization, *CryptoBytes (The technical newsletter of RSA Laboratories)*, 1 (no. 2) (1995), pp. 5–12. Available at http://www.rsa.com/rsalabs/pubs/cryptobytes/ and http://www.research.att.com/~amo.
2. A.M. Odlyzko, Discrete logarithms: The past and the future, *Designs, Codes, and Cryptography* 19 (2000), pp. 129-145. Available at http://www.research.att.com/~amo.

G. Gonnet, D. Panario, and A. Viola (Eds.): LATIN 2000, LNCS 1776, pp. 258–258, 2000.

Communication Complexity and Fourier Coefficients of the Diffie–Hellman Key

Igor E. Shparlinski

Department of Computing, Macquarie University
Sydney, NSW 2109, Australia
igor@comp.mq.edu.au

Abstract. Let p be a prime and let g be a primitive root of the field $\mathbb{F}_p$ of p elements. In the paper we show that the communication complexity of the last bit of the Diffie–Hellman key g^{xy}, is at least $n/24 + o(n)$ where x and y are n-bit integers where n is defined by the inequalities $2^n \leq p \leq 2^{n+1} - 1$. We also obtain a nontrivial upper bound on the Fourier coefficients of the last bit of g^{xy}. The results are based on some new bounds of exponential sums with g^{xy}.

1 Introduction

Let p be a prime and let $\mathbb{F}_p$ be a finite field of p elements which we identify with the set $\{0, \ldots, p-1\}$. We define integer n by the inequalities $2^n \leq p \leq 2^{n+1} - 1$ and denote by $\mathcal{B}_n$ the set of n-bit integers,

$$\mathcal{B}_n = \{x \in \mathbb{Z} \ : \ 0 \leq x \leq 2^n - 1\}.$$

Throughout the paper we do not distinguish between n-bit integers $x \in \mathcal{B}_n$ and their binary expansions. Thus $\mathcal{B}_n$ can be considered as the n-dimensional Boolean cube $\mathcal{B}_n = \{0, 1\}^n$ as well.

Finally, we recall the notion of communication complexity. Given a Boolean function $f(x, y)$ of $2n$ variables

$$x = (x_1, \ldots, x_n) \in \mathcal{B}_n \qquad \text{and} \qquad y = (y_1, \ldots, y_n) \in \mathcal{B}_n,$$

we assume that there are two collaborating parties and the value of x is known to one party and the values of y is known to the other, however each party has no information about the values of the other. The goal is to create a *communication protocol* $\mathbf{P}$ such that, for any inputs $x, y \in \mathcal{B}_n$, at the end at least one party can compute the value of $f(x, y)$. The largest number of bits to exchange by a protocol $\mathbf{P}$, taken over all possible inputs $x, y \in \mathcal{B}_n$, is called the communication complexity $C(\mathbf{P})$ of this protocol. The smallest possible value of $C(\mathbf{P})$, taken over all possible protocols, is called the *communication complexity* $C(f)$ of the function f, see [2,21].

G. Gonnet, D. Panario, and A. Viola (Eds.): LATIN 2000, LNCS 1776, pp. 259–268, 2000.
© Springer-Verlag Berlin Heidelberg 2000

Given two integers $x, y \in \mathcal{B}_n$, the corresponding *Diffie–Hellman key* is defined as g^{xy}. Studying various complexity characteristics of this function is of primal interest for cryptography and complexity theory. Several lower bounds on the various complexity characteristics of this function as well as the discrete logarithm have been obtained in [30]. In particular, for a primitive root g of $\mathbb{F}_p$, one can consider the Boolean function $f(x, y)$ which is defined as the rightmost bit of g^{xy}, that is,

$$f(x_1, \ldots, x_n, y_1, \ldots, y_n) = \begin{cases} 1, & \text{if } g^{xy} \in \{1, 3, \ldots, p - 2\}; \\ 0, & \text{if } g^{xy} \in \{2, 4, \ldots, p - 1\}. \end{cases} \tag{1}$$

In many cases the complexity lower bounds of [30] are as strong as the best known lower bounds for any other function. However, the lower bound $C(f) \geq \log_2 n + O(1)$ of Theorem 9.4 of [30] is quite weak. Here, using a different method, we derive the linear lower bound $C(f) \geq n/24 + o(n)$ on the communication complexity of f.

We also use the same method to obtain an upper bound of the Fourier coefficients of this function, that is,

$$\hat{f}(u, v) = 2^{-2n} \sum_{x \in \mathcal{B}_n} \sum_{y \in \mathcal{B}_n} (-1)^{f(x,y) + \langle ux \rangle + \langle vy \rangle},$$

where $u, v \in \mathcal{B}_n$ and $\langle wz \rangle$ denotes the dot product of the vectors $w, z \in \mathcal{B}_n$. This bound can be combined with many known relations between Fourier coefficients and various complexity characteristics such as such as the circuit complexity, the average sensitivity, the formula size, the average decision tree depth, the degrees of exact and approximate polynomial representations over the reals and several others, see [3,4,8,15,22,23,27] and references therein.

We remark, that although these results do not seem to have any cryptographic implications it is still interesting to study complexity characteristics of such an attractive number theoretic function. Various complexity lower bounds for Boolean functions associated with other natural number theoretic problems can be found in [1,5,6,7,14,30].

Our main tool is exponential sums including a new upper bound of double exponential sums

$$S_a(\mathcal{X}, \mathcal{Y}) = \sum_{x \in \mathcal{X}} \sum_{y \in \mathcal{Y}} \mathbf{e}(ag^{xy}),$$

where $\mathbf{e}(z) = \exp(2\pi i/p)$, with $a \in \mathbb{F}_p$ and arbitrary sets $\mathcal{X}, \mathcal{Y} \subseteq \mathcal{B}_n$. These sums are of independent number theoretic interest. In particular they can be considered as generalizations of the well known sums

$$T_a(\mathcal{U}, \mathcal{V}) = \sum_{u \in \mathcal{U}} \sum_{v \in \mathcal{V}} \mathbf{e}(auv) \qquad \text{and} \qquad Q_a(H) = \sum_{x=1}^{H} \mathbf{e}(ag^x),$$

where $a \in \mathbb{F}_p$, $1 \leq H \leq t$, and $\mathcal{U}, \mathcal{V} \subseteq \mathbb{F}_p$, which are well known in the literature and have proved to be useful for many applications, see [12,17,28,29] as well as Problem 14.a to Chapter 6 of [31] for $T_a(\mathcal{U}, \mathcal{V})$ and [18,19,20,24,25] for $Q_a(H)$.

In this paper we estimate sums $S_a(\mathcal{X}, \mathcal{Y})$ for arbitrary sets $\mathcal{X}$ and $\mathcal{Y}$. Provided that both sets are of the same cardinality $|\mathcal{X}| = |\mathcal{Y}| = N$ our estimates are nontrivial for $N \geq p^{15/16+\delta}$ with any fixed $\delta > 0$.

We remark that the distribution of the triples of (g^x, g^y, g^{xy}) for $x, y \in \mathcal{B}_n$ has been studied in [9,10], see also [30]. In fact this paper relies on an estimate of some double exponential sums from [9].

Throughout the paper the implied constants in symbols 'O', '$\ll$' and '$\gg$' are absolute (we recall that $A \ll B$ and $B \gg A$ are equivalent to $A = O(B)$).

2   Preparations

We say that a set $\mathcal{S} \subseteq \mathcal{B}_n$ is a *cylinder* if there is a set $\mathcal{J} \subseteq \{1, \ldots, n\}$ such that the membership $x \in \mathcal{S}$ does not depend on components x_j, $j \in \mathcal{J}$, of $x = (x_1, \ldots, x_n) \in \mathcal{B}_n$. The *discrepancy* $\Delta(f)$ of f is defined as

$$\Delta(f) = 2^{-2n} \max_{\mathcal{S}, \mathcal{T}} |N_1(\mathcal{S}, \mathcal{T}) - N_0(\mathcal{S}, \mathcal{T})|,$$

where the maximum is taken over all cylinders $\mathcal{S}, \mathcal{T} \subseteq \mathcal{B}_n$ and $N_\mu(\mathcal{S}, \mathcal{T})$ is the number of pairs $(x, y) \in \mathcal{S} \times \mathcal{T}$ with $f(x, y) = \mu$.

The link between the discrepancy and communication complexity is provided by the following statement which is a partial case of Lemma 2.2 from [2].

Lemma 1. *The bound*

$$C(f) \geq \log_2 \left(\frac{1}{\Delta(f)} \right)$$

holds.

We use exponential sums to estimate the discrepancy of the function (1). The following statement has been proved in [9], see the proof of Theorem 8 of that paper.

Lemma 2. *Let $\lambda \in \mathbb{F}_p$ be of multiplicative order t. For any $a, b \in \mathbb{F}_p^*$, the bound*

$$\sum_{u=1}^{t} \left| \sum_{v=1}^{t} \mathbf{e} \left(a\lambda^v + b\lambda^{uv} \right) \right|^4 \ll pt^{11/3}$$

holds.

We recall the well known fact, see see Theorem 5.2 of Chapter 1 of [26], that for any integer $m \geq 2$ the number of integer divisors $\tau(m)$ of m satisfies the bound

$$\log_2 \tau(m) \leq (1 + o(1)) \frac{\ln m}{\ln \ln m}. \tag{2}$$

We now apply Lemma 2 to estimate $S_a(\mathcal{X}, \mathcal{Y})$ for arbitrary sets $\mathcal{X}, \mathcal{Y} \in \mathcal{B}_n$.

Lemma 3. *The bound*

$$\max_{a \in \mathbb{F}_p^*} |S_a(\mathcal{X}, \mathcal{Y})| \ll |\mathcal{X}|^{1/2} |\mathcal{Y}|^{5/6} p^{5/8} \tau(p-1)$$

holds.

Proof. For a divisor $d|p-1$ we denote by $\mathcal{Y}(d)$ the subset of $y \in \mathcal{Y}$ with $\gcd(y, p-1) = d$. Then

$$|S_a(\mathcal{X}, \mathcal{Y})| \le \sum_{d|p-1} |\sigma_d|,$$

where

$$\sigma_d = \sum_{x \in \mathcal{X}} \sum_{y \in \mathcal{Y}(d)} \mathbf{e}\left(ag^{xy}\right).$$

Using the Cauchy inequality, we derive

$$|\sigma_d|^2 \le |\mathcal{X}| \sum_{x \in \mathcal{X}} \left| \sum_{y \in \mathcal{Y}(d)} \mathbf{e}\left(ag^{xy}\right) \right|^2 \le |\mathcal{X}| \sum_{x=1}^{p-1} \left| \sum_{y \in \mathcal{Y}(d)} \mathbf{e}\left(ag^{xy}\right) \right|^2$$

$$= |\mathcal{X}| \sum_{y,z \in \mathcal{Y}(d)} \sum_{x=1}^{p-1} \mathbf{e}\left(a\left(g^{xy} - g^{xz}\right)\right).$$

By the Hölder inequality we have

$$|\sigma_d|^8 \le |\mathcal{X}|^4 |\mathcal{Y}(d)|^6 \sum_{y,z \in \mathcal{Y}(d)} \left| \sum_{x=1}^{p-1} \mathbf{e}\left(a\left(g^{xy} - g^{xz}\right)\right) \right|^4$$

$$\le |\mathcal{X}|^4 |\mathcal{Y}(d)|^6 \sum_{y \in \mathcal{Y}(d)} \sum_{u=1}^{(p-1)/d} \left| \sum_{x=1}^{p-1} \mathbf{e}\left(a\left(g^{xy} - g^{xud}\right)\right) \right|^4.$$

Because each element $y \in \mathcal{Y}(d)$ can be represented in the form $y = dv$ with $\gcd(v, (p-1)/d) = 1$ and $\lambda_d = g^d$ is of multiplicative order $(p-1)/d$, we see that the double sum over u and x does not depend on y. Therefore,

$$|\sigma_d|^8 \le |\mathcal{X}|^4 |\mathcal{Y}(d)|^7 \sum_{u=1}^{(p-1)/d} \left| \sum_{x=1}^{p-1} \mathbf{e}\left(a\left(\lambda_d^x - \lambda_d^{xu}\right)\right) \right|^4$$

$$= |\mathcal{X}|^4 |\mathcal{Y}(d)|^7 d^4 \sum_{u=1}^{(p-1)/d} \left| \sum_{u=1}^{(p-1)/d} \mathbf{e}\left(a\left(\lambda_d^x - \lambda_d^{xu}\right)\right) \right|^4.$$

By Lemma 2 we obtain

$$|\sigma_d|^8 \ll |\mathcal{X}|^4 |\mathcal{Y}(d)|^7 d^4 p \left(\frac{p-1}{d}\right)^{11/3} \le |\mathcal{X}|^4 |\mathcal{Y}(d)|^7 p^{14/3} d^{1/3}. \qquad (3)$$

Using the bound $|\mathcal{Y}(d)| \le |\mathcal{Y}|$ for $d \le p/|\mathcal{Y}|$ and the bound $|\mathcal{Y}(d)| \le p/d$ for $d > p/|\mathcal{Y}|$, we see that

$$|\sigma_d| \ll |\mathcal{X}|^{1/2} |\mathcal{Y}|^{5/6} p^{5/8}$$

for any divisor $d|p-1$ and the desired result follows. $\qquad \square$

Finally, to apply Lemma 3 to the discrepancy we need the following two well known statements which are Problems 11.a and 11.c to Chapter 3 of [31], respectively.

Lemma 4. *For any integers u and $m \ge 2$,*

$$\sum_{\lambda=0}^{m-1} \mathbf{e}_m(\lambda u) = \begin{cases} 0, & \text{if } u \not\equiv 0 \pmod{m}; \\ m, & \text{if } u \equiv 0 \pmod{m}. \end{cases}$$

Lemma 5. *For any integers H and $m \ge 2$,*

$$\sum_{a=1}^{m-1} \left| \sum_{z=0}^{H} \mathbf{e}_m(az) \right| = O(m \ln m).$$

3   Communication Complexity and Fourier Coefficients of the Diffie–Hellman Key

Now we can prove our main results.

Theorem 1. *For the communication complexity of the function $f(x,y)$ given by (1), the bound*

$$C(f) \ge \frac{1}{24} n + o(n)$$

holds.

Proof. We can assume that $p \ge 3$. Put $H = (p-1)/2$. Then for any sets $\mathcal{S}, \mathcal{T} \subseteq \mathcal{B}_n$ (not necessary cylinders), Lemma 4 implies that

$$N_0(\mathcal{S}, \mathcal{T}) = \frac{1}{p} \sum_{a=0}^{p-1} \sum_{x \in \mathcal{S}} \sum_{y \in \mathcal{T}} \sum_{z=1}^{H} \mathbf{e}(a(g^{xy} - 2z)).$$

Separating the term $|\mathcal{S}||\mathcal{T}|H/p$, corresponding to $a = 0$, we obtain

$$\left| N_0(\mathcal{S}, \mathcal{T}) - \frac{|\mathcal{S}||\mathcal{T}|H}{p} \right| \le \frac{1}{p} \sum_{a=1}^{p-1} |S_a(\mathcal{S}, \mathcal{T})| \left| \sum_{z=1}^{H} \mathbf{e}(-2az) \right|.$$

Using Lemma 3 and then Lemma 5, we derive

$$\left| N_0(\mathcal{S}, \mathcal{T}) - \frac{|\mathcal{S}||\mathcal{T}|H}{p} \right|$$

$$\ll |\mathcal{S}|^{1/2} |\mathcal{T}|^{5/6} p^{-3/8} \tau(p-1) \sum_{a=1}^{p-1} \left| \sum_{z=1}^{H} \mathbf{e}(-2az) \right|$$

$$\ll p^{23/24} \tau(p-1) \sum_{a=1}^{p-1} \left| \sum_{z=1}^{H} \mathbf{e}(-2az) \right| = p^{23/24} \tau(p-1) \sum_{a=1}^{p-1} \left| \sum_{z=1}^{H} \mathbf{e}(az) \right|$$

$$\ll p^{47/24} \tau(p-1) \ln p.$$

Because $N_0(\mathcal{S}, \mathcal{T}) + N_1(\mathcal{S}, \mathcal{T}) = |\mathcal{S}||\mathcal{T}|$ and $H/p = 1/2 + O(p^{-1})$ we see that

$$\left| N_0(\mathcal{S}, \mathcal{T}) - \frac{|\mathcal{S}||\mathcal{T}|H}{p} \right| \ll p^{47/24} \tau(p-1) \ln p$$

as well. Therefore the discrepancy of f satisfies the bound

$$\Delta(f) \ll 2^{-2n} p^{47/24} \tau(p-1) \ln p \ll p^{-1/24} \tau(p-1) \ln p.$$

Using (2), we derive that $\Delta(f) \ll 2^{-n/24+o(n)}$. Applying Lemma 1, we obtain the desired result. $\square$

The same considerations also imply the following estimate on the Fourier coefficients.

Theorem 2. *For for the Fourier coefficients of the function $f(x, y)$ given by (1), the bound*

$$\max_{u, v \in \mathcal{B}_n} \left| \hat{f}(u, v) \right| \ll 2^{-n/24+o(n)}$$

holds.

Proof. We fix some nonzero vectors $u, v \in \mathcal{B}_n$ and denote by $\mathcal{X}_0$ and $\mathcal{Y}_0$ the sets of integers $x \in \mathcal{B}_n$ and $y \in \mathcal{B}_n$ for which $\langle ux \rangle = 0$ and $\langle vy \rangle = 0$, respectively. Similarly, we define the sets $\mathcal{X}_1$ and $\mathcal{Y}_1$ by the conditions $\langle ux \rangle = 1$ and $\langle vy \rangle = 1$, respectively. Then we obtain,

$$\hat{f}(u, v) = 2^{-2n} \sum_{x \in \mathcal{X}_0} \sum_{y \in \mathcal{Y}_0} (-1)^{f(x,y)} + 2^{-2n} \sum_{x \in \mathcal{X}_1} \sum_{y \in \mathcal{Y}_1} (-1)^{f(x,y)}$$

$$-2^{-2n} \sum_{x \in \mathcal{X}_1} \sum_{y \in \mathcal{Y}_0} (-1)^{f(x,y)} - 2^{-2n} \sum_{x \in \mathcal{X}_0} \sum_{y \in \mathcal{Y}_1} (-1)^{f(x,y)}.$$

It is easy to see that

$$\sum_{x \in \mathcal{X}_\eta} \sum_{y \in \mathcal{Y}_\mu} (-1)^{f(x,y)} = 2N_0(\mathcal{X}_\eta, \mathcal{Y}_\mu) - |\mathcal{X}_\eta||\mathcal{Y}_\mu|, \qquad \eta, \mu = 0, 1,$$

where, as before, $N_0(\mathcal{X}_\eta, \mathcal{Y}_\mu)$ is the number of pairs $(x, y) \in \mathcal{X}_\eta \times \mathcal{Y}_\mu$ with $f(x, y) = 0$. Using the same arguments as in the proof of Theorem 1, we derive that

$$\left| N_0(\mathcal{X}_\eta, \mathcal{Y}_\mu) - \frac{1}{2}|\mathcal{X}_\eta||\mathcal{Y}_\mu| \right| \ll p^{47/24} \tau(p-1) \ln p, \qquad \eta, \mu = 0, 1,$$

and from (2) we derive the desired result for non-zero vectors u and v.

Now, if $u = \bar{0}$ is a zero vector and v is not then defining $\mathcal{Y}_0$ and $\mathcal{Y}_1$ as before we obtain

$$\hat{f}(\bar{0}, v) = 2^{-2n} \sum_{x \in \mathcal{B}_n} \sum_{y \in \mathcal{Y}_0} (-1)^{f(x,y)} - 2^{-2n} \sum_{x \in \mathcal{B}_n} \sum_{y \in \mathcal{Y}_1} (-1)^{f(x,y)}.$$

As before we derive

$$\left| N_0(\mathcal{B}_n, \mathcal{Y}_\mu) - 2^{n-1}|\mathcal{Y}_\mu| \right| \ll p^{47/24} \tau(p-1) \ln p, \qquad \mu = 0, 1,$$

which implies the desired estimate in this case. The same arguments apply if $v = \bar{0}$ is a zero vector and u is not.

Finally, if both $u = v = \bar{0}$ are zero vectors then

$$\hat{f}(\bar{0}, \bar{0}) = 2^{-2n} \sum_{x \in \mathcal{B}_n} \sum_{y \in \mathcal{B}_n} (-1)^{f(x,y)} = 2^{-2n} \left(2N_0(\mathcal{B}_n, \mathcal{B}_n) - 2^{2n-1} \right)$$

and using the bound

$$\left| N_0(\mathcal{B}_n, \mathcal{B}_n) - 2^{2n-1} \right| \ll p^{47/24} \tau(p-1) \ln p, \qquad \mu = 0, 1,$$

we conclude the proof. $\square$

4   Remarks

Our bound of the discrepancy of f, obtained in the proof of Theorem 1, combined with Lemma 2.2 of [2] can be used to derive a linear lower bound on ε-*distributional communication complexity*, which is defined in a similar way, however the communicating parties are allowed to make mistakes on at most $\varepsilon 2^{2n}$ inputs $x, y \in \mathcal{B}_n$.

Similar considerations can be used to estimate more general sums

$$S_a(\mathcal{X}_1, \ldots, \mathcal{X}_k) = \sum_{x_1 \in \mathcal{X}_1} \cdots \sum_{x_k \in \mathcal{X}_k} \mathbf{e}\left(ag^{x_1 \cdots x_k}\right)$$

with $k \geq 2$ sets $\mathcal{X}_1, \ldots, \mathcal{X}_k \subseteq \mathcal{B}_n$ and thus study the *multi-party communication complexity* of the function (1).

It is obvious that the bound of Lemma 3 can be improved if a nontrivial upper bounds on $|\mathcal{Y}(d)|$ is known and substituted in (3). Certainly one cannot hope to obtain such bounds for arbitrary sets $\mathcal{Y}$ but for cylinders such bounds can be proved. Unfortunately this does not yield any improvement of Theorems 1 and 2. Indeed, nontrivial bounds on $|\mathcal{Y}(d)|$ improve the statement of Lemma 3 for sets of small cardinality, however in our applications sets of cardinality of order p turn out to be most important. But for such sets the trivial bound $|\mathcal{Y}(d)| \leq p/d$, which has been used in the proof of Lemma 3, is the best possible.

The bound of Lemma 3 also implies the same results for modular exponentiation $u^x \pmod{p}$, $u, x \in \mathcal{B}_n$. It would be interesting to extend this result for modular exponentiation modulo arbitrary integers m. In some cases, for example, when m contains a large prime divisor, this can be done within the frameworks of this paper. Other moduli may require some new ideas.

Finally, similar results hold in a more general situation when g is an element of multiplicative order $t \geq p^{3/4+\delta}$ with any fixed $\delta > 0$, rather than a primitive root.

On the other hand, out method does not seem to work for Boolean functions representing middle bits of g^{xy} and obtaining such results is an interesting open question.

References

1. E. Allender, M. Saks and I. E. Shparlinski, 'A lower bound for primality', *Proc. 14 IEEE Conf. on Comp. Compl.*, Atlanta, 1999, IEEE Press, 1999, 10–14.
2. L. Babai, N. Nisan and M. Szegedy, 'Multiparty protocols, pseudorandom generators for logspace and time–space trade-offs', *J. Comp. and Syst. Sci.*, **45** (1992), 204–232.
3. A. Bernasconi, 'On the complexity of balanced Boolean functions', *Inform. Proc. Letters*, **70** (1999), 157–163.
4. A. Bernasconi, 'Combinatorial properties of classes of functions hard to compute in constant depth', *Lect. Notes in Comp. Sci.*, Springer-Verlag, Berlin, **1449** (1998), 339–348.
5. A. Bernasconi, C. Damm and I. E. Shparlinski, 'Circuit and decision tree complexity of some number theoretic problems', *Tech. Report 98-21*, Dept. of Math. and Comp. Sci., Univ. of Trier, 1998, 1–17.
6. A. Bernasconi, C. Damm and I. E. Shparlinski, 'On the average sensitivity of testing square-free numbers', *Lect. Notes in Comp. Sci.*, Springer-Verlag, Berlin, **1627** (1999), 291–299.

7. A. Bernasconi and I. E. Shparlinski, 'Circuit complexity of testing square-free numbers', *Lect. Notes in Comp. Sci.*, Springer-Verlag, Berlin, **1563** (1999), 47–56.

8. R. B. Boppana, 'The average sensitivity of bounded-depth circuits', *Inform. Proc. Letters*, **63** (1997), 257–261.

9. R. Canetti, J. B. Friedlander, S. Konyagin, M. Larsen, D. Lieman and I. E. Shparlinski, 'On the statistical properties of Diffie–Hellman distributions', *Israel J. Math.*, (to appear).

10. R. Canetti, J. B. Friedlander and I. E. Shparlinski, 'On certain exponential sums and the distribution of Diffie–Hellman triples', *J. London Math. Soc.*, (to appear).

11. H. Cohen, *A course in computational algebraic number theory*, Springer-Verlag, Berlin, 1997.

12. J. Friedlander and H. Iwaniec, 'Estimates for character sums', *Proc. Amer. Math. Soc.*, **119** (1993), 363–372.

13. J. von zur Gathen and J. Gerhard, *Modern computer algebra*, Cambridge Univ. Press, Cambridge, 1999.

14. J. von zur Gathen and I. E. Shparlinski, 'The CREW PRAM complexity of modular inversion', *SIAM J. Computing*, (to appear).

15. M. Goldmann, 'Communication complexity and lower bounds for simulating threshold circuits', *Theoretical Advances in Neural Computing and Learning*, Kluwer Acad. Publ., Dordrecht (1994), 85–125.

16. D. M. Gordon, 'A survey of fast exponentiation methods', *J. Algorithms*, **27** (1998), 129–146.

17. H. Iwaniec and A. Sárközy, 'On a multiplicative hybrid problem', *J. Number Theory*, **26** (1987), 89–95.

18. S. Konyagin and I. E. Shparlinski, *Character sums with exponential functions and their applications*, Cambridge Univ. Press, Cambridge, 1999.

19. N. M. Korobov, 'On the distribution of digits in periodic fractions', *Matem. Sbornik*, **89** (1972), 654–670 (in Russian).

20. N. M. Korobov, *Exponential sums and their applications*, Kluwer Acad. Publ., Dordrecht, 1992.

21. E. Kushilevitz and N. Nisan, *Communication complexity*, Cambridge University Press, Cambridge, 1997.

22. N. Linial, Y. Mansour and N. Nisan, 'Constant depth circuits, Fourier transform, and learnability', *Journal of the ACM*, **40** (1993), 607-620.

23. Y. Mansour, 'Learning Boolean functions via the Fourier transform', *Theoretical Advances in Neural Computing and Learning*, Kluwer Acad. Publ., Dordrecht (1994), 391–424.

24. H. Niederreiter, 'Quasi-Monte Carlo methods and pseudo-random numbers', *Bull. Amer. Math. Soc.*, **84** (1978), 957–1041.

25. H. Niederreiter, *Random number generation and Quasi–Monte Carlo methods*, SIAM Press, Philadelphia, 1992.

26. K. Prachar, *Primzahlverteilung*, Springer-Verlag, Berlin, 1957.

27. V. Roychowdhry, K.-Y. Siu and A. Orlitsky, 'Neural models and spectral methods', *Theoretical Advances in Neural Computing and Learning*, Kluwer Acad. Publ., Dordrecht (1994), 3–36.

28. A. Sárközy, 'On the distribution of residues of products of integers', *Acta Math. Hungar.*, **49** (1987), 397–401.

29. I. E. Shparlinski, 'On the distribution of primitive and irreducible polynomials modulo a prime', *Diskretnaja Matem.*, **1** (1989), no.1, 117–124 (in Russian).

30. I. E. Shparlinski, *Number theoretic methods in cryptography: Complexity lower bounds*, Birkhäuser, 1999.
31. I. M. Vinogradov, *Elements of number theory*, Dover Publ., NY, 1954.

Quintic Reciprocity and Primality Test for Numbers of the Form $M = A5^n \pm \omega_n$

Pedro Berrizbeitia[1], Mauricio Odreman Vera[1], and Juan Tena Ayuso[3]

[1] Universidad Simón Bolívar, Departamento de Matemáticas
Caracas 1080-A, Venezuela. {pedrob,odreman}@usb.ve
[2] Facultad de Ciencias, Universidad de Valladolid
Valladolid, España. tena@agt.uva.es

Abstract. The Quintic Reciprocity Law is used to produce an algorithm, that runs in polynomial time, and that determines the primality of numbers M such that $M^4 - 1$ is divisible by a power of 5 which is larger that $\sqrt{M}$, provided that a small prime p, $p \equiv 1 (mod\, 5)$ is given, such that M is not a fifth power modulo p. The same test equations are used for all such M.
If M is a fifth power modulo p, a sufficient condition that determines the primality of M is given.

1 Introduction

Deterministic primality tests that run in polynomial time, for numbers of the form $M = A5^n - 1$, have been given by Williams, [9]. Moreover Williams and Judd [11], also considered primality tests for numbers M, such that $M^2 \pm 1$, have large prime factors. A more general deterministic primality test was developed by Adleman, Pomerance and Rumely [1], improved by H. Cohen and H.W. Lenstra [4], and implemented by H. Cohen and A.K. Lenstra [5]. Although this is more general, for specific families of numbers one may find more efficient algorithms.

This is what we give in this paper, for numbers M, such that $M^4 - 1$ is divisible by a large power of 5. More specifically let $M = A5^n \pm \omega_n$, where $0 < A < 5^n$; $0 < \omega_n < 5^n/2$; $\omega_n^4 \equiv 1 (mod\, 5^n)$.

In the given range there are exactly two possible values for ω_n. One is $\omega_n = 1$ and the other is computed inductively via Hensel's lemma. Thus, given ω_n satisfying $\omega_n^2 \equiv -1 (mod\, 5^n)$, there is a unique $x (mod\, 5)$, such that $(\omega_n + x5^n)^2 \equiv -1 (mod\, 5^{n+1})$.

Once $x (mod\, 5)$ is found select $\omega_{n+1} = \omega_n + x5^n$ or $\omega_{n+1} = 5^n - (\omega_n + x5^n)$ according to which one satisfies $\omega_{n+1} < 5^{n+1}/2$.

For such integers M we use the Quintic Reciprocity Law to produce an algorithm, which runs in polynomial time, that determines the primality of M provided that a small prime p, $p \equiv 1 (mod\, 5)$, is given, such that M is not a fifth power modulo p.

We next describe the theorem that leads naturally to the algorithm.

Let $\zeta = e^{2\pi i/5}$ be a fifth complex primitive root of unity.

G. Gonnet, D. Panario, and A. Viola (Eds.): LATIN 2000, LNCS 1776, pp. 269–279, 2000.

Let $D = \mathbb{Z}[\zeta]$ be the corresponding Cyclotomic Ring. Let π be a primary irreducible element of D lying over p. Let $K = \mathbb{Q}(\zeta + \zeta^{-1}) = \mathbb{Q}(\sqrt{5})$. Let $G = Gal(\mathbb{Q}(\zeta)/\mathbb{Q})$ be the Galois Group of the cyclotomic field $\mathbb{Q}(\zeta)$ over $\mathbb{Q}$. For every integer c denote by σ_c the element of G that sends ζ in ζ^c. For τ in $\mathbb{Z}[G]$ and α in D we denote by α^τ to the action of the element τ of $\mathbb{Z}[G]$ on the element α of D.

Let f be the order of M modulo 5 (f is also the order of M modulo 5^n). Denote by $\Phi_f(x)$ the f-th Cyclotomic Polynomial. We note that $\phi_f(M) \equiv 0 (mod\, 5^n)$.

For $f = 1$ and $f = 2$ let $\gamma = \pi^{1-3\sigma_3}$. For $f = 4$ let $\gamma = \pi$. For all cases let $\alpha = (\gamma/\bar{\gamma})^{\phi_f(M)/5^n}$, where bar indicates complex conjugation.

Let $T_0 = Trace_{K/Q}(\alpha + \bar{\alpha})$,
and $N_0 = Norm_{K/Q}(\alpha + \bar{\alpha})$

For $k \geq 0$ define T_{k+1}, N_{k+1} recursively by the formulas:

$$T_{k+1} = T_k^5 - 5N_kT_k^3 + 5N_k^2T_k + 15N_kT_k - 5T_k^3 + 5T_k \qquad (1.1)$$
$$N_{k+1} = N_k^5 - 5N_k^3(T_k^2 - 2N_k) + 5N_k[(T_k^2 - 2N_k)^2 - 2N_k^2]$$
$$+25N_k^3 - 25N_k(T_k^2 - 2N_k) + 25N_k \qquad (1.2)$$

Let $\prod_{j=1}^{g} P_j(x)$ a the factorization modulo M of the polynomial $\phi_5(x)$ as a product of irreducible polynomials. Let $\mu_j = (M, P_j(\zeta))$ be the ideal of D generated by M and $P_j(\zeta)$.

We prove the following Theorem:

Theorem 1. *Let* M, A, w_n *as before and suppose that* M *is not divisible by any of the solutions of* $x^4 \equiv 1(mod\,5^n)$; $1 < x < 5^n$. *The following statements are equivalent:*

i) M is prime
ii) For each μ_k there is an integer $i_k \not\equiv 0(mod\,5)$ such that

$$\alpha^{5^{(n-1)}} \equiv \zeta^{i_k}(\,mod\,\mu_k) \qquad (1.3)$$

iii)

$$T_{n-1} \equiv N_{n-1} \equiv -1(mod\,M) \qquad (1.4)$$

We note that the equivalence of (i) and (ii) is an extension of Proth Theorem, and the equivalence of (i) and (iii) extends the Lucas-Lehmer Test.

We use the Quintic Reciprocity Law to extend Proth's Theorem, in the same way Guthman [6] and Berrizbeitia-Berry [3] used the Cubic Reciprocity Law to extend Proth's Theorem for numbers of the form $A3^n \pm 1$. From this extension of Proth's Theorem we derive a Lucas-Lehmer type test, by taking Traces and Norms of certain elements in the field $\mathbb{Q}(\sqrt{5})$, in a way which is analogous to Rosen's proof [8] of the Lucas-Lehmer test. Generalization of this scheme to a wider family of numbers is the object of a forthcoming paper.

In section **2** of this paper we introduce the quintic symbol, and state the facts we need from the arithmetic of the ring D, including the Quintic Reciprocity

Law. In section **3** we prove theorem 1. Section **4** is devoted to remarks that have interest on their own, and are useful for implementation. We include a theorem that gives a sufficient condition that determines the primality of M when assumption on p is removed. In section **5** we describe an implementation of the algoritm derived from theorem 1. Work similar to this had been done earlier by Williams [10]. Williams derived his algoritm from properties of some Lucas Sequences. Our algorithm is derived from a generalization of Proth's Theorem, and gives a unified treatment to test primality of numbers M such that $M^4 - 1$ is divisible by a large enough power of 5. In particular, an interesting observation is that the algorithm we use to test numbers M of the form $A5^n + 1$ is the same as the one we use to test numbers of the form $A5^n - 1$, which was not the case for earlier algorithm we found in the literature.

2   The Ring D. Quintic Symbol and Quintic Reciprocity

What we state in this section may be found, among other places, in [7], chapter 12 to 14, from where we borrow the notation and presentation.

Let $D = \mathbb{Z}[\zeta]$ the ring of integer of the cyclotomic field $\mathbb{Q}(\zeta)$. Let p be a rational prime, $p \neq 5$. Let f the order of p modulo 5. Then p factors as the product of $4/f$ prime ideals in D. If $\mathcal{P}$ and $\mathcal{P}'$ are two of these prime ideals, there is a σ in $G = Gal(\mathbb{Q}(\zeta)/\mathbb{Q})$ such that $\sigma(\mathcal{P}) = \mathcal{P}'$. $D/\mathcal{P}$ is a finite field with p^f elements and is called the residue class field mod $\mathcal{P}$. The multiplicative group of units mod $\mathcal{P}$, denoted by $(D/P)^*$ is cyclic of order $(p^f - 1)$. Let α in D an element not in $\mathcal{P}$. There is an integer i, unique modulo 5 such that $\alpha^{(p^f-1)/5} \equiv \zeta^i (mod \mathcal{P})$. The quintic symbol $(\alpha/\mathcal{P})$ is defined to be that unique fith root of unity satisfying

$$\alpha^{(p^f-1)/5} \equiv (\alpha/\mathcal{P})(mod\, \mathcal{P}) \qquad (2.1)$$

The symbol has the following properties:
$(\alpha/\mathcal{P}) = 1$ if, and only if,

$$x^5 \equiv \alpha(mod\mathcal{P}) \qquad (2.2)$$

is solvable in D.
For every $\sigma \in G$

$$(\alpha/\mathcal{P})^\sigma = (\sigma(\alpha)/\sigma(\mathcal{P})) \qquad (2.3)$$

Let $\mathcal{A}$ be an ideal in D, prime to 5. Then $\mathcal{A}$ can be written as a product of prime ideals: $\mathcal{A} = \mathcal{P}_1 \cdots \mathcal{P}_s$. Let $\alpha \in D$ be prime to A. The symbol (α/A) is defined as the product of the symbols $(\alpha/\mathcal{P}_1)\cdots(\alpha/\mathcal{P}_s)$. Let $\beta \in D$ prime to 5 and to α. The symbol (α/β) is defined as $(\alpha/(\beta))$.

D is a Principal Ideal Domain (PID) (see the notes in page 200 of [7] for literature on cyclotomic fields with class number one). An element $\alpha \in D$ is called primary if it is not a unit, is prime to 5 and is congruent to a rational integer modulo $(1 - \zeta)^2$. For each $\alpha \in D$, prime to 5, there is an integer c in $\mathbb{Z}$, unique modulo 5, such that $\zeta^c \alpha$ is primary. In particular, every prime ideal $\mathcal{P}$ in D has a primary generator π.

Quintic Reciprocity Law:

Let M be an integer, prime to 5. Let α be a primary element of D and assume α is prime to M and prime to 5. Then

$$(\alpha/M) = (M/\alpha) \qquad (2.4)$$

3   Proof of the Theorem

The condition imposed on the prime p implies $p \equiv 1 (mod\, 5)$ (otherwise the equation $x^5 \equiv M (mod\, p)$ would have an integer solution). It follows that the ideal (p) factors as the product of four prime ideals in D. These are all principal, since D is a PID. We denote by π a primary generator of one of these prime ideals. The other ideals are generated by the Galois conjugates of π, which are also primary.

We note that $(M/\pi) \neq 1$, otherwise M would be a fifth power modulo each of π's Galois conjugates, hence modulo p. We prove

i) implies ii)

Suppose first $f = 1$.

Let $(M/\pi) = \zeta^{i_1}$. Then $i_1 \not\equiv 0 (mod\, 5)$. Since M is a rational prime, $M \equiv 1 (mod\, 5)$, then (M) factors in D as the product of 4 prime ideals. We write $(M) = (\mu_1)(\sigma_2(\mu_1))(\bar{\mu}_1)(\sigma_2(\bar{\mu}_1))$. We get

$$
\begin{aligned}
\zeta^{i_1} &= (M/\pi) = (\pi/M) && (by\,(2.4)) \\
&= (\pi/\mu_1)(\pi/\sigma_2(\mu_1))(\pi/\bar{\mu}_1)(\pi/\sigma_2(\bar{\mu}_1)) \\
&&& (because\,(M) = (\mu_1)(\sigma_2(\mu_1))(\bar{\mu}_1)(\sigma_2(\bar{\mu}_1))) \\
&= (\frac{\pi}{\bar{\pi}}/\mu_1)(\frac{\pi}{\bar{\pi}}/\sigma_2(\mu_1)) = (\frac{\pi}{\bar{\pi}}/\mu_1)(\sigma_3(\frac{\pi}{\bar{\pi}})/(\mu_1))^{-3} = ((\frac{\pi}{\bar{\pi}})^{1-3\sigma_3}/\mu_1) \\
&&& (by\,(2.3)) \\
&\equiv ((\frac{\pi}{\bar{\pi}})^{1-3\sigma_3})^{(M-1)/5} (mod\, \mu_1) && (by\,(2.1)) \\
&\equiv \alpha^{5^{n-1}} (mod\, \mu_1) && (since\, \phi_1(M) = M - 1)
\end{aligned}
$$

Next suppose $f = 2$. In this case $(M) = (\mu)(\sigma_2(\mu))$. Again we use (2.4), (2.3) and (2.1). This time we get:

There is an integer $i_2 \not\equiv 0 (mod\, 5)$, such that

$$
\begin{aligned}
\zeta^{i_2} &= (\pi/M) = (\pi^{1-3\sigma_3}/\mu) \\
&\equiv (\pi^{1-3\sigma_3})^{(M^2-1)/5} \\
&\equiv (\pi^{1-3\sigma_3})^{(M-1)(M+1)/5} (mod\, \mu).
\end{aligned}
$$

Noting that raising to the Mth power mod μ is same as complex conjugation mod μ and that $\phi_2(M) = M + 1$ we get the result. Finally, if $f = 4$, (M) remains prime in D. We get $(M/\pi) = (\pi/M) \equiv \pi^{(M^4-1)/5}(mod\, M) \equiv \pi^{(M^2-1)(M^2+1)/5}(mod\, M)$. This time raising to the power M^2 is equivalent to complex conjugation and $\phi_4(M) = M^2 + 1$, so we obtain the desired result $\square$.

ii) implies iii)

For $k \geq 0$ let $T_k = Trace_{K/Q}(\alpha^{5^k} + \bar{\alpha}^{5^k})$ and $N_k = Norm_{K/Q}(\alpha^{5^k} + \bar{\alpha}^{5^k})$. We claim that T_k and N_k satisfy the recurrent relations given by (1.1) and (1.2). To see this we let $A_k = \alpha^{5^k} + \bar{\alpha}^{5^k}$ and $B_k = \sigma_2(A_k)$.

So $T_k = A_k + B_k$ and $N_k = A_k B_k$

We first will obtain (1.1).

Raising T_k to the fifth power we get

$$A_k^5 + B_k^5 = T_k^5 - 5N_k(A_k^3 + B_k^3) - 10N_k^2 T_k \tag{3.1}$$

Computing T_k^3 we obtain:

$$A_k^3 + B_k^3 = T_k^3 - 3N_k T_k \tag{3.2}$$

On the other hand, keeping in mind that $\bar{\alpha} = \alpha^{-1}$ inverse one gets:

$$A_k^5 = A_{k+1} + 5((\alpha^{5^k})^3 + (\alpha^{-5^k})^3) + 10A_k \tag{3.3}$$

and

$$A_k^3 = (\alpha^{5^k})^3 + (\alpha^{-5^k})^3 + 3A_k \tag{3.4}$$

Combining (3.3) with (3.4) leads to:

$$A_{k+1} = A_k^5 - 5A_k^3 + 5A_k \tag{3.5}$$

Similarly, one obtains

$$B_{k+1} = B_k^5 - 5B_k^3 + 5B_k \tag{3.6}$$

Adding (3.5) with (3.6) we get

$$T_{k+1} = (A_k^5 + B_k^5) - 5(A_k^3 + B_k^3) + 5T_k \tag{3.7}$$

Substituting (3.1) and (3.2) in (3.7) we obtain (1.1)

To obtain (1.2) we first multiply (3.5) and (3.6). This leads to:

$$N_{k+1} = N_k^5 - 5N_k^3(A_k^2 + B_k^2) + 5N_k(A_k^4 + B_k^4)$$
$$+ \qquad\qquad 25N_k^3 - 25N_k(A_k^2 + B_k^2) + 25N_k \tag{3.8}$$

Next we note:

$$A_k^2 + B_k^2 = T_k^2 - 2N_k \tag{3.9}$$

from where we deduce

$$A_k^4 + B_k^4 = (T_k^2 - 2N_k)^2 - 2N_k^2 \tag{3.10}$$

(1.2) is then obtained by substituting (3.9) and (3.10) in (3.8).

Since we have proved that T_k and N_k satisfy the recurrence relations given by (1.1) and (1.2), ii) implies that $T_{n-1} \equiv (\zeta + \zeta^{-1}) + (\zeta^2 + \zeta^{-2}) \equiv -1 (mod \, \mu)$.

Since T_{n-1} is a rational number then the congruence holds modulo $\mu \cap \mathbb{Q} = M$. Similarly we get $N_{n-1} \equiv -1 (mod \, M) \; \square$.

iii) implies i)

We will show that under the hypothesis every prime divisor Q of M is larger than square root of M. This will imply that M is prime. Let Q be a prime divisor of M. Let $\mathcal{Q}$ be a prime ideal in D lying over Q. Clearly, (1.4) holds modulo $\mathcal{Q}$. We will show that also (1.3) holds modulo $\mathcal{Q}$.

From

$$T_{n-1} = Trace_{K/Q}(\alpha^{5^{n-1}} + \bar{\alpha}^{5^{n-1}}) \equiv -1 (mod \, \mathcal{Q})$$

and

$$N_{n-1} = Norm_{K/Q}(\alpha^{5^{n-1}} + \bar{\alpha}^{5^{n-1}}) \equiv -1 (mod \, \mathcal{Q}),$$

we deduce that $(\alpha^{5^{n-1}} + \bar{\alpha}^{5^{n-1}})$ has the same norm and trace modulo $\mathcal{Q}$ than $(\zeta + \zeta^{-1})$, it follows that $(\alpha^{5^{n-1}} + \bar{\alpha}^{5^{n-1}}) \equiv (\zeta + \zeta^{-1}) (mod \, \mathcal{Q})$ or $(\alpha^{5^{n-1}} + \bar{\alpha}^{5^{n-1}}) \equiv (\zeta^2 + \zeta^{-2}) (mod \, \mathcal{Q})$. This fact, together with the fact $\alpha^{-1} = \bar{\alpha}$ leads to $\alpha^{5^{n-1}} \equiv \zeta^i (mod \, \mathcal{Q})$ for some i $\not\equiv 0 (mod \, 5)$. Hence the class of $\alpha (mod \, \mathcal{Q})$ has order 5^n in the multiplicative group of units $(\mathbb{Q}/\mathcal{Q})^*$. It follows that 5^n divides the order of this group which is a divisor $Q^4 - 1$. In other words, $Q^4 - 1 \equiv 0 (mod \, 5^n)$. Since by hypothesis no solution of this last congruence equation less than 5^n is a divisor of M it follows that Q is larger than 5^n that in turn is larger than square root of M, by the hypothesis made on A. $\square$

4   Remarks on the Implementation

In this section we will make remarks on the Implementation, and find T_0 and N_0. We will also study what happens if M is a fifth power modulo p.

- Although in principle part ii) of theorem 1 provides an algorithm for testing the primality of M, it assumes that a factorization of $\phi_5(x)$ modulo M is given. If M is not a prime the algorithm that finds this factorization may not converge. Part iii) instead gives an algorithm easy to implement provided that N_0 and T_0 are computed.
- Note that the recurrence relations (1.1) and (1.2) are independent of the value of p. This is the case because $\bar{\alpha} = \alpha^{-1}$.

- In practice A is fixed, while n is taken in a range of values which vary from relatively small to as large as possible. In the cases $f = 1$ and $f = 2$ we obtain
$T_0 = Trace_{K/Q}(\alpha + \bar{\alpha}) = Trace_{K/Q}((\gamma/\bar{\gamma})^A + (\bar{\gamma}/\gamma)^A)$ and
$N_0 = Norm_{K/Q}(\alpha + \bar{\alpha}) = Norm_{K/Q}(\gamma/\bar{\gamma})^A + (\bar{\gamma}/\gamma)^A)$. Hence T_0 and N_0 are computable with $O(\log A)$ modular operations.
When $f = 4$ the calculation of $\alpha (mod\, M)$ is longer. In fact, in this case $\alpha = (\gamma/\bar{\gamma})^{(M^2+1)/5^n}$. The exponent this time is very big, and the calculation of T_0 and N_0 in this case involve a lot of work. The calculation is still done with $O(\log M)$ modular operations, but not anymore with $O(\log A)$, as it is in the cases of $f = 1$ and $f = 2$. The following observation reduces somewhat the amount of work involved in the computation of $\alpha (mod\, M)$ for the case $f = 4$.
When dividing $(M^2 + 1)/5^n$ by M one obtains:

$$(M^2 + 1)/5^n = AM + (w_n^2 + 1)/5^n \pm Aw_n$$

The calculation of $\alpha (mod\, M)$ is therefore simplified by keeping in mind that raising $(\gamma/\bar{\gamma})$ to the Mth power modulo M is equivalent to applying σ_2 or σ_3, according to the congruence of M (mod 5).
- If $p \equiv 1(mod\, 5)$ and M is a fith power modulo p, the following proposition provides a sufficient condition to prove the primality of M.

Proposition 1. *If $T_k \equiv N_k \equiv -1(mod\, M)$ for some k such that 5^k is larger than square root of M and if no nontrivial solution of $x^4 \equiv 1(mod\, 5^k)$, $x < 5^k$, is a divisor of M, then M is prime.*

The proof of this proposition goes along the lines of iii) implies i) in theorem 1, the key point being that α has order $5^k mod\, Q$, which obliges Q to be too large or a smaller solution of $x^4 \equiv 1 mod(5^k)$ $\square$.
This proposition may be particularly useful when A is much smaller than 5^n.

5   Implementation

Table 1 below consist of a 2x2 matrix containing all number w_n, $0 < n < 25$, such that $w_n^2 + 1 \equiv 0(mod\, 5^n)$; $0 < w_n < 5^n$; $w_n \equiv \pm 2(mod\, 5)$. The first column contains exactly those w_n which are congruent to $2(mod\, 5)$ and the second column those which are congruent to $3(mod\, 5)$. The term $n + 1$ of the first column, w_{n+1}, is obtained from the nth term of the same column by the following formula:

$$w_{n+1} = w_n + \left[\left(\frac{w_n^2 + 1}{5^n}\right) mod(\,5)\right] 5^n,\ w_1 = 2$$

For the second column we use

$$w_{n+1} = w_n + \left[-\left(\frac{w_n^2 + 1}{5^n}\right) mod(\,5)\right] 5^n,\ w_1 = 3$$

Table 1. ω_n

n	$\omega_n(\omega_1 = 2)$	$\omega_n(\omega_1 = 3)$
1	2	3
2	7	18
3	57	68
4	182	443
5	2057	1068
6	14557	1068
7	45807	32318
8	280182	110443
9	280182	1672943
10	6139557	3626068
11	25670807	23157318
12	123327057	120813568
13	123327057	1097376068
14	5006139557	1097376068
15	11109655182	19407922943
16	102662389557	49925501068
17	407838170807	355101282318
18	3459595983307	355101282318
19	3459595983307	15613890344818
20	79753541295807	15613890344818
21	365855836217682	110981321985443
22	2273204469030182	110981321985443
23	2273204469030182	9647724486047943
24	49956920289342682	9647724486047943
25	109561565064733307	188461658812219818

Table 2 below also consist of a 24x2 matrix, this time the Ath term of the first column contains a list of values of n , $0 < n < 100$, such that $M = A5^n + w_n$, with $w_n \equiv 2 \pmod 5$, is prime, followed by the time it took a Pentium II , 350 mhz, to compute them, using the programm we next describe. Maple was used for implementation.

Table 2. Primes for $w_1 = 2, 3$

A	$w_1 = 2$		$w_1 = 3$	
	$1 \leq n \leq 100$	time	$1 \leq n \leq 100$	time
1	1	28.891	2,3,6,16,17,25	31.563
2	3,20,57,73	39.943	1,4,31	26.087
3	1,22,24	27.705	3,12,73,77,82	34.346
4	2,3,5,17	24.494	1	27.809
5	4,9,64	27.938	5,6	35.504
6	2,5	35.372	15,39	27.162
7	1,34	28.933	2,5,16,35	36.022
8	14	35.883	1,4,24	28.936
9	1,4,29,59	27.788	3,7,55	36.717
10	2,3,10,11,13,43	37.103	1	29.457
11	4,61,86	28.533	2,43,94	36.183
12	2,27,32,63,73	36.900	21	25.896
13	1,8,33,34,56	28.671	3,11,17,18,30,35,37,46,48	37.445
14	7,19,72	36.126	1,24,92	28.857
15	–	44.894	68,72	38.615
16	5,13,17	37.311	1,28,76	29.468
17	28	28.510	2,5,11,27	36.624
18	2,11,54,57	36.766	28,59	30.104
19	1,15,21,23,69	28.971	5,7,35,81	38.568
20	3,14	38.138	1	31.106
21	1	28.237	3,13,14,19,42,57	38.671
22	2,7,12,16,75	36.921	1,8,56	30.001
23	4,8	29.075	2,58,81	38.983
24	2,78	36.275	4	30.680

The first column of the table 3 contains the values of n for which $A5^n + 1$ is prime and the second column those values for which $A5^n - 1$ is prime.

5.1   Description of the Algorithm

Some precomputation is needed.

We fix the primes $p = 11, 31, 41, 61$.

For each of these primes we found a prime element of the ciclotomic ring D, which we will denote by $\Pi_p(\zeta)$, lying over p (this means that $|Norm_{Q(\zeta)/Q}(\Pi_p(\zeta))| = p$).

Table 3. Primes for $\omega_1 = 1, -1$

A	$\omega_1 = 1$		$\omega_1 = -1$	
	$1 \leq n \leq 100$	time	$1 \leq n \leq 100$	time
2	1,3,13,45	12.013	4,6,16,24,30,54,96	12.321
4	2,6,18,50	14.149	1,3,9,13,15,25,39,69	12.497
6	1,2,3,23,27,33,63	12.158	1,2,5,11,28,65,72	13.058
8	1	12.715	2,4,8,10,28	13.219
12	1,5,7,18,19,23,46,51,55,69	12.893	1,3,4,8,9,28,31,48,51,81	13.309
14	1,7,23,33	13.239	2,6,14	13.587
16	2,14,22,26,28,42	13.072	1,3,5,7,13,17,23,33,45,77	13.446
18	3,4,6,10,15,30	13.199	1,2,5,6,9,13,17,24,26,49,66	13.577
22	4,10,40	13.907	1,3,5,7,27,35,89	14.085
24	2,3,8,19,37,47	12.921	2,3,10,14,15,23,27,57,60	13.715

We note that the left side of equation (1.3) does not vary if $\Pi_p(\zeta)$ is replaced by another prime lying over p when $n \geq 2$. Therefore the condition of primary may be disregarded, hence we let $\Pi_{11}(\zeta) = (\zeta + 2)$, $\Pi_{31}(\zeta) = (\zeta - 2)$, $\Pi_{41}(\zeta) = (\zeta^3 + 2\zeta^2 + 3\zeta + 3)$, $\Pi_{61}(\zeta) = (\zeta + 3)$.

For the case $f = 1$ and $f = 2$ ($M = A5^n \pm 1$) we let

$$\beta_{p,f} = \left(\frac{\Pi_p(\zeta)}{\overline{\Pi_p(\zeta)}} \right)^{1-3\sigma}$$

for the case $f = 4$ (or $M = A5^n + w_n$; $w_n = \pm 2$), we let

$$\beta_{p,f} = \left(\frac{\Pi_p(\zeta)}{\overline{\Pi_p(\zeta)}} \right)$$

and

$$T_{p,f,A,n} \equiv Tr_{K/Q} \left(\beta_p^{\phi_f(M)/5^n}(\zeta) + \overline{\beta_p^{\phi_f(M)/5^n}}(\zeta) \right) \ (Mod\, M)$$

$$N_{p,f,A,n} \equiv Norm_{K/Q} \left(\beta_p^{\phi_f(M)/5^n}(\zeta) + \overline{\beta_p^{\phi_f(M)/5^n}}(\zeta) \right) \ (Mod\, M)$$

The program finds the first values of p for which M is not a fith power. If the condition is not satisfied a note is made and these number are later tested by other means.

Otherwise we set $T_0 = T_{p,f,A,n}$ and $N_0 = N_{p,f,A,n}$ and we use the recurrence equation (1.1) and (1.2) to verify if (1.4) holds.

When $f = 1$ or 2 we note that $\phi_f(M)/5^n = A$. Hence $T_{p,f,A,n}$ depends only on A, not on n. In this case, for relatively small values of A, we recommend to compute the value of

$$Tr_{K/Q}\left(\beta_p^{\phi_f(M)/5^n}(\zeta) + \overline{\beta_p^{\phi_f(M)/5^n}(\zeta)}\right)\ (not\ modulo\ M)$$

and

$$Norm_{K/Q}\left(\beta_p^{\phi_f(M)/5^n}(\zeta) + \overline{\beta_p^{\phi_f(M)/5^n}(\zeta)}\right)\ (not\ modulo\ M)$$

These same numbers may be used as the starting number T_0 and N_0 for all numbers n in a given range. If for a fixed value of A the calculation of T_0 and N_0 is counted as part of the precomputation, then the complexity of the primality test for numbers of the form $A5^n \pm 1$, which are not congruent to a fith power modulo p, is simply the complexity of the calculation of the recurrence relations (1.1) and (1.2) n-1 times.

When $f = 4$, $\phi_f(M)/5^n$ is large and depends on A and n. In this case, even for small values of A, the computation of T_0 and N_0 is, for each value of M, of approximately same complexity as the the the computation of T_{n-1}, N_{n-1}, given T_0 and N_0.

Acknowledgements

We are grateful to Daniel Sandonil who implemented the programm.

We are also grateful to the referees, for their useful comments, that helped to improve the paper.

References

1. L. Adleman, C. Pomerance and R. Rumely *On distinguishing prime numbers from composite numbers*, Ann. of Math. 117, (1983), 173-206.
2. D. Bressoud, *Factorization and Primality Testing*, Springer-Verlag, New York 1989.
3. P. Berrizbeitia,T.G. Berry, *Cubic Reciprocity and generalised Lucas-Lehmer test for primality of $A3^n \pm 1$* Proc.AMS.127(1999), 1923-1925.
4. H. Cohen and H.W. Lenstra, *Primality testing and Jacobi sums*, Math. Comp. 42, (1984), 297-330.
5. H. Cohen and A.K. Lenstra, *Implemetation of a new primality test*, Math. Comp. 48, (1987), 103-121.
6. A. Guthmann, *Effective Primality Test for Integers of the Forms $N = K3^n + 1$ and $N = K2^m3^n + 1$*. BIT 32 (1992), 529-534.
7. K. Ireland and M. Rosen, *A Classical Introduction to Modern Number Theory*, 2da ed. Springer-Verlag, New York 1982.
8. M. Rosen, *A Proof of the Lucas-Lehmer Test*, Amer. Math. Monthly 95(1980) 9, 855-856.
9. H.C. Williams, *Effective Primality Tests for some Integers of the Forms $A5^n - 1$ and $A7^n - 1$*, Math. of Comp. 48(1987), 385-403.
10. H.C. Williams, *A Generalization of Lehmer's functions*, Acta Arith 29 (1976), 315-341.
11. H.C. Williams and J.S. Judd, *Determination of the Primality of N by Using Factors of $N^2 \pm 1$*, Math. of Comp. 30(1976), 157-172.

Determining the Optimal Contrast for Secret Sharing Schemes in Visual Cryptography

Matthias Krause[1] and Hans Ulrich Simon[2]

[1] Theoretische Informatik
Universität Mannheim
D-68131 Mannheim, Germany
krause@th.informatik.uni-mannheim.de
[2] Fakultät für Mathematik
Ruhr-Universität Bochum
D-44780 Bochum, Germany
simon@lmi.ruhr-uni-bochum.de

Abstract. This paper shows that the largest possible contrast $C_{k,n}$ in an k-out-of-n secret sharing scheme is approximately $4^{\cdot\ (k\cdot\ 1)}$. More precisely, we show that $4^{\cdot\ (k\cdot\ 1)} \leq C_{k,n} \leq 4^{\cdot\ (k\cdot\ 1)} n^k/(n(n-1)\cdots(n-(k-1)))$. This implies that the largest possible contrast equals $4^{\cdot\ (k\cdot\ 1)}$ in the limit when n approaches infinity. For large n, the above bounds leave almost no gap. For values of n that come close to k, we will present alternative bounds (being tight for $n = k$). The proofs of our results proceed by revealing a central relation between the largest possible contrast in a secret sharing scheme and the smallest possible approximation error in problems occuring in Approximation Theory.

1 Introduction

Visual cryptography and k-out-of-n secret sharing schemes are notions introduced by Naor and Shamir in [10]. A sender wishing to transmit a secret message distributes n transparencies among n recipients, where the transparencies contain seemingly random pictures. A k-out-of-n scheme achieves the following situation: If any k recipients stack their transparencies together, then a secret message is revealed visually. On the other hand, if only $k-1$ recipients stack their transparencies, or analyze them by any other means, they are not able to obtain any information about the secret message. The reader interested in more background information about secret sharing schemes is referred to [10].

An important measures of a scheme is its *contrast*, i.e., the clarity with which the message becomes visible. This parameter lies in interval $[0, 1]$, where contrast 1 means "perfect clarity" and contrast 0 means "invisibility". Naor and Shamir constructed k-out-of-k secret sharing schemes with contrast $2^{-(k-1)}$ and were also able to prove optimality. However, they did not determine the largest possible contrast $C_{k,n}$ for arbitrary k-out-of-n secret sharing schemes.

In the following, there were made several attempts to find accurate estimations for the optimal contrast and the optimal tradeoff between contrast and

G. Gonnet, D. Panario, and A. Viola (Eds.): LATIN 2000, LNCS 1776, pp. 280–291, 2000.

subpixel expansion for arbitrary k-out-of-n secret sharing schemes [4],[5],[1],[2], [3]. For $k = 2$ and arbitrary n this problem was completely solved by Hofmeister, Krause, and Simon in [5]. But the underlying methods, which are based on the theory of linear codes, do not work for $k \geq 3$. Strengthening the approach of Droste [4], the first step in the direction of determining $C_{k,n}$ for some values k and n, where $k \geq 3$, was taken in [5]. They presented a simple linear program $LP(k, n)$ whose optimal solution represents a contrast-optimal k-out-of-n secret sharing scheme. The profit achieved by this solution equals $C_{k,n}$. Although, $C_{k,n}$ was computable in $\text{poly}(n)$ steps this way, and even elemantary formulas were given for $k = 3, 4$, there was still no general formula for $C_{k,n}$ (or for good bounds). Based on computations of $C_{k,n}$ for specific choices of k, n, it was conjectured in [5] that $C_{k,n} \geq 4^{-(k-1)}$ with equality in the limit when n approaches infinity. In [2] and [3], some of the results from [5] concerning $k = 3, 4$ and arbitrary n could be improved. Furthermore, in [3], Blundo, D'Arco, DeSantis and Stinson determine the optimal contrast of k-out-of-n secret sharing schemes for arbitrary n and $k = n - 1$.

In this paper, we confirm the above conjecture of [5] by showing the following bounds on $C_{k,n}$:

$$4^{-(k-1)} \leq C_{k,n} \leq 4^{-(k-1)} \frac{n^k}{n(n-1)\cdots(n-(k-1))}.$$

This implies that the largest possible contrast equals $4^{-(k-1)}$ in the limit when n approaches infinity. For large n, the above bounds leave almost no gap. For values of n that come close to k, we will present alternative bounds (being tight for $n = k$). The proofs of our results proceed by revealing a central relation between the largest possible contrast in a secret sharing scheme and the smallest possible approximation error in problems occuring in Approximation Theory. A similar relation was used in the paper [8] of Linial and Nisan about Approximate Inclusion-Exclusion (although there are also some differences and paper [8] ends-up with problems in Approximation Theory that are different from ours).

2   Definitions and Notations

For the sake of completeness, we recall the definition of visual secret sharing schemes given in [10]. In the sequel, we simply refer to them under the notion *scheme*. For a 0-1–vector v, let $H(v)$ denote the Hamming weight of v, i.e., the number of ones in v.

Definition 1. *A k-out-of-n scheme $\mathcal{C} = (\mathcal{C}_0, \mathcal{C}_1)$ with m subpixels, contrast $\alpha = \alpha(\mathcal{C})$ and threshold d consists of two collections of Boolean $n \times m$ matrices $\mathcal{C}_0 = [C_{0,1}, \ldots, C_{0,r}]$ and $\mathcal{C}_1 = [C_{1,1}, \ldots, C_{1,s}]$, such that the following properties are valid:*

1. *For any matrix $S \in \mathcal{C}_0$, the OR v of any k out of the n rows of S satisfies $H(v) \leq d - \alpha m$.*

2. *For any matrix $S \in C_1$, the OR v of any k out of the n rows of S satisfies $H(v) \geq d$.*

3. *For any $q < k$ and any q-element subset $\{i_1, \ldots, i_q\} \subseteq \{1, \ldots, n\}$, the two collections of $q \times m$ matrices $\mathcal{D}_0$ and $\mathcal{D}_1$ obtained by restricting each $n \times m$ matrix in $\mathcal{C}_0$ and $\mathcal{C}_1$ to rows $i_1, \ldots, i_q$ are indistinguishable in the sense that they contain the same matrices with the same relative frequencies.*

k-out-of-n schemes are used in the following way to achieve the situation described in the introduction. The sender translates every pixel of the secret image into n sets of subpixels, in the following way: If the sender wishes to transmit a white pixel, then she chooses one of the matrices from $\mathcal{C}_0$ according to the uniform distribution. In the case of a black pixel, one of the matrices from $\mathcal{C}_1$ is chosen. For all $1 \leq i \leq n$, recipient i obtains the i-th row of the chosen matrix as an array of subpixels, where a 1 in the row corresponds to a black subpixel and a 0 corresponds to a white subpixel. The subpixels are arranged in a fixed pattern, e.g. a rectangle. (Note that in this model, stacking transparencies corresponds to "computing" the OR of the subpixel arrays.)

The third condition in Definition 1 is often referred to as the "security property" which guarantees that any $k - 1$ of the recipients cannot obtain any information out of their transparencies. The "contrast property", represented by the first two conditions in Definition 1, guarantees that k recipients are able to recognize black pixels visually since any array of subpixels representing a black pixel contains a "significant" amount of black subpixels more than any array representing a white pixel.[1]

In [5], it was shown that the largest possible contrast $C_{k,n}$ in an k-out-of-n scheme coincides with the maximal profit in the following linear program (with variables $\xi_0, \ldots, \xi_n$ and $\eta_0, \ldots, \eta_n$):

Linear Program LP(k, n)

$\max \sum_{j=0}^{n-k} \binom{n-k}{j} \binom{n}{j}^{-1} (\xi_j - \eta_j)$ **subject to**

1. For $j = 0, \ldots, n : \xi_j \geq 0, \eta_j \geq 0$.

2. $\sum_{j=0}^{n} \xi_j = \sum_{j=0}^{n} \eta_j = 1$.

3. For $l = 0, \ldots, k - 1: \sum_{j=l}^{n-k+l+1} \binom{n-k+1}{j-l} \binom{n}{j}^{-1} (\xi_j - \eta_j) = 0$.

The following sections only use this linear program (and do not explicitly refer to Definition 1).

We make the following conventions concerning matrices and vectors. For matrix A, A' denotes its transpose (resulting from A by exchanging rows and

---

[1] The basic notion of a secret sharing scheme, as given in Definition 1, has been generalized in several ways. The generalized schemes in [1], for instance, intend to achieve a situation where certain subsets of recipients can work successfully together, whereas other subsets will gain no information. If the two classes of subsets are the sets of at least k recipients and the sets of at most $k - 1$ recipients, respectively, we obtain (as a special case) the schemes considered in this paper. Another model for 2-out-of-2 schemes involving three colors is presented in [11].

columns). A vector which is denoted by c is regarded as a column vector. Thus, its transpose c' is a row vector. The all-zeros (column) vector is denoted as 0. For matrix A, A_j denotes its j-th row vector. A'_j denotes the j-th row vector of its transpose (as opposed to the transpose of the j'th row vector).

3 Approximation Error and Contrast

In Subsection 3.1, we relate the problem of finding the best k-out-of-n secret sharing scheme to approximation problems of type BAV and BAP. Problem BAV (Best Approximating Vector) asks for the "best approximation" of a given vector c within a vector space V. Problem BAP (Best Approximating Polynomial) asks for the "best approximation" of a given polynomial p of degree k within the set of polynomials of degree $k-1$ or less. It turns out that, choosing c, V, p properly, the largest possible contrast is twice the smallest possible approximation error. In Subsection 3.2, we use this relationship to determine lower and upper bounds. Moreover, the largest possible contrast is determined exactly in the limit (when n approaches infinity). In Subsection 3.3, we derive a criterion that helps to determine those pairs (k, n) for which $C_{k,n}$ coincides with its theoretical upper bound from Subsection 3.2.

3.1 Secret Sharing Schemes and Approximation Problems

As explained in Section 2, the largest possible contrast in an k-out-of-n secret sharing scheme is the maximal profit in linear program $\mathrm{LP}(k, n)$. The special form exhibited by $\mathrm{LP}(k, n)$ is captured by the more abstract definitions of a linear program of type BAV (Best Approximating Vector) or of type BAP (Best Approximating Polynomial).

We start with the discussion of type BAV. We say that a linear program LP is of *type BAV* if there exists a matrix $A \in \Re^{k \times (1+n)}$ and a vector $c \in \Re^{n+1}$ such that LP (with variables $\boldsymbol{\xi} = (\xi_0, \dots, \xi_n)$ and $\boldsymbol{\eta} = (\eta_0, \dots, \eta_n)$) can be written in the following form:

The primal linear program $\mathrm{LP}(A, c)$ of type BAV

max $c'(\boldsymbol{\xi} - \boldsymbol{\eta})$ subject to
(LP1) $\boldsymbol{\xi} \geq 0, \boldsymbol{\eta} \geq 0$
(LP2) $\sum_{j=0}^{n} \xi_j = \sum_{j=0}^{n} \eta_j = 1$
(LP3) $A(\boldsymbol{\xi} - \boldsymbol{\eta}) = 0$

Condition (LP2) implies that

$$\sum_{j=0}^{n}(\xi_j - \eta_j) = 0.$$

Thus, we could add the all-ones row vector $(1, \dots, 1)$ to matrix A in (LP3) without changing the set of legal solutions. For this reason, we assume in the sequel that the following condition holds in addition to (LP1), (LP2), (LP3):

(LP4) The vector space V_A spanned by the row vectors of A contains the all-ones vector.

We aim to show that linear program $LP(A, c)$ can be reformulated as the problem of finding the "best" approximation of c in V_A. To this end, we pass to the dual problem[2] (with variables s, t and $u = (u_0, \ldots, u_{k-1})$):

The dual linear program $DLP(A, c)$ of type BAV

min s+t **subject to**
(DLP1) $A'u + (s, \ldots, s)' \geq c$
(DLP2) $A'u - (t, \ldots, t)' \leq c$

Conditions (DLP1) and (DLP2) are obviously equivalent to

$$s \geq \max_{j=0,\ldots,n} (c_j - A'_j u) \text{ and } t \geq \max_{j=0,\ldots,n} (A'_j u - c_j),$$

and an optimal solution certainly satisfies

$$s = \max_{j=0,\ldots,n} (c_j - A'_j u) \text{ and } t = \max_{j=0,\ldots,n} (A'_j u - c_j).$$

Note that vector $A'u$ is a linear combination of the row vectors of A. Thus, $V_A = \{A'u | \, u \in \Re^k\}$. $DLP(A, c)$ can therefore be rewritten as follows:

$$\min_{v \in V_A} \left[\max_{j=0,\ldots,n} (c_j - v_j) + \max_{j=0,\ldots,n} (v_j - c_j) \right]$$

Consider a vector $v \in V_A$ and let

$$j_-(v) = \arg \max_{j=0,\ldots,n} (c_j - v_j) \text{ and } j_+(v) = \arg \max_{j=0,\ldots,n} (v_j - c_j).$$

Term $S(v) := c_{j_-(v)} - v_{j_-(v)}$ represents the penalty for $v_{j_-(v)}$ being smaller than $c_{j_-(v)}$. Symmetrically, $L(v) := v_{j_+(v)} - c_{j_+(v)}$ represents the penalty for $v_{j_+(v)}$ being larger than $c_{j_+(v)}$. Note that the total penalty $S(v) + L(v)$ does not change if we translate v by a scalar multiple of the all-ones vector $(1, \ldots, 1)'$. According to (LP4), any translation of this form can be performed within V_A. Choosing the translation of v appropriately, we can achieve $S(v) = L(v)$, that is, a perfect balance between the two penalty terms. Consequently, the total penalty for v is twice the distance between c and v measured by the metric induced by the maximum-norm. We thus arrive at the following result.

Theorem 1. *Given linear program $LP(A, c)$ of type BAV, the maximal profit C in $LP(A, c)$ satisfies*
$$C = 2 \cdot \min_{v \in V_A} \max_{j=0,\ldots,n} |c_j - v_j|.$$

---

[2] The rules, describing how the dual linear program is obtained from a given primal, can be looked up in any standard text about linear programming (like [12], for instance).

Thus, the problem of finding an optimal solution to $LP(A, c)$ boils down to the problem of finding a best approximation of c in V_A w.r.t. the maximum-norm.

We now pass to the discussion of linear programs of type BAP. We call $d \in \Re^{1+n}$ *evaluation-vector* of polynomial $p \in \Re[X]$ if $d_j = p(j)$ for $j = 0, \ldots, n$. We say that a linear program $LP(A, c)$ of type BAV is of *type BAP* if, in addition to Conditions (LP1),...,(LP4), the following holds:

(LP5) c is the evaluation vector of a polynomial, say p, of degree k.
(LP6) Matrix $A \in \Re^{k \times (1+n)}$ has rank k, i.e., its row vectors are linearly independent.
(LP7) For $l = 0, \ldots, k-1$, row vector A_l is the evaluation vector of a polynomial,
  say q_l, of degree at most $k - 1$.

Let P_m denote the set of polynomials of degree at most m. Conditions (LP6) and (LP7) imply that V_A is the vector space of evaluation vectors of polynomials from P_{k-1}. Theorem 1 implies that the maximal profit C in a linear program of type BAP satisfies

$$C = 2 \cdot \min_{q \in P_{k-1}} \max_{j=0,\ldots,n} |p(j) - q(j)|.$$

Let λ denote the leading coeffient of p. Thus p can be be written as sum of λX^k and a polynomial in P_{k-1}. Obviously, p is as hard to approximate within P_{k-1} as $|\lambda|X^k$. We obtain the following result:

Corollary 1. *Given linear program $LP(A, c)$ of type BAP, let p denote the poly-nomial of degree k with evaluation vector c, and λ the leading coefficient of p. Then the maximal profit C in $LP(A, c)$ satisfies*

$$C = 2 \cdot \min_{q \in P_{k-1}} \max_{j=0,\ldots,n} \left| |\lambda| j^k - q(j) \right|.$$

We introduce the notation

$$n^{\underline{k}} = n(n - 1) \cdots (n - (k - 1))$$

for so-called "falling powers" and proceed with the following result:

Lemma 1. *The linear program $LP(k, n)$ is of type BAP. The leading coefficient of the polynomial p with evaluation vector c is $(-1)^k / n^{\underline{k}}$.*

The proof of this lemma is obtained by a close inspection of $LP(k, n)$ and a (more or less) straightforward calculation.

Corollary 2. *Let $C_{k,n}$ denote the largest possible contrast in an k-out-of-n se-cret sharing scheme. Then:*

$$C_{k,n} = 2 \cdot \min_{q \in P_{k-1}} \max_{j=0,\ldots,n} |j^k / n^{\underline{k}} - q(j)|.$$

Thus, the largest possible contrast in an k-out-of-n secret sharing scheme is identical to twice the smallest "distance" between polynomial $X^k / n^{\underline{k}}$ and a polynomial in P_{k-1}, where the "distance" between two polynomials is measured as the maximum absolute difference of their evaluations on points $0, 1 \ldots, n$.

3.2   Lower and Upper Bounds

Finding the "best approximating polynomial" of X^k within P_{k-1} is a classical problem in Approximation Theory. Most of the classical results are stated for polynomials defined on interval $[-1, 1]$. In order to recall these results and to apply them to our problem at hand, the definition of the following metric will be useful:

$$d_\infty(f, g) = \max_{x \in [-1,1]} |f(x) - g(x)| \tag{1}$$

This definition makes sense for functions that are continuous on $[-1, 1]$ (in particular for polynomials). The metric implicitly used in Corollaries 1 and 2 is different because distance between polynomials is measured on a finite set of points rather than on a continuous interval. For this reason, we consider sequence

$$z_j = -1 + \frac{2j}{n} \text{ for } j = 0, \dots, n. \tag{2}$$

It forms a regular subdivision of interval $[-1, 1]$ of step width $2/n$. The following metric is a "discrete version" of d_∞:

$$d_n(f, g) = \max_{j=0,\dots,n} |f(z_j) - g(z_j)|. \tag{3}$$

Let $U_k(X) = X^k$ and $U_{k,\infty}^*$ the best approximation of U_k within P_{k-1} w.r.t. d_∞. Analogously, $U_{k,n}^*$ denotes the best approximation of U within P_{k-1} w.r.t. d_n.

$$D_{k,\infty} = d_\infty(U_k, U_{k,\infty}^*) \text{ and } D_{k,n} = d_n(U_k, U_{k,n}^*) \tag{4}$$

are the corresponding approximation errors. It is well known from Approximation Theory[3] that

$$U_{k,\infty}^*(X) = X^k - 2^{-(k-1)} T_k(X) \tag{5}$$

where T_k denotes the Chebyshev polynomial of degree k (defined and visualized in Figure 1). It is well known that $T_k = \cos(k\theta)$ is a polynomial of degree k in $X = \cos(\theta) \in [-1, 1]$ with leading coefficient 2^{k-1}. Thus, $U_{k,\infty}^*$ is indeed from P_{k-1}. Since $\max_{-1 \le x \le 1} |T_k(x)| = 1$, we get

$$D_{k,\infty} = 2^{-(k-1)}. \tag{6}$$

Unfortunately, there is no such simple formula for $D_{k,n}$ (the quantity we are interested in). It is however easy to see that the following inequalities are valid:

$$\left(1 - \frac{k^2}{n}\right) 2^{-(k-1)} \le D_{k,n} \le D_{k,\infty} = 2^{-(k-1)} \tag{7}$$

Inequality $D_{k,n} \le D_{k,\infty}$ is obvious because $d_n(f, g) \le d_\infty(f, g)$ for all f, g. The first inequality can be derived from the fact that the first derivation of T_k is

---

[3] See Chapter 1.2 in [13], for instance.

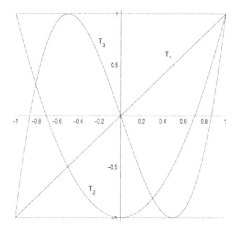

Fig. 1. The Chebyshev polynomial T_k of degree k for $k = 1, 2, 3$. $T_k(X) = \cos(k\theta)$, where $0 \leq \theta \leq \pi$ and $X = \cos(\theta)$.

bounded by k^2 on $[-1, 1]$ (applying some standard tricks). We will improve on this inequality later and present a proof for the improved statement.

Quantities $D_{k,n}$ and $C_{k,n}$ are already indirectly related by Corollary 2. In order to get the precise relation, we have to apply linear transformation $X \rightarrow \frac{n}{2}(X + 1)$, because the values attained by a function $f(X)$ on $X = 0, \dots, n$ coincide with the values attained by function $f\left(\frac{n}{2}(X + 1)\right)$ on $X = z_0, \dots, z_n$. This transformation, applied to a polynomial of degree k with leading coefficient λ, leads to a polynomial of the same degree with leading coefficient $\lambda \left(\frac{n}{2}\right)^k$. The results corresponding to Corollaries 1 and 2 now read as follows:

Corollary 3. *Given a linear program $LP(A, c)$ of type BAP, let p denote the polynomial of degree k with evaluation vector c, and λ the leading coefficient of p. Then the maximal profit C in $LP(A, c)$ satisfies*

$$C = 2 \cdot |\lambda| \left(\frac{n}{2}\right)^k D_{k,n}.$$

Plugging in $(-1)^k / n^{\underline{k}}$ for λ, we obtain

Corollary 4. *The largest possible contrast in an k-out-of-n secret sharing scheme satisfies*

$$C_{k,n} = \frac{n^k}{n^{\underline{k}}} 2^{-(k-1)} D_{k,n}.$$

Since $D_{k,\infty} = 2^{-(k-1)}$, we get the following result:

Corollary 5. *The limit of the largest possible contrast in an k-out-of-n secret sharing scheme, when n approaches infinity, satisfies*

$$C_{k,\infty} = \lim_{n \to \infty} C_{k,n} = 4^{-(k-1)}.$$

The derivation of $C_{k,\infty}$ from $D_{k,\infty}$ profited from the classical Equation (6) from Approximation Theory. For $n = k$, we can go the other way and derive $D_{k,k}$ from the fact (see [10]) that the largest possible contrast in an k-out-of-k secret sharing scheme is $2^{-(k-1)}$:

$$C_{k,k} = 2^{-(k-1)} \tag{8}$$

Applying Corollary 4, we obtain

$$D_{k,k} = \frac{k!}{k^k} \tag{9}$$

According to Stirling's formula, this quantity is asymptotically equal to $\sqrt{2\pi k}e^{-k}$. Equation (9) presents the precise value for the smallest possible approximation error when X^k is approximated by a polynomial of degree $k - 1$ or less, and the distance between polynomials is measured by metric d_k.

Sequence $C_{k,n}$ monotonically decreases with n because the secret sharing scheme becomes harder to design when more people are going to share the secret (and threshold k is fixed). Thus, the unknown value for $C_{k,n}$ must be somewhere between $C_{k,\infty} = 4^{-(k-1)}$ and $C_{k,k} = 2^{-(k-1)}$. We don't expect the sequence $D_{k,n}$ to be perfectly monotonous. However, we know that $D_{k,n} \le D_{k,\infty}$. If n is a multiple of k, the regular subdivision of $[-1, 1]$ with step width $2/n$ is a refinement of the regular subdivision of $[-1, 1]$ with step width $2/k$. This implies $D_{k,n} \ge D_{k,k}$.

Figure 2 presents an overview over the results obtained so far. An edge from a to b with label s should be interpreted as $b = s \cdot a$. For instance, the edges with labels $r_{k,n}, r'_{k,n}, s'_{k,n}, s_{k,n}$ represent the equations

$$C_{k,n} = r_{k,n} \cdot C_{k,\infty} \text{ with } r_{k,n} \ge 1,$$
$$C_{k,k} = r'_{k,n} \cdot C_{k,n} \text{ with } r'_{k,n} \ge 1,$$
$$D_{k,n} = s'_{k,n} \cdot D_{k,k} \text{ with } s'_{k,n} \ge 1 \text{ if } n \text{ is a multiple of } k,$$
$$D_{k,\infty} = s_{k,n} \cdot D_{k,n} \text{ with } s_{k,n} \ge 1,$$

respectively. The edges between $C_{k,n}$ and $D_{k,n}$ explain how $D_{k,n}$ is derived from $C_{k,n}$ and vice versa, i.e., these edges represent Corollary 4. Figure 2 can be used to obtain approximations for the unknown parameters $r_{k,n}, r'_{k,n}, s'_{k,n}, s_{k,n}$. The simple path from $C_{k,\infty} = 4^{-(k-1)}$ to $D_{k,\infty} = 2^{-(k-1)}$ corresponds to equation

$$2^{-(k-1)} = s_{k,n} \cdot 2^{k-1} \frac{n^{\underline{k}}}{n^k} \cdot r_{k,n} \cdot 4^{-(k-1)}.$$

Using $r_{k,n} \ge 1, s_{k,n} \ge 1$ and performing some cancellation, we arrive at

$$r_{k,n} \cdot s_{k,n} = \frac{n^k}{n^{\underline{k}}}. \tag{10}$$

A similar computation associated with the simple path from $D_{k,k}$ to $C_{k,k}$ leads to

$$r'_{k,n} \cdot s'_{k,n} = \frac{k^k n^{\underline{k}}}{k! n^k} = \left(\frac{k}{n}\right)^k \binom{n}{k}. \tag{11}$$

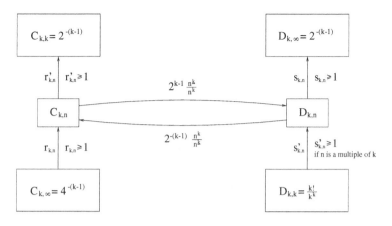

Fig. 2. Sequence $C_{k,n}$, sequence $D_{k,n}$ and relations between them.

The following bounds on $C_{k,n}$ and $D_{k,n}$ are now evident from Figure 2 and (10):

$$4^{-(k-1)} \le C_{k,n} = r_{k,n}4^{-(k-1)} \le \frac{n^k}{n^{\underline{k}}}4^{-(k-1)} \tag{12}$$

$$\frac{n^{\underline{k}}}{n^k}2^{-(k-1)} \le \frac{2^{-(k-1)}}{s_{k,n}} = D_{k,n} \le 2^{-(k-1)} \tag{13}$$

In both cases, the upper bound exceeds the lower bound by factor $n^k/n^{\underline{k}}$ only (approaching 1 when n approaches infinity).[4] An elementary computation[5] shows that $1 - k^2/n < n^{\underline{k}}/n^k \le 1$ holds for all $1 \le k \le n$. Thus, (13) improves on the classical Inequality (7) from Approximation Theory.

Although bounds (12) and (13) are excellent for large n, they are quite poor when n comes close to k. In this case however, we obtain from Figure 2 and (11)

$$\frac{(n/k)^k}{\binom{n}{k}}2^{-(k-1)} \le C_{k,n} \le 2^{-(k-1)} \ , \tag{14}$$

$$\frac{k!}{k^k} \le D_{k,n} \le \frac{n^{\underline{k}}}{n^k} \ , \tag{15}$$

where the first inequality in (15) is only guaranteed if n is a multiple of k. These bounds are tight for $n = k$.

---

[4] Because of (10), the two gaps cannot be maximal simultaneously. For instance, at least one of the upper bounds exceeds the corresponding lower bound at most by factor $\sqrt{n^k/n^{\underline{k}}}$.

[5] making use of $e^{-2x} \le 1 - x \le e^{-x}$, where the first inequality holds for all $x \in [0, 1/2]$ and the second for all $x \in \Re$

3.3 Discussion

Paper [5] presented some explicit formulas for $C_{k,n}$, but only for small values of k. For instance, it was shown that $C_{k,n}$ matches $4^{-(k-1)} \cdot n^k / n^{\underline{k}}$ (which is the theoretical upper bound, see (12)) if $k = 2$ and n even and if $k = 3$ and n is a multiple of 4.

However, computer experiments (computing $C_{k,n}$ by solving $LP(k,n)$ and comparing it with the theoretical upper bound from (12)) support the conjecture that there is no such coincidence for most other choices of k, n. The goal of this subsection is to provide a simple explanation for this phenomenon. Exhibiting basic results from Approximation Theory concerning best approximating polynomials on finite subsets of the real line (see, e.g., Theorem 1.7 and Theorem 1.11 from [13]), it is possible to derive the following

Theorem 2. *It holds that $C_{k,n} = 4^{-(k-1)} \cdot n^k / n^{\underline{k}}$ iff $E_k \subseteq Z_n$, where*

$$E_k = \left\{ \cos\left(\frac{(k-i)\pi}{k}\right) \mid i = 0, \ldots, k \right\},$$

$$Z_n = \{z_0, \ldots, z_n\} = \left\{ -1 + \frac{2i}{n} \mid i = 0, \ldots, n \right\}.$$

Due to lack of space, for the proof of this result we refer to the journal version of this paper. It is quite straightforward to derive that $E_2 \subseteq Z_n$ iff n is even, that $E_3 \subseteq Z_n$ iff n is divisible by 4, and that $E_k \nsubseteq Z_n$ for all n and ≥ 4 as E_k contains irrational numbers. Consequently,

Corollary 6. *It holds that $C_{k,n} = 4^{-(k-1)} \cdot n^k / n^{\underline{k}}$ iff $k = 2$ and n is even or if $k = 3$ and n is a multiple of 4.*

We conclude the paper with a final remark and an open problem. Based on the results of this paper, Kuhlmann and Simon [6] were able to design arbitrary k-out-of-n secret sharing schemes with asymptotically optimal contrast. More precisely, the contrast achieved by their schemes is optimal up to a factor of at most $1 - k^2/n$. For moderate values of k and n, these schemes are satisfactory. For large values of n, they use too many subpixels. It is an open problem to determine (as precise as possible) the tradeoff between the contrast (which should be large) and the number of subpixels (which should be small).

References

1. G. Ateniese, C. Blundo, A. De Santis, D. R. Stinson, *Visual Cryptography for General Access Structures*, Proc. of ICALP 96, Springer, 416-428, 1996.
2. C. Blundo, A. De Santis, D. R. Stinson, *On the contrast in visual cryptography schemes*, Journal of Cryptology 12 (1999), 261-289.
3. C. Blundo, P. De Arco, A. De Santis and D. R. Stinson, *Contrast optimal threshold visual cryptography schemes*, Technical report 1998 (see http://www.cacr.math.uwaterloo.ca/~dstinson/#Visual Cryptography) To appear in SIAM Journal on Discrete Mathematics.

4. S. Droste, *New Results on Visual Cryptography*, in "Advances in Cryptology" - CRYPTO '96, Springer, pp. 401-415, 1996.
5. T. Hofmeister, M. Krause, H. U. Simon, *Contrast-Optimal k out of n Secret Sharing Schemes in Visual Cryptography*, in "Proceedings of the 3rd International Conference on Computing and Combinatorics" - COCOON '97, Springer, pp. 176-186, 1997. Full version will appear in Theoretical Computer Science.
6. C. Kuhlmann, H. U. Simon, *Construction of Visual Secret Sharing Schemes with Almost Optimal Contrast*. Submitted for publication.
7. R. Lidl, H. Niederreiter, *Introduction to finite fields and their applications*, Cambridge University Press, 1994.
8. N. Linial, N. Nisan, *Approximate inclusion-exclusion*, Combinatorica 10, 349-365, 1990.
9. J. H. van Lint, R. M. Wilson, *A course in combinatorics*, Cambridge University Press, 1996.
10. M. Naor, A. Shamir, *Visual Cryptography*, in "Advances in Cryptology - Eurocrypt 94", Springer, 1-12, 1995.
11. M. Naor, A. Shamir, *Visual Cryptography II: Improving the Contrast via the Cover Base*, in Proc. of "Security protocols: international workshop 1996", Springer LNCS 1189, 69-74, 1997.
12. Christos H. Papadimitriou and Kenneth Steiglitz, *Combinatorial Optimization: Algorithms and Complexity*, Prentice Hall, 1982.
13. Theodore J. Rivlin, *An Introduction to the Approximation of Functions*, Blaisdell Publishing Company, 1969.

Average-Case Analysis of Rectangle Packings

E.G. Coffman, Jr.[1], George S. Lueker[2], Joel Spencer[3], and Peter M. Winkler[4]

[1] New Jersey Institute of Technology, Newark, NJ 07102
[2] University of California, Irvine, Irvine, CA 92697
[3] New York University, New York, NY 10003
[4] Bell Labs, Lucent Technologies, Murray Hill, NJ 07974

Abstract. We study the average-case behavior of algorithms for finding a maximal disjoint subset of a given set of rectangles. In the probability model, a random rectangle is the product of two independent random intervals, each being the interval between two points drawn uniformly at random from $[0,1]$. We have proved that the expected cardinality of a maximal disjoint subset of n random rectangles has the tight asymptotic bound $\Theta(n^{1/2})$. Although tight bounds for the problem generalized to $d > 2$ dimensions remain an open problem, we have been able to show that $\Omega(n^{1/2})$ and $O((n \log^{d-1} n)^{1/2})$ are asymptotic lower and upper bounds. In addition, we can prove that $\Theta(n^{d/(d+1)})$ is a tight asymptotic bound for the case of random cubes.

1 Introduction

We estimate the expected cardinality of a maximal disjoint subset of n rectangles chosen at random in the unit square. We say that such a subset is a *packing* of the n rectangles, and stress that a rectangle is specified by its position as well as its sides; it can not be freely moved to any position such as in strip packing or two-dimensional bin packing (see [2] and the references therein for the probabilistic analysis of algorithms for these problems). A random rectangle is the product of two independent random intervals on the coordinate axes; each random interval in turn is the interval between two independent random draws from a distribution G on $[0,1]$.

This problem is an immediate generalization of the one-dimensional problem of packing random intervals [3]. And it generalizes in an obvious way to packing random rectangles (boxes) in $d > 2$ dimensions into the d-dimensional unit cube, where each such box is determined by $2d$ independent random draws from $[0,1]$, two for every dimension. A later section also studies the case of random cubes in $d \geq 2$ dimensions. For this case, to eliminate irritating boundary effects that do not influence asymptotic behavior, we wrap around the dimensions of the unit cube to form a toroid. In terms of an arbitrarily chosen origin, a random cube is then determined by $d+1$ random variables, the first d locating the vertex closest to the origin, and the last giving the size of the cube, and hence the coordinates of the remaining $2^d - 1$ vertices. Each random variable is again an independent random draw from the distribution G.

G. Gonnet, D. Panario, and A. Viola (Eds.): LATIN 2000, LNCS 1776, pp. 292–297, 2000.
© Springer-Verlag Berlin Heidelberg 2000

Applications of our model appear in jointly scheduling multiple resources, where customers require specific "intervals" of a resource or they require a resource for specific intervals of time. An example of the former is a linear communication network and an example of the latter is a reservation system. In a linear network, we have a set S of call requests, each specifying a pair of endpoints (calling parties) that define an interval of the network. If we suppose also that each request gives a future time interval to be reserved for the call, then a call request is a rectangle in the two dimensions of space and time. In an unnormalized and perhaps discretized form, we can pose our problem of finding the expected value of the number of requests in S that can be accommodated.

The complexity issue for the combinatorial version of our problem is easily settled. Consider the two-dimensional case, and in particular a collection of equal size squares. In the associated intersection graph there is a vertex for each square and an edge between two vertices if and only if the corresponding squares overlap. Then our packing problem specialized to equal size squares becomes the problem of finding maximal independent sets in intersection graphs. It is easy to verify that this problem is NP-complete. For example, one can use the approach in [1] which was applied to equal size circles; the approach is equally applicable to equal size squares. We conclude that for any fixed $d \geq 2$, our problem of finding maximal disjoint subsets of rectangles is NP-complete, even for the special case of equal size cubes. As a final note, we point out that, in contrast to higher dimensions, the one-dimensional (interval) problem has a polynomial-time solution [3].

Let S_n be a given set of random boxes, and let C_n be the maximum cardinality of any set of mutually disjoint boxes taken from S_n. After preliminaries in the next section, Section 3 proves that, in the case of cubes in $d \geq 2$ dimensions, $\mathsf{E}[C_n] = \Theta(n^{d/(d+1)})$, and Section 4 proves that, in the case of boxes in d dimensions, $\mathsf{E}[C_n] = \Omega(n^{1/2})$ and $\mathsf{E}[C_n] = O((n \log^{d-1} n)^{1/2})$. Section 5 contains our strongest result, which strengthens the above bounds for $d = 2$ by presenting a $O(n^{1/2})$ tight upper bound. We sketch a proof that relies on a similar result for a reduced, discretized version of the two dimensional problem.

2 Preliminaries

We restrict the packing problem to continuous endpoint distributions G. Within this class, our results are independent of G, because the relevant intersection properties of G depend only on the relative ordering of the points that determine the intervals in each dimension. Thus, for simplicity, we assume hereafter that G is the uniform distribution on $[0, 1]$.

It is also easily verified that we can Poissonize the problem without affecting our results. In this version, the number of rectangles is a Poisson distributed random variable T_n with mean n, and we let $C(n)$ denote the number packed in a maximal disjoint subset. We will continue to parenthesize arguments in the notation of the Poissonized model so as to distinguish quantities like C_n in the model where the number of rectangles to pack is fixed at n.

Let $X_1, \ldots, X_n$ be i.i.d. with a distribution F concentrated on $[0,1]$. We assume that F is *regularly varying* at 0 in that it is strictly increasing and that, for some $\xi > 0$, some constants $K, K' \in (0,1)$, and all $x \in (0, \xi)$, it satisfies $\frac{F(x/2)}{F(x)} \in [K, K']$. For $(s_n \in (0,1]; n \geq 1)$ a given sequence, let $N_n(F, s_n)$ be the maximum number of the X_i that can be chosen such that their sum is at most ns_n on average. Equivalently, in terms of expected values, N_n is such that the sum of the smallest N_n of the X_i is bounded by ns_n, but the sum of the smallest $N_n + 1$ of the X_i exceeds ns_n.

Standard techniques along with a variant of Bernstein's Theorem suffice to prove the following technical lemma.

Lemma 1. *With F and $(s_n, n \geq 1)$ as above, let x_n be the solution to*

$$s_n = \int_0^{x_n} x \, dF(x), \tag{1}$$

and assume the s_n are such that $\lim_{n \to \infty} x_n = 0$. Then if $\lim_{n \to \infty} nF(x_n) = \infty$ and $nF(x_n) = \Omega(\log^2 s_n^{-1})$, we have

$$\mathsf{E}[N_n(F, s_n)] \sim nF(x_n). \tag{2}$$

3   Random Cubes

The optimum packing of random cubes is readily analyzed. We work with a d-dimensional unit cube, and allow (toroidal) wrapping in all axes. The n cubes are generated independently as follows: First a vertex $(v_1, v_2, \ldots, v_d)$ is generated by drawing each c_i independently from the uniform distribution on $[0,1]$. Then one more value w is drawn independently, again uniformly from $[0,1]$. The cube generated is

$$[v_1, v_1 + w) \times [v_2, v_2 + w) \times \cdots \times [v_d, v_d + w),$$

where each coordinate is taken modulo 1. In this set-up, we have the following result.

Theorem 1. *The expected cardinality of a maximum packing of n random cubes is $\Theta(n^{d/(d+1)})$.*

Proof: For the lower bound consider the following simple heuristic. Subdivide the cube into c^{-d} cells with sides

$$c = \alpha n^{-1/(d+1)},$$

where α is a parameter that may be chosen to optimize constants. For each cell $\mathcal{C}$, if there are any generated cubes contained in $\mathcal{C}$, include one of these in the packing. Clearly, all of the cubes packed are nonoverlapping.

One can now show that the probability that a generated cube fits into a particular cell $\mathcal{C}$ is $c^{d+1}/(d+1)$, and so the probability that $\mathcal{C}$ remains empty after generating all n cubes is

$$\left(1 - \frac{c^{d+1}}{d+1}\right)^n = \left(1 - \frac{\alpha^{d+1}}{n(d+1)}\right)^n \sim \exp\left(-\frac{\alpha^{d+1}}{d+1}\right).$$

Since the number of cells is $1/c^d = n^{d/(d+1)}/\alpha^d$, the expectation of the total number of cubes packed is

$$\alpha^{-d}\left(1 - \exp\left(-\frac{\alpha^{d+1}}{d+1}\right)\right) n^{d/(d+1)},$$

which gives the desired lower bound.

The upper bound is based on the simple observation that the sum of the volumes of the packed cubes is at most 1. First we consider the probability distribution of the volume of a single generated cube. The side of this cube is a uniform random variable U over $[0,1]$. Thus the probability that its volume is bounded by z is

$$F(z) = \Pr\left\{U^d \leq z\right\} = \Pr\left\{U \leq z^{1/d}\right\} = z^{1/d}.$$

Then applying Lemma 1 with $s_n = 1/n$, $x_n = ((d+1)/n)^{d/(d+1)}$, and $F(x_n) = ((d+1)/n)^{1/(d+1)}$, we conclude that the expected number of cubes selected before their total volume exceeds 1 is asymptotic to $(d+1)^{1/(d+1)}n^{d/(d+1)}$, which gives the desired matching upper bound. ∎

4   Bounds for $d \geq 2$ Dimensional Boxes

Let $\mathcal{H}_d$ denote the unit hypercube in $d \geq 1$ dimensions. The approach of the last section can also be used to prove asymptotic bounds for the case of random boxes in $\mathcal{H}_d$.

Theorem 2. *Fix d and draw n boxes independently and uniformly at random from $\mathcal{H}_d$. The maximum number that can be packed is asymptotically bounded from below by $\Omega(\sqrt{n})$ and from above by $O(\sqrt{n \ln^{d-1} n})$.*

Proof sketch: The lower bound argument is the same as that for cubes, except that $\mathcal{H}_d$ is partitioned into cells with sides on the order of $1/n^{2d}$. It is easy to verify that, on average, there is a constant fraction of the $n^{1/2}$ cells in which each cell wholly contains at least one of the given rectangles.

To apply Lemma 1 in a proof of the upper bound, one first conducts an asymptotic analysis of the distribution F_d, the volume of a d-dimensional box, which shows that

$$dF_d(x) \sim \frac{2^d}{(d-1)!} \ln^{d-1} x^{-1}.$$

Then, with $s_n = 1/n$, we obtain

$$x_n \sim \sqrt{(d-1)!/(n \ln^{d-1} n)} \quad \text{and} \quad F_d(x_n) \sim 2\sqrt{\ln^{d-1} n/((d-1)!n)}.$$

which together with Lemma 1 yields the desired upper bound. ∎

5   Tight Bound for $d = 2$

Closing the gaps left by the bounds on $\mathsf{E}[C_n]$ for $d \geq 3$ remains an interesting open problem. However, one can show that the lower bound for $d \geq 2$ is tight, i.e., $\mathsf{E}[C_n] = \Theta(n^{1/2})$. To outline the proof of the $O(n^{1/2})$ bound, we first introduce the following reduced, discretized version. A *canonical* interval is an interval that, for some $i \geq 0$, has length 2^{-i} and has a left endpoint at some multiple of 2^{-i}. A canonical rectangle is the product of two canonical intervals. In the reduced, rectangle-packing problem, a Poissonized model of canonical rectangles is assumed in which the number of rectangles of area a is Poisson distributed with mean λa^2, independently for each possible a. Let $C^*(\lambda)$ denote the cardinality of a maximum packing for an instance of the reduced problem with parameter λ.

Note that there are $i+1$ shapes possible for a rectangle of area 2^{-i}, and that for each of these shapes there are 2^i canonical rectangles. The mean number of each of these is $\lambda/2^{2i}$. Thus, the total number $T(\lambda)$ of rectangles in the reduced problem with parameter λ is Poisson distributed with mean

$$\sum_{i=0}^{\infty}(i+1)2^i(\lambda 2^{-2i}) = \lambda \sum_{i=0}^{\infty}(i+1)2^{-i} = 4\lambda. \tag{3}$$

To convert an instance of the original problem to an instance of the reduced problem, we proceed as follows. It can be seen that any interval in $\mathcal{H}_1$ contains either one or two canonical intervals of maximal length. Let the *canonical subinterval* I' of an interval I be the maximal canonical interval in I, if only one exists, and one such interval chosen randomly otherwise. A straightforward analysis shows that a canonical subinterval $I = [k2^{-i}, (k+1)2^{-i})$ has probability 0 if it touches a boundary of $\mathcal{H}_1$, and has probability $\frac{3}{2}2^{-2i}$, otherwise. The *canonical subrectangle* R' of a rectangle R is defined by applying the above separately to both coordinates. Extending the calculations to rectangles, we get $\frac{9}{4}a^2$ as the probability of a canonical subrectangle R of area a, if R does not touch the boundary of $\mathcal{H}_2$, and 0 otherwise. Now consider a random family of rectangles $\{R_i\}$, of which a maximum of $C(n)$ can be packed in $\mathcal{H}_2$. This family generates a random family of canonical subrectangles $\{R_i'\}$. The maximum number $C'(n)$ of the R_i' that can be packed trivially satisfies $C(n) \leq C'(n)$. Since the number of each canonical subrectangle of area a that does not touch a boundary is Poisson distributed with mean $9na^2/4$, we see that an equivalent way to generate a random family $\{R_i'\}$ is simply to remove from a random instance of the reduced problem with parameter $9n/4$ all those rectangles touching a boundary. It follows easily that $\mathsf{E}C(n) \leq \mathsf{E}C'(n) \leq \mathsf{E}C^*(9n/4)$ so if we can prove that $\mathsf{E}C^*(9n/4)$ or more simply $\mathsf{E}C^*(n)$, has the $O(n^{1/2})$ upper bound, then we are done.

The following observations bring out the key recursive structure of maximal packings in the reduced problem. Let Z_1 be the maximum number of rectangles that can be packed if we disallow packings that use rectangles spanning the height of the square. Define Z_2 similarly when packings that use rectangles spanning the width of the square are disallowed. By symmetry, Z_1 and Z_2 have the same distribution, although they may not be independent. To find this distribution, we begin by noting that (i) a rectangle spanning the width of $\mathcal{H}_2$ and one spanning the height of $\mathcal{H}_2$ must intersect and hence can not coexist in a packing; (ii) rectangles spanning the height of $\mathcal{H}_2$ are the only rectangles crossing the horizontal line separating the top and bottom halves of $\mathcal{H}_2$ and rectangles spanning the width of $\mathcal{H}_2$ are the only ones crossing the vertical line separating the left and right halves of $\mathcal{H}_2$. It follows that, if a maximum cardinality packing is not just a single 1×1 square, then it consists of a pair of disjoint maximum cardinality packings, one in the bottom half and one in the top half of $\mathcal{H}_2$, or a similar pair of subpackings, one in the left half and one in the right half of $\mathcal{H}_2$. After rescaling, these subpackings become solutions to our original problem on $\mathcal{H}_2$ with the new parameter λ times the square of half the area of $\mathcal{H}_2$, i.e., $\lambda/4$. We conclude that Z_1 and Z_2 are distributed as the sum of two independent samples of $C^*(\lambda/4)$, and that

$$C^*(\lambda) \leq Z_0 + \max(Z_1, Z_2),$$

where Z_0 is the indicator function of the event that the entire square is one of the given rectangles. Note that Z_0 is independent of Z_1 and Z_2.

To exploit the above recursion, it is convenient to work in terms of the generating function, $S(\lambda) := \mathsf{E}e^{\alpha C^*(\lambda)}$. One can show that $S(\lambda) \leq 2e^{\alpha} \left(S(\lambda/4)\right)^2$, and that a solution to this relation along with the inequality $\mathsf{E}[C^*(\lambda)] \leq \alpha^{-1} \ln \mathsf{E}[e^{\alpha C_*(\lambda)}]$ yields the desired bound, $\mathsf{E}[C^*(\lambda)] = O(n^{1/2})$.

Acknowledgment

In the early stages of this research, we had useful discussions with Richard Weber, which we gratefully acknowledge.

References

1. Clark, B. N., Colburn, C. J., and Johnson, D. S., "Unit Disk Graphs," *Discrete Mathematics*, **86**(1990), 165-177.
2. Coffman, E. G., Jr. and Lueker, G. S., *An Introduction to the Probabilistic Analysis of Packing and Partitioning Algorithms*, Wiley & Sons, New York, 1991.
3. Justicz, J., Scheinermann, E. R., and Winkler, P. M., "Random Intervals," *Amer. Math. Monthly*, **97**(1990), 881-889.

Heights in Generalized Tries and PATRICIA Tries[*]

Charles Knessl[1] and Wojciech Szpankowski[2]

[1] Dept. Mathematics, Statistics & Computer Science, University of Illinois, Chicago, Illinois 60607-7045, USA
[2] Dept. Computer Science, Purdue University, W. Lafayette, IN 47907, USA

Abstract. We consider digital trees such as (generalized) tries and PA-TRICIA tries, built from n random strings generated by an unbiased memoryless source (i.e., all symbols are equally likely). We study limit laws of the height which is defined as the longest path in such trees. For tries, in the region where most of the probability mass is concentrated, the asymptotic distribution is of extreme value type (i.e., double exponential distribution). Surprisingly enough, the height of the PATRICIA trie behaves quite differently in this region: It exhibits an exponential of a Gaussian distribution (with an *oscillating term*) around the most probable value $k_1 = \lfloor \log_2 n + \sqrt{2 \log_2 n} - \frac{3}{2} \rfloor + 1$. In fact, the asymptotic distribution of PATRICIA height concentrates on one or two points. For most n all the mass is concentrated at k_1, however, there exist subsequences of n such that the mass is on the two points $k_1 - 1$ and k_1, or k_1 and $k_1 + 1$. We derive these results by a combination of analytic methods such as generating functions, Mellin transform, the saddle point method and ideas of applied mathematics such as linearization, asymptotic matching and the WKB method.

1 Introduction

Data structures and algorithms on words have experienced a new wave of interest due to a number of novel applications in computer science, communications, and biology. These include dynamic hashing, partial match retrieval of multidimensional data, searching and sorting, pattern matching, conflict resolution algorithms for broadcast communications, data compression, coding, security, genes searching, DNA sequencing, genome maps, IP-addresses lookup on the internet, and so forth. To satisfy these diversified demands various data structures were proposed for these algorithms. Undoubtly, the most popular data structures for algorithms on words are digital trees [9,12] (e.g., tries, PATRICIA tries, digital search trees), and suffix trees [6,18].

The most basic digital tree is known as a *trie* (the name comes from re*trie*val). The primary purpose of a trie is to store a set $\mathcal{S}$ of strings (words, keys), say

---

[*] The work was supported by NSF Grant DMS-93-00136 and DOE Grant DE-FG02-93ER25168, as well as by NSF Grants NCR-9415491, NCR-9804760.

G. Gonnet, D. Panario, and A. Viola (Eds.): LATIN 2000, LNCS 1776, pp. 298–307, 2000.

$\mathcal{S} = \{X_1, \ldots, X_n\}$. Each word $X = x_1 x_2 x_3 \ldots$ is a finite or infinite string of symbols taken from a finite alphabet. Throughout the paper, we deal only with the binary alphabet $\{0, 1\}$, but all our results should be extendable to a general finite alphabet. A string will be stored in a leaf of the trie. The trie over $\mathcal{S}$ is built recursively as follows: For $|\mathcal{S}| = 0$, the trie is, of course, empty. For $|\mathcal{S}| = 1$, $trie(\mathcal{S})$ is a single node. If $|\mathcal{S}| > 1$, $\mathcal{S}$ is split into two subsets $\mathcal{S}_0$ and $\mathcal{S}_1$ so that a string is in $\mathcal{S}_j$ if its first symbol is $j \in \{0, 1\}$. The tries $trie(\mathcal{S}_0)$ and $trie(\mathcal{S}_1)$ are constructed in the same way except that at the k-th step, the splitting of sets is based on the k-th symbol of the underlying strings.

There are many possible variations of the trie. One such variation is the b-trie, in which a leaf is allowed to hold as many as b strings (cf. [12,18]). A second variation of the trie, the *PATRICIA trie* eliminates the waste of space caused by nodes having only one branch. This is done by collapsing one-way branches into a single node. In a *digital search tree* (in short DST) strings are directly stored in nodes, and hence external nodes are eliminated. The branching policy is the same as in tries. The reader is referred to [6,9,12] for a detailed description of digital trees. Here, we consider tries and PATRICIA tries built over n randomly generated strings of binary symbols. We assume that every symbol is equally likely, thus we are within the framework of the so called *unbiased memoryless* model. Our interest lies in establishing asymptotic distributions of the heights for random b-tries, $\mathcal{H}_n^T$, and PATRICIA tries, $\mathcal{H}_n^P$. The height is the longest path in such trees, and its distribution is of considerable interest for several applications.

We now summarize our main results. We obtain asymptotic expansions of the distributions $\Pr\{\mathcal{H}_n^T \le k\}$ (b-tries) and $\Pr\{\mathcal{H}_n^P \le k\}$ (PATRICIA tries) for three ranges of n and k. For b-tries we consider: (i) the "right–tail region" $k \to \infty$ and $n = O(1)$; (ii) the "central region" $n, k \to \infty$ with $\xi = n2^{-k}$ and $0 < \xi < b$; and (iii) the "left–tail region" $k, n \to \infty$ with $n - b2^k = O(1)$. We prove that most probability mass is concentrated in between the right tail and the central region. In particular, for real x

$$\Pr\left\{\mathcal{H}_n^T \le \frac{1+b}{b} \log_2 n + x\right\} \sim \exp\left(-\frac{1}{(b+1)!} 2^{-bx+b\langle \frac{1+b}{b} \log_2 n + x\rangle}\right),$$

where $\langle r \rangle = r - \lfloor r \rfloor$ is the fractional part of r.[1] In words, the asymptotic distribution of b-tries height around its most likely value $\frac{1+b}{b} \log_2 n$ resembles a double exponential (extreme value) distribution. In fact, due to the oscillating term $\langle \frac{1+b}{b} \log_2 n + x \rangle$ the limiting distribution does *not* exist, but one can find lim inf and lim sup of the distribution.

The height of PATRICIA tries behaves differently in the central region (i.e., where most of the probability mass is concentrated). It is concentrated at or near the most likely value $k_1 = \lfloor \log_2 n + \sqrt{2 \log_2 n} - \frac{3}{2} \rfloor + 1$. We shall prove that the asymptotic distribution around k_1 resembles an exponential of a Gaussian distribution, with an oscillating term (cf. Theorem 3). In fact, there exist

---

[1] The fractional part $\langle r \rangle$ is often denoted as $\{r\}$, but in order to avoid confusion we adopt the above notation.

subsequences of n such that the asymptotic distribution of PATRICIA height concentrates only on k_1, or on k_1 and one of the two points $k_1 - 1$ or $k_1 + 1$.

With respect to previous results, Devroye [1] and Pittel [14] established the asymptotic distribution in the central regime for tries and b-tries, respectively, using probabilistic tools. Jacquet and Régnier [7] obtained similar results by analytic methods. The most probable value, $\log_2 n$, of the height for PATRICIA was first proved by Pittel [13]. This was then improved to $\log_2 n + \sqrt{2 \log_2 n}(1 + o(1))$ by Pittel and Rubin [15], and independently by Devroye [2]. No results concerning the asymptotic distribution for PATRICIA height were reported.

The full version of this paper with all proofs can be found on http://www.cs. purdue.edu/people/spa.

2   Summary of Results

As before, we let $\mathcal{H}_n^T$ and $\mathcal{H}_n^P$ denote, respectively, the height of a b-trie and a PATRICIA trie. Their probability distributions are

$$\bar{h}_n^k = \Pr\{\mathcal{H}_n^T \le k\} \qquad \text{and} \qquad h_n^k = \Pr\{\mathcal{H}_n^P \le k\}. \tag{1}$$

We note that for tries $\bar{h}_n^k = 0$ for $n > b2^k$ (corresponding to a balanced tree), while for PATRICIA tries $h_n^k = 0$ for $n > 2^k$. In addition, for PATRICIA we have the following boundary condition: $h_n^k = 1$ for $k \ge n$. It asserts that the height in a PATRICIA trie cannot be bigger than n (due to the elimination of all one-way branches).

The distribution of b-tries satisfies the recurrence relation

$$\bar{h}_n^{k+1} = 2^{-n} \sum_{i=0}^{n} \binom{n}{i} \bar{h}_i^k \bar{h}_{n-i}^k, \qquad k \ge 0 \tag{2}$$

with the initial condition(s)

$$\bar{h}_n^0 = 1, \ n = 0, 1, 2, \ldots, b; \qquad \text{and} \qquad \bar{h}_n^0 = 0, \ n > b. \tag{3}$$

This follows from $\mathcal{H}_n^T = \max\{\mathcal{H}_i^{LT}, \mathcal{H}_{n-i}^{RT}\} + 1$, where $\mathcal{H}_i^{LT}$ and $\mathcal{H}_{n-i}^{RT}$ denote, respectively, the left subtree and the right subtree of sizes i and $n - i$, which happens with probability $2^{-n}\binom{n}{i}$. Similarly, for PATRICIA tries we have

$$h_n^{k+1} = 2^{-n+1} h_n^{k+1} + 2^{-n} \sum_{i=1}^{n-1} \binom{n}{i} h_i^k h_{n-i}^k, \ k \ge 0 \tag{4}$$

with the initial conditions

$$h_0^0 = h_1^0 = 1 \qquad \text{and} \qquad h_n^0 = 0, \ n \ge 2. \tag{5}$$

Unlike b-tries, in a PATRICIA trie the left and the right subtrees cannot be empty (which occurs with probability 2^{-n+1}).

We shall analyze these problems asymptotically, in the limit $n \to \infty$. Despite the similarity between (2) and (4), we will show that even asymptotically the two distributions behave very differently.

We first consider ordinary tries (i.e., 1-tries). It is relatively easy to solve (2) and (3) explicitly and obtain the integral representation

$$\bar{h}_n^k = \frac{n!}{2\pi i} \oint (1 + z2^{-k})^{2^k} z^{-n-1} dz \qquad (6)$$

$$= \begin{cases} 0, & n > 2^k \\ \frac{(2^k)!}{2^{nk}(2^k-n)!}, & 0 \le n \le 2^k. \end{cases}$$

Here the loop integral is for any closed circle surrounding $z = 0$.

Using asymptotic methods for evaluating integrals, or applying Stirling's formula to the second part of (6), we obtain the following.

Theorem 1. *The distribution of the height of tries has the following asymptotic expansions:*

(i) RIGHT-TAIL REGION: $k \to \infty$, $n = O(1)$

$$\Pr\{\mathcal{H}_n^T \le k\} = \bar{h}_n^k = 1 - n(n-1)2^{-k-1} + O(2^{-2k}).$$

(ii) CENTRAL REGION: $k, n \to \infty$ with $\xi = n2^{-k}$, $0 < \xi < 1$

$$\bar{h}_n^k \sim A(\xi)e^{n\phi(\xi)},$$

where

$$\phi(\xi) = \left(1 - \frac{1}{\xi}\right)\log(1 - \xi) - 1,$$

$$A(\xi) = (1 - \xi)^{-1/2}.$$

(iii) LEFT-TAIL REGION: $k, n \to \infty$ with $2^k - n = j = O(1)$

$$\bar{h}_n^k \sim n^j \frac{e^{-n-j}}{j!} \sqrt{2\pi n}.$$

This shows that there are three ranges of k and n where the asymptotic form of $\bar{h}_n^k$ is different.

We next consider the "asymptotic matching" (see [11]) between the three expansions. If we expand (i) for n large, we obtain $1 - \bar{h}_n^k \sim n^2 2^{-k-1}$. For $\xi \to 0$ we have $A(\xi) \sim 1$ and $\phi(\xi) \sim -\xi/2$ so that the result in (ii) becomes

$$A(\xi)e^{n\phi(\xi)} \sim e^{-n\xi/2} = \exp\left(-\frac{1}{2}n^2 2^{-k}\right) \sim 1 - \frac{1}{2}n^2 2^{-k} \qquad (7)$$

where the last approximation assumes that $n, k \to \infty$ in such a way that $n^2 2^{-k} \to 0$. Since (7) agrees precisely with the expansion of (i) as $n \to \infty$, we say that (i) and (ii) asymptotically match. To be precise, we say they match the leading order; higher order matchings can be verified by computing higher order terms in the asymptotic series in (i) and (ii). We can easily show that the expansion of (ii) as $\xi \to 1^-$ agrees with the expansion of (iii) as $j \to \infty$, so that (ii) and (iii) also asymptotically match. The matching verifications imply that, at least to leading order, there are no "gaps" in the asymptotics. In other words, one of the results in (i)-(iii) applies for any asymptotic limit which has k and/or n large. We recall that $\bar{h}_n^k = 0$ for $n > 2^k$ so we need only consider $k \geq \log_2 n$.

The asymptotic limits where (i)-(iii) apply are the three "natural scales" for this problem. We can certainly consider other limits (such as $k, n \to \infty$ with k/n fixed), but the expansions that apply in these limits would necessarily be limiting cases of one of the three results in Theorem 1. In particular, if we let $k, n \to \infty$ with $k - 2\log_2 n = O(1)$, we are led to

$$\bar{h}_n^k \sim \exp\left(-\frac{1}{2}n^2 2^{-k}\right) = \exp\left(-\frac{1}{2}\exp(-k\log 2 + 2\log n)\right). \tag{8}$$

This result is well-known (see [1,7]) and corresponds to a limiting double exponential (or extreme value) distribution. However, according to our discussion, $k = 2\log_2 n + O(1)$ is *not* a natural scale for this problem. The scale $k = \log_2 n + O(1)$ (where (ii) applies) is a natural scale, and the result in (8) may be obtained as a limiting case of (ii), by expanding (ii) for $\xi \to 0$.

We next generalize Theorem 1 to arbitrary b, and obtain the following result whose proof can be found in our full paper available on http://www.cs.purdue.edu/people/spa.

Theorem 2. *The distribution of the height of b-tries has the following asymptotic expansions for fixed b:*

(i) RIGHT-TAIL REGION: $k \to \infty$, $n = O(1)$:

$$\Pr\{\mathcal{H}_n^T \leq k\} = \bar{h}_n^k \sim 1 - \frac{n!}{(b+1)!(n-b-1)!}2^{-kb}.$$

(ii) CENTRAL REGIME: $k, n \to \infty$ with $\xi = n2^{-k}$, $0 < \xi < b$:

$$\bar{h}_n^k \sim A(\xi; b)e^{n\phi(\xi; b)},$$

where

$$\phi(\xi; b) = -1 - \log \omega_0 + \frac{1}{\xi}\left(b\log(\omega_0 \xi) - \log b! - \log\left(1 - \frac{1}{\omega_0}\right)\right),$$

$$A(\xi; b) = \frac{1}{\sqrt{1 + (\omega_0 - 1)(\xi - b)}}.$$

In the above, $\omega_0 = \omega_0(\xi; b)$ is the solution to

$$1 - \frac{1}{\omega_0} = \frac{(\omega_0 \xi)^b}{b!\left(1 + \omega_0 \xi + \frac{\omega_0^2 \xi^2}{2!} + \cdots + \frac{\omega_0^b \xi^b}{b!}\right)}.$$

(iii) LEFT-TAIL REGION: $k, n \to \infty$ with $j = b2^k - n$

$$\bar{h}_n^k \sim \sqrt{2\pi n}\frac{n^j}{j!}b^n \exp\left(-(n+j)\left(1 + b^{-1}\log b!\right)\right)$$

where $j = O(1)$.

When $b = 1$ we can easily show that Theorem 2 reduces to Theorem 1 since in this case $w_0(\xi; 1) = 1/(1 - \xi)$. We also can obtain w_0 explicitly for $b = 2$, namely:

$$w_0(\xi; 2) = \frac{2}{1 - \xi + \sqrt{1 + 2\xi - \xi^2}}. \tag{9}$$

For arbitrary b, we have $w_0 \to \infty$ as $\xi \to b^-$ and $w_0 \to 1$ as $\xi \to 0^+$. More precisely,

$$w_0 = 1 - \frac{\xi^b}{b!} + O(\xi^{b+1}), \qquad \xi \to 0 \tag{10}$$

$$w_0 = \frac{1}{b - \xi} + \frac{b - 1}{b} + O(b - \xi), \quad \xi \to b. \tag{11}$$

Using (10) and (11) we can also show that the three parts of Theorem 2 asymptotically match. In particular, by expanding part (ii) as $\xi \to 0$ we obtain

$$\Pr\{\mathcal{H}_n^T \leq k\} \sim A(\xi)e^{n\phi(\xi)} \sim \exp\left(-\frac{n\xi^b}{(b+1)!}\right) \qquad \xi \to 0$$

$$= \exp\left(-\frac{n^{1+b}2^{-kb}}{(b+1)!}\right). \tag{12}$$

This yields the well-known (see [7,14]) asymptotic distribution of b-tries. We note that, for $k, n \to \infty$, (12) is $O(1)$ for $k - (1 + 1/b)\log_2 n = O(1)$. More precisely, let us estimate the probability mass of $\mathcal{H}_n^T$ around $(1 + 1/b)\log_2 n + x$ where x is a real fixed value. We observe from (12) that

$$\Pr\{\mathcal{H}_n^T \leq (1 + 1/b)\log_2 n + x\} = \Pr\{\mathcal{H}_n^T \leq \lfloor(1 + 1/b)\log_2 n + x\rfloor\}$$

$$\sim \exp\left(-\frac{1}{(1 + b)!}2^{-bx + b\langle(1+b)/b \cdot \log_2 n + x\rangle}\right) \tag{13}$$

where $\langle x \rangle$ is the fractional part of x, that is, $\langle x \rangle = x - \lfloor x \rfloor$.

Corollary 1. *While the limiting distribution of the height for b-tries does not exist, the following lower and upper envelopes can be established*

$$\liminf_{n\to\infty} \Pr\{\mathcal{H}_n^T \leq (1 + 1/b)\log_2 n + x\} = \exp\left(-\frac{1}{(1 + b)!}2^{-b(x-1)}\right),$$

$$\limsup_{n\to\infty} \Pr\{\mathcal{H}_n^T \leq (1 + 1/b)\log_2 n + x\} = \exp\left(-\frac{1}{(1 + b)!}2^{-bx}\right)$$

for fixed real x.

We next turn our attention to PATRICIA tries. Using ideas of applied mathematics, such as linearization and asymptotic matching, we obtain the following. The derivation can be found on `http://www.cs.purdue.edu/people/spa` where we make certain assumptions about the forms of the asymptotic expansions, as well as the asymptotic matching between the various scales.

Theorem 3. *The distribution of PATRICIA tries has the following asymptotic expansions:*
(i) RIGHT-TAIL REGIME: $k, n \to \infty$ *with* $n - k = j = O(1)$, $j \geq 2$

$$\Pr\{\mathcal{H}_n^P \leq n - j\} = h_n^{n-j} \sim 1 - \rho_0 K_j \cdot n! \cdot 2^{-n^2/2+(j-3/2)n},$$

where

$$K_j = \frac{1}{j!} 2^{-j^2/2} 2^{3j/2} C_j, \tag{14}$$

$$C_j = \frac{j!}{2\pi i} \oint \frac{z^{1-j} e^z}{2} \prod_{m=0}^{\infty} \left(\frac{1 - \exp(-z2^{-m-1})}{z2^{-m-1}} \right) dz \tag{15}$$

and $\rho_0 = \prod_{\ell=2}^{\infty} (1 - 2^{-\ell})^{-1} = 1.73137\ldots$
(ii) CENTRAL REGIME: $k, n \to \infty$ *with* $\xi = n2^{-k}$, $0 < \xi < 1$

$$h_n^k \sim \sqrt{1 + 2\xi\Phi'(\xi) + \xi^2\Phi''(\xi)} e^{-n\Phi(\xi)}.$$

We know $\Phi(\xi)$ *analytically only for* $\xi \to 0$ *and* $\xi \to 1$. *In particular, for* $\xi \to 0$

$$\Phi(\xi) \sim \frac{1}{2}\rho_0 e^{\varphi(\log_2 \xi)} \xi^{3/2} \exp\left(-\frac{\log^2 \xi}{2\log 2} \right), \qquad \xi \to 0^+, \tag{16}$$

with

$$\varphi(x) = \frac{\log 2}{2} x(x+1) + \sum_{\ell=0}^{\infty} \log\left(\frac{1 - \exp(-2^{x-\ell})}{2^{x-\ell}} \right) + \sum_{\ell=1}^{\infty} \log(1 - \exp(-2^{x+\ell}))$$

$$= \Psi(x) - \frac{\log 2}{12} + \frac{1}{\log 2} \left(\frac{\gamma^2}{2} + \gamma(1) - \frac{\pi^2}{12} \right), \tag{17}$$

$$\Psi(x) = \sum_{\substack{\ell=-\infty \\ \ell \neq 0}}^{\infty} \frac{1}{2\pi i \ell} \Gamma\left(1 - \frac{2\pi i \ell}{\log 2} \right) \zeta\left(1 - \frac{2\pi i \ell}{\log 2} \right) e^{2\pi i x \ell}. \tag{18}$$

In the above, $\Gamma(\cdot)$ *is the Gamma function,* $\zeta(\cdot)$ *is the Riemann zeta function,* $\gamma = -\Gamma'(1)$ *is the Euler constant, and* $\gamma(1)$ *is defined by the Laurent series* $\zeta(s) = 1/(s-1) + \gamma - \gamma(1)(s-1) + O((s-1)^2)$. *The function* $\Psi(x)$ *is periodic with a very small amplitude, i.e.,* $|\Psi(x)| < 10^{-5}$. *Moreover, for* $\xi \to 1$ *the function* $\Phi(\xi)$ *becomes*

$$\Phi(\xi) \sim D_1 + (1 - \xi)\log(1 - \xi) - (1 - \xi)(1 + \log D_2), \qquad \xi \to 1^-$$

where $D_1 = 1 + \log(K_0^*)$ and $D_2 = K_1^* K_0^*/e$ with

$$K_0^* = \prod_{\ell=1}^{\infty} \left(1 - 2^{-2^{\ell}+1}\right)^{2^{-\ell}} = .68321974\ldots,$$

$$K_1^* = \prod_{\ell=1}^{\infty}\prod_{m=1}^{\infty} \left(1 - 2^{-2^{\ell+1}+2}\right)^{-1}\left[1 - 2^{-2^{\ell+m}+1}\right]^{2^{-m}} = 1.2596283\ldots$$

(iii) LEFT-TAIL REGIME: $k, n \to \infty$ *with* $2^k - n = M = O(1)$

$$h_n^k \sim \frac{\sqrt{2\pi}}{M!} D_2^M n^{M+1/2} e^{-D_1 n}$$

where D_1 and D_2 are defined above.

The expressions for h_n^k in parts (i) and (iii) are completely determined. However, the expression in part (ii) involves the function $\Phi(\xi)$. We have not been able to determine this function analytically, except for its behaviors as ξ approaches 0 or 1. The behavior of $\Phi(\xi)$ as $\xi \to 1^-$ implies the asymptotic matching of parts (ii) and (iii), while the behavior as $\xi \to 0^+$ implies the matching of (i) and (ii). As $\xi \to 0$, this behavior involves the periodic function $\varphi(x)$, which satisfies $\varphi(x+1) = \varphi(x)$. In part (ii) we give two different representations for $\varphi(x)$; the latter (which involves $\Psi(x)$) is a Fourier series.

Since $\Phi(\xi) > 0$, we see that in (ii) and (iii), the distribution is exponentially small in n, while in (i), $1 - h_n^k$ is super–exponentially small (the dominant term in $1 - h_n^k$ is $2^{-n^2/2}$). Thus, (i) applies in the right tail of the distribution while (ii) and (iii) apply in the left tail. We wish to compute the range of k where h_n^k undergoes the transition from $h_n^k \approx 0$ to $h_n^k \approx 1$, as $n \to \infty$. This must be in the asymptotic matching region between (i) and (ii). We can show that C_j, defined in Theorem 3(i), becomes as $j \to \infty$

$$C_j \sim \frac{j^{5/2}}{2} e^{\varphi(\alpha)} \exp\left(-\frac{1}{2}\frac{\log^2 j}{\log 2}\right), \tag{19}$$

where $\alpha = \langle \log_2 j \rangle$. With (19), we can verify the matching between parts (i) and (ii), and the limiting form of (ii) as $\xi \to 0^+$ is

$$h_n^k \sim \exp\left(-\frac{\rho_0}{2} e^{\varphi(\log_2 n)} \exp\left(-\frac{\log 2}{2}\left(\left(k + \frac{3}{2} - \log_2 n\right)^2 - 2\log_2 n - \frac{9}{4}\right)\right)\right)$$

$$= \exp\left(-\rho_0 e^{\varphi(\log_2 n)} 2^{1/8} n \exp\left(-\frac{\log 2}{2}(k + 1.5 - \log_2 n)^2\right)\right) \tag{20}$$

$$= \exp\left(-\rho_0 \cdot n \cdot \exp\left(-\frac{\log 2}{2}(k + 1.5 - \log_2 n)^2 + \theta + \Psi(\log_2 n)\right)\right) \tag{21}$$

where ρ_0 is defined in Theorem 3(i) and

$$\theta = \frac{1}{\log 2}\left(\frac{\gamma^2}{2} + \gamma(1) - \frac{\pi^2}{12}\right) + \frac{\log 2}{24} = -1.022401\ldots$$

while $|\Psi(\log_2 n)| \leq 10^{-5}$. We have written (20) in terms of k and n, recalling that $\xi = n2^{-k}$. We also have used $\sqrt{1 + 2\xi\Phi'(\xi) + \xi^2\Phi''(\xi)} \sim 1$ as $\xi \to 0$.

We now set, for an integer ℓ,

$$k_\ell = \left\lfloor \log_2 n + \sqrt{2\log_2 n} - \frac{3}{2} \right\rfloor + \ell \tag{22}$$

$$= \log_2 n + \sqrt{2\log_2 n} - \frac{3}{2} + \ell - \beta_n,$$

where

$$\beta_n = \left\langle \log_2 n + \sqrt{2\log_2 n} - \frac{3}{2} \right\rangle \in [0, 1). \tag{23}$$

In terms of ℓ and β_n, (21) becomes

$$\Pr\{\mathcal{H}_n^P \leq \lfloor \log_2 n + \sqrt{2\log_2 n} - 1.5 \rfloor + \ell\} \tag{24}$$

$$\sim \exp\left(-\rho_0 e^{\theta + \Psi(\log_2 n)} 2^{-(\ell-\beta_n)^2/2 - (\ell-\beta_n)\sqrt{2\log_2 n}}\right).$$

For $0 < \beta_n < 1$ and $n \to \infty$ the above is small for $\ell \leq 0$, and it is close to one for $\ell \geq 1$. This shows that asymptotically, as $n \to \infty$, all the mass accumulates when $k = k_1$ is given by (22) with $\ell = 1$. Now suppose $\beta_n = 0$ for some n, or more generally that we can find a sequence n_i such that $n_i \to \infty$ as $i \to \infty$ but $\sqrt{2\log_2 n_i}\langle \log_2 n_i + \sqrt{2\log_2 n_i} - \frac{3}{2}\rangle$ remains bounded. Then, the expression in (24) would be $O(1)$ for $\ell = 0$ (since $\beta_n\sqrt{2\log_2 n} = O(1)$). For $\ell = 1$, (24) would then be asymptotically close to 1. Thus, now the mass would accumulate at two points, namely, $k_0 = k_1 - 1$ and k_1. Finally, if $\beta_n = 1 - o(1)$ such that $(1 - \beta_n)\sqrt{2\log_2 n} = O(1)$, then the probability mass is concentrated on k_1 and $k_1 + 1$.

In order to verify the latter assertions, we must either show that $\beta_n = 0$ for an integer n or that there is a subsequence n_i such that $\sqrt{2\log_2 n_i}\beta_{n_i} = O(1)$. The former is false, while the latter is true. To prove that $\beta_n = 0$ is impossible for integer n, let us assume the contrary. If there exists an integer N such that

$$\log_2 n + \sqrt{2\log_2 n} - \frac{3}{2} = N,$$

then

$$n = 2^{N+5/2-\sqrt{4+2N}}.$$

But this is impossible since this would require that $4 + 2N$ is odd. To see that there exists a subsequence such that $R(n_i) = \beta_{n_i}\sqrt{2\log_2 n_i} = O(1)$, we observe that the function $R(n)$ fluctuates from zero to $\sqrt{2\log_2 n}$. We can show that if $n_i = \lfloor 2^{i+5/2-\sqrt{2i+4}} \rfloor + 1$, then $R(n_i) \to 0$ as $i \to \infty$. Note that this subsequence corresponds to the minima of $R(n)$.

Corollary 2. *The asymptotic distribution of PATRICIA height is concentrated among the three points k_1-1, k_1 and k_1+1 where $k_1 = \lfloor \log_2 n + \sqrt{2\log_2 n - \frac{3}{2}} \rfloor + 1$, that is,*

$$\Pr\{\mathcal{H}_n^P = k_1 - 1 \text{ or } k_1 \text{ or } k_1 + 1\} = 1 - o(1)$$

as $n \to \infty$. *More precisely: (i) there are subsequences* n_i *for which* $\Pr\{\mathcal{H}_{n_i}^P = k_1\} = 1 - o(1)$ *provided that*

$$R(n) = \sqrt{2\log_2 n} \left\langle \log_2 n + \sqrt{2\log_2 n} - \frac{3}{2} \right\rangle \to \infty$$

as $i \to \infty$; *(ii) there are subsequences* n_i *for which* $\Pr\{\mathcal{H}_{n_i}^P = k_1 - 1 \text{ or } k_1\} = 1 - o(1)$ *provided that* $R(n_i) = O(1)$; *(iii) finally, there are subsequences* n_i *for which* $\Pr\{\mathcal{H}_{n_i}^P = k_1 \text{ or } k_1+1\} = 1 - o(1)$ *provided that* $\sqrt{2\log_2 n_i} - R(n_i) = O(1)$.

References

1. L. Devroye, A Probabilistic Analysis of the Height of Tries and the complexity of Trie Sort, *Acta Informatica*, 21, 229–237, 1984.
2. L. Devroye, A Note on the Probabilistic Analysis of Patricia Tries, *Random Structures & Algorithms*, 3, 203–214, 1992.
3. L. Devroye, A Study of Trie-Like Structures Under the Density Model, *Ann. Appl. Probability*, 2, 402–434, 1992.
4. P. Flajolet, On the Performance Evaluation of Extendible Hashing and Trie Searching, *Acta Informatica*, 20, 345–369, 1983.
5. N. Froman and P. Froman, *JWKB Approximation*, North-Holland, Amsterdam 1965.
6. D. Gusfield, *Algorithms on Strings, Trees, and Sequences*, Cambridge University Press, 1997.
7. P. Jacquet and M. Régnier, Trie Partitioning Process: Limiting Distributions, Lecture Notes in Computer Science, **214**, 196-210, Springer Verlag, New York 1986.
8. P. Jacquet and W. Szpankowski, Asymptotic Behavior of the Lempel-Ziv Parsing Scheme and Digital Search Trees, *Theoretical Computer Science*, 144, 161-197, 1995.
9. D. Knuth, *The Art of Computer Programming. Sorting and Searching*, Second Edition, Addison-Wesley, 1998.
10. C. Knessl and W. Szpankowski, Quicksort Algorithm Again Revisited, *Discrete Mathematics and Theoretical Computer Science*, 3, 43-64, 1999.
11. P. Lagerstrom, *Matched Asymptotic Expansions: Ideas and Techniques*, Springer-Verlag, New York 1988.
12. H. Mahmoud, *Evolution of Random Search Trees*, John Wiley & Sons, New York 1992.
13. B. Pittel, Asymptotic Growth of a Class of Random Trees, *Ann. of Probab.*, 13, 414–427, 1985.
14. B. Pittel, Path in a Random Digital Tree: Limiting Distributions, *Adv. Appl. Prob.*, 18, 139–155, 1986.
15. B. Pittel and H. Rubin, How Many Random Questions Are Necessary to Identify n Distinct Objects?, *J. Combin. Theory*, Ser. A., 55, 292–312, 1990.
16. W. Szpankowski, Patricia Tries Again Revisited, *Journal of the ACM*, 37, 691–711, 1990.
17. W. Szpankowski, On the Height of Digital Trees and Related Problems, *Algorithmica*, 6, 256–277, 1991.
18. W. Szpankowski, A Generalized Suffix Tree and Its (Un)Expected Asymptotic Behaviors, *SIAM J. Compt.*, 22, 1176–1198, 1993.
19. J. Ziv and A. Lempel, Compression of Individual Sequences via Variable-rate Coding, *IEEE Trans. Information Theory*, 24, 530-536, 1978.

On the Complexity of Routing Permutations on Trees by Arc-Disjoint Paths
Extended Abstract

D. Barth[1], S. Corteel[2], A. Denise[2], D. Gardy[1], and M. Valencia-Pabon[2]

[1] PRiSM, Université de Versailles,
45 Av. des Etats Unis, 78035
VERSAILLES, FR.
[2] L.R.I., Bât 490,
Université Paris-Sud,
91405 ORSAY, FR.

Abstract. In this paper we show that the routing permutation problem is NP-hard even for binary trees. Moreover, we show that in the case of unbounded degree tree networks, the routing permutation problem is NP-hard even if the permutations to be routed are involutions. Finally, we show that the average-case complexity of the routing permutation problem on linear networks is $n/4 + o(n)$.

Keywords: Average-Case Complexity, Routing Permutations, Path Coloring, Tree Networks, NP-completeness.

1 Introduction

Efficient communication is a prerequisite to exploit the performance of large parallel systems. The routing problem on communication networks consists in the efficient allocation of resources to connection requests. In this network, establishing a connection between two nodes requires *selecting* a path connecting the two nodes and *allocating* sufficient resources on all links along the paths associated to the collection of requests. In the case of *all-optical* networks, data is transmitted on lightwaves through optical fiber, and several signals can be transmitted through a fiber link simultaneously provided that different wavelengths are used in order to prevent interference (wavelength-division multiplexing) [4]. As the number of wavelengths is a limited resource, then it is desirable to establish a given set of connection requests with a minimum number of wavelengths. In this context, it is natural to think in wavelengths as colors. Thus the routing problem for all-optical networks can be viewed as a path coloring problem: it consists in finding a desirable collection of paths on the network associated with the collection of connection requests in order to minimize the number of colors needed to color these paths in such a way that any two different paths sharing a same link of the network are assigned different colors. For simple networks, such as trees, the routing problem is simpler, as there is always a unique path for each communication request.

This paper is concerned with routing permutations on trees by arc-disjoint paths,

G. Gonnet, D. Panario, and A. Viola (Eds.): LATIN 2000, LNCS 1776, pp. 308–317, 2000.

that is, the path coloring problem on trees when the collection of connection requests represents a permutation of the nodes of the tree network.

Previous and related work. In [1], Aumann and Rabani have shown that $O(\frac{\log^2 n}{\beta^2})$ colors suffice for routing any permutation on any bounded degree network on n nodes, where β is the *arc expansion* of the network. The result of Aumman and Rabani almost matches the existential lower bound of $\Omega(\frac{1}{\beta^2})$ obtained by Raghavan and Upfal [18]. In the case of specific network topologies, Gu and Tamaki [13] proved that 2 colors are sufficient to route any permutation on any symmetric directed hypercube. Independently, Paterson et al. [17] and Wilfong and Winkler [22] have shown that the routing permutation problem on ring networks is NP-hard. Moreover, in [22] a 2-approximation algorithm is given for this problem on ring networks. To our knowledge, the routing permutation problem on tree networks by arc-disjoint paths has not been studied in the literature.

Our results. In Section 2 we first give some definitions and recall previous results. In Section 3 we show that for arbitrary permutations, the routing permutation problem is NP-hard even for binary trees. Moreover, we show that the routing permutations problem on unbounded degree trees is NP-hard even if the permutations to be routed are involutions, i.e. permutations with cycles of length at most two. In Section 4 we focus on linear networks. In this particular case, since the problem reduces to coloring an interval graph, the routing of any permutation is easily done in polynomial time [14]. We show that the average number of colors needed to color any permutation on a linear network on n vertices is $n/4 + o(n)$. As far as we know, this is the first result on the average-case complexity for routing permutations on networks by arc-disjoint paths. Finally, in Section 5 we give some open problems and future work.

2   Definitions and Preliminary Results

We model the tree network as a rooted labeled symmetric directed tree $T = (V, A)$, where processors and switches are vertices and links are modeled by two arcs in opposite directions. In the sequel, we assume that the labels of the vertices of a tree T on n vertices are $\{1, 2, \ldots, n\}$ and are such that a postfix tree traversal would be exactly $1, 2, \ldots, n$. This implies that for any internal vertex labeled by i the labels of the vertices in his subtree are less than i. Given two vertices i and j of the tree T, we denote by $<i, j>$ the unique path from vertex i to vertex j. The arc from vertex i to its father (resp. from the father of i to i) $(1 \leq i \leq n-1)$ is labeled by i^+ (resp. i^-). See Figure 1(a) for the linear network on $n = 6$ vertices rooted at vertex $i = 6$. We want to route permutations in S_n on any tree T on n vertices. Given a tree T and a vertex i we call $T(i)$ the subtree of T rooted at vertex i.

   We associate with any permutation a graphical representation. To represent the permutation σ we draw an arrow from i to $\sigma(i)$, if $i \neq \sigma(i)$, that is, the path $<i, \sigma(i)>$, $1 \leq i \leq n$. The arrow going from i to $\sigma(i)$ crosses the arc j^+ if and only if i is in $T(j)$ and $\sigma(i)$ is not in $T(j)$ and it crosses the arc j^- if and only if i is not in $T(j)$ and $\sigma(i)$ is in $T(j)$, $1 \leq j \leq n-1$.

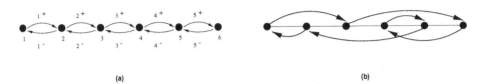

(a)   (b)

Fig. 1. (a) Labeling of the vertices and the arcs for the linear network on $n = 6$ vertices rooted at vertex $i = 6$. (b) representation of permutation $\sigma = (3, 1, 6, 5, 2, 4)$ on the linear network given in (a).

Definition 1. *Let T be a tree on n vertices and σ be a permutation in S_n. We define the* **height of the arc** i^+ *(resp.* **height of the arc** i^-*), $1 \leq i \leq n-1$, denoted $h_T^+(\sigma, i)$ (resp. $h_T^-(\sigma, i)$), as the number of paths crossing the arc i^+ (resp. i^-); that is, $h_T^+(\sigma, i) = |\{j \in T(i) \mid \sigma(j) \notin T(i)\}|$ (resp. $h_T^-(\sigma, i) = |\{j \notin T(i) \mid \sigma(j) \in T(i)\}|$).*

Lemma 1. *Let T be a tree with n vertices. For all σ in S_n and for all $i \in \{1, 2, \ldots, n-1\}$, $h_T^+(\sigma, i) = h_T^-(\sigma, i)$.*

This lemma is straightforward to prove. It tells us that in order to study the height of a permutation on a tree on n vertices, it suffices to consider only the height of the labeled arcs i^+.

Definition 2. *Given a tree T and a permutation σ to be routed on T, the* **height** *of σ, denoted $h_T(\sigma)$, is the maximum number of paths crossing any arc of T: $h_T(\sigma) = \max_i h_T^+(\sigma, i)$.*

For example the permutation $\sigma = (3, 1, 6, 5, 2, 4)$ on the linear network in Figure 1(a) has height 2 (see Figure 1(b)). The maximum is reached in the arcs $4^\pm$.

Definition 3. *Given a tree T and a permutation σ to be routed on T, the* **coloration number** *of σ, denoted $R_T(\sigma)$, is the minimum number of colors assigned to the paths on T associated with σ such that no two paths sharing a same arc of T are assigned the same color.*

Clearly, for any permutation σ of the vertex set of a tree T, we have $R_T(\sigma) \geq h_T(\sigma)$. For linear networks the equality holds, because the conflict graph of the paths associated with σ is an *interval graph* (see [12]). Moreover, optimal vertex coloring for interval graphs can be computed efficiently [14]. However, for arbitrary tree networks, equality does not hold as we will see in the Section 3.3.

3   Complexity of Computing the Coloration Number

We begin this section by showing the NP-completeness of the routing permutations problem in binary trees, and then for the case of routing involutions on unbounded degree trees. Finally, we discuss some polynomial cases of this problem and we show, by an exemple, that in the case of binary trees having at most two vertices with degree equal to 3, the equality between the height and the coloration number of permutations does not hold.

3.1   NP-Completeness Results

Independently, Kumar et al. [15] and Erlebach and Jansen [6] have shown that computing a minimal coloring of any collection of paths on symmetric directed binary trees is NP-hard. However, the construction given in [15,6] does not work when the collection of paths represents a permutation of the vertex set of a binary tree. Thus, by using a reduction similar to the one used in [15,6] we obtain the following result.

Theorem 1. *Let $\sigma \in S_n$ be any permutation.to be routed on a symmetric directed binary tree T on n vertices, then computing $R_T(\sigma)$ is NP-hard.*

Sketch of the proof. We use a reduction from the ARC-COLORING problem [19]. The ARC-COLORING problem can be defined as follows : given a positive integer k, an undirected cycle C_n with vertex set numbered clockwise as $1, 2, \dots, n$, and any collection of paths F on C_n, where each path $<v, w> \in F$ is regarded as the path beginning at vertex v and ending at vertex w again in the clockwise direction, does F can be colored with k colors so that no two paths sharing an edge of C_n are assigned the same color ? It is well known that the ARC-COLORING problem is NP-complete [10]. Let I be an instance of the ARC-COLORING problem. We construct from I an instance I' of the routing permutations problem on binary trees, consisting of a symmetric directed binary tree T and a permutation-set of paths F' on T such that F can be k-colored if and only if F' can be k-colored. Without loss of generality, we may assume that each edge of C_n is crossed by exactly k paths in F. If some edge of C_n is crossed by more than k paths, then this can be discovered in polynomial time, and it implies that the answer in this instance I must be "no". If some edge $[i, i+1]$ of C_n is crossed by $r < k$ paths, then we can add $k - r$ paths of the form $<i, i+1>$ (or $<i, 1>$ if $i = n$) to F without changing its k-colorability.

Let $B(i) \subset F$ (resp. $E(i) \subset F$) be the subcollection of paths of F beginning (resp. ending) at vertex i of C_n, $1 \le i \le n$. Thus, by the previous hypothesis, it is easy to verify that the following property holds for instance I.

Claim. For all vertices i of C_n, $|B(i)| = |E(i)|$.

Construction of the binary tree T of I': first, construct a line on $2k+n$ vertices denoted from left to right by $l_k, l_{k-1}, \dots, l_2, l_1, v_1, v_2, \dots, v_n, r_1, r_2, \dots, r_k$. Next, for each vertex l_i (resp. r_i), $1 \le i \le k$, construct a new different line on $2k + 1$ vertices denoted from left to right by $ll_i^1, ll_i^2, \dots, ll_i^k, wl_i, rl_i^k, rl_i^{k-1}, \dots, rl_i^1$ (resp. $lr_i^1, lr_i^2, \dots, lr_i^k, wr_i, rr_i^k, rr_i^{k-1}, \dots, rr_i^1$) and add to T the arc set $\{(wl_i, l_i), (l_i, wl_i)\}$ (resp. $\{(wr_i, r_i), (r_i, wr_i)\}$). Finally, for each vertex v_i, $1 \le i \le n$, if $|B(i)| > 1$, then construct a new different line on $\alpha_i = |B(i)| - 1$ vertices denoted by $v_i^1, v_i^2, \dots, v_i^{\alpha_i}$ and add to T the arc set $\{(v_i^1, v_i), (v_i, v_i^1)\}$.

The construction of the permutation-set of paths F' of I' is as follows: for each path $<i,j> \in F$, let b_i (resp. e_j) be the first vertex of T in $\{v_i, v_i^1, \dots, v_i^{\alpha_i}\}$ (resp. $\{v_j, v_j^1, \dots, v_j^{\alpha_j}\}$) not already used by any path in F' as beginning-vertex (resp. ending-vertex), then we consider the following two types of paths in F :

- *Type 1* : $i < j$. Then add to F' the path set $\{<b_i, e_j>\}$.
- *Type 2* : $i > j$. Let r_p (resp. l_q) be the first vertex of T in $\{r_1, r_2, \dots, r_k\}$ (resp. $\{l_1, l_2, \dots, l_k\}$) such that the arc (r_p, wr_p) (resp. (l_q, wl_q)) of T has not be already used by any path in F', then add to F' the path set $\{<b_i, rr_p^1>, <lr_p^1, rl_q^1>,$ $<ll_q^1, e_j>\}$. In addition, for each i, $1 \le i \le k$, add to F' the following path sets : $\{<ll_i^j, rl_i^j> : 2 \le j \le k\} \cup \{<rl_i^s, ll_i^s> : 1 \le s \le k\}$ and $\{<lr_i^j, rr_i^j> : 2 \le j \le k\} \cup \{<rr_i^s, lr_i^s> : 1 \le s \le k\}$. The paths $<ll_i^j, rl_i^j>$ and $<lr_i^j, rr_i^j>$, $2 \le j \le k$, $1 \le i \le k$, act as blockers. They make sure that all the three paths in F' corresponding to one path in F of type 2 are colored with the same color in any k-coloration of F'. The other paths that we call *permutation paths*, are used to ensure that the path collection F' represents a permutation of the vertex set of T. In Figure 2 we present an example of this polynomial construction. By

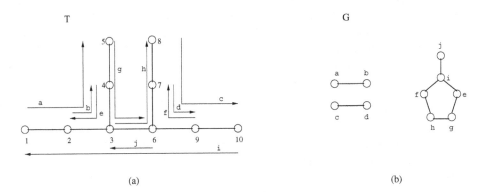

(a)   (b)

Fig. 2. Partial construction of I' from I, where $k = 3$.

our construction, it is easy to check that the set of paths F' on T represents a permutation of the vertex set of T, and that there is a k-coloring of F if and only if there is a k-coloring of F'. □

In the case of unbounded degree symmetric directed trees, Caragiannis et al. [3] have shown that the path coloring problem remains NP-hard even if the collection of paths is symmetric (we call this problem the symmetric path coloring problem), i.e., for each path beginning at vertex v_1 and ending at vertex v_2, there also exists its symmetric, a path beginning at v_2 and ending at v_1. Thus, using a polynomial reduction from the symmetric path coloring problem on trees [3] we have the following result which proof is omitted for lack of space.

Theorem 2. *Let $\sigma \in I_n$ be any involution to be routed on an unbounded degree tree T on n vertices. Then computing $R_T(\sigma)$ is NP-hard.*

3.2   Polynomial Cases

As noticed in Section 2, the coloration number associated to any permutation to be routed on a linear network can be computed efficiently in polynomial time [14]. In the case of *generalized star* networks, i.e., a tree network having only one vertex with degree greater to 2 and the other vertices with degree at most equal to 2, Gargano et al. [11] show that an optimal coloring of any collection of paths on these networks can be computed efficiently in polynomial time. Moreover, in [11] is also showed that the number of colors needed to color any collection of paths on a generalized star network is equal to the height of such a collection of paths. Thus, based on the results given in [11] we obtain the following proposition.

Proposition 1. *Given a generalized star network G on n vertices and a permutation $\sigma \in S_n$ to be routed on G, the coloration number $R_G(\sigma)$ can be computed efficiently in polynomial time. Moreover, $R_G(\sigma) = h_G(\sigma)$ always holds.*

3.3   General Trees

Given any permutation $\sigma \in S_n$ to be routed on a tree T on n vertices, the equality between the heigth $h_T(\sigma)$ and the coloration number $R_T(\sigma)$ does not always hold. In Figure 3(a) we give an exemple of a permutation $\sigma \in S_{10}$ to be routed on a tree T on 10 vertices, which height $h_T(\sigma)$ is equal to 2. Moreover, in Figure 3(b) we present the conflict graph G associated with σ, that is an undirected graph whose vertices are the paths on T associated with σ, and in which two vertices are adjacent if and only if their associated paths share a same arc in T. Thus, clearly the coloration number $R_T(\sigma)$ is equal to the chromatic number of G. Therefore, as the conflict graph G has the cycle C_5 as induced subgraph, then the chromatic number of G is equal to 3, and thus $R_T(\sigma) = 3$.

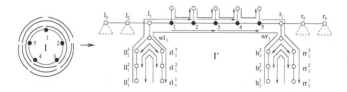

Fig. 3. (a) A tree T on 10 vertices and a permutation $\sigma = (5, 4, 8, 2, 6, 3, 9, 10, 7, 1)$ to be routed on T. (b) The conflict graph G associated with permutation σ in (a).

The best known approximation algorithm for coloring any collection of paths with height h on any tree network is given in [7], which uses at most $\lceil \frac{5}{3}h \rceil$ colors. Therefore it trivially also holds for any permutation-set of paths with height h on any tree.

Proposition 2. *Given a tree T on n vertices and a permutation $\sigma \in S_n$ to be routed on T with heigth $h_T(\sigma)$, there exists a polynomial algorithm for coloring the paths on T associated with σ which uses at most $\lceil \frac{5}{3}h_T(\sigma) \rceil$ colors.*

4   Average Coloration Number on Linear Networks

The main result of this section is the following:

Theorem 3. *The average coloration number of the permutations in S_n to be routed on a linear network on n vertices is*

$$\frac{n}{4} + \frac{\lambda}{2} n^{1/3} + O(n^{1/6})$$

where $\lambda = 0.99615\ldots$.

To prove this result, we use the equality between the height and the coloration number (see Section 2). Then our approach, developed in Subsections 4.1 and 4.2, is as follows: at first we recall a bijection between permutations in S_n and special walks in $\mathbb{N} \times \mathbb{N}$, called "Motzkin walks", which are labeled in a certain way. The bijection is such that the height parameter is "preserved". Then we prove Theorem 3 by studying the asymptotic behaviour of the height of these walks. On the other hand, we get in Subsection 4.3 the generating function of permutations with coloration number k, for any given k. This gives rise to an algorithm to compute exactly the average coloration number of the permutations for any fixed n.

4.1   A Bijection between Permutations and Motzkin Walks

A **Motzkin walk** of length n is a (n+1)-uple $(s_0, s_1, \ldots, s_n)$ of points in $\mathbb{N} \times \mathbb{N}$ satisfying the following conditions:

- For all $0 \leq i \leq n$, $s_i = (i, y_i)$ with $y_i \geq 0$;
- $y_0 = y_n = 0$;
- For all $0 \leq i < n$, $y_{i+1} - y_i$ equals either 1 (North-East step), or 0 (East step), or -1 (South-East step);

The **height** of a Motzkin walk ω is $H(\omega) = \max\limits_{i \in \{0,1,\ldots,n\}} \{y_i\}$.

Labeled Motzkin walks are Motzkin walks in which steps can be labeled by integers. These structures are in relation with several well-studied combinatorial objects [8,20,21] and in particular with permutations. The walks we will deal with are labeled as follows:

- each South-East step $(i, y_i) \rightarrow (i+1, y_i - 1)$ is labeled by an integer between 1 and $y_i{}^2$ (or, equivalently, by a pair of integers, each one between 1 and y_i);
- each East step $(i, y_i) \rightarrow (i+1, y_i)$ is labeled by an integer between 1 and $2y_i + 1$.

Let P_n be the set of such labeled Motzkin walks of length n. We recall that S_n is the set of permutations on $[n]$. The following result was first established by Françon and Viennot [9]:

Theorem (Françon-Viennot) *There is a one-to-one correspondence between the elements of P_n and the elements of S_n.*

Several bijective proofs of this theorem are known. Biane's bijection [2] is particular, in the sense that it preserves the height: to any labeled Motzkin walk of length n and height k corresponds a permutation in S_n with height k (and so with coloration number k). We do not present here the whole Biane's bijection; we just focus on the construction of the (unlabelled) Motzkin walk associated to a permutation, in order to show that the height is preserved. This property, which is not explicitly noticed in Biane's paper, is essential for our purpose.

Biane's correspondence between a permutation $\sigma = (\sigma(1), \sigma(2), \ldots, \sigma(n))$ and a labeled Motzkin walk $\omega = (s_0, s_1, \ldots, s_n)$ is such that, for $1 \le i \le n$:

- step (s_{i-1}, s_i) is a North-East step if and only if $\sigma(i) > i$ and $\sigma^{-1}(i) > i$;
- step (s_{i-1}, s_i) is a South-East step if and only if $\sigma(i) < i$ and $\sigma^{-1}(i) < i$;
- otherwise, step (s_{i-1}, s_i) is an East step.

Now, for any $1 \le i \le n$, the height of point s_i in ω is obviously equal to the number of North-East steps minus the number of South-East steps in the shrinked walk $(s_0, s_1, \ldots, s_i)$. On the other hand, we can prove easily that the height of arc i^+ in σ is equal to the number of integers $j \le i$ such that $\sigma(j) > j$ and $\sigma^{-1}(j) > j$, minus the number of integers $j \le i$ such that $\sigma(j) < j$ and $\sigma^{-1}(j) < j$. This proves the property. We present in Figure 4 an exemple of correspondence. The above description permits to construct the "skeleton" of the permutation, in the center of the figure, given the Motzkin walk on the top. Then the labeling of the path allows to complete the permutation. This is described in detail in [2] and in the full version of this paper, in preparation.

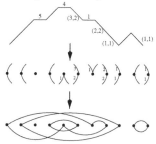

Fig. 4: From a walk to a permutation

4.2   Proof of Theorem 3

In [16], Louchard analyzes some list structures; in particular his "dictionary structure" corresponds to our labeled Motzkin walks. We will use his notation in order to refer directly to his article. From Louchard's theorem 6.2, we deduce the following lemma:

Lemma 2. *The height* $Y^*([nv])$ *of a random labeled Motzkin walk of length* n *after the step* $[nv]$ *($v \in [0,1]$)) has the following behavior*

$$\frac{Y^*([nv]) - nv(1-v)}{\sqrt{n}} \Rightarrow X(v),$$

where "$\Rightarrow$" denotes the weak convergence and X is a Markovian process with mean 0 and covariance $C(s,t) = 2s^2(1-t)^2$, $s \le t$.

Then the work of Daniels and Skyrme [5] gives us a way to compute the maximum of $Y^*([nv])$, that is the height of a random labeled Motzkin walk.

Proposition 3. *The height of a random labeled Motzkin walk Y^* is*

$$\max_v Y^*([nv]) = \frac{n}{4} + m\sqrt{n/2} + O(n^{1/6}), \tag{1}$$

where m is asymptotically Gaussian with mean $E(m) \sim \lambda n^{-1/6}(1/2)^{1/2}$ and variance $\sigma^2(m) \sim 1/8$ and $\lambda = 0.99615\ldots$.

In the formula (1) of the above Proposition 2, the only non-deterministic part is m which is Gaussian. So we just have to replace m by $E(m)$ to prove Theorem 3.

4.3   An Algorithm to Compute Exactly the Average Coloration Number

We just have to look at known results in enumerative combinatorics [8,21] to get the generating function of the permutations of coloration number **exactly** k, that is

$$\frac{(k!)^2 z^{2k}}{P^*_{k+1}(z)P^*_k(z)}$$

with $P_0(z) = 1$, $P_1(z) = z - b_0$ and $P_{n+1}(z) = (t - b_n)P_n(z) - \lambda_n P_{n-1}(z)$ for $n \geq 1$, where P^* is the reprocical polynomial of P, that is $P^*_n(z) = z^n P_n(1/z)$ for $n \geq 0$.

This generating function leads to a recursive algorithm to compute the number of permutations with coloration number k, denoted by $h_{n,k}$.

Proposition 4. *The number of permutations in $\mathcal{S}_{n,k}$ follows the following recurrence*

$$h_{n,k} = \begin{cases} 0 & \text{if } n < 2k \\ (k!)^2 & \text{if } n = 2k \\ -\sum_{i=1}^{2h+1} p(i)h_{n-i,k} & \text{otherwise} \end{cases}$$

where $p(i)$ is the coefficient of z^i in $P^*_{k+1}(z)P^*_k(z)$.

From this result we are able to compute the average height of a permutation as it is $\bar{h}(n) = \sum_{k\geq 0} k h_{n,k}/n!$.

5   Open Problems and Future Work

It remains open the complexity of routing involutions on binary trees by arc-disjoint paths. The average coloration number of permutations to be routed on general trees is also an interesting open problem. Computing the average coloration number of permutations to be routed on arbitrary topology networks seems a very difficult problem.

Acknowledgements. We are very grateful to Philippe Flajolet, Dominique Gouyou-Beauchamps and Guy Louchard for their help.

References

1. Y. Aumann, Y. Rabani. *Improved bounds for all optical routing.* In Proc. of the 6th ACM-SIAM SODA, pp 567-576, 1995.
2. Ph. Biane. *Permutations suivant le type d'excédance et le nombre d'inversions et interprétation combinatoire d'une fraction continue de Heine.* Eur. J. Comb., 14(4):277-284, 1993.
3. I. Caragiannis, Ch. Kaklamanis, P. Persiano. *Wavelength Routing of Symmetric Communication Requests in Directed Fiber Trees.* In Proc. of SIROCCO, 1998.
4. N. K. Cheung, K. Nosu, and G. Winzer, editors. *Special Issue on Dense Wavelength Division Multiplexing Techniques for High Capacity and Multiple Access Communications Systems.* IEEE J. on Selected Areas in Comm., 8(6), 1990.
5. H. E. Daniels, T. H. R. Skyrme. *The maximum of a random walk whose mean path has a maximum.* Adv. Appl. Probab., 17:85-99, 1985.
6. T. Erlebach and K. Jansen. *Call scheduling in trees, rings and meshes.* In Proc. of HICSS-30, vol. 1, pp 221-222. IEEE CS Press, 1997.
7. T. Erlebach, K. Jansen, C. Kaklamanis, M. Mihail, P. Persiano. *Optimal wavelength routing on directed fiber trees.* Theoret. Comput. Sci. 221(1-2):119-137, 1999.
8. Ph. Flajolet. *Combinatorial aspects of continued fractions.* Discrete Math., 32:125-161, 1980.
9. J. Françon, X. Viennot. *Permutations selon leurs pics, creux, doubles montées et doubles descentes, nombres d'Euler et nombres de Genocchi.* Discrete Math., 28:21-35, 1979.
10. M.R. Garey, D.S. Johnson, G.L. Miller, C.H. Papadimitriou. *The complexity of colouring circular arcs and chords.* SIAM J. Alg. Disc. Meth., 1(2):216-227,1980.
11. L. Gargano, P. Hell, S. Perennes. *Coloring all directed paths in a symmetric tree with applications to WDM routing.* In Proc. of ICALP, LNCS 1256, pp 505-515, 1997.
12. M. C. Golumbic. *Algorithmic graph theory and perfect graphs.* Academic Press, New York, 1980.
13. Q.-P. Gu, H. Tamaki. *Routing a permutation in the Hypercube by two sets of edge-disjoint paths.* In Proc. of 10th IPPS. IEEE CS Press, 1996.
14. U. I. Gupta, D. T. Lee, J. Y.-T. Leung. *Efficient algorithms for interval graphs and circular-arc graphs.* Networks, 12:459-467, 1982.
15. S. R. Kumar, R. Panigrahy, A. Russel, R. Sundaram. *A note on optical routing on trees.* Inf. Process. Lett., 62:295-300, 1997.
16. G. Louchard. *Random walks, Gaussian processes and list structures.* Theoret. Comput. Sci., 53:99-124, 1987.
17. M. Paterson, H. Schröder, O. Sýkora, I. Vrto. *On Permutation Communications in All-Optical Rings.* In Proc. of SIROCCO, 1998.
18. P. Raghavan, U. Upfal. *Efficient routing in all optical networks.* In Proc. of the 26th ACM STOC, pp 133-143, 1994.
19. A. Tucker. *Coloring a family of circular arcs.* SIAM J. Appl. Maths., 29(3):493-502, 1975.
20. X. Viennot. *A combinatorial theory for general orthogonal polynomials with extensions and applications.* Lect. Notes Math., 1171:139-157, 1985. Polynômes orthogonaux et applications, Proc. Laguerre Symp., Bar-le-Duc/France 1984.
21. X. Viennot. *Une théorie combinatoire des polynômes orthogonaux généraux.* Notes de conférences, Univ. Quebec, Montréal, 1985.
22. G. Wilfong, P. Winkler. *Ring routing and wavelength translation.* In Proc. of the 9th Annual ACM-SIAM SODA, pp 333-341, 1998.

Subresultants Revisited

Extended Abstract

Joachim von zur Gathen and Thomas Lücking

FB Mathematik-Informatik, Universität Paderborn
33095 Paderborn, Germany
{gathen,luck}@upb.de

1 Introduction

1.1 Historical Context

The *Euclidean Algorithm* was first documented by Euclid (320–275 BC). Knuth (1981), p. 318, writes: *"We might call it the granddaddy of all algorithms, because it is the oldest nontrivial algorithm that has survived to the present day."* It performs division with remainder repeatedly until the remainder becomes zero. With inputs 13 and 9 it performs the following:

$$13 = 1 \cdot 9 + 4,$$
$$9 = 2 \cdot 4 + \boxed{1},$$
$$4 = 4 \cdot 1 + 0.$$

This allows us to compute the *greatest common divisor (gcd)* of two integers as the last non-vanishing remainder. In the example, the gcd of 13 and 9 is computed as 1.

At the end of the 17th century the concept of polynomials was evolving. Researchers were interested in finding the common roots of two polynomials f and g. One question was whether it is possible to apply the Euclidean Algorithm to f and g. In 1707 Newton solved this problem and showed that this always works in $\mathbb{Q}[x]$.

$$x^3 + 2x^2 - x - 2 = (\frac{1}{2}x + \frac{3}{2})(2x^2 - 2x - 4) + \boxed{4x + 4}$$
$$2x^2 - 2x - 4 = (\frac{1}{2}x - 1)(4x + 4) + 0.$$

In this example $f = x^3 + 2x^2 - x - 2$ and $g = 2x^2 - 2x - 4$ have a greatest common divisor $4x + 4$, and therefore the only common root is -1. In a certain sense the Euclidean Algorithm computes all common roots. If you only want to know whether f and g have at least *one* common root, then the whole Euclidean

G. Gonnet, D. Panario, and A. Viola (Eds.): LATIN 2000, LNCS 1776, pp. 318–342, 2000.

Algorithm has to be executed. Thus the next goal was to find an indicator for common roots without using any division with remainder.

The key to success was found in 1748 by Euler, and later by Bézout. They defined the *resultant* of f and g as the smallest polynomial in the coefficients of f and g that vanishes if and only if f and g have a common root. In 1764 Bézout was the first to find a matrix whose determinant is the resultant. The entries of this *Bézout matrix* are quadratic functions of the coefficients of f and g. Today we use the matrix discovered by Sylvester in 1840, known as the *Sylvester matrix*. Its entries are simply coefficients of the polynomials f and g. Sylvester generalized his definition and introduced what we now call *subresultants* as determinants of certain submatrices of the Sylvester matrix. They are nonzero if and only if the corresponding degree appears as a degree of a remainder of the Euclidean Algorithm.

These indicators, in particular the resultant, also work for polynomials in $\mathbb{Z}[x]$. So the question came up whether it is possible to apply the Euclidean Algorithm to f and g in $\mathbb{Z}[x]$ without leaving $\mathbb{Z}[x]$. The answer is no, as illustrated in the example above, since division with remainder is not always defined in $\mathbb{Z}[x]$, although the gcd exists. In the example it is $x + 1$.

However, in 1836 Jacobi found a way out. He introduced *pseudo-division*: he multiplied f with a certain power of the leading coefficient of g before performing the division with remainder. This is always possible in $\mathbb{Z}[x]$. So using pseudo-division instead of division with remainder in every step in the Euclidean Algorithm yields an algorithm with all intermediate results in $\mathbb{Z}[x]$.

About 40 years later Kronecker did research on the *Laurent series* in x^{-1} of g/f for two polynomials f and g. He considered the determinants of a matrix whose entries are the coefficients of the Laurent series of g/f. He obtained the same results as Sylvester, namely that these determinants are nonzero if and only if the corresponding degree appears in the degree sequence of the Euclidean Algorithm. Furthermore Kronecker gave a direct way to compute low degree polynomials s, t and r with $sf + tg = r$ via determinants of matrices derived again from the Laurant series of g/f, and showed that these polynomials are essentially the only ones. He also proved that the polynomial r, if nonzero, agrees with a remainder in the Euclidean Algorithm, up to a constant multiple. This was the first occurrence of *polynomial subresultants*.

In the middle of our century, again 70 years later, the realization of computers made it possible to perform more and more complicated algorithms faster and faster. However, using *pseudo-division* in every step of the Euclidean Algorithm causes *exponential* coefficient growth. This was suspected in the late 1960's. Collins (1967), p. 139 writes: *"Thus, for the Euclidean algorithm, the lengths of the coefficients increases exponentially."* In Brown & Traub (1971) we find: *"Although the Euclidean PRS algorithm is easy to state, it is thoroughly impractical since the coefficients grow exponentially."* An exponential *upper* bound is in Knuth (1981), p. 414: *"Thus the upper bound [...] would be approximately $N^{0.5(2.414)^n}$"*, and experiments show that the simple algorithm does in fact have this behavior; the number of digits in the coefficients grows exponentially at *each step!*". However, we did not find a proof of an exponential *lower* bound; our bound in Theorem 7.3 seems to be new.

One way out of this exponential trap is to make every intermediate result *primitive*, that is, to divide the remainders by the greatest common divisors of their coefficients, the so-called *content*. However, computing the contents seemed to be very expensive since in the worst case the gcd of *all* coefficients has to be computed. So the scientists tried to find divisors of the contents without using any gcd computation. Around 1970, first Collins and then Brown & Traub reinvented the *polynomial subresultants* as determinants of a certain variant of the Sylvester matrix. Habicht had also defined them independently in 1948. Collins and Brown & Traub showed that they agree with the remainders of the Euclidean Algorithm up to a constant factor. They gave simple formulas to compute this factor and introduced the concept of *polynomial remainder sequences (PRS)*, generalizing the concept of Jacobi. The final result is the *subresultant PRS* that features linear coefficient growth with intermediate results in $\mathbb{Z}[x]$.

Since then two further concepts have come up. On the one hand the *fast EEA* allows to compute an arbitrary intermediate line in the Euclidean Scheme directly. Using the fast $O(n \log n \log \log n)$ multiplication algorithm of Schönhage and Strassen, the time for a gcd reduces from $O(n^2)$ to $O(n \log^2 n \log \log n)$ field operations (see Strassen (1983)). On the other hand, the *modular EEA* is very efficient. These two topics are not considered in this thesis; for further information we refer to von zur Gathen & Gerhard (1999), Chapters 6 and 11.

1.2 Outline

After introducing the notation and some well-known facts in Section 2, we start with an overview and comparison of various definitions of subresultants in Section 3. Mulders (1997) describes an error in software implementations of an integration algorithm which was due to the confusion caused by the these various definitions. It turns out that there are essentially two different ways of defining them: the scalar and the polynomial subresultants. Furthermore we show their relation with the help of the Euclidean Algorithm. In the remainder of this work we will mainly consider the scalar subresultants.

In Section 4 we give a formal definition of polynomial remainder sequences and derive the most famous ones as special cases of our general notion. The relation between polynomial remainder sequences and subresultants is exhibited in the Fundamental Theorem 5.1 in Section 5. It unifies many results in the literature on various types of PRS which can be derived as corollaries from this theorem. In Section 6 we apply it to the various definitions of polynomial remainder sequences already introduced. This yields a collection of results from Collins (1966, 1967, 1971, 1973), Brown (1971, 1978), Brown & Traub (1971), Lickteig & Roy (1997) and von zur Gathen & Gerhard (1999). Lickteig & Roy (1997) found a recursion formula for polynomial subresultants not covered by the Fundamental Theorem. We translate it into a formula for scalar subresultants and use it to finally solve an open question in Brown (1971), p. 486. In Section 7 we analyse the coefficient growth and the running time of the various PRS.

Finally in Section 8 we report on implementations of the various polynomial remainder sequences and compare their running times. It turns out that computing contents is quite fast for random inputs, and that the primitive PRS behaves much better than expected.

Much of this Extended Abstract is based on the existing literature. The following results are new:

- rigorous and general definition of division rules and PRS,
- proof that all constant multipliers in the subresultant PRS for polynomials over an integral domain R are also in R,
- exponential lower bound for the running time of the pseudo PRS (algorithm).

2   Foundations

In this chapter we introduce the basic algebraic notions. We refer to von zur Gathen & Gerhard (1999), Sections 2.2 and 25.5, for the notation and fundamental facts about greatest common divisors and determinants. More information on these topics is in Hungerford (1990).

2.1   Polynomials

Let R be a ring. In what follows, this always means a commutative ring with 1.

A basic tool in computer algebra is *division with remainder*. For given polynomials f and g in $R[x]$ of degrees n and m, respectively, the task is to find polynomials q and r in $R[x]$ with

$$f = qg + r \text{ and } \deg r < \deg g. \tag{2.1}$$

Unfortunately such q and r do not always exist.

Example 2.2. It is not possible to divide x^2 by $2x + 3$ with remainder in $\mathbb{Z}[x]$ because $x^2 = (ux + v)(2x + 3) + r$ with $u, v, r \in \mathbb{Q}$ has the unique solution $u = 1/2$, $v = 0$ and $r = -3/2$, which is not over $\mathbb{Z}$.

If defined and unique we call $q = \mathrm{quo}(f, g)$ the *quotient* and $r = \mathrm{rem}(f, g)$ the *remainder*. A ring with a length function (like the degree of polynomials) and where division with remainder is always defined is a *Euclidean domain*. $R[x]$ is a Euclidean domain if and only if R is a field. Moreover a solution of (2.1) is not necessarily unique if the leading coefficient $\mathrm{lc}(g)$ of g is a zero divisor.

Example 2.3. Let $R = \mathbb{Z}_8$ and consider $f = 4x^2 + 2x$ and $g = 2x + 1$. With

$$q_1 = 2x, \qquad r_1 = 0$$
$$q_2 = 2x + 4, \; r_2 = 4$$

we obtain

$$q_1 g + r_1 = 2x(2x + 1) + 0 = 4x^2 + 2x = f,$$
$$q_2 g + r_2 = (2x + 4)(2x + 1) + 4 = 4x^2 + 10x + 8 = 4x^2 + 2x = f.$$

Thus we have two distinct solutions (q_1, r_1) and (q_2, r_2) of (2.1).

A way to get solutions for all commutative rings is the *general pseudo-division* which allows multiplication of f by a ring element α:

$$\alpha f = qg + r, \ \deg r < \deg g. \tag{2.4}$$

If $\alpha = g_m^{n-m+1}$, then this is the *(classical) pseudo-division*. If $\mathrm{lc}(g)$ is not a zero divisor, then (2.4) always has a unique solution in $R[x]$. We call $q = \mathrm{pquo}(f, g)$ the *pseudo-quotient* and $r = \mathrm{prem}(f, g)$ the *pseudo-remainder*.

Example 2.2 continued. For x^2 and $2x + 3$ we get the pseudo-division

$$2^2 \cdot x^2 = (2x - 3)(2x + 3) + 9$$

A simple computation shows that we cannot choose $\alpha = 2$.

Lemma 2.5.

(i) Pseudo-division always yields a solution of (2.4) in $R[x]$.
(ii) If $\mathrm{lc}(g)$ is not a zero divisor, then any solution of (2.4) has $\deg q = n - m$.

Lemma 2.6. *The solution (q, r) of (2.4) is uniquely determined if and only if $\mathrm{lc}(g)$ is not a zero-divisor.*

Let R be a unique factorization domain. We then have $\gcd(f, g) \in R$ for $f, g \in R[x]$, and the *content* $\mathrm{cont}(f) = \gcd(f_0, \dots, f_n) \in R$ of $f = \sum_{0 \le j \le n} f_j x^j$. The polynomial is *primitive* if $\mathrm{cont}(f)$ is a unit. The *primitive part* $\mathrm{pp}(f)$ is defined by $f = \mathrm{cont}(f) \cdot \mathrm{pp}(f)$. Note that $\mathrm{pp}(f)$ is a primitive polynomial.

The *Euclidean Algorithm* computes the gcd of two polynomials by iterating the division with remainder:

$$r_{i-1} = q_i r_i + r_{i+1}. \tag{2.7}$$

3   Various Notions of Subresultants

3.1   The Sylvester Matrix

The various definitions of the subresultant are based on the *Sylvester matrix*. Therefore we first take a look at the historical motivation for this special matrix. Our goal is to decide whether two polynomials $f = \sum_{0 \le j \le n} f_j x^j$ and $g = \sum_{0 \le j \le m} g_j x^j \in R[x]$ of degree $n \ge m > 0$ over a commutative ring R in the indeterminate x have a common root. To find an answer for this question, Euler (1748) and Bézout (1764) introduced the *(classical) resultant* that vanishes if and only if this is true. Although Bézout also succeeded in finding a matrix whose determinant is equal to the resultant, today called *Bézout matrix*, we will follow the elegant derivation in Sylvester (1840). The two linear equations

$$
\begin{aligned}
f_n x_n + f_{n-1} x_{n-1} + \cdots + f_1 x_1 + f_0 x_0 = 0 \\
g_m x_m + g_{m-1} x_{m-1} + \cdots + g_1 x_1 + g_0 x_0 = 0
\end{aligned}
$$

in the indeterminates $x_0, \ldots, x_n$ are satisfied if $x_j = \alpha^j$ for all j, where α is a common root of f and g. For $n > 1$ there are many more solutions of these two linear equations in many variables, but Sylvester eliminates them by adding the $(m-1) + (n-1)$ linear equations that correspond to the following additional conditions:

$$xf(x) = 0, \ldots, x^{m-1}f(x) = 0,$$
$$xg(x) = 0, \ldots, x^{n-1}g(x) = 0.$$

These equations give a total of $n + m$ linear relations among the variables $x_{m+n-1}, \cdots, x_0$:

$$f_n x_{m+n-1} + \cdots + f_0 x_{m-1} \qquad\qquad\qquad = 0$$
$$\vdots$$
$$f_n x_n + f_{n-1} x_{n-1} + \cdots + f_0 x_0 = 0$$
$$g_m x_{m+n-1} + \cdots + g_0 x_{n-1} \qquad\qquad = 0$$
$$\vdots$$
$$g_m x_m + g_{m-1} x_{m-1} + \cdots + g_0 x_0 = 0$$

Clearly $x_j = \alpha^j$ gives a solution for any common root α of f and g, but the point is that (essentially) the converse also holds: a solution of the linear equations gives a common root (or factor). The $(n+m) \times (n+m)$ matrix, consisting of coefficients of f and g, that belongs to this system of linear equations is often called *Sylvester matrix*. In the sequel we follow von zur Gathen & Gerhard (1999), Section 6.3, p. 144, and take its transpose.

Definition 3.1. *Let R be a commutative ring and let $f = \sum_{0 \leq j \leq n} f_j x^j$ and $g = \sum_{0 \leq j \leq m} g_j x^j \in R[x]$ be polynomials of degree $n \geq m > 0$, respectively. Then the $(n + m) \times (n + m)$ matrix*

$$
\mathrm{Syl}(f, g) =
\begin{pmatrix}
f_n & & & & g_m & & & \\
f_{n-1} & f_n & & & g_{m-1} & g_m & & \\
\vdots & \vdots & \ddots & & \vdots & \vdots & \ddots & \\
\vdots & \vdots & & f_n & g_1 & \vdots & & \ddots \\
\vdots & \vdots & & f_{n-1} & g_0 & \vdots & & \ddots \\
\vdots & \vdots & & \vdots & & g_0 & & g_m \\
f_0 & \vdots & & \vdots & & & \ddots & \vdots \\
& f_0 & & \vdots & & & & \vdots \\
& & \ddots & \vdots & & & \ddots & \vdots \\
& & & f_0 & & & & g_0
\end{pmatrix}
\begin{array}{c} \\ \\ \\ \\ \\ \\ \\ \\ \\ \\ \underbrace{}_{m}\underbrace{}_{n} \end{array}
$$

is called the Sylvester matrix *of f and g.*

Remark 3.2. Multiplying the $(n + m - j)$th row by x^j and adding it to the last row for $1 \leq j < n + m$, we get the $(n + m) \times (n + m)$ matrix S^*. Thus $\det(\mathrm{Syl}(f, g)) = \det(\mathrm{Syl}^*(f, g))$.

More details on resultants can be found in Biermann (1891), Gordan (1885) and Haskell (1892). Computations for both the univariate and multivariate case are discussed in Collins (1971).

3.2   The Scalar Subresultant

We are interested in finding out which degrees appear in the degree sequence of the intermediate results in the Euclidean Algorithm. Below we will see that the scalar subresultants provide a solution to this problem.

Definition 3.3. *Let R be a commutative ring and $f = \sum_{0 \leq j \leq n} f_j x^j$ and $g = \sum_{0 \leq j \leq m} g_j x^j \in R[x]$ polynomials of degree $n \geq m > 0$, respectively. The determinant $\sigma_k(f, g) \in R$ of the $(m + n - 2k) \times (m + n - 2k)$ matrix*

$$
S_k(f, g) = \left(
\begin{array}{ccccccccc}
f_n & & & & g_m & & & & \\
f_{n-1} & f_n & & & g_{m-1} & g_m & & & \\
\vdots & & \ddots & & \vdots & & \ddots & & \\
f_{n-m+k+1} & \cdots & \cdots & f_n & g_{k+1} & & \cdots & \cdots & g_m \\
\vdots & & & \vdots & \vdots & & & & \ddots \\
f_{k+1} & \cdots & \cdots & f_m & g_{m-n+k+1} & \cdots & \cdots & \cdots & \cdots & g_m \\
\vdots & & & \vdots & \vdots & & & & \vdots \\
\vdots & & & \vdots & \vdots & & & & \vdots \\
f_{2k-m+1} & \cdots & \cdots & f_k & g_{2k-n+1} & \cdots & \cdots & \cdots & \cdots & g_k
\end{array}
\right)
$$

$$\underbrace{\hphantom{f_{2k-m+1} \cdots \cdots f_k}}_{m-k} \quad \underbrace{\hphantom{g_{2k-n+1} \cdots \cdots g_k}}_{n-k}$$

*is called the kth (**scalar**) **subresultant** of f and g. By convention an f_j or g_j with $j < 0$ is zero. If f and g are clear from the context, then we write S_k and σ_k for short instead of $S_k(f, g)$ and $\sigma_k(f, g)$.*

Sylvester (1840) already contains an explicit description of the (scalar) subresultants. In Habicht (1948), p. 104, σ_k is called *Nebenresultante (minor resultant)* for polynomials f and g of degrees n and $n - 1$. The definition is also in von zur Gathen (1984) and is used in von zur Gathen & Gerhard (1999), Section 6.10, p. 169.

Remark 3.4.

(i) $S_0 = \mathrm{Syl}(f, g)$ and therefore $\sigma_0 = \det(S_0)$ is the *resultant*.
(ii) $\sigma_m = g_m^{n-m}$.

(iii) S_k is the matrix obtained from the Sylvester matrix by deleting the last $2k$ rows and the last k columns with coefficients of f, and the last columns with coefficients of g.

(iv) S_k is a submatrix of S_i if $k \geq i$.

3.3 The Polynomial Subresultant

We now introduce two slightly different definitions of polynomial subresultants. The first one is from Collins (1967), p. 129, and the second one is from Brown & Traub (1971), p. 507 and also in Zippel (1993), Chapter 9.3, p. 150. They yield polynomials that are related to the intermediate results in the Euclidean Algorithm.

Definition 3.5. *Let R be a commutative ring, and $f = \sum_{0 \leq j \leq n} f_j x^j$ and $g = \sum_{0 \leq j \leq m} g_j x^j \in R[x]$ polynomials of degree $n \geq m > 0$. Let $M_{ik} = M_{ik}(f, g)$ be the $(n+m-2k) \times (n+m-2k)$ submatrix of $\mathrm{Syl}(f, g)$ obtained by deleting the last k of the m columns of coefficients of f, the last k of the n columns of coefficients of g and the last $2k+1$ rows except row $(n+m-i-k)$, for $0 \leq k \leq m$ and $0 \leq i \leq n$. The polynomial $R_k(f, g) = \sum_{0 \leq i \leq n} \det(M_{ik}) x^i \in R[x]$ is called the kth **polynomial subresultant** of f and g. In fact Collins (1967) considered the transposed matrices. If f and g are clear from the context, then we write R_k for short instead of $R_k(f, g)$. Note that $\det(M_{ik}) = 0$ if $i > k$ since then the last row of M_{ik} is identical to the $(n+m-i-k)$th row. Thus $R_k = \sum_{0 \leq i \leq k} \det(M_{ik}) x^i$.*

Remark 3.6.

(i) $M_{00} = \mathrm{Syl}(f, g)$ and therefore $R_0 = \det(M_{00})$ is the resultant.

(ii) Remark 3.4(i) implies $\sigma_0 = R_0$.

Definition 3.7. *Let R be a commutative ring and $f = \sum_{0 \leq j \leq n} f_j x^j$ and $g = \sum_{0 \leq j \leq m} g_j x^j \in R[x]$ polynomials of degree $n \geq m > 0$. We consider the determinant $\bar{Z}_k(f, g) = \det(M_k^*) \in R[x]$ of the $(n+m-2k) \times (n+m-2k)$ matrix M_k^* obtained from M_{ik} by replacing the last row with $(x^{m-k-1}f, \cdots, f, x^{n-k-1}g, \cdots, g)$.*

Table 1 gives an overview of the literature concerning these notions. There is a much larger body of work about the special case of the resultant, which we do not quote here.

3.4 Comparison of the Various Definitions

As in Brown & Traub (1971), p. 508, and Geddes *et al.* (1992), Section 7.3, p. 290, we first have the following theorem which shows that the definitions in Collins (1967) and Brown & Traub (1971) describe the same polynomial.

Theorem 3.8.

(i) *If $\sigma_k(f, g) \neq 0$, then $\sigma_k(f, g)$ is the leading coefficient of $R_k(f, g)$. Otherwise, $\deg R_k(f, g) < k$.*

Definition	Authors
$\sigma_k(f,g) = \det(S_k) \in R$	Sylvester (1840), Habicht (1948)
	von zur Gathen (1984)
	von zur Gathen & Gerhard (1999)
$R_k(f,g) = \sum_{0 \cdot i \cdot n} \det(M_{ik})x^i$	Collins (1967), Loos (1982)
	Geddes *et al.* (1992)
$= Z_k(f,g) = \det(M_k^\cdot) \in R[x]$	Brown & Traub (1971)
	Zippel (1993), Lickteig & Roy (1997)
	Reischert (1997)

Table 1. Definitions of subresultants

(ii) $R_k(f,g) = Z_k(f,g)$.

Lemma 3.9. *Let F be a field, f and g in $F[x]$ be polynomials of degree $n \geq m > 0$, respectively, and let r_i, s_i and t_i be the entries in the ith row of the Extended Euclidean Scheme, so that $r_i = s_i f + t_i g$ for $0 \leq i \leq \ell$. Moreover, let $\rho_i = \mathrm{lc}(r_i)$ and $n_i = \deg r_i$ for all i. Then*

$$\frac{\sigma_{n_i}}{\rho_i} \cdot r_i = R_{n_i} \text{ for } 2 \leq i \leq \ell.$$

Remark 3.10. Let f and g be polynomials over an integral domain R, let F be the field of fractions of R, and consider the Extended Euclidean Scheme of f and g in $F[x]$. Then the scalar and the polynomial subresultants are in R and $R[x]$, respectively, and Lemma 3.9 also holds:

$$\frac{\sigma_{n_i}}{\rho_i} \cdot r_i = R_{n_i} \in R[x].$$

Note that r_i is not necessarily in $R[x]$, and ρ_i not necessarily in R.

4   Division Rules and Polynomial Remainder Sequences (PRS)

We cannot directly apply the Euclidean Algorithm to polynomials f and g over an integral domain R since polynomial division with remainder in $R[x]$, which is used in every step of the Euclidean Algorithm, is not always defined. Hence our goal now are definitions modified in such a way that they yield a variant of the Euclidean Algorithm that works over an integral domain. We introduce a generalization of the usual pseudo-division, the concept of *division rules*, which leads to intermediate results in $R[x]$.

Definition 4.1. *Let R be an integral domain. A* one-step division rule *is a partial mapping*

$$\mathcal{R}\colon R[x]^2 \rightharpoonup R^2$$

such that for all $(f,g) \in \mathrm{def}(\mathcal{R})$ there exist $q,r \in R[x]$ satisfying

(i) $\mathcal{R}(f,g) = (\alpha,\beta)$,
(ii) $\alpha f = qg + \beta r$ and $\deg r < \deg g$.

Recall that $\mathrm{def}(\mathcal{R}) \subseteq R[x]^2$ is the *domain of definition* of $\mathcal{R}$, that is, the set of $(f,g) \in R[x]^2$ at which $\mathcal{R}$ is defined. In particular, $\mathcal{R}\colon \mathrm{def}(\mathcal{R}) \longrightarrow R^2$ is a total map. In the examples below, we will usually define one-step division rules by starting with a (total or partial) map $\mathcal{R}_0\colon R[x]^2 \rightharpoonup R^2$ and then taking $\mathcal{R}$ to be the maximal one-step division rule consistent with $\mathcal{R}_0$. Thus

$$\mathrm{def}(\mathcal{R}) = \{(f,g) \in R[x]^2 : \exists \alpha, \beta \in R, \exists q,r \in R[x]$$
$$(\alpha,\beta) = \mathcal{R}_0(f,g) \text{ and (ii) holds}\},$$

and $\mathcal{R}$ is $\mathcal{R}_0$ restricted to $\mathrm{def}(\mathcal{R})$. Furthermore $(f,0)$ is never in $\mathrm{def}(\mathcal{R})$ ("you can't divide by zero"), so that

$$\mathrm{def}(\mathcal{R}) \subseteq \mathcal{D}_{\max} = R[x] \times (R[x] \setminus \{0\}).$$

We are particularly interested in one-step division rules $\mathcal{R}$ with $\mathrm{def}(\mathcal{R}) = \mathcal{D}_{\max}$. In our examples, $(0,g)$ will always be in $\mathrm{def}(\mathcal{R})$ if $g \neq 0$.

We may consider the usual remainder as a partial function $\mathrm{rem}\colon R[x]^2 \rightharpoonup R[x]$ with $\mathrm{rem}(f,g) = r$ if there exist $q,r \in R[x]$ with $f = qg + r$ and $\deg r < \deg g$, and $\mathrm{def}(\mathrm{rem})$ maximal. Recall from Section 2 the definitions of rem, prem and cont.

Example 4.2. Let f and g be polynomials over an integral domain R of degrees n and m, respectively, and let $f_n = \mathrm{lc}(f)$, $g_m = \mathrm{lc}(g) \neq 0$ be their leading coefficients. Then the three most famous types of division rules are as follows:

- *classical division rule:* $\mathcal{R}(f,g) = (1,1)$.
- *monic division rule:* $\mathcal{R}(f,g) = (1, \mathrm{lc}(\mathrm{rem}(f,g)))$.
- *Sturmian division rule:* $\mathcal{R}(f,g) = (1,-1)$.

Examples are given below. When R is a field, these three division rules have the largest possible domain of definition $\mathrm{def}(\mathcal{R}) = \mathcal{D}_{\max}$, but otherwise, it may be smaller; we will illustrate this in Example 4.7. Hence they do not help us in achieving our goal of finding rules with maximal domain $\mathcal{D}_{\max}$. But there exist two division rules which, in contrast to the first examples, always yield solutions in $R[x]$:

- *pseudo-division rule:* $\mathcal{R}(f,g) = (g_m^{n-m+1}, 1)$.

In case R is a unique factorization domain, we have the

- *primitive division rule:* $\mathcal{R}(f,g) = (g_m^{n-m+1}, \mathrm{cont}(\mathrm{prem}(f,g)))$.

For algorithmic purposes, it is then useful for R to be a Euclidean domain.

The disadvantage of the pseudo-division rule, however, is that in the Euclidean Algorithm it leads to exponential coefficient growth; the coefficients of the intermediate results are usually enormous, their bit length may be exponential in the bit length of the input polynomials f and g. If R is a UFD, we get the smallest intermediate results if we use the primitive division rule, but the computation of the content in every step of the Euclidean Algorithm seems to be expensive. Collins (1967) already observed this in his experiments. Thus he tries to avoid the computation of contents and to keep the intermediate results "small" at the same time by using information from *all* intermediate results in the EEA, not only the two previous remainders. Our concept of one-step division rules does not cover his method. So we now extend our previous definition, and will actually capture all the "recursive" division rules from Collins (1967, 1971, 1973), Brown & Traub (1971) and Brown (1971) under one umbrella.

Definition 4.3. *Let R be an integral domain. A division rule is a partial mapping*

$$\mathcal{R} \colon R[x]^2 \dashrightarrow (R^2)^*$$

associating to $(f, g) \in \mathrm{def}(\mathcal{R})$ a sequence $((\alpha_2, \beta_2), \ldots, (\alpha_{\ell+1}, \beta_{\ell+1}))$ of arbitrary length ℓ such that for all $(f, g) \in \mathrm{def}(\mathcal{R})$ there exist $\ell \in \mathbb{N}_{\geq 0}$, $q_1, \ldots, q_\ell \in R[x]$ and $r_0, \ldots, r_{\ell+1} \in R[x]$ satisfying

(i) $r_0 = f, r_1 = g$,
(ii) $\mathcal{R}_i(f, g) = \mathcal{R}(f, g)_i = (\alpha_i, \beta_i)$,
(iii) $\alpha_i r_{i-2} = q_{i-1} r_{i-1} + \beta_i r_i$ and $\deg r_i < \deg r_{i-1}$

for $2 \leq i \leq \ell + 1$. The integer $\ell = |\mathcal{R}(f, g)|$ is the length *of the sequence.*

A division rule where $\ell = 1$ for all values is the same as a one-step division rule, and from an arbitrary division rule we can obtain a one-step division rule by projecting to the first coordinate (α_2, β_2) if $\ell \geq 2$. Using Lemma 2.6, we find that for all $(f, g) \in \mathrm{def}(\mathcal{R})$, q_{i-1} and r_i are unique for $2 \leq i \leq \ell + 1$. If we have a one-step division rule $\mathcal{R}^*$ which is defined at all (r_{i-2}, r_{i-1}) for $2 \leq i \leq \ell + 1$ (defined recursively), then we obtain a division rule $\mathcal{R}$ by using $\mathcal{R}^*$ in every step:

$$\mathcal{R}_i(f, g) = \mathcal{R}^*(r_{i-2}, r_{i-1}) = (\alpha, \beta).$$

If we truncate $\mathcal{R}$ at the first coordinate, we get $\mathcal{R}^*$ back. But the notion of division rules is strictly richer than that of one-step division rules; for example the first step in the reduced division rule below is just the pseudo-division rule, but using the pseudo-division rule repeatedly does not yield the reduced division rule.

Example 4.2 continued. Let $f = r_0, g = r_1, r_2, \ldots, r_\ell \in R[x]$ be as in Definition 4.3, let $n_i = \deg r_i$ be their degrees, $\rho_i = \mathrm{lc}(r_i)$ their leading coefficients, and $d_i = n_i - n_{i+1} \in \mathbb{N}_{\geq 0}$ for $0 \leq i \leq \ell$ (if $n_0 \geq n_1$). We now present two different types of recursive division rules. They are based on polynomial sub-resultants. It is not obvious that they have domain of definition $\mathcal{D}_{\max}$, since divisions occur in their definitions. We will show that this is indeed the case in Remarks Remark 6.8 and Remark 6.12.

– *reduced division rule:* $\mathcal{R}_i(f,g) = (\alpha_i, \beta_i)$ for $2 \leq i \leq \ell + 1$,
  where we set $\alpha_1 = 1$ and

$$(\alpha_i, \beta_i) = (\rho_{i-1}^{d_{i-2}+1}, \alpha_{i-1}) \text{ for } 2 \leq i \leq \ell + 1.$$

– *subresultant division rule:* $\mathcal{R}_i(f,g) = (\alpha_i, \beta_i)$ for $2 \leq i \leq \ell + 1$,
  where we set $\rho_0 = 1$, $\psi_2 = -1$, $\psi_3, \ldots, \psi_{\ell+1} \in R$ with

$$(\alpha_i, \beta_i) = (\rho_{i-1}^{d_{i-2}+1}, -\rho_{i-2}\psi_i^{d_{i-2}}) \text{ for } 2 \leq i \leq \ell + 1,$$
$$\psi_i = (-\rho_{i-2})^{d_{i-3}}\psi_{i-1}^{1-d_{i-3}} \text{ for } 3 \leq i \leq \ell + 1.$$

The subresultant division rule was invented by Collins (1967), p. 130. He tried to find a rule such that the r_i's agree with the polynomial subresultants up to a small constant factor. Brown (1971), p. 486, then provided a recursive definition of the α_i and β_i as given above. Brown (1971) also describes an "improved division rule", where one has some magical divisor of ρ_i.

We note that the exponents in the recursive definition of the ψ_i's in the subresultant division rule may be negative. Hence it is not clear that the β_i's are in R. However, we will show this in Theorem 6.15, and so answer the following open question that was posed in Brown (1971), p. 486:

Question 4.4. *"At the present time it is not known whether or not these equations imply $\psi_i, \beta_i \in R$."*

By definition, a division rule $\mathcal{R}$ defines a sequence $(r_0, \ldots, r_\ell)$ of remainders; recall that they are uniquely defined. Since it is more convenient to work with these "polynomial remainder sequences", we fix this notion in the following definition.

Definition 4.5. *Let $\mathcal{R}$ be a division rule. A sequence $(r_0, \ldots, r_\ell)$ with each $r_i \in R[x] \setminus \{0\}$ is called a polynomial remainder sequence (PRS) for (f,g) according to $\mathcal{R}$ if*

(i) $r_0 = f, r_1 = g$,
(ii) $\mathcal{R}_i(f,g) = (\alpha_i, \beta_i)$,
(iii) $\alpha_i r_{i-2} = q_{i-1}r_{i-1} + \beta_i r_i$,

for $2 \leq i \leq \ell + 1$, where ℓ is the length *of $\mathcal{R}(f,g)$. The PRS is* complete *if $r_{\ell+1} = 0$. It is called* normal *if $d_i = \deg r_i - \deg r_{i+1} = 1$ for $1 \leq i \leq \ell - 1$ (Collins (1967), p. 128/129).*

In fact the remainders for PRS according to arbitrary division rules over an integral domain only differ by a nonzero constant factor.

Proposition 4.6. *Let R be an integral domain, $f, g \in R[x]$ and $r = (r_0, \ldots, r_\ell)$ and $r^* = (r_0^*, \ldots, r_{\ell*}^*)$ be two PRS for (f, g) according to two division rules $\mathcal{R}$ and $\mathcal{R}^*$, respectively, none of whose results $\alpha_i, \beta_i, \alpha_i^*, \beta_i^*$ is zero. Then $r_i^* = \gamma_i r_i$ with*

$$\gamma_i = \prod_{0 \le k \le i/2 - 1} \frac{\alpha_{i-2k}^* \beta_{i-2k}}{\alpha_{i-2k} \beta_{i-2k}^*} \in F \setminus \{0\}$$

for $0 \le i \le \min\{\ell, \ell^*\}$, where F is the field of fractions of R.

The proposition yields a direct way to compute the PRS for (f, g) according to $\mathcal{R}^*$ from the PRS for (f, g) according to $\mathcal{R}$ and the $\alpha_i, \beta_i, \alpha_i^*, \beta_i^*$. In particular, the degrees of the remainders in any two PRS are identical.

In Example 4.2 we have seen seven different division rules. Now we consider the different polynomial remainder sequences according to these rules. Each PRS will be illustrated by the following example.

Example 4.7. We perform the computations on the polynomials

$$f = r_0 = 9x^6 - 27x^4 - 27x^3 + 72x^2 + 18x - 45 \text{ and}$$
$$g = r_1 = 3x^4 - 4x^2 - 9x + 21$$

over $R = \mathbb{Q}$ and, wherever possible, also over $R = \mathbb{Z}$.

i	classical	monic	Sturmian	pseudo
0	$9x^6 - 27x^4 - 27x^3 + 72x^2 + 18x - 45$			
1	$3x^4 - 4x^2 - 9x + 21$			
2	$-11x^2 - 27x + 60$	$x^2 + \frac{27}{11}x - \frac{60}{11}$	$11x^2 - 27x + 60$	$-297x^2 - 729x + 1620$
3	$-\frac{164\,880}{1331}x + \frac{248\,931}{1331}$	$x - \frac{27\,659}{18\,320}$	$\frac{164\,880}{1331}x + \frac{248\,931}{1331}$	$3\,245\,333\,040x - 4\,899\,708\,873$
4	$-\frac{1\,959\,126\,851}{335\,622\,400}$	1	$-\frac{1\,959\,126\,851}{335\,622\,400}$	$-1\,659\,945\,865\,306\,233\,453\,993$

i	primitive	reduced	subresultant
0	$9x^6 - 27x^4 - 27x^3 + 72x^2 + 18x - 45$		
1	$3x^4 - 4x^2 - 9x + 21$		
2	$-11x^2 - 27x + 60$	$-297x^2 - 729x + 1620$	$297x^2 + 729x - 1620$
3	$18\,320x - 27\,659$	$120\,197\,520x - 181\,470\,699$	$13\,355\,280x - 20\,163\,411$
4	-1	$86\,915\,463\,129$	$9\,657\,273\,681$

1. **Classical PRS.** The most familiar PRS for (f, g) is obtained according to the *classical division rule*. Collins (1973), p. 736, calls this the *natural Euclidean PRS (algorithm)*. The intermediate results of the classical PRS and of the Euclidean Algorithm coincide.

2. **Monic PRS.** In Collins (1973), p. 736, the PRS for (f, g) according to the *monic division rule* is called *monic PRS (algorithm)*. The r_i are monic for $2 \le i \le \ell$, and we get the same intermediate results as in the *monic Euclidean Algorithm* in von zur Gathen & Gerhard (1999), Section 3.2, p. 47.

3. **Sturmian PRS.** We choose the PRS for (f, g) according to the *Sturmian division rule* as introduced in Sturm (1835). Kronecker (1873), p. 117, Habicht (1948), p. 102, and Loos (1982), p. 119, deal with this *generalized Sturmian PRS (algorithm)*. Kronecker (1873) calls it *Sturmsche Reihe (Sturmian sequence)*, and in Habicht (1948) it is the *verallgemeinerte Sturmsche Kette (generalized Sturmian chain)*. If $g = \partial f / \partial x$ as in Habicht (1948), p. 99, then this is the *classical Sturmian PRS (algorithm)*. Note that the Sturmian PRS agrees with the classical PRS up to sign.

If R is not a field, then Example 4.7 shows that the first three types of PRS may not have $\mathcal{D}_{max}$ as their domain of definition. In the example they are only of length 1. But fortunately there are division rules that have this property.

4. **Pseudo PRS.** If we choose the PRS according to the *pseudo-division rule*, then we get the so-called *pseudo PRS*. Collins (1967), p. 138, calls this the *Euclidean PRS (algorithm)* because it is the most obvious generalization of the Euclidean Algorithm to polynomials over an integral domain R that is not a field. Collins (1973), p. 737, also calls it *pseudo-remainder PRS*.

5. **Primitive PRS.** To obtain a PRS over R with minimal coefficient growth, we choose the PRS according to the *primitive division rule* which yields primitive intermediate results. Brown (1971), p. 484, calls this the *primitive PRS (algorithm)*.

6. **Reduced PRS.** A perceived drawback of the primitive PRS is the (seemingly) costly computation of the content; recently the algorithm of Cooperman *et al.* (1999) achieves this with an expected number of less than two integer gcd's. In fact, in our experiments in Section 8, the primitive PRS turns out to be most efficient among those discussed here. But Collins (1967) introduced his *reduced PRS (algorithm)* in order to avoid the computation of contents completely. His algorithm uses the *reduced division rule* and keeps the intermediate coefficients reasonably small but not necessarily as small as with the primitive PRS.

7. **Subresultant PRS.** The reduced PRS is not the only way to keep the coefficients small without computing contents. We can also use the *subresultant division rule*. According to Collins (1967), p. 130, this is the *subresultant PRS (algorithm)*.

5   Fundamental Theorem on Subresultants

Collins' Fundamental Theorem on subresultants expresses an arbitrary subresultant as a power product of certain data in the PRS, namely the multipliers α and β and the leading coefficients of the remainders in the Euclidean Algorithm. In this section our first goal is to prove the Fundamental Theorem on subresultants for polynomial remainder sequences according to an arbitrary division rule $\mathcal{R}$.

The following result is shown for PRS in Brown & Traub (1971), p. 511, Fundamental theorem, and for reduced PRS in Collins (1967), p. 132, Lemma 2, and p. 133, Theorem 1.

Fundamental Theorem 5.1. *Let f and $g \in R[x]$ be polynomials of degrees $n \geq m > 0$, respectively, over an integral domain R, let $\mathcal{R}$ be a division rule and $(r_0, \ldots, r_\ell)$ be the PRS for (f, g) according to $\mathcal{R}$, $(\alpha_i, \beta_i) = \mathcal{R}_i(f, g)$ the constant multipliers, $n_i = \deg r_i$ and $\rho_i = \mathrm{lc}(r_i)$ for $0 \leq i \leq \ell$, and $d_i = n_i - n_{i+1}$ for $0 \leq i \leq \ell - 1$.*

(i) For $0 \leq j \leq n_1$, the jth subresultant of (f, g) is

$$\sigma_j(f, g) = (-1)^{b_i} \rho_i^{n_{i-1} - n_i} \prod_{2 \leq k \leq i} \left(\frac{\beta_k}{\alpha_k}\right)^{n_{k-1} - n_i} \rho_{k-1}^{n_{k-2} - n_k}$$

if $j = n_i$ for some $1 \leq i \leq \ell$, otherwise 0, where $b_i = \sum_{2 \leq k \leq i}(n_{k-2} - n_i)(n_{k-1} - n_i)$.

(ii) The subresultants satisfy for $1 \leq i < \ell$ the recursive formulas

$$\sigma_{n_1}(f, g) = \rho_1^{d_0} \text{ and}$$
$$\sigma_{n_{i+1}}(f, g) = \sigma_{n_i}(f, g) \cdot (-1)^{d_i(n_0 - n_{i+1} + i + 1)} (\rho_{i+1} \rho_i)^{d_i} \prod_{2 \leq k \leq i+1} \left(\frac{\beta_k}{\alpha_k}\right)^{d_i}.$$

Corollary 5.2. *Let $\mathcal{R}$ be a division rule and $(r_0, \ldots, r_\ell)$ be the PRS for (f, g) according to $\mathcal{R}$, let $n_i = \deg r_i$ for $0 \leq i \leq \ell$ be the degrees in the PRS, and let $0 \leq k \leq n_1$. Then*

$$\sigma_k \neq 0 \iff \exists i \colon k = n_i.$$

6   Applications of the Fundamental Theorem

Following our program, we now derive results for the various PRS for polynomials $f, g \in R[x]$ of degrees $n \geq m \geq 0$, respectively, over an integral domain R, according to the division rules in Example 4.2.

Corollary 6.1. *Let $(r_0, \ldots, r_\ell)$ be a **classical PRS** and $1 \leq i \leq \ell$. Then*

(i)
$$\sigma_{n_i}(f, g) = (-1)^{b_i} \rho_i^{d_i - 1} \prod_{2 \leq k \leq i} \rho_{k-1}^{n_{k-2} - n_k}.$$

(ii) The subresultants satisfy the recursive formulas

$$\sigma_{n_1}(f, g) = \rho_1^{d_0}, \text{ and}$$
$$\sigma_{n_{i+1}}(f, g) = \sigma_{n_i}(f, g) \cdot (-1)^{d_i(n_0 - n_{i+1} + i + 1)} (\rho_{i+1} \rho_i)^{d_i}.$$

If the PRS is normal, then this simplifies to:

(iii)
$$\sigma_{n_i}(f, g) = (-1)^{(d_0 + 1)(i+1)} \rho_i \rho_1^{d_0 + 1} \prod_{3 \leq k \leq i} \rho_{k-1}^2 \text{ for } i \geq 2.$$

(iv) The subresultants satisfy the recursive formulas

$$\sigma_{n_1}(f, g) = \rho_1^{d_0}, \text{ and}$$
$$\sigma_{n_{i+1}}(f, g) = \sigma_{n_i}(f, g) \cdot (-1)^{d_0 + 1} \rho_{i+1} \rho_i.$$

The following is the Fundamental Theorem 11.13 in Gathen & Gerhard (1999), Chapter 11.2, p. 307.

Corollary 6.2. *Let $(r_0, \ldots, r_\ell)$ be a* **monic PRS***, and $1 \leq i \leq \ell$. Then*

(i)
$$\sigma_{n_i}(f, g) = (-1)^{b_i} \rho_0^{n_1 - n_i} \rho_1^{n_0 - n_i} \prod_{2 \leq k \leq i} \beta_k^{n_{k-1} - n_i}.$$

(ii) *The subresultants satisfy the recursive formulas*

$$\sigma_{n_1}(f, g) = \rho_1^{d_0}, \text{ and}$$
$$\sigma_{n_{i+1}}(f, g) = \sigma_{n_i}(f, g) \cdot (-1)^{d_i(n_0 - n_{i+1} + i + 1)} (\rho_0 \rho_1 \beta_2 \cdots \beta_{i+1})^{d_i}.$$

If the PRS is normal, then this simplifies to:

(iii)
$$\sigma_{n_i}(f, g) = (-1)^{(d_0+1)(i+1)} \rho_0^{i-1} \rho_1^{d_0 + i - 1} \prod_{2 \leq k \leq i} \beta_k^{i - (k-1)} \text{ for } i \geq 2.$$

(iv) *The subresultants satisfy for $1 \leq i < \ell$ the recursive formulas*

$$\sigma_{n_1}(f, g) = \rho_1^{d_0}, \text{ and}$$
$$\sigma_{n_{i+1}}(f, g) = \sigma_{n_i}(f, g) \cdot (-1)^{d_0 + 1} \rho_0 \rho_1 \beta_2 \cdots \beta_{i+1}.$$

Corollary 6.3. *Let $(r_0, \ldots, r_\ell)$ be a* **Sturmian PRS***, and $1 \leq i \leq \ell$. Then*

(i)
$$\sigma_{n_i}(f, g) = (-1)^{b_i + \sum_{2 \leq k \leq i}(n_{k-1} - n_i)} \rho_i^{d_i - 1} \prod_{2 \leq k \leq i} \rho_{k-1}^{n_{k-2} - n_k}.$$

(ii) *The subresultants satisfy the recursive formulas*

$$\sigma_{n_1}(f, g) = \rho_1^{d_0}, \text{ and}$$
$$\sigma_{n_{i+1}}(f, g) = \sigma_{n_i}(f, g) \cdot (-1)^{d_i(n_0 - n_{i+1} + 1)} (\rho_{i+1} \rho_i)^{d_i}.$$

If the PRS is normal, then this simplifies to:

(iii)
$$\sigma_{n_i}(f, g) = (-1)^{(d_0+1)(i+1)} \rho_1^{d_0 + 1} \rho_i \prod_{3 \leq k \leq i} \rho_{k-1}^2 \text{ for } i \geq 2.$$

(iv) *The subresultants satisfy the recursive formulas*

$$\sigma_{n_1}(f, g) = \rho_1^{d_0}, \text{ and}$$
$$\sigma_{n_{i+1}}(f, g) = \sigma_{n_i}(f, g) \cdot (-1)^{d_0 + i + 1} \rho_{i+1} \rho_i.$$

The following corollary can be found in Collins (1966), p. 710, Theorem 1, for polynomial subresultants.

Corollary 6.4. *Let $(r_0, \ldots, r_\ell)$ be a* **pseudo PRS***, and $1 \leq i \leq \ell$. Then*

(i)
$$\sigma_{n_i}(f, g) = (-1)^{b_i} \rho_i^{d_i - 1} \prod_{2 \leq k \leq i} \rho_{k-1}^{n_{k-2} - n_k - (n_{k-1} - n_i)(d_{k-2} + 1)}.$$

(ii) The subresultants satisfy the recursive formulas

$$\sigma_{n_1}(f,g) = \rho_1^{do}, \text{ and}$$

$$\sigma_{n_{i+1}}(f,g) = \sigma_{n_i}(f,g) \cdot (-1)^{d_i(n_0-n_{i+1}+i+1)}(\rho_{i+1}\rho_i)^{d_i} \prod_{2 \leq k \leq i+1} \rho_{k-1}^{-(d_{k-2}+1)d_i}.$$

If the PRS is normal, then this simplifies to:

(iii) $$\sigma_{n_i}(f,g) = (-1)^{(d_0+1)(i+1)}\rho_1^{(d_0+1)(2-i)}\rho_i \prod_{3 \leq k \leq i-1} \rho_{k-1}^{2(k-i)} \text{ for } i \geq 2.$$

(iv) The subresultants satisfy the recursive formulas

$$\sigma_{n_1}(f,g) = \rho_1^{do}, \text{ and}$$

$$\sigma_{n_{i+1}}(f,g) = \sigma_{n_i}(f,g) \cdot (-1)^{d_0+1}\rho_1^{-(d_0+1)}\rho_{i+1}\rho_i \prod_{3 \leq k \leq i+1} \rho_{k-1}^{-2}.$$

Remark 6.5. If the PRS is normal, then Corollary 6.4(iii) implies that

$$\sigma_{n_i}(f,g)(-1)^{(\delta_0+1)(i+1)}\rho_1^{(d_0+1)(i-2)} \prod_{3 \leq k \leq i-1} \rho_{k-1}^{2(i-k)} = \rho_i.$$

Thus $\sigma_{n_i}(f,g)$ divides ρ_i. This result is also shown for polynomial subresultants in Collins (1966), p. 711, Corollary 1.

Since the content of two polynomials cannot be expressed in terms of our parameters ρ_i and n_i, we do not consider the Fundamental theorem for **primitive PRS**.

The following is shown for polynomial subresultants in Collins (1967), p. 135, Corollaries 1.2 and 1.4.

Corollary 6.6. *Let $(r_0, \ldots, r_\ell)$ be a reduced PRS, and $1 \leq i \leq \ell$. Then*

(i) $$\sigma_{n_i}(f,g) = (-1)^{b_i}\rho_i^{d_i-1} \prod_{2 \leq k \leq i} \rho_{k-1}^{d_{k-2}(1-d_{k-1})}$$

(ii) The subresultants satisfy for the recursive formulas

$$\sigma_{n_1}(f,g) = \rho_1^{do}, \text{ and}$$

$$\sigma_{n_{i+1}}(f,g) = \sigma_{n_i}(f,g) \cdot (-1)^{d_i(n_0-n_{i+1}+i+1)}\rho_{i+1}^{d_i}\rho_i^{-d_{i-1}d_i}.$$

If the PRS is normal, then this simplifies to:

(iii) $$\sigma_{n_i}(f,g) = (-1)^{(d_0+1)(i+1)}\rho_i \text{ for } i \geq 2.$$

(iv) The subresultants satisfy the recursive formulas

$$\sigma_{n_1}(f,g) = \rho_1^{do}, \text{ and}$$

$$\sigma_{n_{i+1}}(f,g) = \sigma_{n_i}(f,g) \cdot (-1)^{d_0+1}\rho_{i+1}\rho_i^{-1}.$$

Remark 6.7. We obtain from Corollary 6.6(i)

$$\sigma_{n_i}(f,g) \prod_{2 \leq k \leq i} (-1)^{(n_{k-2}-n_i)(n_{k-1}-n_i)} \rho_{k-1}^{d_{k-2}(d_{k-1}-1)} = \rho_i^{d_{i-1}}.$$

Thus $\sigma_{n_i}(f,g)$ divides $\rho_i^{d_{i-1}}$. This result can also be found in Collins (1967), p.135, Corollary 1.2.

Remark 6.8. For every reduced PRS, r_i is in $R[x]$ for $2 \leq i \leq \ell$. Note that Corollary 6.6(iii) implies $r_i = (-1)^{(d_0+1)(i+1)} R_i(f,g)$. So the normal case is clear. An easy proof for the general case based on polynomial subresultants is in Collins (1967), p. 134, Corollary 1.1, and Brown (1971), p. 485/486.

Lemma 6.9. *Let* $e_{i,j} = d_{j-1} \prod_{j \leq k \leq i} (1-d_k)$, *and let* ψ_i *be as in the subresultant division rule. Then*

$$\psi_i = - \prod_{1 \leq j \leq i-2} \rho_j^{e_{i-3,j}} \ \text{for } 2 \leq i \leq \ell.$$

Corollary 6.10. *Let* $(r_0, \ldots, r_\ell)$ *be a* **subresultant PRS***, and* $1 \leq i \leq \ell$*. Then*

(i)
$$\sigma_{n_i}(f,g) = \prod_{1 \leq k \leq i} \rho_k^{e_{i-1,k}}.$$

(ii) The subresultants satisfy the recursive formulas

$$\sigma_{n_1}(f,g) = \rho_1^{d_0}, \ \text{and}$$
$$\sigma_{n_{i+1}}(f,g) = \sigma_{n_i}(f,g) \cdot \rho_{i+1}^{d_i} \prod_{1 \leq k \leq i} \rho_k^{-d_i e_{i-1,k}}.$$

If the PRS is normal, then this simplifies to:

(iii)
$$\sigma_{n_i}(f,g) = \rho_i \ \text{for } i \geq 2.$$

(iv) The subresultants satisfy the recursive formulas

$$\sigma_{n_1}(f,g) = \rho_1^{d_0}, \ \text{and}$$
$$\sigma_{n_{i+1}}(f,g) = \sigma_{n_i}(f,g) \cdot \rho_{i+1} \rho_i^{-1}.$$

Now we have all tools to prove the relation between normal reduced and normal subresultant PRS which can be found in Collins (1967), p. 135, Corollary 1.3, and Collins (1973), p. 738.

Corollary 6.11. *Let* $(r_0, \ldots, r_\ell)$ *be a normal reduced PRS and* $(a_0, \ldots, a_\ell)$ *a normal subresultant PRS for the polynomials* $r_0 = a_0 = f$ *and* $r_1 = a_1 = g$. *Then the following holds for* $2 \leq i \leq \ell$:

$$\mathrm{lc}(r_i) = (-1)^{(n_0-n_i)(n_1-n_i)} \cdot \mathrm{lc}(a_i).$$

Remark 6.12. For every subresultant PRS the polynomials r_i are in $R[x]$ for $2 \leq i \leq \ell$. Note that Corollary 6.10(iii) implies $r_i = R_i(f, g)$. So the normal case is clear. An easy proof for the general case based on polynomial subresultants is in Collins (1967), p. 130, and Brown (1971), p. 486.

Corollary 6.10 does not provide the only recursive formula for subresultants. Another one is based on an idea in Lickteig & Roy (1997), p. 12, and Reischert (1997), p. 238, where the following formula has been proven for polynomial subresultants. The translation of this result into a theorem on scalar subresultants leads us to an answer to Question 4.4.

Theorem 6.13. *Let $(r_0, \dots, r_\ell)$ be a subresultant PRS. Then the subresultants satisfy for $1 \leq i < \ell$ the recursive formulas*

$$\sigma_{n_1}(f, g) = \rho_1^{d_0} \text{ and}$$
$$\sigma_{n_{i+1}}(f, g) = \sigma_{n_i}(f, g)^{1-d_i} \cdot \rho_{i+1}^{d_i}.$$

The proof of the conjecture now becomes pretty simple:

Corollary 6.14. *Let $\psi_2 = -1$ and $\psi_i = (-\rho_{i-2})^{d_{i-3}} \psi_{i-1}^{1-d_{i-3}}$ for $3 \leq i \leq \ell$. Then*

$$\psi_i = -\sigma_{n_{i-2}}(f, g) \text{ for } 3 \leq i \leq \ell.$$

Since all subresultants are in R, this gives an answer to Question 4.4:

Theorem 6.15. *The coefficients ψ_i and β_i of the subresultant PRS are always in R.*

7   Analysis of Coefficient Growth and Running Time

We first estimate the running times for normal PRS. A proof for an exponential *upper* bound for the pseudo PRS is in Knuth (1981), p. 414, but our goal is to show an exponential *lower* bound. To this end, we prove two such bounds on the bit length of the leading coefficients ρ_i in this PRS. Recall that $\rho_1 = \mathrm{lc}(g)$ and $\sigma_{n_1} = \rho_1^{\delta_0 + 1}$.

Lemma 7.1. *Suppose that $(f, g) \in \mathbb{Z}[x]^2$ have a normal pseudo PRS. Then*

$$|\rho_i| \geq |\rho_1|^{2^{i-3}} \text{ for } 3 \leq i \leq \ell.$$

Lemma 7.2. *Suppose that $(f, g) \in \mathbb{Z}[x]^2$ have a normal pseudo PRS, and that $|\rho_1| = 1$. Then*

$$|\rho_i| \geq |\sigma_{n_i} \prod_{2 \leq k \leq i-2} \sigma_{n_k}(f, g)^{2^{i-k-1}}| \text{ for } 3 \leq i \leq \ell.$$

Theorem 7.3. *Computing the pseudo PRS takes exponential time, at least 2^n, in some cases with input polynomials of degrees at most n.*

We have the following running time bound for the normal reduced PRS algorithm.

Theorem 7.4. *Let $\|f\|_\infty$, $\|g\|_\infty \leq A$, $B = (n+1)^n A^{n+m}$, and let $(r_0, \ldots, r_\ell)$ be the normal reduced PRS for f, g. Then the max-norm of the r_i is at most $4B^3$, and the algorithm uses $O(n^3 m \log^2(nA))$ word operations.*

Corollary 7.5. *Since Corollary 6.11 shows that normal reduced PRS and normal subresultant PRS agree up to sign, the estimates in Theorem 7.4 are also true for normal subresultant PRS.*

We conclude the theoretical part of our comparison with an overview of all worst-case running times for the various normal PRS in Table 2. The length of the coefficients of f and g are assumed to be at most n. The estimations that are not proven here can be found in von zur Gathen & Gerhard (1999).

PRS	time	proven in
classical/Sturmian	n^8	von zur Gathen & Gerhard (1999)
monic	n^6	von zur Gathen & Gerhard (1999)
pseudo	c^n with $c \geq 2$	Theorem 7.3
primitive	n^6	von zur Gathen & Gerhard (1999)
reduced/subresultant	n^6	Theorem 7.4

Table 2. Comparison of various normal PRS. The time in bit operations is for polynomials of degree at most n and with coefficients of length at most n and ignores logarithmic factors.

8   Experiments

We have implemented six of the PRS for polynomials with integral coefficients in C++, using Victor Shoup's "Number Theory Library" NTL 3.5a for integer and polynomial arithmetic. Since the Sturmian PRS agrees with the classical PRS up to sign, it is not mentioned here. The contents of the intermediate results in the primitive PRS are simply computed by successive gcd computations. Cooperman *et al.* (1999) propose a new algorithm that uses only an expected number of two gcd computations, but on random inputs it is slower than the naïve approach. All timings are the average over 10 pseudorandom inputs. The software ran on a Sun Sparc Ultra 1 clocked at 167MHz.

In the first experiment we pseudorandomly and independently chose three polynomials $f, g, h \in \mathbb{Z}[x]$ of degree $n-1$ with nonnegative coefficients of length

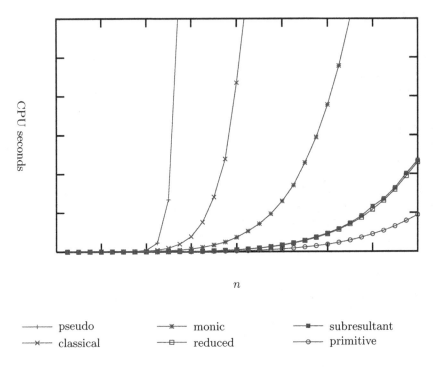

Fig. 1. Computation of polynomial remainder sequences for polynomials of degree $n-1$ with coefficients of bit length less than n for $1 \leq n \leq 32$.

less than n, for various values of n. Then we used the various PRS algorithms to compute the gcd of fh and gh of degrees less than $2n$. The running times are shown in Figures Figure 1 and Figure 2.

As seen in Table 2 the pseudo PRS turns out to be the slowest algorithm. The reason is that for random inputs with coefficients of length at most n the second polynomial is almost never monic. Thus Theorem 7.3 shows that for random inputs the running time for pseudo PRS is mainly exponential. A surprising result is that the primitive PRS, even implemented in a straightforward manner, turns out to be the fastest PRS. Collins and Brown & Traub only invented the subresultant PRS in order to avoid the primitive PRS since it seemed too expensive, but our tests show that for our current software this is not a problem.

Polynomial remainder sequences of random polynomials tend to be normal. Since Corollary 6.11 shows that reduced and subresultant PRS agree up to signs in the normal case, their running times also differ by little.

We are also interested in comparing the reduced and subresultant PRS, so we construct PRS which are not normal. To this end, we pseudorandomly and independently choose six polynomials f, f_1, g, g_1, h, h_1 for various degrees n as follows:

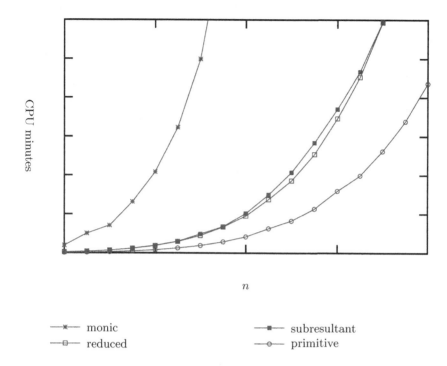

──✳── monic	──■── subresultant
──□── reduced	──○── primitive

Fig. 2. Computation of polynomial remainder sequences for polynomials of degree $n-1$ with coefficients of bit length less than n for $32 \leq n \leq 96$. Time is now measured in minutes.

polynomial	degree	coefficient length
f, g	$n/6$	$n/4$
f_1, g_1	$n/3$	n
h	$n/2$	$3n/4$
h_1	n	n

So the polynomials

$$F = (fh \cdot x^n + f_1)h_1$$
$$G = (gh \cdot x^n + g_1)h_1$$

have degrees less than $2n$ with coefficient length less than n, and every polynomial remainder sequence of F and G has a degree jump of $\frac{n}{3}$ at degree $2n - \frac{n}{6}$. Then we used the various PRS algorithms to compute the gcd of F and G. The running times are illustrated in Figures Figure 3 and Figure 4.

As in the first test series the pseudo PRS turns out to be the slowest, and the primitive PRS is the fastest. Here the monic PRS is faster than the reduced PRS. Since the PRS is non-normal, the α_i's are powers of the leading coefficients of the intermediate results, and their computation becomes quite expensive.

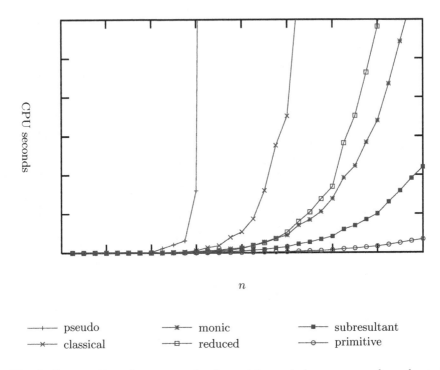

Fig. 3. Computation of non-normal polynomial remainder sequences for polynomials of degree $2n - 1$ with coefficient length less than n and a degree jump of $\frac{n}{3}$ at degree $2n - \frac{n}{6}$, for $1 \leq n \leq 32$.

References

ÉTIENNE BÉZOUT, Recherches sur le degré des équations résultantes de l'évanouissement des inconnues. *Histoire de l'académie royale des sciences* (1764), 288–338. Summary 88–91.

OTTO BIERMANN, Über die Resultante ganzer Functionen. *Monatshefte fuer Mathematik und Physik* (1891), 143–146. II. Jahrgang.

W. S. BROWN, On Euclid's Algorithm and the Computation of Polynomial Greatest Common Divisors. *Journal of the ACM* **18**(4) (1971), 478–504.

W. S. BROWN, The Subresultant PRS Algorithm. *ACM Transactions on Mathematical Software* **4**(3) (1978), 237–249.

W. S. BROWN AND J. F. TRAUB, On Euclid's Algorithm and the Theory of Subresultants. *Journal of the ACM* **18**(4) (1971), 505–514.

G. E. COLLINS, Polynomial remainder sequences and determinants. *The American Mathematical Monthly* **73** (1966), 708–712.

GEORGE E. COLLINS, Subresultants and Reduced Polynomial Remainder Sequences. *Journal of the ACM* **14**(1) (1967), 128–142.

GEORGE E. COLLINS, The Calculation of Multivariate Polynomial Resultants. *Journal of the ACM* **18**(4) (1971), 515–532.

G. E. COLLINS, Computer algebra of polynomials and rational functions. *The American Mathematical Monthly* **80** (1973), 725–755.

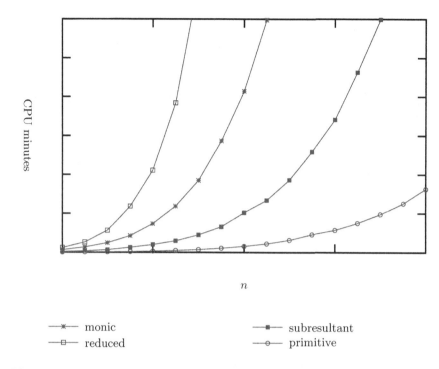

—✳— monic   —■— subresultant

—□— reduced   —○— primitive

Fig. 4. Computation of non-normal polynomial remainder sequences for polynomials of degree $2n - 1$ with coefficient length less than n and a degree jump of $\frac{n}{3}$ at degree $2n - \frac{n}{6}$, for $32 \leq n \leq 96$. Time is now measured in minutes.

GENE COOPERMAN, SANDRA FEISEL, JOACHIM VON ZUR GATHEN, AND GEORGE HAVAS, Gcd of many integers. In *COCOON '99*, ed. T. ASANO ET AL., Lecture Notes in Computer Science **1627**. Springer-Verlag, 1999, 310–317.

LEONHARD EULER, Démonstration sur le nombre des points où deux lignes des ordres quelconques peuvent se couper. *Mémoires de l'Académie des Sciences de Berlin* **4** (1748), 1750, 234–248. Eneström 148. *Opera Omnia*, ser. 1, vol. 26, Orell Füssli, Zürich, 1953, 46–59.

JOACHIM VON ZUR GATHEN, Parallel algorithms for algebraic problems. *SIAM Journal on Computing* **13**(4) (1984), 802–824.

JOACHIM VON ZUR GATHEN AND JÜRGEN GERHARD, *Modern Computer Algebra*. Cambridge University Press, 1999.

K. O. GEDDES, S. R. CZAPOR, AND G. LABAHN, *Algorithms for Computer Algebra*. Kluwer Academic Publishers, 1992.

PAUL GORDAN, *Vorlesungen über Invariantentheorie. Erster Band: Determinanten*. B. G. Teubner, Leipzig, 1885. Herausgegeben von GEORG KERSCHENSTEINER.

WALTER HABICHT, Eine Verallgemeinerung des Sturmschen Wurzelzählverfahrens. *Commentarii Mathematici Helvetici* **21** (1948), 99–116.

M. W. HASKELL, Note on resultants. *Bulletin of the New York Mathematical Society* **1** (1892), 223–224.

THOMAS W. HUNGERFORD, *Abstract Algebra: An Introduction*. Saunders College Publishing, Philadelphia PA, 1990.

C. G. J. JACOBI, De eliminatione variabilis e duabus aequationibus algebraicis. *Journal für die Reine und Angewandte Mathematik* **15** (1836), 101–124.

DONALD E. KNUTH, *The Art of Computer Programming, vol.2, Seminumerical Algorithms*. Addison-Wesley, Reading MA, 2nd edition, 1981.

L. KRONECKER, Die verschiedenen *Sturm*schen Reihen und ihre gegenseitigen Beziehungen. *Monatsberichte der Königlich Preussischen Akademie der Wissenschaften, Berlin* (1873), 117–154.

L. KRONECKER, Zur Theorie der Elimination einer Variabeln aus zwei algebraischen Gleichungen. *Monatsberichte der Königlich Preussischen Akademie der Wissenschaften, Berlin* (1881), 535–600. *Werke*, Zweiter Band, ed. K. HENSEL, Leipzig, 1897, 113–192. Reprint by Chelsea Publishing Co., New York, 1968.

THOMAS LICKTEIG AND MARIE-FRANÇOISE ROY, Cauchy Index Computation. *Calcolo* **33** (1997), 331–357.

R. LOOS, Generalized Polynomial Remainder Sequences. *Computing* **4** (1982), 115–137.

THOM MULDERS, A note on subresultants and the Lazard/Rioboo/Trager formula in rational function integration. *Journal of Symbolic Computation* **24**(1) (1997), 45–50.

ISAAC NEWTON, *Arithmetica Universalis, sive de compositione et resolutione arithmetica liber*. J. Senex, London, 1707. English translation as *Universal Arithmetick: or, A Treatise on Arithmetical composition and Resolution*, translated by the late Mr. Raphson and revised and corrected by Mr. Cunn, London, 1728. Reprinted in: DEREK T. WHITESIDE, *The mathematical works of Isaac Newton*, Johnson Reprint Co, New York, 1967, p. 4 ff.

DANIEL REISCHERT, Asymptotically Fast Computation of Subresultants. In *Proceedings of the 1997 International Symposium on Symbolic and Algebraic Computation ISSAC '97*, Maui HI, ed. WOLFGANG W. KÜCHLIN. ACM Press, 1997, 233–240.

V. STRASSEN, The computational complexity of continued fractions. *SIAM Journal on Computing* **12**(1) (1983), 1–27.

C. STURM, Mémoire sur la résolution des équations numériques. *Mémoires présentés par divers savants à l'Acadèmie des Sciences de l'Institut de France* **6** (1835), 273–318.

J. J. SYLVESTER, A method of determining by mere inspection the derivatives from two equations of any degree. *Philosophical Magazine* **16** (1840), 132–135. *Mathematical Papers* **1**, Chelsea Publishing Co., New York, 1973, 54–57.

RICHARD ZIPPEL, *Effective polynomial computation*. Kluwer Academic Publishers, 1993.

A Unifying Framework for the Analysis of a Class of Euclidean Algorithms

Brigitte Vallée

GREYC, Université de Caen, F-14032 Caen (France)
Brigitte.Vallee@info.unicaen.fr

Abstract. We develop a general framework for the analysis of algorithms of a broad Euclidean type. The average-case complexity of an algorithm is seen to be related to the analytic behaviour in the complex plane of the set of elementary transformations determined by the algorithms. The methods rely on properties of transfer operators suitably adapted from dynamical systems theory. As a consequence, we obtain precise average-case analyses of four algorithms for evaluating the Jacobi symbol of computational number theory fame, thereby solving conjectures of Bach and Shallit. These methods provide a unifying framework for the analysis of an entire class of gcd-like algorithms together with new results regarding the probable behaviour of their cost functions.

1 Introduction

Euclid's algorithm, discovered as early as 300BC, was analysed first in the worst case in 1733 by de Lagny, then in the average-case around 1969 independently by Heilbronn [8] and Dixon [5], and finally in distribution by Hensley [9] who proved in 1994 that the Euclidean algorithm has Gaussian behaviour; see Knuth's and Shallit's vivid accounts [12,20]. The first methods used range from combinatorial (de Lagny, Heilbronn) to probabilistic (Dixon). In parallel, studies by Lévy, Khinchin, Kuzmin and Wirsing had established the metric theory of continued fractions by means of a specific density transformer. The more recent works rely for a good deal on *transfer operators*, a far-reaching generalization of density transformers, originally introduced by Ruelle [17,18] in connection with the thermodynamic formalism and dynamical systems theory [1]. Examples are Mayer's studies on the continued fraction transformation [14], Hensley's work [9] and several papers of the author [21,22] including her analysis of the Binary GCD Algorithm [23].

In this paper, we provide new analyses of several classical and semi-classical variants of the Euclidean algorithm. A strong motivation of our study is a group of gcd-like algorithms that compute the Jacobi symbol whose relation to quadratic properties of numbers is well-known.

Methods. Our approach consists in viewing an algorithm of the broad gcd type as a dynamical system, where each iterative step is a linear fractional transformation (LFT) of the form $z \to (az+b)/(cz+d)$. The system control may be simple, what we call generic below, but also multimodal, what we call Markovian. A specific set of transformations is then associated to each algorithm. It will appear from our treatment that the computational complexity of an algorithm is in fact dictated by the collective dynamics of its associated set of transformations. More precisely, two factors intervene: (i) the characteristics of the LFT's in the complex domain; (ii) their *contraction properties*, notably

G. Gonnet, D. Panario, and A. Viola (Eds.): LATIN 2000, LNCS 1776, pp. 343–354, 2000.

near fixed points. There results a classification of gcd-like algorithms in terms of the average number of iterations: some of them are "fast", that is, of logarithmic complexity $\Theta(\log N)$, while others are "slow", that is, of the log-squared type $\Theta(\log^2 N)$.

It is established here that strong contraction properties of the elementary transformations that build up a gcd-like algorithm entail logarithmic cost, while the presence of an indifferent fixed-point leads to log-squared behaviour. In the latter case, the analysis requires a special twist that takes its inspiration from the study of intermittency phenomena in physical systems that was introduced by Bowen [2] and is nicely exposed in a paper of Prellberg and Slawny [15]. An additional benefit of our approach is to open access to characteristics of the distribution of costs, including information on moments: the fast algorithms appear to have concentration of distribution—the cost converges in probability to its mean—while the slow ones exhibit an extremely large dispersion of costs.

Technically, this paper relies on a description of relevant parameters by means of generating functions, a by now common tool in the average-case of algorithms [7]. As is usual in number theory contexts, the generating functions are Dirichlet series. They are first proved to be algebraically related to specific operators that encapsulate all the important informations relative to the "dynamics" of the algorithm. Their analytical properties depend on spectral properties of the operators, most notably the existence of a "spectral gap" that separates the dominant eigenvalue from the remainder of the spectrum. This determines the singularities of the Dirichlet series of costs. The asymptotic extraction of coefficients is then achieved by means of Tauberian theorems [4], one of several ways to derive the prime number theorem. Average complexity estimates finally result. The main thread of the paper is thus adequately summarized by the chain:

Euclidean algorithm $\rightsquigarrow$ Associated transformations $\rightsquigarrow$ Transfer operator
$\rightsquigarrow$ Dirichlet series of costs $\rightsquigarrow$ Tauberian inversion $\rightsquigarrow$ Average-case complexity.

This chain then leads to effective and simple criteria for distinguishing slow algorithms from fast ones, for establishing concentration of distribution, for analysing various cost parameters of algorithms, etc. The constants relative to the sloŽw algorithms are all explicit, while the constants relative to the fast algorithms are closely related to the entropy of the associated dynamical system: they are computable numbers; however, except in two classical cases, they do not seem be related to classical constants of analysis.

Motivations. We study here eight algorithms: the first four algorithms are variations of the classical Euclidean algorithm and are called *Classical* (G), *By-Excess* (L), *Classical-Centered* (K), and *Subtractive* (T). The last four algorithms serve to compute the Jacobi symbol introduced in Section 2, and are called *Even* (E), *Odd* (O), *Ordinary* (U) and *Centered* (C).

The complexity of the first four algorithms is now known: The two classical algorithms (G) and (K) have been analysed by Heilbronn, Dixon and Rieger [16]. The Subtractive algorithm (T) was studied by Knuth and Yao [25], and Vardi [24] analysed the By-Excess Algorithm (L) by comparing it to the Subtractive Algorithm. The methods used are rather disparate, and their applicability to new situations is somewhat unclear. Here, we design a unifying framework that also provides new results on the distribution of costs.

Two of the Jacobi Symbol Algorithms, the Centered (C) and Even (E) algorithms, have been introduced respectively by Lebesgue [13] in 1847 and Eisenstein [6] in 1844. Three of them, the Centered, Ordinary and Even algorithms, have been studied by Shallit [19] who provided a complete worst-case analysis. The present paper solves completely a conjecture of Bach and Shallit. Indeed, in [19], Shallit writes: "Bach has also suggested that one could investigate the average number of division steps in computing the Jacobi symbol [...]. This analysis is probably feasible to carry out for the Even Algorithm, and it seems likely that the average number of division steps is $\Theta(\log^2 N)$. However, determining the average behaviour for the two other algorithms seems quite hard."

Results and plan of the paper. Section 3 is the central technical section of the paper. There, we develop the line of attack outlined earlier and introduce successively Dirichlet generating functions, transfer operators of the Ruelle type, and the basic elements of Tauberian theory that are adequate for our purposes. The main results of this section are summarized in Theorem 1 and Theorem 2 that imply a general criterion for logarithmic versus log-squared behaviour, while providing a framework for higher moment analyses.

In Section 4, we return to our eight favorite algorithms—four classical variations and four Jacobi symbol variations. The corresponding analyses are summarized in Theorems 3 and 4 where we list our main results, some old and some new, that fall as natural consequences of the present framework. It results from the analysis (Theorem 3) that the Fast Class contains two classic algorithms, the Classical Algorithm (G), and the Classical Centered Algorithm (K), together with three Jacobi Symbol algorithms: the Odd (O), Ordinary (U) and Centered (C) Algorithms. Their respective average-case complexities on pairs of integers less than N are of the form $H_N \sim A_H \log N$ for $H \in \{G, K, O, U, C\}$.

The five constants are effectively characterized in terms of entropies of the associated dynamical system, and the constants related to the two classical algorithms are easily obtained, $A_G = (12/\pi^2) \log 2$, $A_K = (12/\pi^2) \log \phi$.

Theorem 4 proves that the Slow Class contains the remaining three algorithms, the By-Excess Algorithm (L), the Subtractive Algorithm (T), and one of the Jacobi Symbol Algorithm, the Even Algorithm (E). They all have a complexity of the log-squared type,

$$L_N \sim (3/\pi^2) \log^2 N, \quad T_N \sim (6/\pi^2) \log^2 N, \quad E_N \sim (2/\pi^2) \log^2 N.$$

Finally, Theorem 5 provides new probabilistic characterizations of the distribution of the costs: in particular, the approach applies to the analysis of the subtractive GCD algorithms for which we derive the order of growth of higher moments, which appears to be new. We also prove that concentration of distribution holds in the case of the the five fast algorithms (G, K, O, U, C).

Finally, apart from specific analyses, our main contributions are the following:

(a) We show how transfer operator method may be extended to cope with complex situations where the associated dynamical system may be either random or Markovian (or both!).

(b) An original feature in the context of analysis of algorithms is the encapsulation of the method of inducing (related to intermittency as evoked above).

(c) Our approch opens access to information on higher moments of the distribution of costs.

2 Eight Variations of the Euclidean Algorithm

We present here the eight algorithms to be analysed; the first four are classical variants of the Euclidean Algorithm, while the last four are designed for computing the Jacobi Symbol.

2.1. Variations of the classical Euclidean Algorithm. There are two divisions between u and v ($v > u$), that produce a positive remainder r such that $0 \le r < u$: the classical division (by-default) of the form $v = cu + r$, and the division by-excess, of the form $v = cu - r$. The centered division between u and v ($v > u$), of the form $v = cu + \varepsilon r$, with $\varepsilon = \pm 1$ produces a positive remainder r such that $0 \le r < u/2$. There are three Euclidean algorithms associated to each type of division, respectively called the Classical Algorithm (G), the By-Excess Algorithm (L), and the Classical Centered Algorithm (K). Finally, the Subtractive Algorithm (T) uses only subtractions and no divisions, since it replaces the classical division $v = cu + r$ by exactly c subtractions of the form $v := v - u$.

2.2. Variations for computing Jacobi symbol. The Jacobi symbol, introduced in [11], is a very important tool in algebra, since it is related to quadratic characteristics of modular arithmetic. Interest in its efficient computation has been reawakened by its utilisation in primality tests and in some important cryptographic schemes.

For two integers u and v (v odd), the possible values for the Jacobi symbol $J(u, v)$ are $-1, 0, +1$. Even if the Jacobi symbol can be directly computed from the classical Euclidean algorithm, thanks to a formula due to Hickerson [10], quoted in [24], we are mainly interested in specific algorithms that run faster. These algorithms are fundamentally based on the following two properties,

Quadratic Reciprocity law: $J(u, v) = (-1)^{(u-1)(v-1)/4} J(v, u)$ for $u, v \ge 0$ odd,
Modulo law: $J(v, u) = J(v - bu, u)$,

and they perform, like the classical Euclidean algorithm, a sequence of Euclidean-like divisions and exchanges. However, the Quadratic Reciprocity law being only true for odd integers, the standard Euclidean division has to be transformed into a pseudo–euclidean division of the form

$$v = bu + \varepsilon 2^k s \qquad \text{with } \varepsilon = \pm 1, s \text{ odd and strictly less than } u,$$

that creates another pair (s, u) for the following step. Then the symbol $J(u, v)$ is easily computed from the symbol $J(s, u)$.

The binary division, used in the Binary GCD algorithm, can also be used for computing the Jacobi symbol. However, it is different since the pseudo–division that it uses is NOT a modification of the classical euclidean division. We consider here four main algorithms according to the kind of pseudo–euclidean division that is performed. They are called the Even, Odd, Ordinary and Centered Algorithms, and their inputs are odd integers. The Even algorithm (E) performs divisions with even pseudo–quotients, and thus odd pseudo–remainders. The Odd algorithm (O) performs divisions with odd pseudo–quotients, and thus even pseudo-remainders from which powers of 2 are removed. The Ordinary (U) and Centered (C) Algorithms perform divisions where the pseudo–quotients are equal to the ordinary quotients or to the centered quotients; then, remainders may be even or odd, and, when they are even, powers of two are removed for obtaining the pseudo–remainders.

Alg., Type	Division	LFT's	Conditions.
(G) (1, all, 0)	$v = cu + r$ $0 \le r < u$	$\dfrac{1}{c+x}$ $c \ge 1$	$\mathcal{F} : c \ge 2$
(L) (1, all, 1)	$v = cu - r$ $0 \le r < u$	$\dfrac{1}{c-x}$ $c \ge 2$	$\mathcal{F} : c \ge 3$
(K) ($\frac{1}{2}$, all, 0)	$v = cu + \varepsilon r$ $c \ge 2, \varepsilon = \pm 1,$ $0 \le r < \frac{u}{2}$	$\dfrac{1}{c+\varepsilon x}$ $c \ge 2, \varepsilon = \pm 1,$ $(c, \varepsilon) \ne (2, -1)$	$\mathcal{F} : \varepsilon = 1$
(T) (1, all, 0)	$v = u + (v - u)$	$\mathcal{T} = \{q = \dfrac{1}{1+x}, p = \dfrac{x}{1+x}\}$	Finishes with pq
(E) (1, odd, 1)	$v = cu + \varepsilon s$ c even, $\varepsilon = \pm 1$ s odd, $0 < s < u$	$\dfrac{1}{c+\varepsilon x}$ c even, $\varepsilon = \pm 1$	$\mathcal{F} : \varepsilon = 1$
(O) (1, odd, 0)	$v = cu + \varepsilon 2^k s$ c odd, $\varepsilon = \pm 1$, s odd $k \ge 1, 0 \le 2^k s < u$	$\dfrac{2^k}{c+\varepsilon x}$ $k \ge 1, c$ odd $\ge 2^k + 1$	$\mathcal{J} : k = 0$
(U) (1, odd, 0)	$v = cu + 2^k s$ $s = 0$ or s odd, $k \ge 0$ $0 \le 2^k s < u$	$\mathcal{U}_0 = \{\dfrac{1}{c+x}, c \ge 1\}$ $\mathcal{U}_1 = \{\dfrac{2^k}{c+x}, \ge 1, c \ge 2^k\}$ $\mathcal{U}_{i\mid j} = \mathcal{U}_j \cap \{c \equiv i \bmod 2\}$	initial state: 0 final state: 1
(C) ($\frac{1}{2}$, odd, 0)	$v = cu + \varepsilon 2^k s$ $s = 0$ or s odd, $k \ge 0$ $0 \le 2^k s < \frac{u}{2}$	$\mathcal{C}_0 = \{\dfrac{1}{c+\varepsilon x}\}, \varepsilon = \pm 1,$ $c \ge 2, (c, \varepsilon) \ne (2, -1)$ $\mathcal{C}_1 = \{\dfrac{2^k}{c+\varepsilon x}\}, k \ge 1, \varepsilon = \pm 1$ $c \ge 2^{k+1}, (c, \varepsilon) \ne (2^{k+1}, -1)$ $\mathcal{C}_{i\mid j} = \mathcal{C}_j \cap \{c \equiv i \bmod 2\}$	initial state: 0 final state: 1

2.3. The sets of linear fractional transformations. When performing ℓ (pseudo)-euclidean divisions on the input (u, v), each of the eight algorithms builds a specific continued fraction of height ℓ that decomposes the rational $x = u/v$ as

$$u/v = h_1 \circ h_2 \circ \ldots \circ h_\ell(a),$$

where the h_i's are linear fractional transformations (LFT's) and a is the last value of the rational. The value a equals 1 for the Even Algorithm (E) and By-Excess Algorithm (L), and equals 0 for the other six algorithms. The rational inputs of all the algorithms always belong to a basic interval of the form $\mathcal{I} = [0, \rho]$ with $\rho = 1/2$ for the two centered algorithms (K) and (C) and $\rho = 1$ in the other six cases. For the first four algorithms, the valid inputs are all the rationals of $\mathcal{I}$, while the valid inputs of the last four algorithms are only the odd rationals of $\mathcal{I}$. The variable valid has two possible values $\{$all, odd$\}$, and finally, the type of the algorithm is defined as the value of the triple (ρ, valid, a).

The precise form of the possible LFT's depends on the algorithm, and there are two classes of algorithms, the generic class and the Markovian class:

In the case of the first six algorithms, there may exist special sets of LFT's in the initial step $(\mathcal{J})$ or in the final step $(\mathcal{F})$. However, all the other steps are generic, in the sense that they use the same set of LFT's, that we call the generic set. These algorithms are called themselves generic.

On the contrary, the last two algorithms –the Ordinary Algorithm (U) and the Centered Algorithm (C)– have a Markovian flavour. If the quotient b is odd, then the remainder is even, and thus k satisfies $k \geq 1$; if b is even, then the remainder is odd, and thus k satisfies $k = 0$. This link is of Markovian type, and we consider two states: the 0 state, which means "the remainder of (u, v) is odd", i.e. $k = 0$, and the 1 state, which means "the remainder of (u, v) is even ", i.e. $k \geq 1$. Denoting by $\mathcal{U}_j$, resp. $\mathcal{C}_j$ the set of LFT's which can be used in state j, we obtain four different sets, $\mathcal{U}_{i|j}$, resp. $\mathcal{C}_{i|j}$, each of them brings rationals from state j to state i. The initial state is the 0 state and the final state is the 1 state.

3   Dynamical Operators and Tauberian Theorems

Here, we describe the general tools for analysing algorithms of the Euclidean type that are based on some division-like operation and exchanges. We first introduce the generating functions relative to the height of the continued fraction and we relate them to the dynamical operator associated to the algorithm. This operator can be generic or Markovian, according to the structure of the algorithm. In this way, the two Dirichlet series that intervene in the analysis, called $F(s)$ and $G(s)$, are expressed in terms of the Ruelle operator. The average number of steps involves partial sums of coefficients of these two Dirichlet series, and Tauberian Theorems are a classical tool that transfers analytical properties of Dirichlet series into asymptotic behaviour of their coefficients.

3.1. Generating functions. We consider the following sets relative to $\mathcal{I} := [0, \rho]$,

$$\widetilde{\Omega} := \{(u, v); \ u, v \text{ valid}, \ (u/v) \in \mathcal{I}\}, \quad \widetilde{\Omega}_N := \{(u, v) \in \widetilde{\Omega}, v \leq N\},$$

$$\Omega := \{(u, v); \ u, v \text{ valid}, \gcd(u, v) = 1, (u/v) \in \mathcal{I}\}, \quad \Omega_N := \{(u, v) \in \Omega, v \leq N\},$$

for the possible inputs of an algorithm, and we denote by $\Omega^{[\ell]}$, $\widetilde{\Omega}^{[\ell]}$, $\Omega_N^{[\ell]}$, $\widetilde{\Omega}_N^{[\ell]}$ the subsets of Ω, $\widetilde{\Omega}$, Ω_N, $\widetilde{\Omega}_N$ for which the algorithm performs exactly ℓ pseudo–divisions. Equivalently, the height of the continued fraction is equal to ℓ. We study the average number of steps S_N, $\widetilde{S}_N$ of the algorithm on Ω_N, $\widetilde{\Omega}_N$

$$S_N := \frac{1}{|\Omega_N|} \sum_{\ell \geq 0} \ell \, |\Omega_N^{[\ell]}| \qquad \widetilde{S}_N := \frac{1}{|\widetilde{\Omega}_N|} \sum_{\ell \geq 0} \ell \, |\widetilde{\Omega}_N^{[\ell]}| \qquad (1)$$

and we wish to evaluate their asymptotic behaviour (for $N \to \infty$). We first consider pairs (u, v) with fixed $v = n$, and we denote by $\nu_n^{[\ell]}$ (resp. $\widetilde{\nu}_n^{[\ell]}$) the number of such elements of $\Omega^{[\ell]}$ (resp. $\widetilde{\Omega}^{[\ell]}$). We introduce the double generating functions $S(s, w)$ and $\widetilde{S}(s, w)$ of the sequences $(\nu_n^{[\ell]})$ and $(\widetilde{\nu}_n^{[\ell]})$,

$$S(s, w) := \sum_{\ell \geq 1} w^\ell \sum_{n > 1} \frac{\nu_n^{[\ell]}}{n^s}, \qquad \widetilde{S}(s, w) := \sum_{\ell \geq 1} w^\ell \sum_{n > 1} \frac{\widetilde{\nu}_n^{[\ell]}}{n^s}. \qquad (2)$$

The Riemann series $\widehat{\zeta}$ relative to valid numbers $\widehat{\zeta}(s) := \sum_{v \text{ valid}} v^{-s}$ relates the two generating functions via the equality $\widetilde{S}(s, w) = \widehat{\zeta}(s) S(s, w)$. It is then sufficient to study $S(s, w)$. We introduce the two sequences (a_n) and (b_n) together with their associated Dirichlet series $F(s), G(s)$,

$$a_n := \sum_{\ell \geq 1} \nu_n^{[\ell]} \qquad b_n := \sum_{\ell \geq 1} \ell\, \nu_n^{[\ell]}, \qquad F(s) = \sum_{n > 1} \frac{a_n}{n^s}, \qquad G(s) = \sum_{n > 1} \frac{b_n}{n^s}. \qquad (3)$$

Now, $F(s)$ and $G(s)$ can be easily expressed in terms of $S(s, w)$ since

$$F(s) = S(s, 1), \qquad G(s) = \frac{d}{dw} S(s, w)|_{w=1}, \qquad (4)$$

and intervene, via partial sums of their coefficients, in the quantity S_N,

$$S_N = \frac{\sum_{n \leq N} \sum_{\ell \geq 0} \ell\, \nu_n^{[\ell]}}{\sum_{n \leq N} \sum_{\ell \geq 0} \nu_n^{[\ell]}} = \frac{\sum_{n \leq N} b_n}{\sum_{n \leq N} a_n}. \qquad (5)$$

3.2. Ruelle operators. We show now how the Ruelle operators associated to the algorithms intervene in the evaluation of the generating function $S(s, w)$. We denote by $\mathcal{L}$ a set of LFT's. For each $h \in \mathcal{L}$, $D[h]$ denotes the denominator of the linear fractional transformation (LFT) h, defined for $h(x) = (ax + b)/(cx + d)$ with a, b, c, d coprime integers by $D[h](x) := |cx + d| = |\det h|^{1/2} |h'(x)|^{-1/2}$. The Ruelle operator $\mathbf{L}_s$ relative to the set $\mathcal{L}$ depends on a complex parameter s

$$\mathbf{L}_s[f](x) := \sum_{h \in \mathcal{L}} \frac{1}{D[h](x)^s} f \circ h(x). \qquad (6)$$

More generally, when given two sets of LFT's, $\mathcal{L}$ and $\mathcal{K}$, the set $\mathcal{L}\mathcal{K}$ is formed of all $h \circ g$ with $h \in \mathcal{L}$ and $g \in \mathcal{K}$, and the multiplicative property of denominator D, i.e., $D[h \circ g](x) = D[h](g(x))\, D[g](x)$, implies that the operator $\mathbf{K}_s \circ \mathbf{L}_s$ uses all the LFT's of $\mathcal{L}\mathcal{K}$

$$\mathbf{K}_s \circ \mathbf{L}_s[f](x) := \sum_{h \in \mathcal{L}\mathcal{K}} \frac{1}{D[h](x)^s} f \circ h(x). \qquad (7)$$

3.3. Ruelle operators and generating functions. The first six algorithms are generic, since they use the same set $\mathcal{H}$ at each generic step. In this case, the ℓ-th iterate of $\mathbf{H}_s$ generates all the LFT's used in ℓ (generic) steps of the algorithm. The last two algorithms are Markovian. There are four sets of LFT's, and each of these sets, denoted by $\mathcal{U}_{i|j}$ "brings" rationals from state j to state i. We denote by $\mathbf{U}_{s,i|j}$ the Ruelle operator associated to set $\mathcal{U}_{i|j}$, and by $\mathbf{U}_s$ the "matrix operator"

$$\mathbf{U}_s = \begin{pmatrix} \mathbf{U}_{s,0|0} & \mathbf{U}_{s,0|1} \\ \mathbf{U}_{s,1|0} & \mathbf{U}_{s,1|1} \end{pmatrix}. \tag{8}$$

By multiplicative properties (7), the ℓ-th iterate of $\mathbf{U}_s$ generates all the elements of $\mathcal{U}^{<\ell>}$, i.e., all the possible LFT's of height ℓ. More precisely, the coefficient of index (i,j) of the matrix $\mathbf{U}_s^\ell$ is the Ruelle operator relative to the set $\mathcal{U}_{i|j}^{<\ell>}$ that brings rationals from state j to state i in ℓ steps.

In both cases, the Ruelle operator is then a "generating" operator, and generating functions themselves can be easily expressed with the Ruelle operator:

Proposition 1. *The double generating function $S(s,w)$ of the sequence $(\nu_n^{[\ell]})$ can be expressed as a function of the Ruelle operators associated to the algorithm. In the generic case,*

$$S(s,w) = w\mathbf{K}_s[1](a) + w^2\,\mathbf{F}_s \circ (I - w\mathbf{H}_s)^{-1} \circ \mathbf{J}_s[1](a).$$

Here, the Ruelle operators $\mathbf{H}_s$, $\mathbf{F}_s$, $\mathbf{J}_s$, $\mathbf{K}_s$ are relative to the generic set $\mathcal{H}$, final set $\mathcal{F}$, initial set $\mathcal{J}$ or mixed set $\mathcal{K} := \mathcal{J} \cap \mathcal{F}$; the value a is the final value of the rational u/v. In the Markovian case, the Ruelle operator $\mathbf{U}_s$ is a matrix operator, and

$$S(s,w) = \begin{pmatrix} 0 & 1 \end{pmatrix} w\mathbf{U}_s\,(I - w\mathbf{U}_s)^{-1} \begin{bmatrix} 1 \\ 0 \end{bmatrix}(0).$$

In both cases, the Dirichlet series $F(s)$ and $G(s)$ involve powers of quasi-inverse of the Ruelle operator of order 1 for $F(s)$, and of order 2 for $G(s)$.

3.4. Tauberian Theorems. Finally, we have shown that the average number of steps S_N of the four Algorithms on Ω_N is a ratio where the numerators and the denominators involve the partial sums of the Dirichlet series $F(s)$ and $G(s)$. Thus, the asymptotic evaluation of S_N, $\tilde{S}_N$ (for $N \to \infty$) is possible if we can apply the following Tauberian theorem [4] to the Dirichlet series $F(s)$, $\widehat{\zeta}(s)F(s)$, $G(s)$, $\widehat{\zeta}(s)G(s)$.

Tauberian Theorem. *[Delange] Let $F(s)$ be a Dirichlet series with non negative coefficients such that $F(s)$ converges for $\Re(s) > \sigma > 0$. Assume that (i) $F(s)$ is analytic on $\Re(s) = \sigma$, $s \neq \sigma$, and (ii) for some $\beta \geq 0$, one has $F(s) = A(s)(s-\sigma)^{-\beta-1} + C(s)$, where A, C are analytic at σ, with $A(\sigma) \neq 0$. Then, as $N \to \infty$,* $\displaystyle\sum_{n \leq N} a_n =$

$$\frac{A(\sigma)}{\sigma\Gamma(\beta+1)}\,N^\sigma \log^\gamma N\,[1 + \varepsilon(N)], \quad \varepsilon(N) \to 0.$$

In the remainder of the paper, we show that the Tauberian Theorem applies to $F(s), G(s)$ with $\sigma = 2$. For $F(s)$, it applies with $\beta = 0$. For $G(s)$, it applies with $\beta = 1$ or $\beta = 2$. For the slow algorithms, β equals 2, and the average number of steps will be of order $\log^2 N$. For the fast algorithms, β equals 1, and this will prove the logarithmic behaviour of the average number of steps. First, the function $F(s)$ is closely linked to the $\widehat{\zeta}$ function relative to valid numbers. Then, the Tauberian Theorem applies to $F(s)$ and $\widehat{\zeta}(s)F(s)$, with $\sigma = 2$ and $\beta = 0$.

3.5. Functional Analysis. Here, we consider the following conditions on a set $\mathcal{H}$ of LFTs that will entail that the Ruelle operator $\mathbf{H}_s$ relative to $\mathcal{H}$ fulfills all the properties that we need for applying the Tauberian Theorem to Dirichlet series $G(s)$.

Conditions $\mathcal{Q}(\mathcal{H})$. *There exist an open disk $\mathcal{V} \subset \mathcal{I}$ and a real $\alpha > 2$ such that*

(C_1) *Every LFT $h \in \mathcal{H}$ has an analytic continuation on $\mathcal{V}$, and h maps the closure $\bar{\mathcal{V}}$ of disk $\mathcal{V}$ inside $\mathcal{V}$. Every function $|h'|$ has an analytic continuation on $\mathcal{V}$ denoted by $\tilde{h}$.*

(C_2) *For each $h \in \mathcal{H}$, there exists $\delta(h) < 1$ for which $0 < |\tilde{h}(z)| \leq \delta(h)$ for all $z \in \mathcal{V}$ and such that the series $\displaystyle\sum_{h \in \mathcal{H}} |\frac{\delta(h)}{\det(h)}|^{s/2}$ converges on the plane $\Re(s) > \alpha$.*

(C_3) *Let $\mathcal{H}_{[k]}$ denote the subset of $\mathcal{H}$ defined as $\mathcal{H}_{[k]} := \{h \in \mathcal{H} \mid \det(h) = 2^k\}$. One of two conditions (i) or (ii) holds: (i) $\mathcal{H} = \mathcal{H}_{[0]}$,*
(ii) For any $k \geq 1$, $\mathcal{H}_{[k]}$ is not empty and $\mathcal{H} = \bigcup_{k \geq 1} \mathcal{H}_{[k]}$.
Moreover, for any $k \geq 0$ for which $\mathcal{H}_{[k]} \neq \emptyset$, the intervals $(h(\mathcal{I}), h \in \mathcal{H}_{[k]})$ form a pseudo-partition of $\mathcal{I}$.

(C_4) *For some integer A, the set $\mathcal{H}$ contains a subset $\mathcal{D}$ of the form*
$$\mathcal{D} := \{h \mid h(x) = A/(c+x) \quad \text{with integers } c \to \infty\}.$$

We denote by $\mathbf{H}_s$ a Ruelle operator associated to a generic set $\mathcal{H}$, and by $\mathbf{U}_s$ a Ruelle Markovian operator (associated to a Markovian process with two states). In this case, the subset $\mathcal{U}_i$ denotes the set relative to state i. Here $\mathcal{I}$ denotes the basic interval $[0, \rho]$. We consider that conditions $\mathcal{Q}(\mathcal{H})$ and $\mathcal{Q}(\mathcal{U}_i)$ (for $i = 0, 1$) hold. Then, we can prove the following: Under conditions (C_1) and (C_2), the Ruelle operator $\mathbf{H}_s$ acts on $\mathcal{A}_\infty(\mathcal{V})$ for $\Re(s) > \sigma$ and the operator $\mathbf{U}_s$ acts on $\mathcal{A}_\infty(\mathcal{V})^2$ for $\Re(s) > \alpha$. They are compact (even nuclear in the sense of Grothendieck). Furthermore, for real values of parameter s, they have dominant spectral properties: there exist a unique dominant eigenvalue $\lambda(s)$ positive, analytic for $s > \alpha$, a unique dominant eigenfunction ψ_s, and a unique dominant projector e_s such that $e_s[\psi_s] = 1$. Then, there is a spectral gap between the dominant eigenvalue and the remainder of the spectrum. Under conditions (C_3), the operators $\mathbf{H}_2$, $\mathbf{U}_2$ are density transformers; thus one has $\lambda(2) = 1$ and $e_2[f] = \int_0^1 f(t)dt$ (generic case) or $e_2[f_0, f_1] = \int_0^1 [f_0(t) + f_1(t)]dt$ (Markovian case). Finally, condition (C_4) implies that the operators $\mathbf{H}_s$, $\mathbf{U}_s$ have no eigenvalue equal to 1 on the line $\Re(s) = 2, s \neq 2$. Finally, the powers of the quasi–inverse of the Ruelle operator which intervene in the expression of generating functions $F(s)$ and $G(s)$ fulfill all the hypotheses of Tauberian Theorem:

Theorem 1. *Let $\mathcal{H}$ be a generic set that satisfies $\mathcal{Q}(\mathcal{H})$ and $\mathcal{U}$ be a Markovian set that satisfies $\mathcal{Q}(\mathcal{U}_i)$ $(i = 0, 1)$. Then, for any $p \geq 1$, the p-th powers $(I - \mathbf{H}_s)^{-p}$, $(I - \mathbf{U}_s)^{-p}$ of the quasi–inverse of the Ruelle operators relative to $\mathcal{H}$ and $\mathcal{U}$ are analytic on the punctured plane $\{\Re(s) \geq 2, s \neq 2\}$ and have a pole of order p at $s = 2$. Near $s = 2$, one has, for any function f positive on $\mathcal{J}$, and any $x \in \mathcal{J}$,*

$$(I - \mathbf{H}_s)^{-p}[f](x) \quad \text{or} \quad (I - \mathbf{U}_s)^{-p}[f](x) \sim \frac{1}{(s-2)^p} \left(\frac{-1}{\lambda'(2)}\right)^p \psi_2(x)\, e_2[f],$$

Here, $\lambda(s)$ is the dominant eigenvalue, ψ_s is the dominant eigenfunction and e_s the dominant projector with the condition $e_s[\psi_s] = 1$.

Then the Euclidean Algorithm associated to $\mathcal{H}$ or to $\mathcal{U}$ performs an average number of steps on valid rationals of $\mathcal{I}$ with denominator less than N that is asymptotically logarithmic, $H_N \sim A_H \, \log N$, $U_N \sim A_U \, \log N$. The constants A_H, A_U involve the entropy of the associated dynamical systems.

In the case when the set $\mathcal{H}$ is only almost well-behaved –it contains one "bad" LFT p, but the set $\mathcal{Q} := \mathcal{H} \setminus \{p\}$ is well-behaved– we adapt the *method of inducing* that originates from dynamical systems theory.

Theorem 2. *Let $\mathcal{H}$ be a generic set of LFTs for which the following holds: (i) There exists a element p of $\mathcal{H}$ which possesses an indifferent point, i.e., a fixed point where the absolute value of the derivative equals 1, (ii) The LFT p does not belong to the final set $\mathcal{F}$, (iii) If $\mathcal{Q}$ denotes the set $\mathcal{H} \setminus \{p\}$, and $\mathcal{M}$, $\mathcal{M}^+$ the sets $p^\star \mathcal{Q}$, $p^\star \mathcal{F}$, then conditions $\mathcal{Q}(\mathcal{M})$, $\mathcal{Q}(\mathcal{M}^+)$ are fullfilled.*

Then the Euclidean Algorithm associated to $\mathcal{H}$ performs an average number of steps on valid rationals of $\mathcal{I}$ with denominator less than N that is asymptotically of log–squared type, $H_N \sim \widetilde{H}_N \sim A_H \, \log^2 N$.

The average number Q_N of good steps (i.e., steps that use elements of $\mathcal{Q}$) performed by the Euclidean Algorithm on valid rationals of $\mathcal{I}$ with denominator less than N satisfies $Q_N \sim \widetilde{Q}_N \sim A_Q \log N$, and the constant A_Q involves the entropy of the dynamical system relative to set $\mathcal{M}$.

4   Average-Case Analysis of the Algorithms

We now come back to the analysis of the eight algorithms, and we study successively the fast algorithms, and the slow algorithms

4.1. The Fast Algorithms. We consider the generic sets $\mathcal{G}, \mathcal{K}, \mathcal{O}$ relative to the Classical Algorithm, the Classical Centered Algorithm and the Odd Algorithm, or the Markovian sets $\mathcal{U}$ or $\mathcal{C}$ relative to the Ordinary or the Centered Algorithm. It is easy to verify that the conditions $\mathcal{Q}(\mathcal{G}), \mathcal{Q}(\mathcal{K}), \mathcal{Q}(\mathcal{O}), \mathcal{Q}(\mathcal{U}_i)(i = 0, 1), \mathcal{Q}(\mathcal{C}_i)(i = 0, 1)$, hold.

Moreover, at $s = 2$, the Ruelle operators can be viewed as density transformers. However, the dynamical systems to which they are associated may be complex objects, since they are random for the Odd Algorithm, and are both random and Markovian for the Ordinary and Centered Algorithm. The reason is that the three pseudo-divisions (odd, ordinary, centered) are related to dyadic valuation, so that continued fractions expansions are only defined for rationals numbers. However, one can define random continued fraction for real numbers when choosing in a suitable random way the dyadic valuation of a real number. Then, the Ruelle operator relative to each algorithm can be viewed as the transfer operator relative to this random dynamical system. Now, the application of Theorem 1 gives our first main result:

Theorem 3. *Consider the five algorithms, the Classical Algorithm (G), the Classical Centered Algorithm (K), the Odd Algorithm (O), the Ordinary Algorithm (U) or the Centered Algorithm (C). The average numbers of division steps performed by each of these five algorithms, on the set of valid inputs of denominator less than N are of asymptotic logarithmic order. They all satisfy*

$$H_N \sim \widetilde{H}_N \sim (2/h(\mathcal{H})) \log N \qquad for \qquad H \in \{G, K, O, U, C\},$$

where $h(\mathcal{H})$ is the entropy of the dynamical system relative to the algorithm. For the two first algorithms, the entropies are explicit,

$$h(\mathcal{G}) = \pi^2/(6 \log 2), \qquad h(\mathcal{K}) = \pi^2/(6 \log \phi).$$

Each of the previous entropies can be computed, by adapting methods developed in previous papers [3,22]. What we have at the moment is values from simulations that already provide a consistent picture of the relative merits of the Centered, Odd, and Ordinary Algorithms, namely,

$$A_C \approx 0.430 \pm 0.005 \qquad A_0 \approx 0.435 \pm 0.005 \qquad A_U \approx 0.535 \pm 0.005.$$

It is to be noted that the computer algebra system MAPLE makes use of the Ordinary Algorithm, (perhaps on the basis that only unsigned integers need to be manipulated), although this algorithm appears to be from our analysis the fast one that has the *worst* convergence rate.

4.2. The Slow Algorithms. For the Even algorithm and the By-Excess Algorithm, the "bad" LFT is defined by $p(x) := 1/(2 - x)$, with an indifferent point in 1. For the Subtractive Algorithm, the "bad" LFT is defined by $p(x) := x/(1+x)$, with an indifferent point in 0. In the latter case, the induced set $\mathcal{M}$ coincides with the set $\mathcal{G}$ relative to the Classical Algorithm (G).

When applying Theorem 2 to sets $\mathcal{E}, \mathcal{L}, \mathcal{T}$, we obtain our second main result:

Theorem 4. *Consider the three algorithms, the By-Excess Algorithm (L), the Subtractive Algorithm (T), the Even Algorithm (E). The average numbers of steps performed by each of the three algorithms, on the set of valid inputs of denominator less than N are of asymptotic log–squared order. They satisfy*

$$L_N \sim \widetilde{L}_N \sim A_L \log^2 N, \quad T_N \sim \widetilde{T}_N \sim A_T \log^2 N, \quad E_N \sim \widetilde{E}_N \sim A_E \log^2 N,$$
$$\text{with } A_E = (3/\pi^2), A_T = (6/\pi^2), A_E = (2/\pi^2).$$

The average numbers of good steps performed by the algorithms on the set of valid inputs of denominator less than N satisfy

$$P_N \sim \widetilde{P}_N \sim A_P \log N \qquad G_N \sim \widetilde{G}_N \sim A_G \log N, \qquad M_N \sim \widetilde{M}_N \sim A_M \log N.$$
$$\text{with } A_P = (6 \log 2)/\pi^2, A_G = (12 \log 2)/\pi^2, A_M = (4 \log 3)/\pi^2.$$

4.3. Higher moments. Our methods apply to other parameters of continued fraction. On the other hand, by using successive derivatives of the double generating function $S(s, w)$, we can easily evaluate higher moments of the random variable "number of iterations".

Theorem 5. (i) *For any integer $\ell \geq 1$ and any of the five fast algorithms, the ℓ-th moment of the cost function is asymptotic to the ℓth power of the mean. In particular the standard deviation is $o(\log N)$. Consequently the random variable expressing the cost satisfies the concentration of distribution property.*

(ii) *For any integer $\ell \geq 2$ and any of the three slow algorithms, the ℓth moment of the cost function is of order $N^{\ell-1}$ and the standard deviation is $\Theta(\sqrt{N})$.*

Acknowledgements. I wish to thank Pierre Ducos for earlier discussions of 1992, Charlie Lemée and Jérémie Bourdon for their master's theses closely related to this work, Thomas Prellberg for his explanations of the inducing method, Dieter Mayer for his introduction to random dynamical systems, Ilan Vardi for Hickerson's formula, and Philippe Flajolet for the experimentations.

References

1. BEDFORD, T., KEANE, M., AND SERIES, C., Eds. *Ergodic Theory, Symbolic Dynamics and Hyperbolic Spaces*, Oxford University Press, 1991.
2. BOWEN, R. Invariant measures for Markov maps of the interval, *Commun. Math. Phys.* 69 (1979) 1–17.
3. DAUDÉ, H., FLAJOLET, P., AND VALLÉE, B. An average-case analysis of the Gaussian algorithm for lattice reduction, *Combinatorics, Probability and Computing* (1997) 6 397–433
4. DELANGE, H. Généralisation du Théorème d'Ikehara, *Ann. Sc. ENS, (1954) 71, pp 213–242.*
5. DIXON, J. D. The number of steps in the Euclidean algorithm, *Journal of Number Theory 2* (1970), 414–422.
6. EISENSTEIN, G. Einfacher Algorithmus zur Bestimmung der Werthes von $\left(\frac{a}{b}\right)$, *J. für die Reine und Angew. Math.* 27 (1844) 317-318.
7. FLAJOLET, P. AND SEDGEWICK, R. Analytic Combinatorics, Book in preparation (1999), see also INRIA Research Reports 1888, 2026, 2376, 2956.
8. HEILBRONN, H. On the average length of a class of continued fractions, Number Theory and Analysis, ed. by P. Turan, New-York, Plenum, 1969, pp 87-96.
9. HENSLEY, D. The number of steps in the Euclidean algorithm, *Journal of Number Theory* 49, 2 (1994), 142–182.
10. HICKERSON, D. Continued fractions and density results for Dedekind sums, *J. Reine Angew. Math.* 290 (1977) 113-116.
11. JACOBI, C.G.J. Uber die Kreistheilung und ihre Anwendung auf die Zalhentheorie, *J. für die Reine und Angew. Math.* 30 (1846) 166–182.
12. KNUTH, D.E. The art of Computer programming, Volume 2, 3rd edition, Addison Wesley, Reading, Massachussets, 1998.
13. LEBESGUE V. A. Sur le symbole (a/b) et quelques unes de ses applications, *J. Math. Pures Appl.* 12(1847) pp 497–517
14. MAYER, D. H. Continued fractions and related transformations, In *Ergodic Theory, Symbolic Dynamics and Hyperbolic Spaces*, T. Bedford, M. Keane, and C. Series, Eds. Oxford University Press, 1991, pp. 175–222.
15. PRELLBERG, T. AND SLAWNY, J. Maps of intervals with Indifferent fixed points: Thermodynamic formalism and Phase transitions. *Journal of Statistical Physics* 66 (1992) 503-514
16. RIEGER, G. J. Uber die mittlere Schrittazahl bei Divisionalgorithmen, *Math. Nachr.* (1978) pp 157-180
17. RUELLE, D. *Thermodynamic formalism,* Addison Wesley (1978)
18. RUELLE, D. *Dynamical Zeta Functions for Piecewise Monotone Maps of the Interval*, vol. 4 of *CRM Monograph Series*, American Mathematical Society, Providence, 1994.
19. SHALLIT, J. On the worst–case of the three algorithmss for computing the Jacobi symbol, *Journal of Symbolic Computation* 10 (1990) 593–610.
20. SHALLIT, J. Origins of the analysis of the Euclidean Algorithm, *Historia Mathematica* 21 (1994) pp 401-419
21. VALLÉE, B. Opérateurs de Ruelle-Mayer généralisés et analyse des algorithmes d'Euclide et de Gauss, *Acta Arithmetica* 81.2 (1997) 101–144.
22. VALLÉE, B. Fractions continues à contraintes périodiques, *Journal of Number Theory* 72 (1998) pp 183–235.
23. VALLÉE, B. Dynamics of the Binary Euclidean Algorithm: Functional Analysis and Operators., *Algorithmica* (1998) vol 22 (4) pp 660–685.
24. VARDI, I. Continued fractions, Preprint, chapter of a book in preparation.
25. YAO, A.C., AND KNUTH, D.E. Analysis of the subtractive algorithm for greatest common divisors. *Proc. Nat. Acad. Sc. USA* 72 (1975) pp 4720-4722.

Worst–Case Complexity of the Optimal LLL Algorithm

Ali Akhavi

GREYC - Université de Caen, F-14032 Caen Cedex, France
ali.akhavi@info.unicaen.fr

Abstract. In this paper, we consider the open problem of the complexity of the LLL algorithm in the case when the approximation parameter of the algorithm has its extreme value 1. This case is of interest because the output is then the strongest Lovász–reduced basis. Experiments reported by Lagarias and Odlyzko [13] seem to show that the algorithm remain polynomial in average. However no bound better than a naive exponential order one is established for the worst–case complexity of the optimal LLL algorithm, even for fixed small dimension (higher than 2). Here we prove that, for any *fixed dimension* n, the number of iterations of the LLL algorithm is *linear* with respect to the size of the input. It is easy to deduce from [17] that the linear order is optimal. Moreover in 3 dimensions, we give a tight bound for the maximum number of iterations and we characterize precisely the output basis. Our bound also improves the known one for the usual (non–optimal) LLL algorithm.

1 Introduction

A Euclidean lattice is a set of all integer linear combinations of p linearly independent vectors in $\mathbb{R}^n$. Any lattice can be generated by many bases (all of them of cardinality p). The lattice basis reduction problem is to find bases with good Euclidean properties, that is sufficiently short vectors and almost orthogonal. The problem is old and there exist numerous notions of reduction; the most natural ones are due to Minkowski or to Korkhine–Zolotarev. For a general survey, see for example [8,16]. Both of these reduction processes are "strong", since they build reduced bases with in some sense best Euclidean properties. However, they are also computationally hard to find, since they demand the first vector of the basis should be a shortest one in the lattice. It appears that finding such an element in a lattice is likely to be NP-hard [18,1,5].

Fortunately, even approximate answers to the reduction problem have numerous theoretical and practical applications in computational number theory and cryptography: Factoring polynomials with rational coefficients [12], finding linear diophantine approximations (Lagarias, 1980), breaking various cryptosystems [15] and integer linear programming [7,11]. In 1982, Lenstra, Lenstra and Lovász [12] gave a powerful approximation reduction algorithm. It depends on a real approximation parameter $\delta \in [1,2[$ and is called LLL(δ). It is a possible generalization of its 2–dimensional version, which is the famous Gauss algorithm. The celebrated LLL algorithm seems difficult to analyze precisely, both in the worst–case and in average–case. The original paper [12] gives an upper bound for the number of iterations of LLL(δ), which is polynomial in the data

G. Gonnet, D. Panario, and A. Viola (Eds.): LATIN 2000, LNCS 1776, pp. 355–366, 2000.

size, for all values of δ except the optimal value 1: When given n input vectors of $\mathbb{R}^n$ of length at most M, the data size is $O(n^2 \log M)$ and the upper bound is $n^2 \log_\delta M + n$. When the approximation parameter δ is 1, the only known upper–bound is M^{n^2}, which is *exponential* even for fixed dimension. It was still an open problem whether the optimal LLL algorithm is polynomial. In this paper, we prove that the number of iterations of the algorithm is *linear* for any fixed dimension. More precisely, it is $O(A^{n^3} \log M)$ where A is any constant strictly greater than $(2/\sqrt{3})^{(1/6)}$. We prove also that under a quite reasonable heuristic principle, the number of iterations is $O((2/\sqrt{3})^{n^2/2} \log M)$. In the 3–dimensional case (notice that the problem was totally open even in this case), we provide a precise linear bound, which is even better than the usual bounds on the non–optimal versions of the LLL algorithm. Several reasons motivate our work on the complexity of the optimal LLL algorithm.

1. This problem is cited as an open question by respected authors [4,17] and I think that the answer will bring at least a better understanding of the lattice reduction process. Of course, this paper is just an insight to the general answer of the question.

2. The optimal LLL algorithm provides the strongest Lovász–reduced basis in a lattice (the best bounds on the classical length defects and orthogonality defect). In many applications, people seem to be interested by such a basis [13], and sometimes even in fixed low dimension [14].

3. We believe that the complexity of finding an optimal Lovász–reduced basis is of great interest and the LLL algorithm is the most natural way to find an optimal Lovász–reduced basis in a lattice (we develop it more in the conclusion).

Plan of the paper. Section 2 presents the LLL algorithm and introduces some definitions and notations. In Section 3, we recall some known results in 2 dimensions. Section 4 deals with the worst–case complexity of the optimal LLL algorithm in 3-dimensional case. Finally, in Section 5, we prove that in any fixed dimension, the number of iterations of the LLL algorithm is linear with respect to the length of the input.

2   General Description of the LLL Algorithm

Let $\mathbb{R}^p$ be endowed with the usual scalar product $(\cdot, \cdot)$ and Euclidean length $|u| = (u, u)^{1/2}$. The notation $(u)_{\perp H}$ will denote the projection of the vector u in the classical orthogonal space $H^\perp$ of H in $\mathbb{R}^p$. The set $\langle u_1, u_2, ..., u_r \rangle$ denotes the vector space spanned by a family of vectors $(u_1, u_2, ..., u_r)$. A lattice of $\mathbb{R}^p$ is the set of all integer linear combinations of a set of linearly independent vectors. Generally it is given by one of its bases $(b_1, b_2, \ldots, b_n)$ and the number n is the dimension of the lattice. So, if M is the maximum length of the vectors b_i, the data-size is $(n^2 \log M)$ and when working in fixed dimension, the data-size is $O(\log M)$. The determinant $\det(L)$ of the lattice L is the volume of the n–dimensional parallelepiped spanned by the origin and the vectors of any basis. Indeed it does not depend on the choice of a basis. The usual Gram–Schmidt orthogonalization process, builds in polynomial–time from a basis $b = (b_1, b_2, \ldots, b_n)$ an orthogonal basis $b^* = (b_1^*, b_2^*, \ldots, b_n^*)$ (which is generally not a basis for the lattice generated by b) and a lower–triangular matrix $m = (m_{ij})$ that expresses the system b

into the system b^*. By construction,

$$
\begin{cases}
b_1^* = b_1 \\
b_2^* = b_{2\perp\langle b_1\rangle} \\
\vdots \\
b_i^* = b_{i\perp\langle b_1,\cdots,b_{i-1}\rangle} \\
\vdots \\
b_n^* = b_{n\perp\langle b_1,\cdots,b_{n-1}\rangle}
\end{cases}
, \quad
m =
\begin{array}{c}
b_1 \\ \vdots \\ b_i \\ b_{i+1} \\ \vdots \\ b_n
\end{array}
\begin{pmatrix}
1 & \cdots & \cdots & \cdots & \cdots & 0 \\
\vdots & \ddots & & \ddots & & \vdots \\
\cdots & \cdots & 1 & 0 & & \vdots \\
\cdots & \cdots & m_{i+1,i} & 1 & \ddots & \vdots \\
\vdots & \vdots & \vdots & & \ddots & 0 \\
m_{n1} & \cdots & \cdots & \cdots & \cdots & 1
\end{pmatrix}. \tag{1}
$$

We recall that if L is the lattice generated by the basis b, its determinant $\det(L)$ is expressed in term of the lengths $|b_i^*|$: $\det(L) = \prod_{i=1}^n |b_i^*|$.

The ordered basis b is called *proper* if $|m_{ij}| \leq 1/2$, for $1 \leq j < i \leq n$. There exists a simple polynomial-time algorithm which makes any basis proper by means of integer translations of each b_i in the directions of b_j, for j decreasing from $i - 1$ to 1.

Definition 1. *[12] For any $\delta \in [1, 2[$, the basis $(b_1, \ldots, b_n)$ is called δ–reduced (or LLL(δ)-reduced or δ-Lovász–reduced) if it fullfils the two conditions:*

(i) $(b_1, ..., b_n)$ is proper.
(ii) $\forall i \in \{1, 2, \cdots, n-1\}$ $(1/\delta)\, |(b_i)_{\perp\langle b_1,\cdots,b_{i-1}\rangle}| < |(b_{i+1})_{\perp\langle b_1,\cdots,b_{i-1}\rangle}|.$

The optimal LLL algorithm ($\delta = 1$) is a possible generalization of its 2–dimensional version, the famous Gauss' algorithm, whose precise analysis is already done both in the worst–case [10,17,9] and in the average–case [6]. In the sequel, a reduced basis denotes always an optimal LLL-reduced one. When talking about the algorithm without other precision, we always mean the optimal LLL algorithm.

For all integer i in $\{1, 2, \cdots, n-1\}$, we call B_i the two–dimensional basis formed by the two vectors $(b_i)_{\perp\langle b_1,\cdots,b_{i-1}\rangle}$ and $(b_{i+1})_{\perp\langle b_1,\cdots,b_{i-1}\rangle}$. Then, by the previous Definition $(b_1, \ldots, b_n)$ is reduced iff it is proper and if all bases B_i are reduced.

Definition 2. *Let t be a real parameter such that $t > 1$. We call a basis $(b_1, \ldots, b_n)$ t–quasi–reduced if it satisfies the following conditions:*

(i) the basis $(b_1, \ldots, b_{n-1})$ is proper.
(ii) For all $1 \leq i \leq n-2$, the bases B_i are reduced.
(iii) The last basis B_{n-1} is not reduced but it is t–reduced: $|m_{n,n-1}| < 1/2$ and

$$(1/t)\, |(b_{n-1})_{\perp\langle b_1,\cdots,b_{n-2}\rangle}| \leq |(b_n)_{\perp\langle b_1,\cdots,b_{n-2}\rangle}| < |(b_{n-1})_{\perp\langle b_1,\cdots,b_{n-2}\rangle}|.$$

In other words, whenever the beginning basis $(b_1, \cdots, b_{n-1})$ is reduced, but the whole basis $b = (b_1, \cdots, b_n)$ is not, then for all $t > 1$ such that the last two–dimensional basis B_i is t–reduced, the basis b is called t–quasi-reduced.

Here is a simple enunciation of the LLL(δ) algorithm:

The LLL(δ)-reduction algorithm:
Input: A basis $b = (b_1, \ldots, b_n)$ of a lattice L.
Output: A $LLL(1)$-reduced basis b of the lattice L.
Initialization: Compute the orthogonalized system b^* and the matrix m.
i := 1;
While i < n do
$b_{i+1} := b_{i+1} - \lfloor m_{i+1,i} \rceil b_i$ ($\lfloor x \rceil$ is the integer nearest to x).
Test: Is the two-dimensional basis B_i δ–reduced?
    If true, make $(b_1, \ldots, b_{i+1})$ proper by translations; **set i := i + 1;**
    If false, swap b_i and b_{i+1}; update b^* and m; if $i \neq 1$ then **set i := i − 1;**

During an execution of the algorithm, the index i variates in $\{1, \ldots, n\}$. It is called the *current index.* When i equals some $k \in \{1, \ldots, n-1\}$, the beginning lattice generated by $(b_1, \ldots, b_k)$ is already reduced. Then, the reduction of the basis B_k is tested. If the test is positive, the basis $(b_1, \ldots, b_{k+1})$ is made proper and the beginning lattice generated by $(b_1, \ldots, b_{k+1})$ is then reduced. So, i is incremented. Otherwise, the vectors b_k and b_{k+1} are swapped. At this moment, nothing guarantees that $(b_1, \ldots, b_k)$ "remains" reduced. So, i is decremented and the algorithm updates b^* and m, translates the new b_k in the direction of b_{k-1} and tests the reduction of the basis B_{k-1}. Thus the index i may fall down to 1. Finally when i equals n, the whole basis is reduced and the algorithm terminates. An example of the variation of the index i is shown by Figure 1.

value of the current index *i*

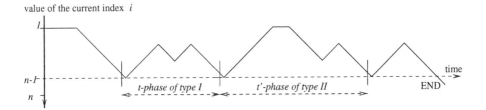

Fig. 1. Variation of the index i presented as a walk.

In the sequel an iteration of the LLL algorithm is precisely an iteration of the "while" loop in the previous enunciation. So each iteration has exactly one test (Is the two–dimensional basis B_i reduced ?) and the number of steps is exactly the number of tests. Notice that whenever a test at a level i is negative, i.e. the basis B_i is not reduced, after the swap of b_i and b_{i+1} the determinant d_i of the lattice $(b_1, \ldots, b_i)$ is decreased. Moreover, for any $t > 1$, if at the moment of the test the basis B_i is even not t–reduced, the determinant d_i is decreased by a factor at least $1/t$. This explains the next definition.

Definition 3. *For a real parameter $t > 1$, a step of index i is called t–decreasing if at the moment of the test, the basis B_i is not t–reduced. Else, it is called t–non–decreasing.*

[12] pointed out that during the execution of a non–optimal LLL algorithm, say LLL(δ) for some $\delta > 1$, all steps with negative tests are δ–decreasing. Similarly, we assert the next lemma, based on the decrease of the integer quantity

$$D := \prod_{i=1}^{n-1} d_i^2 := \prod_{i=1}^{n-1} \prod_{j=1}^{i} |b_j^*|^2, \qquad (2)$$

by a factor $1/t^2$, whenever a step is t–decreasing (other steps do not make D increase).

Lemma 1. *Let the LLL(1) algorithm run on an integer input $(b_1, \ldots, b_n)$ of length $\log M$. For any $t > 1$, the number of t-decreasing steps is less than $n(n-1)/2 \log_t M$.*

Definition 4. A phase *is a sequence of steps that occur between two successive tests of reduction of the last two–dimensional lattice B_{n-1}. For a real $t > 1$, we say that a phase is a t–phase if at the beginning of the phase the basis $(b_1, \ldots b_n)$ is t–quasi–reduced. Phases are classified in two groups: A phase is called of type I if during the next phase the first vector b_1 is never swapped. Else, it is called of of type II (see Figure 1).*

3 Some Known Results in 2 Dimensions: Gauss' Algorithm

In two dimensions a phase of the algorithm is an iteration (of the "while" loop) and the only positive test occurs at the end of the algorithm. Thus the number of steps equals the number of negative tests plus one. For any $t > 1$, before the input is t-quasi-reduced, each step is t–decreasing. So by Lemma 1 any input basis (b_1, b_2) will be t-quasi-reduced within at most $\log_t M$ steps. Then the next Lemma leads to the bound $\log_{\sqrt{3}} M + 2$ for the total number of steps of Gauss' algorithm. This bound is not optimal [17]. However, in next sections we generalize this argumentation to the case of an arbitrary fixed dimension. Notice that the Lemma does not suppose that the input basis is integral and this fact is used in the sequel (proof in [2]).

Lemma 2. *For any $t \in]1, \sqrt{3}]$, during the execution of Gauss' algorithm on any input basis (not necessarily integral), there are at most 2 t–non-decreasing steps.*

4 The 3–Dimensional Case

Let t be a real parameter such that $1 < t \leq 3/2$. Here we count separately the iterations that are inside t–phases and the iterations that are not inside t–phases.
First we show that the total number of steps that are not inside t–phases is *linear* with respect to the input length $\log M$ (Lemma 3). Second we prove that the total number of iterations inside t–phases is always less than nine (Lemma 4). Thus we exhibit for the first time a linear bound for the number of iterations of the LLL algorithm in 3–dimensional space (the naive bound is M^6.) In addition, our argumentation gives a precise characterization of a reduced basis in the three–dimensional space.

Theorem 1. *The number of iterations of the LLL(1) algorithm on an input integer basis (b_1, b_2, b_3) of length $\log M$ is less than $\log_{\sqrt{3}} M + 6 \log_{3/2} M + 9$.*

The linear order for our bound is in fact optimal since it is so in dimension 2 [17] and one can obviously build from a basis b of $n - 1$ vectors of maximal length M, another basis b' of n vectors of the same maximal length such the number of iterations of the

LLL algorithm on the b' is strictly greater than on b. Moreover, even if we have not tried here to give the best coefficient of linearity in dimension 3, our bound is quite acceptable since [19] exhibits a family of bases of lengths $\log M$, for which the number of iterations of the algorithm is *greater* than $2.6 \log_2 M + 4$. Our bound is $13.2 \log_2 M + 9$. Observe that the classical bound on the usual non–optimal $LLL(2/\sqrt{3})$ is $28.9 \log_2 M + 2$, and even computed more precisely as in Lemma 3 it remains $24.1 \log_2 M + 2$. So our bound, which is also valid for $LLL(\delta)$ with $\delta < 3/2$, improves the classical upper–bound on the number of steps of $LLL(\delta)$ provided that $\delta < 1.3$.

The next Lemma (proof in [2]) is a more precise version of Lemma 1 in the particular case of 3 dimensions. (It is used in the sequel for $t \le 3/2$.)

Lemma 3. *Let the LLL algorithm run on a integer basis (b_1, b_2, b_3) of length $\log M$. Let t be a real parameter such that $1 < t \le \sqrt{3}$. The number of steps that are not inside any t–phase is less than $\log_{\sqrt{3}} M + 6 \log_t M$.*

4.1 From a Quasi–reduced Basis to a Reduced One

Lemma 4. *Let t be a real parameter in $]1, 3/2]$. When the dimension is fixed at 3, there are at most three t–phases during an execution of the algorithm. The total number of steps inside t–phases is less than nine.*

The proof is based on Lemma 5 and Corollaries 2 and 1. Lemma 5 shows that a t–phase of type I is necessarily an ending phase and has exactly 3 iterations.

The central role in the proof is played by Lemma 7 and its Corollary 2 asserting that in 3 dimensions, there are at most 2 t–phases of type II during an execution.

Finally, Lemma 6 and Corollary 1 show that any t–phase of type II has a maximum number of 3 iterations[1].

Remarks. (1) For t chosen closer to 1 ($\sqrt{6/5}$ rather than $3/2$), if during an execution, a t-reduced basis is obtained, then a reduced one will be obtained after at most 9 steps (see [2]). (A t–phase is necessarily followed by another t–phase.) **(2)** Of course, the general argumentation of the next section (for an arbitrary fixed dimension) holds here. But both of these directions lead to a less precise final upper bound.

Lemma 5. *For all $t \in]1, \sqrt{3}]$, a t-phase of type I has 3 steps and is an ending phase.*

Proof. The vector b_1 will not be modified during a phase of type I. Then, by Lemma 2, the basis $((b_2)_{\perp b_1}, (b_3)_{\perp b_1})$ will be reduced after only two iterations[2]. But here, there is one additional step (of current index 1, with a positive test) between these two iterations.

Lemma 6. *For any $t > 1$, if a basis $(b_1, \ldots, b_{n-1}, b_n)$ is t–quasi–reduced, then the basis $(b_1, \ldots, b_{n-2}, b_n)$ is t'–quasi–reduced, with $t' > (2/\sqrt{3})t$.*

---

[1] These facts are used to make the proof clearer but they are not essential in the proof: actually, if a phase has more than 3 iterations, then these additional steps (which are necessarily with negative tests and with the index i equal to 1) are t–decreasing and all t–decreasing steps are already counted by Lemma 3.

[2] Lemma 2 does not demand the input basis to be integral.

Corollary 1. *In* 3 *dimensions, for all* $t \in]1, 3/2]$*, a* t*–phase of type II has* 3 *steps.*

Proof. Since (b_1, b_2, b_3) is $3/2$–quasi–reduced, by the previous Lemma, (b_1, b_3) is $\sqrt{3}$–quasi–reduced. Then by Lemma 2, (b_1, b_3) will be reduced after 2 steps.

The next Lemma plays a central role in the whole proof. This result which remains true when (b_1, b_2, b_3) is reduced gives also a precise characterization of a 1-Lovász reduced basis in dimension 3. A detailed proof is available in [2].

Lemma 7. *For all* $t \in]1, 3/2]$*, if the basis* (b_1, b_2, b_3) *is* t*–quasi–reduced and proper, then among all the vectors of the lattice that are not in the plan* $\langle b_1, b_2 \rangle$*, there is at most one pair of vectors* $\pm u$ *whose lengths are strictly less than* $|b_3|$*.*

Proof (sketch). Let $u := x\, b_1 + y\, b_2 + z\, b_3$ be a vector of the lattice $((x, y, z) \in \mathbb{Z}^3)$. The vector u is expressed in the orthogonal basis b^* defined by (1) and its length satisfies $|u|^2 = (x + y\, m_{21} + z\, m_{31})^2 \, |b_1^*|^2 + (y + z\, m_{32})^2 \, |b_2^*|^2 + |b_3^*|^2$.
First, since (b_1, b_2, b_3) is $3/2$–quasi reduced, one gets easily that if $|z| > 1$ or $|y| > 1$ or $|x| > 1$, then $|u| > |b_3|$. Now, if $z = 1$, by considering the ratio $|u|^2/|b_3|^2$, one shows that there exits at most one pair $(x, y) \in \{0, 1, -1\}^2 \backslash \{(0, 0)\}$ such that $|u| < |b_3|$. This unique vector depends on the signs of m_{21}, m_{31} et m_{32} as recapitulated by Table 1.

u	$b_3 - b_2$	$b_3 - b_1 + b_2$	$b_3 + b_1 - b_2$	$b_3 + b_2$	$b_3 - b_1 - b_2$	$b_3 + b_2$	$b_3 - b_2$	$b_3 + b_1 + b_2$
m_{21}	+	+	+	+	−	−	−	−
m_{31}	+	+	−	−	+	+	−	−
m_{32}	+	−	+	−	+	−	+	−

Table 1. The unique vector possibly strictly shorter than b_3, as a function of signs of m_{ij}.

Corollary 2. *During an execution of LLL(1) algorithm in three dimensions, for all* $t \in]1, 3/2]$*, there are at most two* t*–phases of type II.*

Proof. Assume (b_1, b_2, b_3) is the t–quasi–reduced basis at the beginning of a first t–phase of type II and and let (b_1, b_2, b_3') denote the basis obtained from (b_1, b_2, b_3) by making the latter proper. Since the t-phase is of type II, $|b_3'| < |b_1|$ and the algorithm swaps $|b_1|$ and $|b_3'|$. As (b_1, b_2) is Gauss-reduced, b_1 is a shortest vector of sub-lattice generated by (b_1, b_2). Thus the fact $|b_3'| < |b_1|$ together with the previous Lemma show that in the whole lattice there is at most one pair of vectors $\pm u$ strictly shorter than b_3'. So the vector b_3' can be swapped only once. In particular, only one new t'–phase (for any $t' > 1$) of type II may occur before the end of the algorithm and after the first t–phase ($t \leq 3/2$) of type II, all phases except eventually one have exactly 2 iterations.

5 Arbitrary Fixed Dimension n

In the previous section, we argued in an *additive* way: We chose a tractable value t_0 ($3/2$ in the 3 dimensions) such that for $1 < t \leq t_0$ we could easily count the maximum number of steps inside all t–phases. Then we *added* the last bound (9 in the last Section) to the total number of iterations that were not inside t–phases.

Here we argue differently. On one hand, the total number of t–decreasing steps is classically upper bounded by $n^2 \log_t M$ (Lemma 1). Now, for all $t > 1$, we call a t–*non–decreasing* sequence, a sequence of t–non–decreasing consecutive steps. During such a sequence just before any negative test of index i, the basis $(b_1, \ldots, b_{i+1})$ is t–quasi–reduced. The problem is that during a t–*non–decreasing* sequence, we cannot *quantify* efficiently the decreasing of the usual integer potential function[3] D (whose definition is recalled in (2)). The crucial point here (Proposition 1) is that for all $t \in]1, \sqrt{3}]$, there exists some integer $c(n, t)$ such that any t–non–decreasing sequence of the LLL(1) algorithm – when it works on an arbitrary input basis $(b_1, \ldots, b_n)$ (no matter the lengths of the vectors)– has strictly less than $c(n, t)$ steps. In short, any sequence of iterations, which is longer than $c(n, t)$, has a t–decreasing step.

Hence, our argumentation is in some sense *multiplicative* since the total number of iterations with negative tests is thus bounded from above by $c(n, t)n^2 \log_t M$. We deduce the following theorem which for the first time exhibits a linear bound on the number of iterations of the LLL algorithm in fixed dimension.

Theorem 2. *For any fixed dimension n, let the optimal LLL algorithm run on an integer input $(b_1, \ldots, b_n)$ of length $\log M$. The maximum number of iterations K satisfies:*

(i) for all constant $A > (2/\sqrt{3})^{(1/6)}$, K is $O(A^{n^3} \log M)$.

(ii) under a very plausible heuristic, K is also $O\left((2/\sqrt{3})^{n^2/2} \log M\right)$.

The first formulation (i) is based on Proposition 1, and Lemmata 1, 8. For the second formulation (ii) we use also Lemma 9 (proved under a very plausible heuristic).

The next Lemma is an adaptation of counting methods used by Babai, Kannan and Schnorr [3,7,14] when finding a shortest vector in a lattice with a Lovász–reduced basis on hand. For a detailed proof, see [2].

Lemma 8. *Let $t \in]1, 2[$ be a real parameter and $\mathcal{L}$ be a lattice generated by a basis $b := (b_1, \ldots, b_n)$, which is not necessarily integral and whose vectors are of arbitrary length. If b is proper and t–quasi-reduced then there exists an integer $\alpha(n, t)$ such that the number of vectors of the lattice $\mathcal{L}$ whose lengths are strictly less than $|b_1|$ is strictly less than $\alpha(n, t)$. Moreover,*

$$\alpha(n, t) < \sqrt{3t^2/(4 - t^2)} \; 3^{n-1} \, (2/\sqrt{3})^{\frac{n(n-1)}{2}}. \tag{3}$$

Remark. The sequence $\alpha(n, t)$ is increasing with n (and also with t).

Proposition 1. *Let n be a fixed dimension and t a real parameter in $]1, \sqrt{3}]$. There exists an integer $c(n, t)$ such that the length of any t–non–decreasing sequence of the LLL(1) algorithm – on any input basis $(b_1, \ldots, b_n)$, no matter the lengths of its vectors and no matter the basis is integral – is strictly less than $c(n, t)$.*

---

[3] The naive bound is obtained using only the fact that D is a strictly positive integer less than $M^{n(n-1)}$ and it is strictly decreasing at each step with a negative test.

Proof (sketch). By induction on n. The case $n = 2$ is trivial and $c(2, t) = 2$ (Lemma 2). Suppose that the assertion holds for any basis of $n - 1$ vectors and let the algorithm run on a basis $b := (b_1, \ldots, b_n)$. Let us consider the longest possible t–non–decreasing sequence. After at most $c(n - 1, t)$ t–non–decreasing steps, b is t–quasi–reduced [4].

If the next phase if of type I, then the algorithm works actually with the basis of cardinality $(n - 1)$, $b_{\perp b_1} := ((b_2)_{\perp b_1}, \ldots, (b_n)_{\perp b_1})$, which is also t–quasi–reduced. Then by the induction hypothesis, the t–non–decreasing sequence will be finished after at most $c(n - 1, t) + \alpha(n - 1, t)$ more steps[5].

On the other hand, there are at most $\alpha(n, t)$ successive phases of type II, since Lemma 8 asserts that the first vector of the t–quasi–reduced basis $(b_1, \ldots, b_n)$ can be modified at most $\alpha(n, t)$ times. Each of them has no more than $c(n - 1, t)$ steps, because the algorithm works actually on $(b_1, \ldots, b_{n-1})$.

After the last t–phase of type II, it may be one more t-phase of type I. Finally, since $\alpha(n, t)$ is increasing with respect to n, the quantity $c(n, t)$ is less than

$$c(n-1,t)+c(n-1,t)\alpha(n,t)+c(n-1,t)+\alpha(n-1,t) < (c(n-1,t)+1)(\alpha(n,t)+2),$$

and finally $\quad c(n, t) + 1 \leq (c(2, t) + 1) \prod_{i=2}^{n} (\alpha(i, t) + 2) \,.$ \hfill (4)

Proof ((i) of Theorem 2). Each sequence of $c(n, t)$ steps contains at least one t–decreasing step. At each t–decreasing step, the quantity D, which is always in the interval $[1, M^{n^2}]$, is decreasing by at least $1/t$. So the total number of iterations of the algorithm is always less than $c(n, t) n^2 \log_t M$. Now by choosing a fixed $t \in]1, \sqrt{3}]$, relations (3) and (4) together show that the quantity $n^2 c(n, t)$ is bounded from above by A^{n^3}, where A is any constant greater than $(2/\sqrt{3})^{1/6}$.

In the first proof, we choose for t an arbitrary value in the interval $]1, \sqrt{3}]$. Now, we improve our bound by choosing t as a function of n. What we really need here is to evaluate the number of possible successive t-phases of type II. So the main question is: When a basis $(b_1, \ldots, b_n)$ of a lattice L is t–quasi–reduced, how many lattice points u are satisfying $(1/t)|b_1| < |u| < |b_1|$? More precisely, is it possible to choose t, as a function of dimension n, such that the open volume between the two n–dimensional balls of radii $|b_1|$ and $(1/t)|b_1|$ does not contain any lattice point?

Now, we answer these questions under a quite reasonable heuristic principle which is often satisfied. So the bound on $C(n)$ and on the maximum number of iterations will be improved. This heuristic is due to Gauss. Consider a lattice of determinant $\det(L)$. The heuristic claims that the number of lattice points inside a ball $\mathcal{B}$ is well approximated by $\text{volume}(\mathcal{B})/\det(L)$. More precisely the error is of the order of the surface of the ball $\mathcal{B}$. This principle holds for very large class of lattices, in particular those used in applications (for instance "regular lattices" where the minima are close to each other and where the fundamental parallelepiped is close to a hyper–cube). Moreover, notice that this heuristic also leads to the result of Lemma 8.

---

[4] Otherwise, there would be a t–non–decreasing sequence of more than $c(n - 1, t)$ steps while the algorithm runs on the basis $(b_1, \ldots, b_{n-1})$.

[5] During the $c(n - 1, t)$ steps on $b_{\perp b_1}$, each change of the first vector $(b_2)_{\perp b_1}$ (no more than $\alpha(n - 1, t)$, by Lemma 8) is followed by one step (of current index one) with a positive test which has not been counted yet.

Under this assumption and if $\gamma_n = \pi^{n/2}/\Gamma(1 + n/2)$ denotes the volume of the n–dimensional unit ball, then the number of lattice points $\beta(n, t)$ that lie strictly between balls of radii $|b_1|$ and $(1/t)|b_1|$ satisfies (at least asymptotically)

$$\beta(n, t_n) \le \gamma_n \ (|b_1|^n/\det(L)) \ (1 - 1/t^n). \tag{5}$$

Now if $(b_1, \ldots, b_n)$ is t–quasi–reduced with $t \le \sqrt{3}$,

$$|b_1|^n/\det(L) = |b_1|^n / \prod_{i=1}^n |b_i^*| \le 3^n \ (2/\sqrt{3})^{\frac{n(n-1)}{2}}.$$

Then, using the classical Stirling approximation, $\beta(n, t)$ is bounded from above:

$$\beta(n, t) < 3 \ (\pi e/n)^{n/2} \ (2/\sqrt{3})^{\frac{n(n-1)}{2}} \ (1 - 1/t^n).$$

By routine computation, we deduce the following Lemma.

Lemma 9. *Suppose that there exists n_0 such that for $n \ge n_0$, relation (5) is true. Then there exists a sequence $t_n > 1$ satisfying:*
(i) (t_n) is decreasing and tends to 1 with respect to n.
(ii) for all $n \ge n_0$, $\beta(n, t_n) < 1$ and $1/\log t_n < 3n \ (\pi e/n)^{n/2} \ (2/\sqrt{3})^{\frac{n(n-1)}{2}}$.

Remark: One deduces from (i) that if $\beta(n - 1, t_n) = 0$, then $\beta(n - 1, t_{n-1}) = 0$.

Proof (sketch for (ii) of Theorem 2).
The quantities t_n and n_0 are defined by the previous Lemma. First we prove that for $n > n_0$ and with the notations of Proposition 1,

$$c(n, t_n) \le c(n - 1, t_n) + (c(n_0, t_n) + \alpha((n_0, t_n))). \tag{6}$$

Indeed, after $c(n - 1, t)$ steps, $(b_1, \ldots, b_n)$ is t_n–quasi–reduced and if H denotes the vector space $\langle b_1, \ldots, b_{n-n_0-1} \rangle$, the (n_0)–dimensional basis $b_{\perp H} := ((b_{n-n_0})_{\perp H}, \ldots, (b_n)_{\perp H})$ is t_n–quasi–reduced as well. Thus by the previous Lemma during the t_n–non–decreasing sequence, its first vector cannot be modified. So from the first time that $(b_1, \ldots, b_n)$ is t_n–quasi–reduced until the end of the t_n–non–decreasing sequence the current index i will always be in the integral interval $\{n - n_0, \ldots, n\}$. Then by Proposition 1 the sequence of t_n–non–decreasing iterations may continue for at most $c(n_0, t_n) + \alpha(n_0, t_n)$ more iterations.[6] This ends the proof of relation (6).

So for $n > n_0$, $c(n, t_n) \le (n - n_0 + 1)(c(n_0, t_n) + \alpha(n_0, t_n))$.

Since $t_n < t_{n_0}$, the basis $b_{\perp H}$ is also t_{n_0}–quasi–reduced and by Lemma 8, $\alpha(n_0, t_n) \le \alpha(n_0, t_{n_0})$. (The same relation is true for $k < n_0$.) Finally the quantity $c(n_0, t_n) + \alpha(n_0, t_n)$ is a constant B that depends only on n_0. We have then $c(n, t_n) < nB$. So a sequence longer than nB contains always a t_n–decreasing step and the total number of iterations is less than $nBn^2 \log_{t_n} M$. Finally Lemma 9 gives an upper–bound for $1/\log t_n$ and leads to the (ii) of Theorem 2.

---

[6] The quantity $\alpha(n_0, t_n)$ corresponds to the maximum number of positive tests with the current index $i = n - n_0$, after the the first time $b_{\perp H}$ is t_n–quasi–reduced.

6   Conclusion

Our paper gives for the first time linear bounds for the the maximum number of iterations of the optimal LLL algorithm, in fixed dimension. I believe that the complexity of finding an optimal Lovász–reduced basis is of great interest and not well–known.

Kannan presented [7] an algorithm which uses as sub–routine the non–optimal LLL algorithm ($\delta > 1$) and outputs a Korkine–Zolotarev basis of the lattice in $O(n^n) \log M$ steps. Such an output is also an optimal Lovász–reduced basis (Actually it is stronger). Thus Kannan's algorithm provides an upper–bound on the complexity of finding an optimal Lovász–reduced basis[7]. For the future, one of the two following possibilities (or both) has to be considered.

(1) Our upper–bound is likely to be improved. However, observe that in this paper we have already improved notably the naive bound for fixed dimension (the exponential order is replaced by linear order). For the moment our bound remains worse than the one Kannan exhibits for his algorithm.

(2) The LLL algorithm which is the most natural way to find an optimal Lovász–reduced basis is not the best way (and then the same phenomenon may be possible for finding a non–optimal Lovász–reduced basis: more efficient algorithms than the classical LLL algorithm may output the same reduced basis).

Acknowledgments

I am indebted to Brigitte Vallée for drawing my attention to algorithmic problems in lattice theory and for regular helpful discussions.

References

1. M. Ajtai. The shortest vector problem in L_2 is NP-hard for randomized reduction. *ECCC-TR97-047*, 1997. http://www.eccc.uni-trier.de/eccc/.
2. A. Akhavi. The worst case complexity of the optimal LLL algorithm. Preprint. http://www.info.unicaen.fr/~akhavi/publi.html, Caen, 1999.
3. L. Babai. On Lovász' lattice reduction and the nearest lattice point problem. *Combinatorica*, 6(1):1–13, 1986.
4. A. Bachem and R. Kannan. Lattices and the basis reduction algorithm. *CMU-CS*, pages 84–112, 1984.
5. J. Cai. Some recent progress on the complexity of lattice problems. *ECCC-TR99-006*, 1999. http://www.eccc.uni-trier.de/eccc/
6. H. Daudé, Ph. Flajolet, and B. Vallée. An average-case analysis of the Gaussian algorithm for lattice reduction. *Comb., Prob. & Comp.*, 123:397–433, 1997.
7. R. Kannan. Improved algorithm for integer programming and related lattice problems. In *15th Ann. ACM Symp. on Theory of Computing*, pages 193–206, 1983.
8. R. Kannan. Algorithmic geometry of numbers. *Ann. Rev. Comp. Sci.*, 2:231–267, 1987.

---

[7] Moreover, so far as I know, there exists no polynomial time algorithm for finding a Korkine–Zolotarev reduced basis from an an optimal Lovász–reduced one. Finding an optimal Lovász–reduced basis seems to be *strictly* easier than finding a Korkine–Zolotarev reduced one.

9. M. Kaib and C. P. Schnorr. The generalized Gauss reduction algorithm. *J. of Algorithms*, 21:565–578, 1996.

10. J. C. Lagarias. Worst-case complexity bounds for algorithms in the theory of integral quadratic forms. *J. Algorithms*, 1:142–186, 1980.

11. H.W. Lenstra. Integer programming with a fixed number of variables. *Math. Oper. Res.*, 8:538–548, 1983.

12. A. K. Lenstra, H. W. Lenstra, and L. Lovász. Factoring polynomials with rational coefficients. *Math. Ann.*, 261:513–534, 1982.

13. J. C. Lagarias and A. M. Odlyzko. Solving low-density subset sum problems. In *24th IEEE Symposium FOCS*, pages 1–10, 1983.

14. C. P. Schnorr. A hierarchy of polynomial time lattice basis reduction algorithm. *Theoretical Computer Science*, 53:201–224, 1987.

15. A. Joux and J. Stern. Lattice reduction: A toolbox for the cryptanalyst. *J. of Cryptology*, 11:161–185, 1998.

16. B. Vallée. Un problème central en géométrie algorithmique des nombres: la réduction des réseaux. *Inf. Th. et App.*, 3:345–376, 1989.

17. B. Vallée. Gauss' algorithm revisited. *J. of Algorithms*, 12:556–572, 1991.

18. P. van Emde Boas. Another NP-complete problem and the complexity of finding short vectors in a lattice. *Rep. 81-04 Math. Inst. Amsterdam*, 1981.

19. O. von Sprang. *Basisreduktionsalgorithmen für Gitter kleiner Dimension*. PhD thesis, Universität des Saarlandes, 1994.

Iteration Algebras Are Not Finitely Axiomatizable
Extended Abstract

Stephen L. Bloom and Zoltán Ésik*

1 Stevens Institute of Technology
Department of Computer Science
Hoboken, NJ 07030
bloom@cs.stevens-tech.edu
2 A. József University
Department of Computer Science
Szeged, Hungary
esik@inf.u-szeged.hu

Abstract. Algebras whose underlying set is a complete partial order and whose term-operations are continuous may be equipped with a least fixed point operation $\mu x.t$. The set of all equations involving the μ-operation which hold in all continuous algebras determines the variety of iteration algebras. A simple argument is given here reducing the axiomatization of iteration algebras to that of Wilke algebras. It is shown that Wilke algebras do not have a finite axiomatization. This fact implies that iteration algebras do not have a finite axiomatization, even by "hyperidentities".

1 Introduction

For a fixed signature Σ, a μ/Σ-algebra $\mathfrak{A} = (A, \sigma)_{\sigma \in \Sigma}$, is a Σ-algebra equipped with an operation $(\mu x.t)^A$, for each μ/Σ-term t. Algebras whose underlying set A is equipped with a complete partial order and whose basic operations $\sigma : A^n \to A$ are continuous, determine μ/Σ-algebras in which $\mu x.t$ is defined using least fixed points (see below). The variety of μ/Σ-algebras generated by these continuous algebras is the variety of μ/Σ-iteration algebras. Such algebras have been used in many studies in theoretical computer science (for only a few of many references, see [14,8,15,9,11,12,1].)

The main theorem in the current paper shows that the identities satisfied by continuous algebras are not finitely based. This result has been known for some

* Partially supported by grant no. FKFP 247/1999 from the Ministry of Education of Hungary and grant no. T22423 from the National Foundation of Hungary for Scientific Research.

G. Gonnet, D. Panario, and A. Viola (Eds.): LATIN 2000, LNCS 1776, pp. 367–376, 2000.

time, [6], but only in an equivalent form for **iteration theories**. In this note we give an argument which may have independent interest. We show how to translate "scalar" iteration algebra identities into Wilke algebra identities, [17]. Since the identities of Wilke algebras are not finitely based, as we show, the same property holds for iteration algebras.

In fact, we prove a stronger result. We show there is no finite number of *hyperidentities* which axiomatize iteration algebras. Our notion of "hyperidentity" is stronger than that introduced by Taylor [19]. (In this extended abstract, we will omit many of the proofs.)

2 μ/Σ-Terms and Algebras

In this section, we formulate the notion of a μ/Σ-algebra, where Σ is a signature, i.e., a ranked set. We do **not** assume that the underlying set of an algebra is partially ordered. Let $V = \{x_1, x_2, \dots\}$ be a countably infinite set of "variables", and let $\Sigma = (\Sigma_0, \Sigma_1, \dots)$ be a ranked alphabet. The set of μ/Σ-**terms**, denoted T_Σ, is the smallest set of expressions satisfying the following conditions:

- each variable is in T_Σ;
- if $\sigma \in \Sigma_n$ and $t_1, \dots, t_n \in T_\Sigma$, then $\sigma(t_1, \dots, t_n)$ is in T_Σ;
- if $x \in V$ and $t \in T_\Sigma$, then $\mu x.t$ is in T_Σ.

Every occurrence of the variable x is **bound** in $\mu x.t$. The free variables occurring in a term are defined as usual. We use the notation $t = t[x_1, \dots, x_n]$ to indicate that t is a term whose free variables are among the distinct variables $x_1, \dots, x_n$, *and no bound variable is among* $x_1, \dots, x_n$. Perhaps confusingly, we write

$$t[t_1/x_1, \dots, t_n/x_n]$$

to indicate the term obtained by simultaneously substituting the terms t_i for the free occurrences of x_i in t, for $i = 1, \dots, n$. (By convention, we assume no variable free in t_i becomes bound in $t[t_1/x_1, \dots, t_n/x_n]$.) But here, we do not rule out the possibility that there are other free variables in t not affected by this substitution.

Definition 1. *A μ/Σ-algebra consists of a Σ-algebra $\mathfrak{A} = (A, \sigma^A)_{\sigma \in \Sigma}$ and an assignment of a function $t^A : A^n \to A$ to each μ/Σ-term $t = t[x_1, \dots, x_n]$ which satisfies the (somewhat redundant) requirements that*

1.

$$x_i^A(a_1, \dots, a_n) = a_i, \ i = 1, 2, \dots, n;$$

2. for each $\sigma \in \Sigma_n$,

$$(\sigma(t_1, \ldots, t_n))^A(a_1, \ldots, a_n) = \sigma^A(t_1^A(a_1, \ldots, a_n), \ldots, t_n^A(a_1, \ldots, a_n));$$

3. if s and t differ only in their bound variables, then $s^A = t^A$;
4. if $s = s[x_1, \ldots, x_n]$, $t = t[x_1, \ldots, x_n]$ and if

$$s^A(a_1, \ldots, a_n) = t^A(a_1, \ldots, a_n),$$

for all $a_1, \ldots, a_n \in A$, then, for each $i \in [n]$, all $a_j \in A$, $1 \leq j \leq n$, $j \neq i$,

$$(\mu x_i.s)^A(a_1, \ldots a_{i-1}, a_{i+1}, \ldots, a_n) = (\mu x_i.t)^A(a_1, \ldots a_{i-1}, a_{i+1}, \ldots, a_n);$$

5. if $t = t[x_1, \ldots, x_n]$, then the function t^A depends on at most the arguments corresponding to the variables occurring freely in t.

If $\mathfrak{A}$ and $\mathfrak{B}$ are μ/Σ-algebras, a **morphism** $\varphi : \mathfrak{A} \to \mathfrak{B}$ is a function $A \to B$ such that for all terms $t = t[x_1, \ldots, x_n]$, and all $a_1, \ldots, a_n \in A$,

$$\varphi(t^A(a_1, \ldots, a_n)) = t^B(\varphi(a_1), \ldots, \varphi(a_n)).$$

In particular, a morphism of μ/Σ-algebras is also a Σ-algebra morphism.

As usual, we say that a μ/Σ-algebra $\mathfrak{A}$ **satisfies an identity** $s \approx t$ between μ/Σ-terms if the functions s^A and t^A are the same.

3   Conway and Iteration Algebras

Definition 2. *A μ/Σ-Conway algebra is a μ/Σ-algebra satisfying the **double iteration identities** (1) and **composition identities** (2)*

$$\mu x.\mu y.t \approx \mu z.t[z/x, z/y] \tag{1}$$
$$\mu x.s[r/x] \approx s[\mu x.r[s/x]/x], \tag{2}$$

for all terms $t = t[x, y, z_1, \ldots, z_p]$, $s = s[x, z_1, \ldots, z_p]$, and $r = r[x, z_1, \ldots, z_p]$ in T_Σ. A **morphism** of Conway algebras is just a morphism of μ/Σ-algebras.

The class of **Conway algebras** is the class of all μ/Σ-Conway algebras, as Σ varies over all signatures.

Letting r be the variable x in the composition identity, we obtain the following well known fact.

Lemma 1. *Any μ/Σ-Conway algebra satisfies the **fixed point identities***

$$\mu x.s \approx s[\mu x.s/x].$$

In particular,

$$\mu x.s \approx s,$$

whenever x does not occur free in s.

We will be mostly concerned with scalar signatures. A signature Σ is **scalar** if $\Sigma_n = \emptyset$ for $n > 1$. The class of **scalar Conway algebras** is the collection of all μ/Σ-Conway algebras, as Σ varies over all scalar signatures.

Proposition 1. *Suppose that Σ is a scalar signature. If $\mathfrak{A}$ is a μ/Σ-algebra satisfying the composition identities, then the double iteration identities (1) hold in $\mathfrak{A}$. Moreover, a μ/Σ-algebra $\mathfrak{A}$ is a Conway-algebra iff (2) holds in $\mathfrak{A}$ for all terms $r = r[x]$ and $s = s[x]$.*   □

Note again that unlike most treatments of μ/Σ-algebras, we do **not** assume that such an algebra comes with a partial order with various completeness properties guaranteeing that all functions which are either monotone and/or continuous have least fixed points. We say that a μ/Σ-algebra $\mathfrak{A} = (A, \sigma^A)_{\sigma \in \Sigma}$ is **continuous** if the underlying set A is equipped with a directed complete partial order, and each basic function $\sigma_A : A^n \to A$ preserves all sups of directed sets; the μ-operator is defined via least fixed points. (See [8].) For example, when a term $t[x, y]$ denotes such a function $t^A : A \times A \to A$, say, then $\mu x.t$ denotes the function $A \to A$ whose value at $b \in A$ is the *least* a in A such that $a = t^A(a, b)$. For us, $\mu x.t$ is interpreted as a function which has no particular properties. Of course, we will be interested in classes of algebras in which these functions are required to satisfy certain identities.

Definition 3. *A μ/Σ-algebra is a μ/Σ-**iteration algebra** if it satisfies all identities satisfied by the continuous Σ-algebras or, equivalently, the regular Σ-algebras of [15], or the iterative Σ-algebras of [11,16]. A **morphism** of iteration algebras is a morphism of μ/Σ-algebras.*

For axiomatic treatments of iteration algebras, see [2]. It is proved in [7] that an μ/Σ-algebra is an iteration algebra iff it is a Conway algebra satisfying certain "group-identities".

When Σ is either clear from context or not important, we say only **iteration algebra** instead of "μ/Σ-iteration algebra". The class of **scalar iteration algebras** is the class of all μ/Σ-iteration algebras, as Σ varies over all scalar signatures. The following is well known [2].

Proposition 2. *If $\mathfrak{A}$ is an iteration algebra, $\mathfrak{A}$ satisfies the **power identities**: for each term $t = t[x, y_1, \dots, y_p]$,*

$$\mu x.t \approx \mu x.t^n, \quad n \geq 1,$$

where $t^1 = t$ and $t^{k+1} := t[t^k/x]$.   □

4   Scalar μ/Σ-Iteration Algebras

We have already pointed out in Proposition 2 that any iteration algebra satisfies the power identities. It turns out that for scalar signatures, the composition and power identities are complete.

Theorem 1. *When Σ is scalar, a μ/Σ-algebra $\mathfrak{A}$ is an iteration algebra iff $\mathfrak{A}$ satisfies the composition identities and the power identities.*

Proof. We need prove only that if $\mathfrak{A}$ satisfies the composition and power identities, then $\mathfrak{A}$ satisfies all iteration algebra identities. The idea of the proof is to show that in fact $\mathfrak{A}$ is a quotient of a free iteration algebra. $\square$

5   Wilke Algebras

A **Wilke algebra** is a two-sorted algebra $A = (A_f, A_\omega)$ equipped with an associative operation $A_f \times A_f \to A_f$, written $u \cdot v$, a binary operation $A_f \times A_\omega \to A_\omega$, written $u \cdot x$, and a unary operation $A_f \to A_\omega$, written $u^\dagger$, which satisfies the following identities:

$$(u \cdot v) \cdot w = u \cdot (v \cdot w), \quad u, v, w \in A_f \tag{3}$$
$$(u \cdot v) \cdot x = u \cdot (v \cdot x), \quad u, v \in A_f, x \in A_\omega, \tag{4}$$
$$(u \cdot v)^\dagger = u \cdot (v \cdot u)^\dagger, \quad u, v \in A_f, \tag{5}$$
$$(u^n)^\dagger = u^\dagger, \quad u \in A_f, n \geq 2. \tag{6}$$

(See [17], where these structures were called "binoids". In [18,13], it is shown how Wilke algebras may be used to characterize regular sets of ω-words.)

A **morphism** $h = (h_f, h_\omega) : A \to B$ of Wilke algebras $A = (A_f, A_\omega)$ and $B = (B_f, B_\omega)$, is a pair of functions

$$h_f : A_f \to B_f$$
$$h_\omega : A_\omega \to B_\omega$$

which preserve all of the structure, so that h_f is a semigroup morphism, and

$$h_\omega(u^\dagger) = h_f(u)^\dagger, \quad u \in A_f$$
$$h_\omega(u \cdot x) = h_f(u) \cdot h_\omega(x), \quad u \in A_f, x \in A_\omega$$

The function $A_f \times A_\omega \to A_\omega$ is called the **action**. We refer to the two equations (3) and (4) as the **associativity conditions**; equation (5) is the **commutativity condition**, and (6) are the **power identities**.

We will also consider **unitary Wilke algebras**, which are Wilke algebras $A = (A_f, A_\omega)$ in which A_f is a monoid with neutral element 1, which satisfies the **unit conditions:**

$$1 \cdot u = u = u \cdot 1, \quad u \in A_f$$
$$1 \cdot x = x, \quad x \in A_\omega.$$

A **morphism of unitary Wilke algebras** $h : A \to B$ is a morphism of Wilke algebras which preserves the neutral element. (It follows that $h(1^\dagger) = 1^\dagger$.)

We will need a notion weaker than that of a unitary Wilke algebra.

Definition 4. *A **unitary Wilke prealgebra** (A_f, A_ω) is an algebra with the operations and constant of a unitary Wilke algebra whose operations need to satisfy only the associativity and unit conditions. A **morphism** $h = (h_f, h_\omega) : (A_f, A_\omega) \to (B_f, B_\omega)$ is defined as for unitary Wilke algebras.*

6   Axiomatizing Wilke Algebras

We adopt an argument from [4] to show that unitary Wilke algebras do not have a finite axiomatization. Thus, neither do Wilke algebras. In particular, we prove the following fact.

Proposition 3. *For each prime $p > 2$ there is a unitary Wilke prealgebra $M_p = (M_f, M_\omega)$ which satisfies the commutativity condition (5) and all power identities (6) for integers $n < p$. However, for some $u \in M_f$, $(u^p)^\dagger \neq u^\dagger$. Thus, (unitary) Wilke algebras have no finite axiomatization.*

Proof. Let $\perp, \top$ be distinct elements not in $\mathbf{N}$, the set of nonnegative integers. We define a function $\rho_p : \mathbf{N} \to \{\perp, \top\}$ as follows.

$$\rho_p(n) = \begin{cases} \top & \text{if } p \text{ divides } n \\ \perp & \text{otherwise.} \end{cases}$$

Let $M_f = \mathbf{N}$, and $M_\omega = \{\perp, \top\}$. The monoid operation $u \cdot v$ on M_f is *addition*, $u + v$, and the action of M_f on M_ω is trivial: $u \cdot x = x$, for $u \in M_f$, $x \in M_\omega$. Lastly, define $u^\dagger = \rho_p(u)$. It is clear that (M_f, M_ω) is a unitary Wilke prealgebra satisfying the commutativity condition.

Now, p divides $u^n = nu$ iff p divides n or p divides u. Thus, in particular, for $n < p$, $(u^n)^\dagger = u^\dagger$. Also, $(uv)^\dagger = u(vu)^\dagger$, since $uv = vu$ and the action is trivial. But if $u = 1$, then $u^p = p$ and $u^\dagger = \perp \neq \top = p^\dagger = (u^p)^\dagger$.

Now if there were any finite axiomatization of unitary Wilke algebras, then, by the compactness theorem, there would be a finite axiomatization consisting of the associativity, commutativity and unit conditions, together with some finite subset of the power identities. This has just been shown impossible.   □

7   From Iteration Algebras to Wilke Algebras and Back

Suppose that Σ is a signature, not necessarily scalar, and that $\mathfrak{B}$ is a μ/Σ-algebra. Let A_f denote the set of all functions $t^B : B \to B$ for Σ-terms $t = t[x]$ having at most the one free variable x, and let A_ω denote the set of elements of the form t^B, for μ/Σ-terms t having no free variables. We give $A = (A_f, A_\omega)$ the structure of a unitary Wilke prealgebra as follows.

For $a_i = t_i^B \in A_f$, $i = 1, 2$, and $w = s^B \in A_\omega$,

$$a_1 \cdot a_2 := (t_1[t_2/x])^B; \quad a_1 \cdot w := (t_1[s/x])^B; \quad a_1^\dagger := (\mu x.t_1)^B.$$

Proposition 4. *With the above definitions, (A_f, A_ω) is a unitary Wilke prealgebra. Moreover, (A_f, A_ω) is a unitary Wilke algebra iff $\mathfrak{B}$ is an iteration algebra.* $\square$

Notation: We write $\mathfrak{B}\mathbf{W}$ for the unitary Wilke algebra determined by the iteration algebra $\mathfrak{B}$.

We want to give a construction of a scalar μ/Σ-algebra $\mathfrak{A}[B]$ from a unitary Wilke prealgebra $B = (B_f, B_\omega)$. So let $B = (B_f, B_\omega)$ be a unitary Wilke prealgebra. Define the scalar signature Σ as follows:

$$\Sigma_1 := \{\sigma_b : b \in B_f\}$$
$$\Sigma_0 := \{\sigma_z : z \in B_\omega\}.$$

Let the underlying set A of $\mathfrak{A}[B]$ be $B_f \cup B_\omega$. The functions σ^A for $\sigma \in \Sigma_1$ are defined as follows:

$$\sigma_b^A(w) := b \cdot w.$$

For $z \in B_\omega$, $\sigma_z^A := z$. Lastly, we define the functions t^A for all μ/Σ-terms $t = t[x]$, by induction on the structure of the term t. By Lemma 1, we need only consider the case $t = \mu x.s[x]$ and x occurs free in s. In this case, s is $\sigma_{b_1}(\ldots \sigma_{b_k}(x) \ldots)$, for some $k \geq 0$ and $b_j \in B_f$. If $k = 0$, $(\mu x.x)^A := 1^\dagger$; otherwise,

$$(\mu x.\sigma_{b_1}(\ldots \sigma_{b_k}(x) \ldots))^A := (b_1 \cdot \ldots \cdot b_k)^\dagger.$$

Proposition 5. *$\mathfrak{A}[B]$ is a Conway-algebra iff B satisfies the commutativity condition.* $\square$

Lemma 2. *For each $n \geq 2$, the Conway algebra $\mathfrak{A}[B]$ satisfies the power identity $\mu x.t \approx \mu x.t^n$, for all terms $t = t[x]$ iff B satisfies the identity $x^\dagger \approx (x^n)^\dagger$.* $\square$

Corollary 1. $\mathfrak{A}[B]$ *is an iteration algebra iff* B *is a unitary Wilke algebra.*

Corollary 2. *For any finite subset* X *of the power identities, there is a scalar Conway algebra* $\mathfrak{A}$ *which satisfies all of the identities in* X, *but fails to satisfy all power identities.*

Proof. This follows from the previous lemma and Proposition 3. □

Remark 1. When $B = (B_f, B_\omega)$ is generated by a set $B_0 \subseteq B_f$, one may reduce the signature of $\mathfrak{A}[B]$ to contain only the letters associated with the elements of B_0. By the same argument one obtains the following stronger version of Corollary 2: For any finite subset X of the power identities, there is a scalar Conway μ/Σ-algebra $\mathfrak{A}$ having a single operation which satisfies all of the identities in X, but fails to satisfy all power identities.

Corollary 3. *Each unitary Wilke algebra* B *is isomorphic to* $\mathfrak{A}[B]\mathbf{W}$. □

8   Hyperidentities

A 'hyperidentity' differs from a standard identity only in the way it is interpreted. Suppose that Δ is a fixed ranked signature in which Δ_n is countably infinite, for each $n \geq 0$. For any signature Σ, an identity $s \approx t$ between μ/Δ terms is said to be a **hyperidentity** of the μ/Σ-algebra $\mathfrak{A}$, if for each way of substituting a μ/Σ term $t_\delta[x_1, \ldots, x_n, \overline{y}]$ for each letter $\delta \in \Delta_n$ in the terms s, t, the resulting μ/Σ-identity is true in $\mathfrak{A}$. Thus, the operation symbols in Δ may be called "meta-operation symbols". This definition of "hyperidentity" extends the notion in Taylor [19], in that terms with more than n variables are allowed to be substituted for n-ary function symbols.

For example, if $F, G \in \Delta_1$, then

$$\mu x.F(G(x)) \approx F(\mu x.G(F(x))) \tag{7}$$

is a hyperidentity of the class of all iteration algebras. Indeed, this is just a restatement of the composition identity.

The following proposition follows from our definition of hyperidentity.

Proposition 6. *The two hyperidentities (7) and*

$$\mu x.\mu y.F(x, y) \approx \mu z.F(z, z)$$

axiomatize the Conway algebras. □

Each group identity mentioned above may be formulated as a hyperidentity (containing no free variables), as well. Thus, iteration algebras can be axiomatized by an infinite set of hyperidentities.

Theorem 2. *There is no finite set of hyperidentities which axiomatize iteration algebras.*

Proof idea. Suppose that there is a finite set E of hyperidentities that axiomatize iteration algebras. Then there is a finite set E' of scalar hyperidentities such that a scalar μ/Σ-algebra is an iteration algebra iff it is a model of the hyperidentities E'. The equations in E' are obtained from those in E by replacing each meta-operation symbol $F(x_1, \ldots, x_n)$, $n > 1$, by unary meta-operation symbols $f_i(x_j)$ in all possible ways. (For example, $F(x, y) \approx F(y, x)$ gets replaced by the two equivalent equations $f_1(x_1) \approx f_1(x_2)$ and and $f_2(x_1) = f_2(x_2)$.) When the rank n of F is zero, F remains unchanged. Now if a scalar hyperidentity $s \approx t$ holds in all iteration algebras, then either no variable occurs free in either s or t, or both sides contain the same free variable. We then translate each such scalar hyperidentity $s \approx t$ in the finite set E' into an identity $tr(s) \approx tr(t)$ between two unitary Wilke prealgebra terms. The translation $t \mapsto tr(t)$ is by induction on the structure of the term t. For example, if x has a free occurrence in t, the translation of $\mu x.t$ is $tr(t)^\dagger$; otherwise $tr(\mu x.t)$ is $tr(t)$. We then show the resulting set of identities $tr(s) \approx tr(t)$ together with the axioms for unitary Wilke prealgebras gives a finite axiomatization of unitary Wilke algebras, contradicting Proposition 3. $\square$

9   Conclusion

Although most equational theories involving a fixed point operation which are of interest in theoretical computer science are nonfinitely based, several of them have a finite *relative* axiomatization over iteration algebras. Examples of such theories are the equational theory of Kleene algebras of binary relations, or (regular) languages, the theory of bisimulation or tree equivalence classes of processes equipped with the regular operations, etc. See [10,3,7] and [5]. Since the nonfinite equational axiomatizability of these theories is caused by the nonfinite axiomatizability of iteration algebras, the constructions of this paper may be used to derive simple proofs of the nonfinite axiomatizability of these theories as well.

References

1. S.L. Bloom and Z. Ésik. Iteration algebras. *International Journal of Foundations of Computer Science*, 3(3):245–302, 1992. Extended abstract in *Colloq. on Trees in Algebra and Programming*, volume 493 of *Lecture Notes in Computer Science*, 264–274, 1991.

2. S.L. Bloom and Z. Ésik. *Iteration Theories: The Equational Logic of Iterative Processes.* EATCS Monographs on Theoretical Computer Science. Springer–Verlag, 1993.

3. S.L. Bloom and Z. Ésik. Equational axioms for regular sets. *Mathematical Structures in Computer Science*, 3:1–24, 1993.

4. S.L. Bloom and Z. Ésik. Shuffle binoids. *Theoretical Informatics and Applications*, 32(4-5-6):175–198, 1998.

5. S.L. Bloom, Z. Ésik and D. Taubner. Iteration theories of synchronization trees. *Information and Computation*, 102:1–55, 1993.

6. Z. Ésik. Independence of the equational axioms of iteration theories. *Journal of Computer and System Sciences*, 36:66–76, 1988.

7. Z. Ésik. Group axioms for iteration. *Information and Computation*, 148:131–180, 1999.

8. J. Goguen, J. Thatcher, E. Wagner, and J. Wright. Initial algebra semantics and continuous algebras. *Journal of the ACM*, 24:68–95, 1977.

9. I. Guessarian. Algebraic Semantics. Lecture Notes in Computer Science 99, Springer, Berlin-New York, 1981.

10. D. Krob. Complete systems of B-rational identities. *Theoretical Computer Science*, 89:207–343, 1991.

11. E. Nelson. Iterative algebras. *Theoretical Computer Science* 25:67–94, 1983.

12. D. Niwinski. Equational mu-calculus. *Computation theory* (Zaborów, 1984), Lecture Notes in Comput. Sci., 208:169–176, Springer, Berlin-New York, 1985.

13. D. Perrin and J-E. Pin. Semigroups and automata on infinite words. in J. Fountain (ed.), Semigroups, Formal Languages and Groups, pages 49–72, Kluwer Academic Pub., 1995.

14. D. Scott. Data types as lattices. *SIAM Journal of Computing*, 5:522–587, 1976.

15. J. Tiuryn. Fixed points and algebras with infinitely long expressions, I. Regular algebras. *Fundamenta Informaticae*, 2:103–127, 1978.

16. J. Tiuryn. Unique fixed points vs. least fixed points. *Theoretical Computer Science*, 12:229–254, 1980.

17. T. Wilke. An Eilenberg Theorem for ∞-languages. In "Automata, Languages and Programming", Proc. of 18th ICALP Conference, vol. 510 of *Lecture Notes in Computer Science*, 588–599, 1991.

18. T. Wilke. An algebraic theory for regular languages of finite and infinite words. *International Journal of Algebra and Computation*, 3:447–489, 1993.

19. W. Taylor. Hyperidentities and hypervarieties. *Aequationes Mathematicae*, 21:30–49, 1981.

Undecidable Problems in Unreliable Computations

Richard Mayr *

Department of Computer Science, University of Edinburgh,
JCMB, Edinburgh EH9 3JZ, UK. e-mail: `mayrri@dcs.ed.ac.uk`

Abstract. Lossy counter machines are defined as Minsky n-counter machines where the values in the counters can spontaneously decrease at any time. While termination is decidable for lossy counter machines, structural termination (termination for every input) is undecidable. This undecidability result has far reaching consequences. Lossy counter machines can be used as a general tool to prove the undecidability of many problems, for example (1) The verification of systems that model communication through unreliable channels (e.g. model checking lossy fifo-channel systems and lossy vector addition systems). (2) Several problems for reset Petri nets, like structural termination, boundedness and structural boundedness. (3) Parameterized problems like fairness of broadcast communication protocols.

1   Introduction

Lossy counter machines (LCM) are defined just like Minsky counter machines [19], but with the addition that the values in the counters can spontaneously decrease at any time. This is called 'lossiness', since a part of the counter is lost. (In a different framework this corresponds to lost messages in unreliable communication channels.) There are many different kinds of lossiness, i.e. different ways in which the counters can decrease. For example, one can define that either a counter can only spontaneously decrease by 1, or it can only become zero, or it can change to any smaller value. All these different ways are described by different *lossiness relations* (see Section 2).

The addition of lossiness to counter machines weakens their computational power. Some types of lossy counter machines (with certain lossiness relations) are not Turing-powerful, since reachability and termination are decidable for them. Since lossy counter machines are *weaker* than normal counter machines, any *undecidability* result for lossy counter machines is particularly interesting.

The main result of this paper is that *structural termination* (termination for every input) is undecidable for every type of lossy counter machine (i.e. for every lossiness relation).

This result can be applied to prove the undecidability of many problems. To prove the undecidability of a problem X, it suffices to choose a suitable lossiness relation L and reduce the structural termination problem for lossy counter machines with lossiness relation L to the problem X. The important and nice point here is that problem X does *not* need to simulate a counter machine perfectly. Instead, it suffices if X can simulate a counter machine imperfectly, by simulating only a lossy counter machine. Furthermore, one can choose the right type of imperfection (lossiness) by choosing the lossiness relation L.

---

* Work supportet by DAAD Post-Doc grant D/98/28804.

G. Gonnet, D. Panario, and A. Viola (Eds.): LATIN 2000, LNCS 1776, pp. 377–386, 2000.

Thus lossy counter machines can be used as a general tool to prove the undecidability of problems. Firstly, they can be used to prove new undecidability results, and secondly they can be used to give more elegant, simpler and much shorter proofs of existing results (see Section 5).

2   Definitions

Definition 1. *A* n-*counter machine [19]* M *is described by a finite set of states* Q, *an initial state* $q_0 \in Q$, *a final state* accept $\in Q$, n *counters* $c_1, \ldots, c_n$ *and a finite set of instructions of the form* $(q :\ c_i := c_i + 1; \mathsf{goto}\ q')$ *or* $(q :\ \mathsf{If}\ c_i = 0\ \mathsf{then\ goto}\ q'\ \mathsf{else}\ c_i := c_i - 1; \mathsf{goto}\ q'')$ *where* $i \in \{1, \ldots, n\}$ *and* $q, q', q'' \in Q$. *A configuration of* M *is described by a tuple* $(q, m_1, \ldots, m_n)$ *where* $q \in Q$ *and* $m_i \in \mathbb{N}$ *is the content of the counter* c_i $(1 \leq i \leq n)$. *The size of a configuration is defined by* $size((q, m_1, \ldots, m_n)) := \sum_{i=1}^{n} m_i$. *The possible computation steps are defined by*

1. $(q, m_1, \ldots, m_n) \to (q', m_1, \ldots, m_i + 1, \ldots, m_n)$
   if there is an instruction $(q :\ c_i := c_i + 1; \mathsf{goto}\ q')$.
2. $(q, m_1, \ldots, m_n) \to (q', m_1, \ldots, m_n)$ *if there is an instruction* $(q :\ \mathsf{If}\ c_i = 0\ \mathsf{then\ goto}\ q'\ \mathsf{else}\ c_i := c_i - 1; \mathsf{goto}\ q'')$ *and* $m_i = 0$.
3. $(q, m_1, \ldots, m_n) \to (q'', m_1, \ldots, m_i - 1, \ldots, m_n)$ *if there is an instruction* $(q :\ \mathsf{If}\ c_i = 0\ \mathsf{then\ goto}\ q'\ \mathsf{else}\ c_i := c_i - 1; \mathsf{goto}\ q'')$ *and* $m_i > 0$.

A *run* of a counter machine is a (possibly infinite) sequence of configurations $s_0, s_1, \ldots$ with $s_0 \to s_1 \to s_2 \to s_3 \to \ldots$. *Lossiness relations* describe spontaneous changes in the configurations of lossy counter machines.

Definition 2. *Let* $\xrightarrow{s}$ *(for 'sum') be a relation on configurations of* n-*counter machines*
$$(q, m_1, \ldots, m_n) \xrightarrow{s} (q', m_1', \ldots, m_n') :\Leftrightarrow (q, m_1, \ldots, m_n) = (q', m_1', \ldots, m_n') \vee$$
$$\left(q = q' \wedge \sum_{i=1}^{n} m_i > \sum_{i=1}^{n} m_i' \right)$$
This relation means that either nothing is changed or the sum of all counters strictly decreases. Let id *be the identity relation. A relation* $\xrightarrow{l}$ *is a* lossiness relation *iff* $id \subseteq \xrightarrow{l} \subseteq \xrightarrow{s}$. *A* lossy counter machine *(LCM) is given by a counter machine* M *and a lossiness relation* $\xrightarrow{l}$. *Let* $\to$ *be the normal transition relation of* M. *The lossy transition relation* $\Longrightarrow$ *of the LCM is defined by* $s_1 \Longrightarrow s_2 :\Leftrightarrow \exists s_1', s_2'. s_1 \xrightarrow{l} s_1' \to s_2' \xrightarrow{l} s_2$. *An arbitrary lossy counter machine is a lossy counter machine with an arbitrary (unspecified) lossiness relation. The following relations are lossiness relations:*

Perfect *The relation* id *is a lossiness relation. Thus arbitrary lossy counter machines subsume normal counter machines.*

Classic Lossiness *The classic lossiness relation* $\xrightarrow{cl}$ *is defined by* $(q, m_1, \ldots, m_n) \xrightarrow{cl} (q', m_1', \ldots, m_n') :\Leftrightarrow q = q' \wedge \forall i. m_i \geq m_i'$. *Here the contents of the counters can become any smaller value. A relation* $\xrightarrow{l}$ *is called a* subclassic lossiness relation *iff* $id \subseteq \xrightarrow{l} \subseteq \xrightarrow{cl}$.

Bounded Lossiness A counter can loose at most $x \in \mathbb{N}$ before and after every computation step. Here the lossiness relation $\xrightarrow{l(x)}$ is defined by $(q, m_1, \ldots, m_n) \xrightarrow{l(x)} (q', m'_1, \ldots, m'_n) :\Leftrightarrow q = q' \wedge \forall i.\, m_i \geq m'_i \geq max\{0, m_i - x\}$. Note that $\xrightarrow{l(x)}$ is a subclassic lossiness relation.

Reset Lossiness If a counter is tested for zero, then it can suddenly become zero. The lossiness relation $\xrightarrow{rl}$ is defined as follows: $(q, m_1, \ldots, m_n) \xrightarrow{rl} (q', m'_1, \ldots, m'_n)$ iff $q = q'$ and for all i either $m_i = m'_i$ or $m'_i = 0$ and there is an instruction $(q: \text{ If } c_i = 0 \text{ then goto } q' \text{ else } c_i := c_i - 1; \text{goto } q'')$. Note that $\xrightarrow{rl}$ is subclassic.

The definition of these lossiness relations carries over to other models like Petri nets [21], where places are considered instead of counters.

Definition 3. *For any arbitrary lossy n-counter machine and any configuration s let $runs(s)$ be the set of runs that start at configuration s. (There can be more than one run if the counter machine is nondeterministic or lossy.) Let $runs^\omega(s)$ be the set of infinite runs that start at configuration s. A run $r = \{(q^i, m^i_1, \ldots, m^i_n)\}^\infty_{i=0} \in runs^\omega(s)$ is space-bounded iff $\exists c \in \mathbb{N}. \forall i. \sum^n_{j=1} m^i_j \leq c$. Let $runs^\omega_b(s)$ be the space-bounded infinite runs that start at s. An (arbitrary lossy) n-counter machine M is*

zero-initializing iff in the initial state q_0 it first sets all counters to 0.
space-bounded iff the space used by M is bounded by a constant c. $\exists c \in \mathbb{N}. \forall r \in runs((q_0, 0, \ldots, 0)). \forall s \in r.\, size(s) \leq c$
input-bounded iff in every run from any configuration the size of every reached configuration is bounded by the input. $\forall s. \forall r \in runs(s). \forall s' \in r.\, size(s') \leq size(s)$
strongly-cyclic iff every infinite run from any configuration visits the initial state q_0 infinitely often. $\forall q \in Q, m_1, \ldots, m_n \in \mathbb{N}. \forall r \in runs^\omega((q, m_1, \ldots, m_n))$. $\exists m'_1, \ldots, m'_n \in \mathbb{N}.\, (q_0, m'_1, \ldots, m'_n) \in r$.
bounded-strongly-cyclic iff every space-bounded infinite run from any configuration visits the initial state q_0 infinitely often. $\forall q \in Q, m_1, \ldots, m_n \in \mathbb{N}$. $\forall r \in runs^\omega_b((q, m_1, \ldots, m_n)). \exists m'_1, \ldots, m'_n \in \mathbb{N}.\, (q_0, m'_1, \ldots, m'_n) \in r$

If M is input-bounded then it is also space-bounded. If M is strongly-cyclic then it is also bounded-strongly-cyclic. If M is input-bounded and bounded-strongly-cyclic then it is also strongly-cyclic.

3   Decidable Properties

Since arbitrary LCM subsume normal counter machines, nothing is decidable for them. However, some problems are decidable for classic LCM (with the classic lossiness relation). They are not Turing-powerful. The following results in this section are special cases of positive decidability results in [4,5,2].

Lemma 1. *Let M be a classic LCM and s a configuration of M. The set $pre^*(s) := \{s' \mid s' \Longrightarrow^* s\}$ of predecessors of s is effectively constructible.*

Theorem 1. *Reachability is decidable for classic LCM.*

Lemma 2. *Let M be a classic LCM with initial configuration s_0. It is decidable if there is an infinite run that starts at s_0, i.e. if $runs^\omega(s_0) \neq \emptyset$.*

Theorem 2. *Termination is decidable for classic LCM.*

It has been shown in [4] that even model checking classic LCM with the temporal logics EF and EG (natural fragments of computation tree logic (CTL) [7,10]) is decidable.

4   The Undecidability Result

We show that structural termination is undecidable for LCM for every lossiness relation. We start with the problem CM, which was shown to be undecidable by Minsky [19].

CM

Instance: A 2-counter machine M with initial state q_0.
Question: Does M accept $(q_0, 0, 0)$?

BSC-ZI-CM$_b^\omega$

Instance: A bounded-strongly-cyclic, zero-initializing 3-counter machine M with initial state q_0.
Question: Does M have an infinite space-bounded run from $(q_0, 0, 0, 0)$,
    i.e. $runs_b^\omega((q_0, 0, 0, 0)) \neq \emptyset$?

Lemma 3. *BSC-ZI-CM$_b^\omega$ is undecidable.*

Proof. We reduce CM to BSC-ZI-CM$_b^\omega$. Let M be a 2-counter machine with initial state q_0. We construct a 3-counter machine M' as follows: First M' sets all three counters to 0. Then it does the same as M, except that after every instruction it increases the third counter c_3 by 1. Every instruction of M of the form $(q : c_i := c_i + 1;$ goto $q')$ with $(1 \le i \le 2)$ is replaced by $q : c_i := c_i + 1;$ goto q_2 and $q_2 : c_3 := c_3 + 1;$ goto q', where q_2 is a new state. Every instruction of the form $(q :$ If $c_i = 0$ then goto q' else $c_i := c_i - 1;$ goto $q'')$ with $(1 \le i \le 2)$ is replaced by three instructions: $q :$ If $c_i = 0$ then goto q_2 else $c_i := c_i - 1;$ goto q_3, $q_2 : c_3 := c_3 + 1;$ goto q', $q_3 : c_3 := c_3 + 1;$ goto q'' where q_2, q_3 are new states.
    Finally, we replace the accepting state *accept* of M by the initial state q_0' of M', i.e. we replace every instruction (goto *accept*) by (goto q_0'). M' is zero-initializing by definition. M' is bounded-strongly-cyclic, because c_3 is increased after every instruction and only set to zero at the initial state q_0'.

$\Rightarrow$ If M is a positive instance of CM then it has an accepting run from $(q_0, 0, 0)$. This run has finite length and is therefore space-bounded. Then M' has an infinite space-bounded cyclic run that starts at $(q_0', 0, 0, 0)$. Thus M' is a positive instance of BSC-ZI-CM$_b^\omega$.

$\Leftarrow$ If M' is a positive instance of BSC-ZI-CM$_b^\omega$ then there exists an infinite space-bounded run that starts at the configuration $(q_0', 0, 0, 0)$. By the construction of M' this run contains an accepting run of M from the configuration $(q_0, 0, 0)$. Thus M is a positive instance of CM. $\square$

$\exists n$LCM$^\omega$

Instance: A strongly-cyclic, input-bounded 4-counter LCM M with initial state q_0.
Question: Does there exist an $n \in \mathbb{N}$ s.t. $runs^\omega((q_0, 0, 0, 0, n)) \neq \emptyset$?

Theorem 3. *$\exists n$LCM$^\omega$ is undecidable for every lossiness relation.*

Proof. We reduce BSC-ZI-CM$_b^\omega$ to $\exists n$LCM$^\omega$ with any lossiness relation $\overset{l}{\to}$. For any bounded-strongly-cyclic, zero-initializing 3-counter machine M we construct a strongly-cyclic, input-bounded lossy 4-counter machine M' with initial state q_0' and lossiness relation $\overset{l}{\to}$ as follows: The 4-th counter c_4 holds the 'capacity'. In every operation it is changed in a way s.t. the sum of all counters never increases. (More exactly, the sum of all counters can increase by 1, but only if it was decreased by 1 in the previous step.) Every instruction of M of the form $(q : c_i := c_i + 1;$ goto $q')$ with $(1 \leq i \leq 3)$ is replaced by two instructions: $q :$ If $c_4 = 0$ then goto $fail$ else $c_4 := c_4 - 1;$ goto q_2, $q_2 : c_i := c_i + 1;$ goto q', where $fail$ is a special final state and q_2 is a new state. Every instruction of the form $(q :$ If $c_i = 0$ then goto q' else $c_i := c_i - 1;$ goto $q'')$ with $(1 \leq i \leq 3)$ is replaced by two instructions: $q :$ If $c_i = 0$ then goto q' else $c_i := c_i - 1;$ goto q_2, $q_2 : c_4 := c_4 + 1;$ goto q'', where q_2 is a new state.

M' is bounded-strongly-cyclic, because M is bounded-strongly-cyclic. M' is input-bounded, because every run from a configuration $(q, m_1, \ldots, m_4)$ is space-bounded by $m_1 + m_2 + m_3 + m_4$. Thus M' is also strongly-cyclic.

$\Rightarrow$ If M is a positive instance of BSC-ZI-CM$_b^\omega$ then there exists a $n \in \mathbb{N}$ and an infinite run of M that starts at $(q_0, 0, 0, 0)$, visits q_0 infinitely often and always satisfies $c_1 + c_2 + c_3 \leq n$. Since $id \subseteq \xrightarrow{l}$, there is also an infinite run of M' that starts at $(q_0, 0, 0, 0, n)$, visits q_0 infinitely often and always satisfies $c_1 + c_2 + c_3 + c_4 \leq n$. Thus M' is a positive instance of $\exists n \mathrm{LCM}^\omega$.

$\Leftarrow$ If M' is a positive instance of $\exists n \mathrm{LCM}^\omega$ then there exists an $n \in \mathbb{N}$ s.t. there is an infinite run that starts at the configuration $(q'_0, 0, 0, 0, n)$. This run is space-bounded, because it always satisfies $c_1 + c_2 + c_3 + c_4 \leq n$. By the construction of M', the sum of all counters can only increase by 1 if it was decreased by 1 in the previous step. By the definition of lossiness (see Def. 2) we get the following: If lossiness occurs (when the contents of the counters spontaneously change) then this strictly and permanently decreases the sum of all counters. It follows that lossiness can only occur at most n times in this infinite run and the sum of all counters is bounded by n. Thus there is an infinite suffix of this run of M' where lossiness does not occur. Thus there exist $q' \in Q, m'_1, \ldots, m'_4 \in \mathbb{N}$ s.t. an infinite suffix of this run of M' without lossiness starts at $(q', m'_1, \ldots, m'_4)$. It follows that there is an infinite space-bounded run of M that starts at $(q', m'_1, \ldots, m'_3)$. Since M is bounded-strongly-cyclic, this run must eventually visit q_0. Thus there exist $m''_1, \ldots, m''_3 \in \mathbb{N}$ s.t. an infinite space-bounded run of M starts at $(q_0, m''_1, \ldots, m''_3)$. Since M is zero-initializing, there is an infinite space-bounded run of M that starts at $(q_0, 0, 0, 0)$. Thus M is a positive instance of BSC-ZI-CM$_b^\omega$. $\qquad\square$

Note that this undecidability result even holds under the additional condition that the LCMs are strongly-cyclic and input-bounded. It follows immediately that model checking LCM with the temporal logics CTL (computation-tree logic [7,10]) and LTL (linear-time temporal logic [22]) is undecidable, since the question of $\exists n \mathrm{LCM}^\omega$ can be expressed in these logics. There are two variants of the structural termination problem:

STRUCTTERM-LCM, VARIANT 1

Instance: A strongly-cyclic, input-bounded 4-counter LCM M with initial state q_0.
Question: Does M terminate for all inputs from q_0 ?
    Formally: $\forall n_1, \ldots, n_4 \in \mathbb{N}. runs^\omega((q_0, n_1, n_2, n_3, n_4)) = \emptyset$?

STRUCTTERM-LCM, VARIANT 2

Instance: A strongly-cyclic, input-bounded 4-counter LCM M with initial state q_0.
Question: Does M terminate for all inputs from every control state q ?
    Formally: $\forall n_1, \ldots, n_4 \in \mathbb{N}. \forall q \in Q. runs^\omega((q, n_1, n_2, n_3, n_4)) = \emptyset$?

Theorem 4. *Structural termination is undecidable for lossy counter machines. Both variants of STRUCTTERM-LCM are undecidable for every lossiness relation.*

Proof. The proof of Theorem 3 carries over, because the LCM is strongly-cyclic and the 3-CM in BSC-ZI-CM$_b^\omega$ is zero-initializing.   □

SPACE-BOUNDEDNESS FOR LCM

Instance: A strongly-cyclic 4-counter LCM M with initial configuration $(q_0, 0, 0, 0, 0)$
Question: Is M space-bounded ?

Theorem 5. *Space-boundedness for LCM is undecidable for all lossiness relations.*

Theorem 6. *Structural space-boundedness for LCM is undecidable for every lossiness relation.*

Proof. The proof is similar to Theorem 4. An extra counter is used to count the length of the run. It is unbounded iff the run is infinite. All other counters are bounded.   □

5   Applications

5.1   Lossy Fifo-Channel Systems

Fifo-channel systems are systems of finitely many finite-state processes that communicate with each other by sending messages via unbounded fifo-channels (queues, buffers). In lossy fifo-channel systems these channels are lossy, i.e. they can spontaneously loose (arbitrarily many) messages. This can be used to model communication via unreliable channels. While normal fifo-channel systems are Turing-powerful, some safety-properties are decidable for lossy fifo-channel systems [2,5,1]. However, liveness properties are undecidable even for lossy fifo-channel systems. In [3] Abdulla and Jonsson showed the undecidability of the *recurrent-state problem* for lossy fifo-channel systems. This problem is if certain states of the system can be visited infinitely often. The undecidable core of the problem is essentially if there exists an initial configuration of a lossy fifo-channel system s.t. it has an infinite run. The undecidability proof in [3] was done by a long and complex reduction from a variant of Post's correspondence problem (2-permutation PCP [23], which is (wrongly) called cyclic PCP in [3]).

Lossy counter machines can be used to give a much simpler proof of this result. The lossiness of lossy fifo-channel systems is classic lossiness, i.e. the contents of a fifo-channel can change to any substring at any time. A lossy fifo-channel system can simulate a classic LCM (with some additional deadlocks) in the following way: Every lossy fifo-channel contains a string in X^* (for some symbol X) and is used as a classic lossy counter. The only problem is the test for zero. We test the emptiness of a channel by adding a special symbol Y and removing it in the very next step. If it can be done then the channel is empty (or has become empty by lossiness). If this cannot be done, then the channel was not empty or the symbol Y was lost. In this case we get a deadlock. These additional deadlocks do not affect the existence of infinite runs, and thus the results of Section 4 carry over. Thus the problem $\exists n$LCM$^\omega$ (for the classic lossiness relation) can be reduced to the problem above for lossy fifo-channel systems and the undecidability follows immediately from Theorem 3.

5.2   Model Checking Lossy Basic Parallel Processes

Petri nets [21] (also described as 'vector addition systems' in a different framework) are a widely known formalism used to model concurrent systems. They can also be seen as counter machines without the ability to test for zero, and are not Turing-powerful, since the reachability problem is decidable for them [17]. Basic Parallel Processes correspond to communication-free nets, the (very weak) subclass of labeled Petri nets where every transition has exactly one place in its preset. They have been studied intensively in the framework of model checking and semantic equivalences (e.g. [12,18,6,15,20]).

An instance of the *model checking problem* is given by a system S (e.g. a counter machine, Petri net, pushdown automaton,...) and a temporal logic formula φ. The question is if the system S has the properties described by φ, denoted $S \models \varphi$.

The branching-time temporal logics EF, EG and EG_ω are defined as extensions of Hennessy-Milner Logic [13,14,10] by the operators EF, EG and EG_ω, respectively. $s \models EF\varphi$ iff there exists an s' s.t. $s \xrightarrow{*} s'$ and $s' \models \varphi$. $s_0 \models EG_\omega\varphi$ iff there exists an infinite run $s_0 \to s_1 \to s_2 \to \ldots$ s.t. $\forall i.\, s_i \models \varphi$. EG is similar, except that it also includes finite runs that end in a deadlock. Alternatively, EF and EG can be seen as fragments of computation-tree logic (CTL [7,10]), since $EF\varphi = true\,\mathcal{U}\,\varphi$ and $EG\varphi = \varphi\,w\mathcal{U}\,false$.

Model checking Petri nets with the logic EF is undecidable, but model checking Basic Parallel Processes with EF is $PSPACE$-complete [18]. Model checking Basic Parallel Processes with EG is undecidable [12]. It is different for lossy systems: By induction on the nesting-depth of the operators EF, EG and EG_ω, and constructions similar to the ones in Lemma 1 and Lemma 2, it can be shown that model checking classic LCM with the logics EF, EG and EG_ω is decidable. Thus it is also decidable for classical lossy Petri nets and classical lossy Basic Parallel Processes (see [4]).

However, model checking lossy Basic Parallel Processes with nested EF and EG/EG_ω operators is still undecidable for every subclassic lossiness relation. This is quite surprising, since lossy Basic Parallel Processes are an extremely weak model of infinite-state concurrent systems and the temporal logic used is very weak as well.

Theorem 7. *Model checking lossy Basic Parallel Processes (with any subclassic lossiness relation) with formulae of the form $EF\,EG_\omega\Phi$, where Φ is a Hennessy-Milner Logic formula, is undecidable.*

Proof. Esparza and Kiehn showed in [12] that for every counter machine M (with all counters initially 0) a Basic Parallel Processes P and a Hennessy-Milner Logic formula φ can be constructed s.t. M does not halt iff $P \models EG_\omega\varphi$. The construction carries over to subclassic LCM and subclassic lossy Basic Parallel Processes. The control-states of the counter machine are modeled by special places of the Basic Parallel Processes. In every infinite run that satisfies φ exactly one of these places is marked at any time.

We reduce $\exists n\mathrm{LCM}^\omega$ to the model checking problem. Let M be a subclassic LCM. Let P be the corresponding Basic Parallel Processes as in [12] and let φ be the corresponding Hennessy-Milner Logic formula as in [12]. We use the same subclassic lossiness relation on M and on P. P stores the contents of the 4-th counter in a place Y. Thus $P\|Y^n$ corresponds to the configuration of M with n in the 4-th counter (and 0 in the others). We define a new initial state X and transitions $X \xrightarrow{a} X\|Y$ and $X \xrightarrow{b} P$, where a and b do not occur in P. Let $\Phi := \varphi \wedge \neg\langle b\rangle true$. Then M is a positive instance of $\exists n\mathrm{LCM}^\omega$ iff $X \models EF\,EG_\omega\Phi$. The result follows from Theorem 3.   □

For Petri nets and Basic Parallel Processes, the meaning of Hennessy-Milner Logic formulae can be expressed by boolean combinations of constraints of the form $p \geq k$ (at least k tokens on place p). Thus the results also hold if boolean combinations of such constraints are used instead of Hennessy-Milner Logic formulae. Another consequence of Theorem 7 is that model checking lossy Petri nets with CTL is undecidable.

5.3   Reset/Transfer Petri Nets

Reset Petri nets are an extension of Petri nets by the addition of reset-arcs. A reset-arc between a transition and a place has the effect that, when the transition fires, all tokens are removed from this place, i.e. it is reset to zero. Transfer nets and transfer arcs are defined similarly, except that all tokens on this place are moved to some different place. It was shown in [8] that termination is decidable for 'Reset Post G-nets', a more general extension of Petri nets that subsumes reset nets and transfer nets. (For normal Petri nets termination is $EXPSPACE$-complete [24]). While boundedness is trivially decidable for transfer nets, the same question for reset nets was open for some time (and even a wrong decidability proof was published). Finally, it was shown in [8,9] that boundedness (and structural boundedness) is undecidable for reset Petri nets. The proof in [8] was done by a complex reduction from Hilbert's 10th problem (a simpler proof was later given in [9]).

Here we generalize these results by using lossy counter machines. This also gives a unified framework and considerably simplifies the proofs.

Lemma 4. *Reset Petri nets can simulate lossy counter machines with reset-lossiness.*

Theorem 8. *Structural termination, boundedness and structural boundedness are undecidable for lossy reset Petri nets with every subclassic lossiness relation.*

Proof. It follows from Lemma 4 that a lossy reset Petri net with subclassic lossiness relation $\xrightarrow{l}$ can simulate a lossy counter machine with lossiness relation $\xrightarrow{l} \cup \xrightarrow{rl}$. The results follow from Theorem 4, Theorem 5 and Theorem 6.   □

The undecidability result on structural termination carries over to transfer nets (instead of a reset the tokens are moved to a special 'dead' place), but the others don't. Note that for normal Petri nets structural termination and structural boundedness can be decided in polynomial time (just check if there is a positive linear combination of effects of transitions). Theorem 7 and Theorem 8 also hold for arbitrary lossiness relations, but this requires an additional argument. The main point is that Petri nets (unlike counter machines) can increase a place/counter and decrease another in the same step.

5.4   Parameterized Problems

We consider verification problems for systems whose definition includes a parameter $n \in \mathbb{N}$. Intuitively, n can be seen as the size of the system. Examples are

- Systems of n indistinguishable communicating finite-state processes.
- Systems of communicating pushdown automata with n-bounded stack.
- Systems of (a fixed number of) processes who communicate through (lossy) buffers or queues of size n.

Let $P(n)$ be such a system with parameter n. For every fixed n, $P(n)$ is a system with finitely many states and thus (almost) every verification problem is decidable for it. So the problem $P(n) \models \Phi$ is decidable for any temporal logic formula Φ from any reasonable temporal logic, e.g. modal μ-calculus [16] or monadic second-order theory. The *parameterized verification problem* is if a property holds independently of the parameter n, i.e. for any size. Formally, the question is if for given P and Φ we have $\forall n \in \mathbb{N}. P(n) \models \Phi$ (or $\neg \exists n \in \mathbb{N}. P(n) \models \neg \Phi$). Many of these parameterized problems are undecidable by the following meta-theorem.

Theorem 9. *A parameterized verification problem is undecidable if it satisfies the following conditions:*

1. *It can encode an n-space-bounded lossy counter machine (for some lossiness relation) in such a way that $P(n)$ corresponds to the initial configuration with n in one counter and 0 in the others.*
2. *It can check for the existence of an infinite run.*

Proof. By a reduction of $\exists n$LCM$^\omega$ and Theorem 3. The important point is that in the problem $\exists n$LCM$^\omega$ one can require that the LCM is input-bounded.   □

The technique of Theorem 9 is used in [11] to show the undecidability of the model checking problem for linear-time temporal logic (LTL) and broadcast communication protocols. These are systems of n indistinguishable communicating finite-state processes where a 'broadcast' by one process can affect all other $n - 1$ processes. Such a broadcast can be used to set a simulated counter to zero. However, there is no test for zero. One reduces $\exists n$LCM$^\omega$ with lossiness relation $\overset{rl}{\to}$ to the model checking problem. In the same way, similar results can be proved for parameterized problems about systems with bounded buffers, stacks, etc.

6   Conclusion

Lossy counter machines can be used as a general tool to show the undecidability of many problems. It provides a unified way of reasoning about many quite different classes of systems. For example the recurrent-state problem for lossy fifo-channel systems, the boundedness problem for reset Petri nets and the fairness problem for broadcast communication protocols were previously thought to be completely unrelated. Yet lossy counter machines show that the principles behind their undecidability are the same. Moreover, the undecidability proofs for lossy counter machines are very short and much simpler than previous proofs of weaker results [3,8].

Lossy counter machines have also been used in this paper to show that even for very weak temporal logics and extremely weak models of infinite-state concurrent systems, the model checking problem is undecidable (see Subsection 5.2). We expect that many more problems can be shown to be undecidable with the help of lossy counter machines, especially in the area of parameterized problems (see Subsection 5.4).

Acknowledgments

Thanks to Javier Esparza and Petr Jančar for fruitful discussions.

References

1. P. Abdulla, A. Bouajjani, and B. Jonsson. On-the-fly Analysis of Systems with Unbounded, Lossy Fifo Channels. In *10th Intern. Conf. on Computer Aided Verification (CAV'98)*. LNCS 1427, 1998.

2. P. Abdulla and B. Jonsson. Verifying Programs with Unreliable Channels. In *LICS'93*. IEEE, 1993.

3. P. Abdulla and B. Jonsson. Undecidable verification problems for programs with unreliable channels. *Information and Computation*, 130(1):71–90, 1996.

4. A. Bouajjani and R. Mayr. Model checking lossy vector addition systems. In *Proc. of STACS'99*, volume 1563 of *LNCS*. Springer Verlag, 1999.

5. G. Cécé, A. Finkel, and S.P. Iyer. Unreliable Channels Are Easier to Verify Than Perfect Channels. *Information and Computation*, 124(1):20–31, 1996.

6. S. Christensen, Y. Hirshfeld, and F. Moller. Bisimulation equivalence is decidable for Basic Parallel Processes. In E. Best, editor, *Proceedings of CONCUR 93*, volume 715 of *LNCS*. Springer Verlag, 1993.

7. E.M. Clarke and E.A. Emerson. Design and synthesis of synchronization skeletons using branching time temporal logic. volume 131 of *LNCS*, pages 52–71, 1981.

8. C. Dufourd, A. Finkel, and Ph. Schnoebelen. Reset nets between decidability and undecidability. In *Proc. of ICALP'98*, volume 1443 of *LNCS*. Springer Verlag, 1998.

9. C. Dufourd, P. Jančar, and Ph. Schnoebelen. Boundedness of Reset P/T Nets. In *Proc. of ICALP'99*, volume 1644 of *LNCS*. Springer Verlag, 1999.

10. E.A. Emerson. Temporal and modal logic. In J. van Leeuwen, editor, *Handbook of Theoretical Computer Science : Volume B, FORMAL MODELS AND SEMANTICS*. Elsevier, 1994.

11. J. Esparza, A. Finkel, and R. Mayr. On the verification of broadcast protocols. In *Proc. of LICS'99*. IEEE, 1999.

12. J. Esparza and A. Kiehn. On the model checking problem for branching time logics and Basic Parallel Processes. In *CAV'95*, volume 939 of *LNCS*. Springer Verlag, 1995.

13. M. Hennessy and R. Milner. On observing nondeterminism and concurrency. volume 85 of *LNCS*, pages 295–309, 1980.

14. M. Hennessy and R. Milner. Algebraic laws for nondeterminism and concurrency. *Journal of Association of Computer Machinery*, 32:137–162, 1985.

15. Y. Hirshfeld, M. Jerrum, and F. Moller. A polynomial-time algorithm for deciding bisimulation equivalence of normed Basic Parallel Processes. *Journal of Mathematical Structures in Computer Science*, 6:251–259, 1996.

16. D. Kozen. Results on the propositional μ-calculus. *TCS*, 27:333–354, 1983.

17. E. Mayr. An algorithm for the general Petri net reachability problem. *SIAM Journal of Computing*, 13:441–460, 1984.

18. R. Mayr. Weak bisimulation and model checking for Basic Parallel Processes. In *Foundations of Software Technology and Theoretical Computer Science (FST&TCS'96)*, volume 1180 of *LNCS*. Springer Verlag, 1996.

19. M.L. Minsky. *Computation: Finite and Infinite Machines*. Prentice-Hall, 1967.

20. F. Moller. Infinite results. In Ugo Montanari and Vladimiro Sassone, editors, *Proceedings of CONCUR'96*, volume 1119 of *LNCS*. Springer Verlag, 1996.

21. J.L. Peterson. *Petri net theory and the modeling of systems*. Prentice-Hall, 1981.

22. A. Pnueli. The temporal logic of programs. In *FOCS'77*. IEEE, 1977.

23. K. Ruohonen. On some variants of Post's correspondence problem. *Acta Informatica*, 19:357–367, 1983.

24. H. Yen. A unified approach for deciding the existence of certain Petri net paths. *Information and Computation*, 96(1):119–137, 1992.

Equations in Free Semigroups with Anti-involution and Their Relation to Equations in Free Groups

Claudio Gutiérrez[1,2]

[1] Computer Science Group, Dept. of Mathematics, Wesleyan University
[2] Departamento de Ingeniería Matemática, D.I.M., Universidad de Chile
(Research funded by FONDAP, Matemáticas Aplicadas)
cgutierrez@wesleyan.edu

Abstract. The main result of the paper is the reduction of the problem of satisfiability of equations in free groups to the satisfiability of equations in free semigroups with anti-involution (SGA), by a non-deterministic polynomial time transformation.
A free SGA is essentially the set of words over a given alphabet plus an operator which reverses words. We study equations in free SGA, generalizing several results known for equations in free semigroups, among them that the exponent of periodicity of a minimal solution of an equation E in free SGA is bounded by $2^{\mathcal{O}(|E|)}$.

1 Introduction

The study of the problem of solving equations in free SGA (unification in free SGA) and its computational complexity is a problem closely related to the problem of solving equations in free semigroups and in free groups, which lately have attracted much attention of the theoretical computer science community [3], [12], [13], [14].

Free semigroups with anti-involution is a structure which lies in between that of free semigroups and free groups. Besides the relationship with semigroups and groups, the axioms defining SGA show up in several important theories, like algebras of binary relations, transpose in matrices, inverse semigroups.

The problem of solving equations in free semigroups was proven to be decidable by Makanin in 1976 in a long paper [10] . Some years later, in 1982, again Makanin proved that solving equations in free groups was a decidable problem [11]. The technique used was similar to that of the first paper, although the details are much more involved. He reduced equations in free groups to solving equations in free SGA with special properties ('non contractible'), and showed decidability for equation of this type. For free SGA (without any further condition) the decidability of the problem of satisfiability of equations is still open, although we conjecture it is decidable.

Both of Makanin's algorithms have received very much attention. The enumeration of all unifiers was done by Jaffar for semigroups [6] and by Razborov

G. Gonnet, D. Panario, and A. Viola (Eds.): LATIN 2000, LNCS 1776, pp. 387–396, 2000.

for groups [15]. Then, the complexity has become the main issue. Several authors have analyzed the complexity of Makanin's algorithm for semigroups [6], [16], [1], being EXPSPACE the best upper-bound so far [3]. Very recently Plandowski, without using Makanin's algorithm, presented an upper-bound of PSPACE for the problem of satisfiability of equations in free semigroups [14]. On the other hand, the analysis of the complexity of Makanin's algorithm for groups was done by Koscielski and Pacholski [8], who showed that it is not primitive recursive.

With respect to lower bounds, the only known lower bound for both problems is NP-hard, which seems to be weak for the case of free groups. It is easy to see that this lower bound works for the case of free SGA as well.

The main result of this paper is the reduction of equations in free groups to equations in free SGA (Theorem 9 and Corollary 10). This is achieved by generalizing to SGA several known results for semigroups, using some of Makanin's results in [11], and proving a result that links these results (Proposition 3). Although we do not use it here, we show that the standard bounds on the exponent of periodicity of minimal solutions to word equations also hold with minor modifications in the case of free SGA (Theorem 5).

For concepts of word combinatorics we will follow the notation of [9]. By ϵ we denote the empty word.

2   Equations in Free SGA

A *semigroup with anti-involution* (SGA) is an algebra with a binary associative operation (written as concatenation) and a unary operation $(\ )^{-1}$ with the equational axioms

$$(xy)z = x(yz), \qquad (xy)^{-1} = y^{-1}x^{-1}, \qquad x^{-1-1} = x. \tag{1}$$

A *free* semigroup with anti-involution is an initial algebra for this variety. It is not difficult to check that for a given alphabet C, the set of words over $C \cup C^{-1}$ together with the operator $(\ )^{-1}$, which reverses a word and changes every letter to its twin (e.g. a to a^{-1} and conversely) is a free algebra for SGA over A.

Equations and Solutions. Let C and V be two disjoint alphabets of constants and variables respectively. Denote by $C^{-1} = \{c^{-1} : c \in C\}$. Similarly for V^{-1}. An *equation* E in free SGA with constants C and variables V is a pair (w_1, w_2) of words over the alphabet $\mathcal{A} = C \cup C^{-1} \cup V \cup V^{-1}$. The number $|E| = |w_1| + |w_2|$ is the *length* of the equation E and $|E|_V$ will denote the number of occurrences of variables in E. These equations are also known as *equations in a paired alphabet*.

A map $S : V \longrightarrow (C \cup C^{-1})^*$ can be uniquely extended to a SGA-homomorphism $\bar{S} : \mathcal{A}^* \longrightarrow (C \cup C^{-1})^*$ by defining $S(c) = c$ for $c \in C$ and $S(u^{-1}) = (S(u))^{-1}$ for $u \in C \cup V$. We will use the same symbol S for the map S and the SGA-homomorphism $\bar{S}$. A *solution* S of the equation $E = (w_1, w_2)$ is (the unique SGA-homomorphism defined by) a map $S : V \longrightarrow (C \cup C^{-1})^*$ such that $S(w_1) = S(w_2)$. The length of the solution S is $|S(w_1)|$. By $S(E)$ we denote the word $S(w_1)$ (which is the same as $S(w_2)$). Each occurrence of a symbol $u \in \mathcal{A}$ in E with $S(u) \neq \epsilon$ determines a unique factor in $S(E)$, say $S(E)[i,j]$, which we will denote by $S(u,i,j)$ and call simply an *image* of u in $S(E)$.

The Equivalence Relation (S,E). Let S be a solution of E and P be the set of positions of $S(E)$. Define the binary relation $(S,E)'$ in $P \times P$ as follows: given positions $p, q \in P$, $p(S,E)'q$ if and only if one of the following hold:

1. $p = i + k$ and $q = i' + k$, where $S(x,i,j)$ and $S(x,i',j')$ are images of x in $S(E)$ and $0 \leq k < |S(x)|$.
2. $p = i + k$ and $q = j' - k$, where $S(x,i,j)$ and $S(x^{-1},i',j')$ are images of x and x^{-1} in $S(E)$ and $0 \leq k < |S(x)|$.

Then define (S,E) as the transitive closure of $(S,E)'$. Observe that (S,E) is an equivalence relation.

Contractible Words. A word $w \in \mathcal{A}^*$ is called *non-contractible* if for every $u \in \mathcal{A}$ the word w contains neither the factor uu^{-1} nor $u^{-1}u$. An equation (w_1, w_2) is called non-contractible if both w_1 and w_2 are non-contractible. A solution S to an equation E is called non-contractible if for every variable x which occurs in E, the word $S(x)$ is non-contractible.

Boundaries and Superpositions. Given a word $w \in \mathcal{A}^*$, we define a *boundary* of w as a pair of consecutive positions $(p, p+1)$ in w. We will write simply p_w, the subindex denoting the corresponding word. By extension, we define $i(w) = 0_w$ and $f(w) = |w|_w$, the *initial* and *final* boundaries respectively. Note that the boundaries of w have a natural linear order ($p_w \leq q_w$ iff $p \leq q$ as integers).

Given an equation $E = (w_1, w_2)$, a *superposition* (of the boundaries of the left and right hand sides) of E is a linear order $\leq$ of the set of boundaries of w_1 and w_2 extending the natural orders of the boundaries of w_1 and w_2, such that $i(w_1) = i(w_2)$ and $f(w_1) = f(w_2)$ and possibly identifying some p_{w_1} and q_{w_2}.

Cuts and Witnesses. Given a superposition $\leq$ of $E = (w_1, w_2)$, a *cut* is a boundary j of w_2 (resp. w_1) such that $j \neq b$ for all boundaries b of w_1 (resp. w_2). Hence a cut determines at least three symbols of E, namely $w_2[j]$, $w_2[j+1]$ and $w_1[i+1]$, where i is such that $i_{w_1} < j_{w_2} < (i+1)_{w_1}$ in the linear order, see Figure 1. The triple of symbols $(w_2[j], w_2[j+1], w_1[i])$ is called a *witness* of the cut. A superposition is called *consistent* if $w_1[i+1]$ is a variable.

Observe that every superposition gives rise to a system of equations $(E, \leq)$, which codifies the constraints given by $\leq$, by adding the corresponding equations and variables $x = x'y$ which the cuts determine. Also observe that every solution S of E determines a unique consistent superposition, denoted $\leq_S$. Note finally that the cut j determines a boundary $(r, r+1)$ in $S(E)$; if $p \leq r < q$, we say that the subword $S(E)[p,q]$ of $S(E)$ *contains* the cut j.

Lemma 1 *Let E be an equation in free SGA. Then E has a solution if and only if $(E, \leq)$ has a solution for some consistent superposition $\leq$. There are no more than $|E|^{4|E|_V}$ consistent superpositions.*

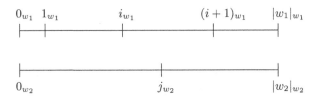

Fig. 1. The cut j_w.

Proof. Obviously if for some consistent superposition $\le$, $(E, \le)$ has a solution, then E has a solution. Conversely, if E has a solution S, consider the superposition generated by S.

As for the bound, let $E = (w_1, w_2)$ and write v for $|E|_V$. First observe that if w_2 consists only of constants, then there are at most $|w_2|^v$ consistent superpositions. To get a consistent superposition in the general case, first insert each initial and final boundary of each variable in w_2 in the linear order of the boundaries of w_1 (this can be done in at most $|E| + v$ ways). Then it rest to deal with the subwords of w_2 in between variables (hence consisting only of constants and of total length $\le |E| - v$). Summing up, there are no more than $(|E| + v)^{2v}(|E| - v)^v \le |E|^{4v}$ consistent superpositions.

Lemma 2 (Compare Lemma 6, [12]) *Assume S is a minimal (w.r.t. length) solution of E. Then*

1. *For each subword $w = S(E)[i, j]$ with $|w| > 1$, there is an occurrence of w or w^{-1} which contains a cut of $(E, \le_S)$.*
2. *For each letter $c = S(E)[i]$ of $S(E)$, there is an occurrence of c or c^{-1} in E.*

Proof. Let $1 \le p \le q \le |S(E)|$. Suppose neither $w = S(E)[p, q]$ nor w^{-1} have occurrences in $S(E)$ which contain cuts. Consider the position p in $S(E)$ and its (S, E)-equivalence class P, and define for each variable x occurring in E,

$S'(x) =$ the subsequence of some image $S(x, i, j)$ of x consisting of all positions which are not in the set P. (i.e. "cut off" from $S(x, i, j)$ all the positions in P).

It is not difficult to see that S' is well defined, *i.e.*, it does not depend on the particular image $S(x, i, j)$ of x chosen, and that $S'(w_1) = S'(w_2)$ (these facts follow from the definition of (S, E)-equivalence). Now, if P does not contain any images of constants of E, it is easy to see that S' is a solution of the equation E. But $|S'(E)| < |S(E)|$, which is impossible because S was assumed to be minimal.

Hence, for each word $w = S[p, q]$, its first position must in the same (S, E)-class of the position of the image of a constant c of E. If $p < q$ the right (resp. left) boundary of that constant is a cut in w (resp. w^{-1}) which is neither initial nor final (check definition of (S, E)-equivalence for $S(E)[p+1]$, etc.), and we are in case 1. If $p = q$ we are in case 2.

Proposition 3 *For each non-contractible equation E there is a finite list of systems of equations $\Sigma_1, \ldots, \Sigma_k$ such that the following conditions hold:*

1. *E has a non-contractible solution if and only if one Σ_i has a solution.*
2. *$k \le |E|^{8|E|_V}$.*
3. *There is $c > 0$ constant such that $|\Sigma_i| \le c|E|$ and $|\Sigma_i|_V \le c|E|_V$ for each $i = 1, \ldots, k$.*

Proof. Let $\le$ be a consistent superposition of E, and let

$$(x_1, y_1, z_1), \ldots, (x_r, y_r, z_r) \tag{2}$$

be a list of those witnesses of the cuts of $(E, \le)$ for which at least one of the x_i, y_i is a variable. Let

$$D = \{(c, d) \in (C \cup C^{-1})^2 : c \ne d^{-1} \wedge d \ne c^{-1}\},$$

and define for each r-tuple $\langle (c_i, d_i) \rangle_i$, of pairs of D the system

$$\Sigma_{\langle (c_i, d_i) \rangle_i} = (E, \le) \cup \{(x_i, x_i' c_i), (y_i, d_i y_i') : i = 1, \ldots, r\}.$$

Now, if S is a non-contractible solution of $(E, \le)$ then S define a solution of some Σ_i, namely the one defined by the r-tuple defined by the elements $(c_i, d_i) = (S(x_i)[|S(x_i)|], S(y_i)[1])$, for $i = 1, \ldots, r$. Note that because E and S are non-contractible, each (c_i, d_i) is in D.

On the other direction, suppose that S is a solution of some Σ_i. Then obviously S is a solution of $(E, \le)$. We only need to prove that the $S(z)$ is non-contractible for all variables z occurring in E. Suppose some z has a factor cc^{-1}, for $c \in C$. Then by Lemma 2 there is an occurrence of cc^{-1} (its converse is the same) which contains a cut of $(E, \le)$. But because E is non-contractible, we must have that one of the terms in (2), say (x_j, y_j, z_j), witnesses this occurrence, hence $x_j = x_j' c$ and $y_j = c^{-1} y_j'$, which is impossible by the definition of the Σ_i's.

The bound in 2. follows by simple counting: observe that $r \le 2|E|_V$ and $|D| \le |C|^{2r} \le |E|^{4|E|_V}$, and the number k of systems is no bigger than the number of superpositions times $|D|$. For the bounds in 3. just sum the corresponding numbers of the new equations added.

The following is an old observation of Hmelevskii [5] for free semigroups which extends easily to free SGA:

Proposition 4 *For each system of equations Σ in free SGA with generators C, there is an equation E in free SGA with generators $C \cup c$, $c \notin (C \cup C^{-1})$, such that*

1. *S is a solution of E if and only if S is a solution of Σ.*
2. *$|E| \le 4|\Sigma|$ and $|E|_V = |\Sigma|_V$.*

Moreover, if the equations in Σ are non-contractible, the E is non-contractible.

Proof. Let $(v_1, w_1), \ldots, (v_n, w_n)$ the system of equations Σ. Define E as

$$(v_1 c v_2 c \cdots c v_n c v_1 c^{-1} v_2 c^{-1} \cdots c^{-1} v_n, \; w_1 c w_2 c \cdots c w_n c w_1 c^{-1} w_2 c^{-1} \cdots c^{-1} w_n).$$

Clearly E is non-contractible because so was each equation (v_i, w_i), and c is a fresh letter. Also if S is a solution of Σ, obviously it is a solution of E. Conversely, if S is a solution of E, then

$$|S(v_1 c v_2 c \cdots c v_n)| = |S(v_1 c^{-1} v_2 c^{-1} \cdots c^{-1} v_n)|,$$

hence

$$|S(v_1 c v_2 c \cdots c v_n)| = |S(w_1 c w_2 c \cdots c w_n)|,$$

and the same for the second pair of expressions with c^{-1}. Now it is easy to show that $S(v_i) = S(w_i)$ for all i: suppose not, for example $|S(v_1)| < |S(w_1)|$. Then $S(w_1)[|S(v_1)| + 1] = c$ and $S(w_1)[|S(v_1)| + 1] = c^{-1}$, impossible. Then argue the same for the rest.

The bounds are simple calculations.

The next result is a very important one, and follows from a straightforward generalization of the result in [7], where it is proved for semigroups.

Theorem 5 *Let E be an equation in free SGA. Then, the exponent of periodicity of a minimal solution of E is bounded by $2^{\mathcal{O}(|E|)}$.*

Proof. It is not worth reproducing here the ten-pages proof in [7] because the changes needed to generalize it to free SGA are minor ones. We will assume that the reader is familiar with the paper [7].

The proof there consist of two independent parts: (1) To obtain from the word equation E a linear Diophantine equation, and (2) To get good bound for it. We will sketch how to do step (1) for free SGA. The rest is completely identical.

First, let us sketch how the system of linear equations is obtained from a word equation E. Let S be a solution of E. Recall that a P-stable presentation of $S(x)$, for a variable x, has the form

$$S(x) = w_0 P^{\mu_1} w_1 P^{\mu_2} \ldots w_{n-1} P^{\mu_{n-1}} w_n.$$

¿From here, for a suitable P (which is the word that witnesses the exponent of periodicity of $S(E)$), a system of linear Diophantine equations $LD_P(E)$ is built, roughly speaking, by replacing the μ_i by variables x_{μ_i} in the case of variables, plus some other pieces of data. Then it is proved that if S is a minimal solution of E, the solution $x_{\mu_i} = \mu_i$ is a minimal solution of $LD_P(E)$.

For the case of free SGA, the are two key points to note. First, for the variables of the form x^{-1}, the solution $S(x^{-1})$ will have the following P^{-1}-stable presentation (same P, w_i, μ_i as before):

$$S(x^{-1}) = w_n^{-1} (P^{-1})^{\mu_{n-1}} w_{n-1}^{-1} (P^{-1})^{\mu_{n-2}} \ldots w_1^{-1} (P^{-1})^{\mu_1} w_0^{-1}.$$

Second, note that P^{-1} is a subword of PP if and only if P is a subword of $P^{-1}P^{-1}$. Call a repeated occurrence of P in w, say $w = uP^k v$, maximal, if P is neither the suffix of u nor a prefix of v. So it holds that maximal occurrences of P and P^{-1} in w either (1) do not overlap each other, or (2) overlap almost completely (exponents will differ at most by 1).

In case (1), consider the system $LD_P(E') \cup LD_{P^{-1}}(E')$ (each one constructed exactly as in the case of word equations) where E' is the equation E where we consider the pairs of variables x^{-1}, x as independent for the sake of building the system of linear Diophantine equations. And, of course, the variables x_{μ_i} obtained from the same μ_i in $S(x)$ and $S(x^{-1})$ are the same.

In case (2), notice that P-stable and P^{-1}-stable presentations for a variable x differ very little. So it is enough to consider $LD_P(E')$, taking care of using for the P-presentation of $S(x^{-1})$ the same set of Diophantine variables (adding 1 or -1 where it corresponds) used for the P-presentation of $S(x)$.

It must be proved then that if S is a minimal solution of the equation in free SGA E, then the solution $x_{\mu_i} = \mu_i$ is a minimal solution of the corresponding system of linear Diophantine equations defined as above. This can be proved easily with the help of Lemma 2.

Finally, as for the the parameters of the system of Diophantine equations, observe that $|E'| = |E|$, hence the only parameters that grow are the number of variables and equations, and by a factor of at most 2. So the asymptotic bound remains the same as for the case of E', which is $2^{\mathcal{O}(|E|)}$.

The last result concerning equations in free SGA we will prove follows from the trivial observation that every equation in free semigroups is an equation in free SGA. Moreover:

Proposition 6 *Let M be a free semigroup on the set of generators C, and N be a free SGA on the set of generators C, and E an equation in M. Then E is satisfiable in M if and only if it is satisfiable in N.*

Proof. An equation in free SGA which does no contain $(\ )^{-1}$ has a solution if and only if it has a solution which does not contain $(\ )^{-1}$. So the codification of equations in free semigroups into free SGA is straightforward: the same equation.

We get immediately a lower bound for the problem of satisfiability of equations in free SGA by using the corresponding result for the free semigroup case.

Corollary 7 *Satisfiability of equations in free SGA is NP-hard.*

3   Reducing the Problem of Satisfiability of Equations in Free Groups to Satisfiability of Equations in Free SGA

A *group* is an algebra with a binary associative operation (written as concatenation), a unary operation $(\ )^{-1}$, and a constant 1, with the axioms (1) plus

$$xx^{-1} = 1, \quad x^{-1}x = 1, \quad 1x = x1 = 1. \tag{3}$$

As in the case of free SGA, is not hard to see that the set of non-contractible words over $C \cup C^{-1}$ plus the empty word, and the operations of composition and reverse suitable defined, is a free group with generators C.

Equations in free groups. The formal concept of *equation in free groups* is almost exactly the same as that for free SGA, hence we will not repeat it here. The difference comes when speaking of solutions. A *solution S* of the equation E is (the unique group-homomorphism $S : \mathcal{A} \longrightarrow (C \cup C^{-1})^*$ defined by) a map $S : V \longrightarrow (C \cup C^{-1})^*$ extended by defining $S(c) = c$ for each $c \in C$ and $S(w^{-1}) = (S(w))^{-1}$, which satisfy $S(w_1) = S(w_2)$. Observe that the only difference with the case of SGA is that now we possibly have 'simplifications' of subexpressions of the form ww^{-1} or $w^{-1}w$ to 1, *i.e.* the use of the equations (3).

Proposition 8 (Makanin, Lemma 1.1 in [11]) *For any non-contractible equation E in the free group G with generators C we can construct a finite list $\Sigma_1, \ldots, \Sigma_k$ of systems of non-contractible equations in the free SGA G' with generators C such that the following conditions are satisfied:*

1. *E has a non-contractible solution in G if and only if $k > 0$ and some system Σ_j has a non-contractible solution in G'.*
2. *There is $c > 0$ constant such that $|\Sigma_i| \leq |E| + c|E|_V^2$ and $|\Sigma_i|_V \leq c|E|_V^2$ for each $i = 1, \ldots, k$.*
3. *There is $c > 0$ constant such that $k \leq (|E|_V)^{c|E|_V^2}$.*

Proof. This is essentially the proof in [11] with the bounds improved. Let E be the equation

$$C_0 X_1 C_1 X_2 \cdots C_{v-1} X_v C_v = 1, \tag{4}$$

where C_i are non-contractible, $v = |E|_V$, and X_i are meta-variables representing the actual variables in E.

Let S be a non-contractible solution of E. By a known result (see [11], p. 486), there is a set W of non-contractible words in the alphabet C, $|W| \leq 2v(2v+1)$, such that each C_i and $S(X_i)$ can be written as a concatenation of no more than $2v$ words in W, and after replacement Equation (4) holds in the free group with generators W.

Let Z be a set of $2v(2v+1)$ fresh variables. Then choose words $y_0, x_1, y_1, x_1, \ldots, x_v, y_v \in (Z \cup Z^{-1})^*$, each of length at most $2v$, non-contractible, and define the system of equations

1. $C_j = y_j$, $j = 0, \ldots, v$,
2. $X_j = x_j$, $j = 1, \ldots, v$.

Each such set of equations, for which Equation (4) holds in the free group with generators Z when replacing C_i and X_i by the corresponding words in $(Z \cup Z^{-1})^*$, defines one system Σ_i.

It is clear from the result mentioned earlier, that E has a solution if and only if there is some Σ_i which has a non-contractible solution. How many Σ_i are there? No more than $[(2v(2v+1))^{2v}]^{2v+1}$.

Theorem 9 *For each equation E in a free group G with generators C there is a finite set Q of equations in a free semigroup with anti-involution G' with generators $C \cup \{c_1, c_2\}$, $c_1, c_2 \notin C$, such that the following hold:*

1. *E is satisfiable in G if and only if one of the equations in Q is satisfiable in G'.*
2. *There is $c > 0$ constant, such that for each $E' \in Q$, it holds $|E'| \leq c|E|^2$.*
3. *$|Q| \leq |E|^{c|E|_V^3}$, for $c > 0$ a constant.*

Proof. By Proposition 8, there is a list of systems of non-contractible equations $\Sigma_1, \ldots, \Sigma_k$ which are equivalent to E (w.r.t. non-contractible satisfiability). By Proposition 4, each such system Σ_j is equivalent (w.r.t. to satisfiability) to a non-contractible equation E'. Then, by Proposition 3, for each such non-contractible E', there is a system of equations (now without the restriction of non-contractibility) $\Sigma'_1, \ldots, \Sigma'_{k'}$ such that E' has a non-contractible solution if and only if one of the Σ'_j has a solution (not necessarily non-contractible). Finally, by Proposition 4, for each system Σ', we have an equation E'' which have the same solutions (if any) of Σ'. So we have a finite set of equations (the E'''s) with the property that E is satisfiable in G if and only if one of the E'' is satisfiable in G'.

The bounds in 2. and 3. follow by easy calculations from the bounds in the corresponding results used above.

Remark. It is not difficult to check that the set Q in the previous theorem can be generated non-deterministically in polynomial time.

Corollary 10 *Assume that f_T is an upper bound for the deterministic TIME-complexity of the problem of satisfiability of equations in free SGA. Then*

$$\max\{f_T(c|E|^2), |E|^{c|E|_V^3}\},$$

for $c > 0$ a constant, is an upper bound for the deterministic TIME-complexity of the problem of satisfiability of equations in free groups.

4   Conclusions

Our results show that solving equations in free SGA comprises the cases of free groups and free semigroups, the first with an exponential reduction (Theorem 9), and the latter with a linear reduction (Proposition 6). This suggest that free SGA, due to its simplicity, is the 'appropriate' theory to study when seeking algorithms for solving equations in those theories.

In a preliminary version of this paper we stated the following conjectures:

1. Satisfiability of equations in free groups is PSPACE-hard.
2. Satisfiability of equations in free groups is in EXPTIME.
3. Satisfiability of equations in free SGA is decidable.

In the meantime the author proved that satisfiability of equations in free SGA is in PSPACE, hence answering positively (2) and (3). Also independently, Diekert and Hagenah announced the solution of (3) [2].

Acknowledgements

Thanks to Volker Diekert for useful comments.

References

1. V. Diekert, *Makanin's Algorithm for Solving Word Equations with Regular Constraints*, in forthcoming M. Lothaire, *Algebraic Combinatorics on Words*. Report Nr. 1998/02, Fakultät Informatik, Universität Stuttgart.
2. V. Diekert, Personal communication, 8 Oct. 1999.
3. C. Gutiérrez, *Satisfiability of Word Equations with Constants is in Exponential Space*, in Proc. FOCS 98.
4. C. Gutiérrez, *Solving Equations in Strings: On Makanin's Algorithm*, in Proceedings of the Third Latin American Symposium on Theoretical Informatics, LATIN'98, Campinas, Brazil, 1998. In LNCS 1380, pp. 358-373.
5. J.I. Hmelevskii, *Equations in a free semigroup*, Trudy Mat. Inst. Steklov 107(1971), English translation: Proc. Steklov Inst. Math. 107(1971).
6. J. Jaffar, *Minimal and Complete Word Unification*, Journal of the ACM, Vol. 37, No.1, January 1990, pp. 47-85.
7. A. Kościelski, L. Pacholski, *Complexity of Makanin's algorithm*, J. Assoc. Comput. Mach. 43 (1996) 670-684.
8. A. Kościelski, L. Pacholski, *Makanin's algorithm is not primitive recursive*, Theoretical Computer Science 191 (1998) 145-156.
9. Lothaire, M. *Combinatorics on Words*, Cambridge Mathematical Texts, reprinted 1998.
10. G.S. Makanin, *The problem of satisfiability of equations in a free semigroup*, Mat. Sbornik 103, 147-236 (in Russian). English translation in Math. USSR Sbornik 32, 129-198.
11. G.S. Makanin. *Equations in a free group*. Izvestiya NA SSSR 46, 1199-1273, 1982 (in Russian). English translation in Math USSR Izvestiya, Vol. 21 (1983), No. 3.
12. W. Rytter and W. Plandowski, *Applications of Lempel-Ziv encodings to the solution of word equations*, In Proceedings of the 25th. ICALP, 1998.
13. Plandowski, W., *Satisfiability of word equations with constants is in NEXPTIME*, in Proc. STOC'99.
14. Plandowski, W., *Satisfiability of word equations with constants is in PSPACE*, in Proc. FOCS'99.
15. A.A. Razborov, *On systems of equations in a free group*, Izvestiya AN SSSR 48 (1984) 779-832 (in Russian). English translation in Math. USSR Izvestiya 25 (1985) 115-162.
16. K. Schulz, *Word Unification and Transformation of Generalized Equations*, Journal of Automated Reasoning 11:149-184, 1993.

Squaring Transducers:
An Efficient Procedure for Deciding
Functionality and Sequentiality of Transducers

Marie-Pierre Béal[1], Olivier Carton[1], Christophe Prieur[2], and Jacques
Sakarovitch[3]

[1] Institut Gaspard Monge, Université Marne-La Vallée
[2] LIAFA, Université Paris 7 / CNRS
[3] Laboratoire Traitement et Communication de l'Information, ENST / CNRS

Abstract. We described here a construction on transducers that give a
new conceptual proof for two classical decidability results on transducers:
it is decidable whether a finite transducer realizes a functional relation,
and whether a finite transducer realizes a sequential relation. A better
complexity follows then for the two decision procedures.

In this paper we give a new presentation and a conceptual proof for two
classical decision results on finite transducers.

Transducers are finite automata with input and output; they realize thus
relations between words, the so-called *rational relations*. Eventhough they are a
very simple model of machines that compute relations — they can be seen as
2-tape *1-way* Turing machines — most of the problems such as equivalence or
intersection are easily shown to be equivalent to the Post Correspondence Problem and thus undecidable. The situation is drastically different for transducers
that are *functional*, that is, transducers that realize functions, and the above
problems become then easily decidable. And this is of interest because of the
following result.

Theorem 1. *[12]   Functionality is a decidable property for finite transducers.*

Among the functional transducers, those which are *deterministic in the input* (they are called *sequential*) are probably the most interesting, both from a
pratical and from a theoretical point of view: they correspond to machines that
can really and easily be implemented. A rational function is *sequential* if it can
be realized by a sequential transducer. Of course, a non sequential transducer
may realize a sequential function and this occurrence is known to be decidable.

Theorem 2. *[7]   Sequentiality is a decidable property for rational functions.*

The original proofs of these two theorems are based on what could be called a
"pumping" principle, implying that a word which contradicts the property may
be chosen of a bounded length, and providing thus directly decision procedures

G. Gonnet, D. Panario, and A. Viola (Eds.): LATIN 2000, LNCS 1776, pp. 397–406, 2000.
© Springer-Verlag Berlin Heidelberg 2000

of exponential complexity. Theorem 1 was published again in [4], with exactly
the same proof, hence the same complexity.

Later, it was proved that the functionality of a transducer can be decided
in polynomial time, as a particular case of a result obtained by reduction to
another decision problem on another class of automata ([10, Theorem 2]).

With this communication, we shall see how a very natural construction performed
on the *square of the transducer* yields a decision procedure for the two
properties, that is, it can be read on the result of the construction whether the
property holds or not.

The size of the object constructed for deciding functionality is *quadratic* in
the size of the considered transducer. In the case of sequentiality, one has to be
more subtle for the constructed object may be too large. But it is shown that it
can be decided in *polynomial* time whether this object has the desired property.

Due to the short space available on the proceedings, the proofs of the results
are omited here and will be published in a forthcoming paper.

1 Preliminaries

We basically follow the definitions and notation of [9,2] for automata.

The set of *words* over a finite alphabet A, *i.e. the free monoid* over A, is
denoted by A^*. Its identity, or *empty word* is denoted by 1_{A^*}.

An *automaton* $\mathcal{A}$ *over a finite alphabet* A, noted $\mathcal{A} = \langle Q, A, E, I, T \rangle$, is a
directed graph labelled by elements of A; Q is the set of vertices, called *states*,
$I \subset Q$ is the set of *initial* states, $T \subset Q$ is the set of *terminal* states and
$E \subset Q \times A \times Q$ is the set of labelled *edges* called *transitions*. The automaton $\mathcal{A}$
is *finite* if Q is finite.

The definition of automata as labelled graphs extends readily to automata
over any monoid: an *automaton* $\mathcal{A}$ *over* M, noted $\mathcal{A} = \langle Q, M, E, I, T \rangle$, is a
directed graph the edges of which are labelled by elements of the monoid M. A
computation is a path in the graph $\mathcal{A}$; its label is the product of the label of
its transitions. A computation is *successful* if it begins with an initial state and
ends with a final state. The *behaviour* of $\mathcal{A}$ is the subset of M consisting of the
labels of the successful computations of $\mathcal{A}$.

A state of $\mathcal{A}$ is said to be *accessible* if it belongs to a computation that begins
with an initial state; it is *useful* if it belongs to a successful computation. The
automaton $\mathcal{A}$ is *trim* if all of its states are useful. The accessible part and the
useful part of a finite automaton $\mathcal{A}$ are easily computable from $\mathcal{A}$.

An automaton $\mathcal{T} = \langle Q, A^* \times B^*, E, I, T \rangle$ over a direct product $A^* \times B^*$ of two
free monoids is called *transducer* from A^* to B^*. The behaviour of a transducer $\mathcal{T}$
is thus (the graph of) a relation α from A^* into B^*: α is said to be *realized* by $\mathcal{T}$.
A relation is *rational* (*i.e.* its graph is a rational subset of $A^* \times B^*$) if and only
if it is realized by a finite transducer.

It is a slight generalization — that does not increase the generating power of
the model — to consider transducers $\mathcal{T} = \langle Q, A^* \times B^*, E, I, T \rangle$ where I and T
are not *subsets* of Q (*i.e.* functions from Q into $\{0, 1\}$) but *functions* from Q

into $B^* \cup \emptyset$ (the classical transducers are those for which the image of a state by I or T is either $\emptyset$ or 1_{B^*}).

A transducer is said to be *real-time* if the label of every transition is a pair (a, v) where a is letter of A, the *input* of the transition, and v a word over B, the *output* of the transition, and if for any states p and q and any letter a there is at most one transition from p to q whose input is a. Using classical algorithms from automata theory, any transducer $\mathcal{T}$ can be transformed into a transducer that is real-time if $\mathcal{T}$ realizes a function ([9, Th. IX.5.1], [2, Prop. III.7.1]).

If $\mathcal{T} = \langle Q, A^* \times B^*, E, I, T \rangle$ is a real-time transducer, the *underlying input automaton* of $\mathcal{T}$ is the automaton $\mathcal{A}$ over A obtained from $\mathcal{T}$ by forgetting the second component of the label of every transition and by replacing the functions I and T by their respective domains. The language recognized by $\mathcal{A}$ is the domain of the relation realized by $\mathcal{T}$.

We call *sequential* a transducer that is real-time, functional, and whose underlying input automaton is *deterministic*. A function α from A^* into B^* is *sequential* if it can be realized by a sequential transducer. It has to be acknowlegded that *this is not the usual terminology*: what we call "sequential" (transducers or functions) have been called "*subsequential*" since the seminal paper by Schützenberger [13] — *cf.* [2,5,7,8,11, *etc.*]. There are good reasons for this change of terminology that has already been advocated by V. Bruyère and Ch. Reutenauer: "the word *subsequential* is unfortunate since these functions should be called simply *sequential*" ([5]). Someone has to make the first move.

2 Squaring Automata and Ambiguity

Before defining the square of a transducer, we recall what is the *square of an automaton* and how it can be used to decide whether an automaton is unambiguous or not. A trim automaton $\mathcal{A} = \langle Q, A, E, I, T \rangle$ is *unambiguous* if any word it accepts is the label of a unique successful computation in $\mathcal{A}$.

Let $\mathcal{A}' = \langle Q', A, E', I', T' \rangle$ and $\mathcal{A}'' = \langle Q'', A, E'', I'', T'' \rangle$ be two automata on A. The *Cartesian product* of $\mathcal{A}'$ and $\mathcal{A}''$ is the automaton $\mathcal{C}$ defined by

$$\mathcal{C} = \mathcal{A}' \times \mathcal{A}'' = \langle Q' \times Q'', A, E, I' \times I'', T' \times T'' \rangle$$

where E is the set of transitions defined by

$$E = \{((p', p''), a, (q', q'')) \mid (p', a, q') \in E' \quad \text{and} \quad (p'', a, q'') \in E''\} \ .$$

Let $\mathcal{A} \times \mathcal{A} = \langle Q \times Q, A, F, I \times I, T \times T \rangle$ be the Cartesian product of the automaton $\mathcal{A} = \langle Q, A, E, I, T \rangle$ with itself; the set F of transitions is defined by:

$$F = \{((p, r), a, (q, s)) \mid (p, a, q), (r, a, s) \in E\} \ .$$

Let us call *diagonal* of $\mathcal{A} \times \mathcal{A}$ the sub-automaton $\mathcal{D}$ of $\mathcal{A} \times \mathcal{A}$ determined by the diagonal D of $Q \times Q$, *i.e.* $D = \{(q, q) \mid q \in Q\}$, as set of states. The states and transitions of $\mathcal{A}$ and $\mathcal{D}$ are in bijection, hence $\mathcal{A}$ and $\mathcal{D}$ are equivalent.

Lemma 1. *[3, Prop. IV.1.6]   A trim automaton $\mathcal{A}$ is unambiguous *if and only if the* trim part of $\mathcal{A} \times \mathcal{A}$ is equal to $\mathcal{D}$.*   ∎

Remark that as (un)ambiguity, *determinism* can also be described in terms of Cartesian square, by a simple rewording of the definition: *a trim automaton $\mathcal{A}$ is* deterministic *if and only if the* accessible part *of $\mathcal{A} \times \mathcal{A}$ is equal to $\mathcal{D}$.*

3   Product of an Automaton by an Action

We recall now what is an *action*, how an action can be seen as an automaton, and what can be then defined as the product of a (normal) automaton by an action. We end this section with the definition of the specific action that will be used in the sequel.

Actions.   A (right) *action of a monoid M on a set S* is a mapping $\delta \colon S \times M \to S$ which is consistent with the multiplication in M:

$$\forall s \in S, \ \forall m, m' \in M \qquad \delta(s, 1_M) = s \quad \text{and} \quad \delta(\delta(s, m), m') = \delta(s, m\, m') \ .$$

We write $s \cdot m$ rather than $\delta(s, m)$ when it causes no ambiguity.

Actions as automata.   An action δ of M on a set S with s_0 as distinguished element may then be seen as an automaton on M (without terminal states):

$$\mathcal{G}_\delta = \langle\, S, M, E, s_0 \,\rangle$$

is defined by the set of transitions $E = \{(s, m, s \cdot m) \mid s \in S, \ m \in M\}$.

Note that, as both S and M are usually infinite, the automaton $\mathcal{G}_\delta$ is "doubly" infinite: the *set of states* is infinite, and, for every state s, the *set of transitions* whose origin is s is infinite as well.

Product of an automaton by an action.   Let $\mathcal{A} = \langle\, Q, M, E, I, T \,\rangle$ be a (finite trim) automaton on a monoid M and δ an action of M on a (possibly infinite) set S. The product of $\mathcal{A}$ and $\mathcal{G}_\delta$ is the automaton on M:

$$\mathcal{A} \times \mathcal{G}_\delta = \langle\, Q \times S, M, F, I \times \{s_0\}, T \times S \,\rangle$$

the transitions of which are defined by

$$F = \{((p, s), m, (q, s \cdot m)) \mid s \in S, \ (p, m, q) \in E\} \ .$$

We shall call *product of $\mathcal{A}$ by δ*, and denote by $\mathcal{A} \times \delta$, the *accessible part* of $\mathcal{A} \times \mathcal{G}_\delta$.

The projection on the first component induces a bijection between the transitions of $\mathcal{A}$ whose origin is p and the transitions of $\mathcal{A} \times \delta$ whose origin is (p, s), for any p in Q and any (p, s) in $\mathcal{A} \times \delta$. The following holds (by induction on the length of the computations):

$$(p, s) \xrightarrow[\mathcal{A} \times \delta]{m} (q, t) \quad \Longrightarrow \quad t = s \cdot m \ .$$

We call *value of a state* (p, s) of $\mathcal{A} \times \delta$ the element s of S. We shall say that the product $\mathcal{A} \times \delta$ itself *is a valuation* if the projection on the first component is a 1-to-1 mapping between the states of $\mathcal{A} \times \delta$ and the states of $\mathcal{A}$.

Remark 1.   Let us stress again the fact that $\mathcal{A} \times \delta$ is the *accessible part* of $\mathcal{A} \times \mathcal{G}_\delta$. This makes possible that it may happen that $\mathcal{A} \times \delta$ is finite eventhough $\mathcal{G}_\delta$ is infinite (*cf.* Theorem 5).

The "Advance or Delay" action.   Let B^* be a free monoid and let us denote by H_B the set $H_B = (B^* \times 1_{B^*}) \cup (1_{B^*} \times B^*) \cup \{\mathbf{0}\}$. A mapping $\psi \colon B^* \times B^* \to H_B$ is defined by:

$$\forall u, v \in B^* \qquad \psi(u, v) = \begin{cases} (v^{-1}u, 1_{B^*}) & \text{if} \quad v \text{ is a prefix of } u \\ (1_{B^*}, u^{-1}v) & \text{if} \quad u \text{ is a prefix of } v \\ \mathbf{0} & \text{otherwise} \end{cases}$$

Intuitively, $\psi(u, v)$ tells either how much the first component u is ahead of the second component v, or how much it is late, or if u and v are not prefixes of a common word. In particular, $\psi(u, v) = (1_{B^*} \times 1_{B^*})$ if, and only if, $u = v$.

Lemma 2.   *The mapping ω_B from $H_B \times (B^* \times B^*)$ into H_B defined by:*

$$\forall (f, g) \in H_B \setminus \mathbf{0} \qquad \omega_B((f, g), (u, v)) = \psi(f\,u, g\,v) \quad and \quad \omega_B(\mathbf{0}, (u, v)) = \mathbf{0}$$

is an action, which will be called the "Advance or Delay" (or "AD") action (relative to B^*) and will thus be denoted henceforth by a dot.   ■

Remark 2.   The transition monoid of ω_B is isomorphic to $B^* \times B^*$ if B has at least two letters, to $\mathbb{Z}$ if it has only one letter. (We have denoted by $\mathbf{0}$ the absorbing element of H_B under ω_B in order to avoid confusion with 0, the identity element of the monoid $\mathbb{Z}$).

4   Deciding Functionality

Let $\mathcal{T} = \langle Q, A^* \times B^*, E, I, T \rangle$ be a real-time trim transducer such that the output of every transition is a single word of B^* — recall that this is a necessary condition for the relation realized by $\mathcal{T}$ to be a function. The transducer $\mathcal{T}$ is not functional if and only if there exist two *distinct* computations:

$$q_0' \xrightarrow[\mathcal{T}]{a_1/u_1'} q_1' \cdots \xrightarrow[\mathcal{T}]{a_n/u_n'} q_n' \qquad \text{and} \qquad q_0'' \xrightarrow[\mathcal{T}]{a_1/u_1''} q_1'' \cdots \xrightarrow[\mathcal{T}]{a_n/u_n''} q_n''$$

with $u_1' u_2' \ldots u_n' \neq u_1'' u_2'' \ldots u_n''$. There exists then at least one i such that $u_i' \neq u_i''$, and thus such that $q_i' \neq q_i''$.

This implies, by projection on the first component, that the underlying input automaton $\mathcal{A}$ of $\mathcal{T}$ is *ambiguous*. But it may be the case that $\mathcal{A}$ is ambiguous and $\mathcal{T}$ still functional, as it is shown for instance with the transducer $\mathcal{Q}_1$ represented on the top of Figure 1 (*cf.* [2]). We shall now carry on the method of Cartesian square of section 2 from automata to transducers.

Cartesian square of a real-time transducer.   By definition, the *Cartesian product* of T by itself is the transducer $T \times T$ from A^* into $B^* \times B^*$:

$$T \times T = \langle Q \times Q, A^* \times (B^* \times B^*), F, I \times I, T \times T \rangle$$

whose transitions set F is defined by:

$$F = \{((p, r), (a, (u', u'')), (q, s)) \mid \quad (p, (a, u'), q) \quad \text{and} \quad (r, (a, u''), s) \in E\} .$$

The underlying input automaton of $T \times T$ is the square of the underlying input automaton $\mathcal{A}$ of T. If $\mathcal{A}$ is unambiguous, then T is functional, and the trim part of $\mathcal{A} \times \mathcal{A}$ is reduced to its diagonal.

An effective characterization of functionality.   The transducer $T \times T$ is an automaton on the monoid $M = A^* \times (B^* \times B^*)$. We can consider that the AD action is an action of M on H_B, by forgetting the first component. We can thus make the product of $T \times T$, or of any of its subautomata, by the AD action ω_B.

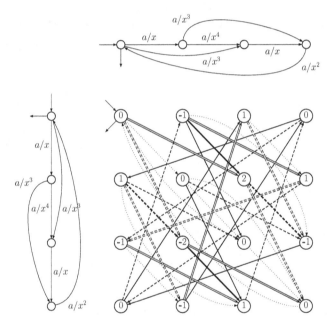

Fig. 1. Cartesian square of $\mathcal{Q}_1$, valued by the product with the action $\omega_{\{x\}}$.
As the output alphabet has only one letter, $H_{\{x\}}$ is identified with $\mathbb{Z}$ and the states are labelled by an integer. Labels of transitions are not shown: the input is always a and is kept implicit; an output of the form (x^n, x^m) is coded by the integer $n - m$ which is itself symbolised by the drawing of the arrow: a dotted arrow for 0, a simple solid arrow for +1, a double one for +2 and a bold one for +3; and the corresponding dashed arrows for the negative values.

Theorem 3.   *A transducer $\mathcal{T}$ from A^* into B^* is functional if and only if the product of the trim part $\mathcal{U}$ of the Cartesian square $\mathcal{T} \times \mathcal{T}$ by the AD action ω_B is a valuation of $\mathcal{U}$ such that the value of any final state is $(1_{B^*}, 1_{B^*})$.*   ∎

Figure 1 shows the product of the Cartesian square of a transducer $\mathcal{Q}_1$ by the AD action[1].

Let us note that if α is the relation realized by $\mathcal{T}$, the transducer obtained from $\mathcal{T} \times \mathcal{T}$ by forgetting the first component is a transducer from B^* into itself that realizes the composition product $\alpha \circ \alpha^{-1}$. The conditon expressed may then seen as a condition for $\alpha \circ \alpha^{-1}$ being the identity, which is clearly a condition for the functionality of α.

5   Deciding Sequentiality

The original proof of Theorem 2 goes indeed in three steps: first, sequential functions are characterized by a property expressed by means of a *distance function*, then this property (on the function) is proved to be equivalent to a property on the transducer, and finally a pumping-lemma like procedure is given for deciding the latter property (*cf.* [7,2]). We shall see how the last two steps can be replaced by the computation of the product of the Cartesian square of the transducer by the AD action. We first recall the first step.

5.1   A Quasi-Topological Characterization of Sequential Functions

If f and g are two words, we denote by $f \wedge g$ the *longuest prefix common* to f and g. The free monoid is then equipped with the *prefix distance*

$$\forall f, g \in A^* \qquad \mathsf{d_p}(f, g) = |f| + |g| - 2|f \wedge g| \ .$$

In other words, if $f = h\,f'$ and $g = h\,g'$ with $h = f \wedge g$, then $\mathsf{d_p}(f, g) = |f'| + |g'|$.

Definition 1.   *A function $\alpha \colon A^* \to B^*$, is said to be* uniformly diverging[2] *if for every integer n there exists an integer N which is greater than the prefix distance of the images by α of any two words (in the domain of α) whose prefix distance is smaller than n, i.e.*

$$\forall n \in \mathbb{N}, \ \exists N \in \mathbb{N}, \ \forall f, g \in \mathsf{Dom}\,\alpha \qquad \mathsf{d_p}(f, g) \leqslant n \implies \mathsf{d_p}(f\alpha, g\alpha) \leqslant N \ .$$

Theorem 4. *[7,13]*   *A rational function is sequential if, and only if it is uniformly diverging.*

Remark 3.   The characterization of sequential functions by uniform divergence holds in the larger class of functions whose inverse preserves rationality. This is a generalization of a theorem of Ginsburg and Rose due to Choffrut, a much stronger result, the full strength of which will not be of use here (*cf.* [5,8]).

---

[1] It turns out that, in this case, the trim part is equal to the whole square.

[2] After [7] and [2], the usual terminology is "function with *bounded variation*". We rather avoid an expression that is already used, with an other meaning, in other parts of mathematics.

5.2   An Effective Characterization of Sequential Functions

Theorem 5.   *A transducer $\mathcal{T}$ realizes a sequential function if, and only if the product of the accessible part $\mathcal{V}$ of $\mathcal{T} \times \mathcal{T}$ by the AD action ω_B*

i)   *is finite;*

ii)   *has the property that if a state with value $\mathbf{0}$ belongs to a cycle in $\mathcal{V}$, then the label of that cycle is $(1_{B^*}, 1_{B^*})$.*   ∎

The parallel between automata and transducers is now to be emphasized. Unambiguous (resp. deterministic) automata are characterized by a condition on the trim (resp. accessible) part of the Cartesian square of the automaton whereas functional transducers (resp. transducers that realize sequential functions) are characterized by a condition on the product by ω_B of the trim (resp. accessible) part of the Cartesian square of the transducer.

Figure 2 shows two cases where the function is sequential: in (a) since the accessible part of the product is finite and no state has value $\mathbf{0}$; in (b) since the accessible part of the product is finite as well and the states whose value is $\mathbf{0}$ all belong to a cycle every transition of which is labelled by $(1_{B^*}, 1_{B^*})$.

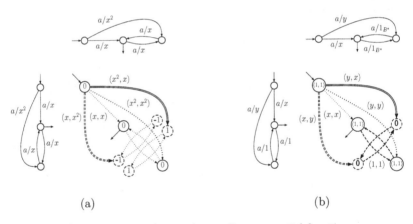

(a)   (b)

Fig. 2. Two transducers that realize sequential functions.

Figure 3 shows two cases where the function is not sequential: in (a) since the accessible part of the product is infinite; in (b) since although the accessible part of the product is finite some states whose value is $\mathbf{0}$ belong to a cycle whose label is different from $(1_{B^*}, 1_{B^*})$.

The following lemma is the key to the proof of Theorem 5 as well as to its effectivity.

Lemma 3.   *Let $w = (1_{B^*}, z)$ be in $H_B \setminus \mathbf{0}$ and (u, v) in $B^* \times B^* \setminus (1_{B^*}, 1_{B^*})$. Then the set $\{w \cdot (u, v)^n \mid n \in \mathbb{N}\}$ is finite and does not contain $\mathbf{0}$ if, and only if, u and v are congugate words and z is a prefix of a power of u.*   ∎

Remark 4.   The original proof of Theorem 2 by Ch. Choffrut goes by the definition of the so-called *twinning property* (*cf.* [2, p. 128]). It is not difficult to check that two states p and q of a real-time transducer $\mathcal{T}$ are (non trivially) *twinned* when: i) (p,q) is accessible in $\mathcal{T} \times \mathcal{T}$;   ii) (p,q) belongs to a cycle in $\mathcal{V}$ every transition of which is not labelled by $(1_{B^*}, 1_{B^*})$;   iii) (p,q) has not the value $\mathbf{0}$ in the product of $\mathcal{V}$ by ω_B.

It is then shown that a transducer realizes a sequential function if, and only if, every pair of its states has the twinning property.

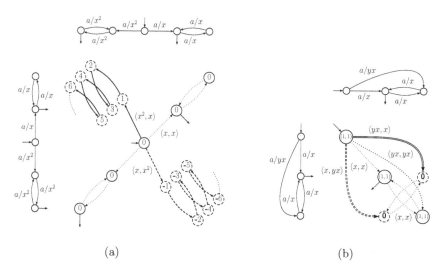

(a)   (b)

Fig. 3. Two transducers that realize non sequential functions.

6   The Complexity Issue

The "*size*" of an automaton $\mathcal{A}$ (on a free monoid A^*) is measured by the number m of transitions. (The size $|A| = k$ of the (input) alphabet is seen as a constant.) The size of a transducer $\mathcal{T}$ will be measured by *the sum of the sizes* of its transitions where the size of a transition $(p, (u,v), q)$ is the length $|u\,v|$. It is denoted by $|\mathcal{T}|$.

The size of the transducer $\mathcal{T} \times \mathcal{T}$ is $|\mathcal{T}|^2$ and the complexity to build it is proportional to that size. The complexity of determining the trim part as well as the accessible part is linear in the size of the transducer.

Deciding whether the product of the *trim part* $\mathcal{U}$ of $\mathcal{T} \times \mathcal{T}$ by the AD action ω_B is a valuation of $\mathcal{U}$ (and if the value of any final state is $(1_{B^*}, 1_{B^*})$) is again linear in the size of $\mathcal{U}$. Hence deciding whether a transducer $\mathcal{T}$ is functional is quadratic in the size of the transducer. Note that the same complexity is also established in [6].

The complexity of a decision procedure for the sequentiality of a function, based on Theorem 5, is polynomial. However, this is less straightforward to establish than functionality, for the size of the product $\mathcal{V} \times \omega_B$ may be exponential.

One first checks whether the label of every cycle in $\mathcal{V}$ is of the form (u, v) with $|u| = |v|$. It suffices to check it on a base of simple cycles and this can be done by a deep-first search in $\mathcal{V}$. Let us call *true cycle* a cycle which is not labelled by $(1_{B^*}, 1_{B^*})$ and let $\mathcal{W}$ be the subautomaton of $\mathcal{V}$ consisting of states from which a true cycle is accessible. By Theorem 5, if suffices to consider the product $\mathcal{W} \times \omega_B$. This product may still be of exponential size. However one does not construct it entirely. For every state of $\mathcal{W}$, the number of values which are to be considered in $\mathcal{W} \times \omega_B$ may be bounded by the size of $\mathcal{T}$. This yields an algorithm of polynomial complexity in order to decide the sequentiality of the function realized by $\mathcal{T}$.

In [1], it is shown directly that the twinning property is decidable in polynomial time.

References

1. M.-P. Béal and O. Carton: Determinization of transducers over finite and infinite words, *to appear*.
2. J. Berstel: *Transductions and context-free languages*, Teubner, 1979.
3. J. Berstel and D. Perrin: *Theory of codes*, Academic Press, 1985.
4. M. Blattner and T. Head: Single valued a-transducers, *J. Computer System Sci.* **7** (1977), 310–327.
5. V. Bruyère and Ch. Reutenauer: A proof of Choffrut's theorem on subsequential functions, *Theoret. Comput. Sci.* **215** (1999), 329–335.
6. O. Carton, Ch. Choffrut and Ch. Prieur: How to decide functionality of rational relations on infinite words, *to appear*.
7. Ch. Choffrut: Une caractérisation des fonctions séquentielles et des fonctions sous-séquentielles en tant que relations rationnelles, *Theoret. Comput. Sci.* **5** (1977), 325–337.
8. Ch. Choffrut: A generalization of Ginsburg and Rose's characterization of g-s-m mappings, *in Proc. of ICALP'79* (H. Maurer, Ed.), *Lecture Notes in Comput. Sci.* **71** (1979), 88–103.
9. S. Eilenberg: *Automata, Languages and Machines* vol. A, Academic Press, 1974.
10. E. M. Gurari and O. H. Ibarra: Finite-valued and finitely ambiguous transducers, *Math. Systems Theory* **16** (1983), 61-66.
11. Ch. Reutenauer: Subsequential functions: characterizations, minimization, examples, *Lecture Notes in Comput. Sci.* **464** (1990), 62–79.
12. M. P. Schützenberger: Sur les relations rationnelles, *in Automata Theory and Formal Languages* (H. Brackhage, Ed.), *Lecture Notes in Comput. Sci.* **33** (1975), 209–213.
13. M. P. Schützenberger: Sur une variante des fonctions séquentielles, *Theoret. Comput. Sci.* **4** (1977), 47–57.

Unambiguous Büchi Automata

Olivier Carton[1] and Max Michel[2]

[1] Institut Gaspard Monge, 5, boulevard Descartes F-77454 Marne-la-Vallée cedex 2
Olivier.Carton@univ-mlv.fr
[2] CNET, 38, rue du Général Leclerc F-92131 Issy-les-Moulineaux
Max.Michel@cnet.francetelecom.fr

Abstract. In this paper, we introduce a special class of Büchi automata called unambiguous. In these automata, any infinite word labels exactly one path going infinitely often through final states. The word is accepted by the automaton if this path starts at an initial state. The main result of the paper is that any rational set of infinite words is recognized by such an automaton. We also provide two characterizations of these automata. We finally show that they are well suitable for boolean operations.

1 Introduction

Automata on infinite words have been introduced by Büchi [3] in order to prove the decidability of the monadic second-order logic of the integers. Since then, automata on infinite objects have often been used to prove the decidability of numerous problems. From a more practical point of view, they also lead to efficient decision procedures as for temporal logic [12]. Therefore, automata of infinite words or infinite trees are one of the most important ingredients in model checking tools [14]. The complementation of automata is then an important issue since the systems are usually modeled by logical formulas which involve the negation operator.

There are several kinds of automata that recognize sets of infinite words. In 1962, Büchi [3] introduced automata on ω-words, now referred to as *Büchi automata*. These automata have initial and final states and a path is successful if it starts at an initial state and goes infinitely often through final states. However, not all rational sets of infinite words are recognized by a deterministic Büchi automaton [5]. Therefore, complementation is a rather difficult operation on Büchi automata [12].

In 1963, Muller [9] introduced automata, now referred to as *Muller automata*, whose accepting condition is a family of accepting subsets of states. A path is then successful if it starts at the unique initial state and if the set of states which occurs infinitely in the path is accepting. A deep result of McNaughton [6] shows that any rational set of infinite words is recognized by a deterministic Muller automaton. A deterministic automaton is unambiguous in the following sense. With each word is associated a canonical path which is the unique path starting at the initial state. A word is then accepted iff its canonical path is successful. In a deterministic Muller automaton, the unambiguity is due to the uniqueness of the initial state and to the determinism of the transitions. Independently, the acceptance condition determines if a path is successful or not. The unambiguity of a deterministic Muller automaton makes it easy to complement. It suffices to exchange accepting and non-accepting subsets of states. However, the main drawback of using deterministic Muller automata is that the acceptance condition is much more complicated. It is a family of subsets of states instead of a simple set of final states. There are other kinds of deterministic automata recognizing all rational sets of infinite words like Rabin automata [11], Street automata or parity automata [8]. In all these automata, the acceptance condition is more complicated than a simple set of final states.

In this paper, we introduce a class of Büchi automata in which any infinite word labels exactly one path going infinitely often through final states. A canonical path can then be associated with each infinite word and we call these automata unambiguous. In these automata, the unambiguity is due to the transitions and to the final states whereas the initial states determine if a path is successful. An infinite word is then accepted iff its canonical path starts at an initial state. The main result is that any rational set of infinite words is recognized by such an automaton. It turns out that these unambiguous Büchi automata are codeterministic, i.e., reverse deterministic. Our result is thus the counterpart of McNaughton's result for codeterministic automata. It has already been proved independently in [7] and [2] that any rational set of infinite words is recognized by a codeterministic automaton but the construction given in [2] does not provide unambiguous automata. We also show that unambiguous automata are well suited for boolean operations and especially complementation. In particular, our construction can be used to find a Büchi automaton which recognizes the complement of the set recognized by another Büchi automaton. For a Büchi automaton with n states, our construction provides an unambiguous automaton which has at most $(12n)^n$ states.

The unambiguous automata introduced in the paper recognize right-infinite words. However, the construction can be adapted to bi-infinite words. Two unambiguous automata on infinite words can be joined to make an unambiguous automaton on bi-infinite words. This leads to an extension of McNaughton's result to the realm of bi-infinite words.

The main result of this paper has been first obtained by the second author and his proof has circulated as a hand-written manuscript among a bunch of people. It was however never published. Later, the first author found a different proof of the same result based on algebraic constructions on semigroups. Both authors have decided to publish their whole work on this subject together.

The paper is organized as follows. Section 2 is devoted to basic definitions on words and automata. Unambiguous Büchi automata are defined in Sect. 3. The main result (Theorem 1) is stated there. The first properties of these automata are presented in Sect. 4. Boolean Operations are studied in Sect. 5.

2   Automata

We recall here some elements of the theory of rational sets of finite and infinite words. For further details on automata and rational sets of finite words, see [10] and for background on automata and rational sets of infinite words, see [13]. Let A be a set called an *alphabet* and usually assumed to be finite. We respectively denote by A^* and A^+ the set of finite words and the set of nonempty finite words. The set of right-infinite words, also called ω-words, is denoted by A^ω.

A *Büchi automaton* $\mathcal{A} = (Q, A, E, I, F)$ is a non-deterministic automaton with a set Q of states, subsets $I, F \subset Q$ of *initial* and *final* states and a set $E \subset Q \times A \times Q$ of *transitions*. A transition (p, a, q) of $\mathcal{A}$ is denoted by $p \xrightarrow{a} q$. A *path* in $\mathcal{A}$ is an infinite sequence

$$\gamma : q_0 \xrightarrow{a_0} q_1 \xrightarrow{a_1} q_2 \cdots$$

of consecutive transitions. The *starting* state of the path is q_0 and the ω-word $\lambda(\gamma) = a_0 a_1 \ldots$ is called the *label* of γ. A *final* path is a path γ such that at least one of the final states of the automaton is infinitely repeated in γ. A *successful* path is a final path which starts at an initial state.

As usual, an ω-word is accepted by the automaton if it is the label of a successful path. The set of accepted ω-words is said to be recognized by the automaton and is denoted by $L(\mathcal{A})$. It is well known that a set of ω-words is rational iff it is recognized by some automaton.

A state of a Büchi automaton $\mathcal{A}$ is said to be *coaccessible* if it is the starting state of a final path. A Büchi automaton is said to be *trim* if all states are coaccessible. Any state which occurs in a final path is coaccessible and thus non-coaccessible states of an automaton can be removed. In the sequel, automata are usually assumed to be trim.

An automaton $\mathcal{A} = (Q, A, E, I, F)$ is said to be *codeterministic* if for any state q and any letter a, there is at most one incoming transition $p \xrightarrow{a} q$ for some state p. If this condition is met, for any state q and any finite word w, there is at most one path $p \xrightarrow{w} q$ ending in q.

3   Unambiguous Automata

In this section, we introduce the concept of unambiguous Büchi automata. We first give the definition and we state one basic property of these automata. We then establish a characterization of these automata. We give some examples and we state the main result.

Definition 1. *A Büchi automaton $\mathcal{A}$ is said to be* unambiguous *(respectively* complete*) iff any ω-word labels at most (respectively at least) one final path in $\mathcal{A}$.*

The set of final paths is only determined by the transitions and the final states of $\mathcal{A}$. Thus, the property of being unambiguous or complete does not depend on the set of initial states of $\mathcal{A}$. In the sequel, we will freely say that an automaton $\mathcal{A}$ is unambiguous or complete without specifying its set of initial states.

The definition of the word "complete" we use here is not the usual definition given in the literature. A deterministic automaton is usually said to be complete if for any state q and letter a there is at least an outgoing transition labeled by a. This definition implies that any finite or infinite word labels at least a path starting at the initial state. It will stated in Proposition 1 that the unambiguity implies that the automaton is codeterministic. Thus, we should reverse the definition and we should say that for any state q and letter a there is at least an incoming transition labeled by a. However, since the words are right-infinite, this condition does not imply anymore that any ω-word labels a path going infinitely often though final states as it is shown in Example 3. Thus, the definition chosen in this paper really insures that any ω-word is the label of a final path. It will be stated in Proposition 1 that this condition is actually stronger that the usual one.

In the sequel, we write UBA for Unambiguous Büchi Automaton and CUBA for Complete Unambiguous Büchi Automaton. The following example is the simplest CUBA.

Example 1. The automaton $(\{0\}, A, E, I, \{0\})$ with $E = \{0 \overset{a}{\rightarrow} 0 \mid a \in A\}$ is obviously a CUBA. It recognizes the set A^ω of all ω-words if the state 0 is initial and recognizes the empty set otherwise. It is called the *trivial* CUBA.

The following proposition states that an UBA must be codeterministic. Such an automaton can be seen as a deterministic automaton which reads infinite words from right to left. It starts at infinity and ends at the beginning of the word. Codeterministic automata on infinite words have already been considered in [2]. It is proved in that paper that any rational set of ω-words is recognized by a codeterministic automata. Our main theorem generalizes this results. It states that any rational set of ω-words is recognized by a CUBA.

Proposition 1. *Let $\mathcal{A} = (Q, A, E, I, F)$ be a trim Büchi automaton. If $\mathcal{A}$ is unambiguous, then $\mathcal{A}$ is codeterministic. If $\mathcal{A}$ is complete, then for any state q and any letter a, there is at least one incoming transition $p \overset{a}{\rightarrow} q$ for some state p.*

The second statement of the proposition says that our definition of completeness implies the usual one. Example 3 shows that the converse does not hold. However, Proposition 3 provides some additional condition on the automaton to ensure that it is unambiguous and complete.

Before giving some other examples of CUBA, we provide a simple characterization of CUBA which makes it easy to verify that an automaton is unambiguous and complete. This proposition also shows that it can be effectively checked if a given automaton is unambiguous or complete.

Let $\mathcal{A} = (Q, A, E, I, F)$ be a Büchi automaton and let q be a state of $\mathcal{A}$. We denote by $\mathcal{A}_q = (Q, A, E, \{q\}, F)$ the new automaton obtained by taking the singleton $\{q\}$ as set of initial states. The set $L(\mathcal{A}_q)$ is then the set of ω-words labeling a final path starting at state q.

Proposition 2. *Let $\mathcal{A} = (Q, A, E, I, F)$ be a Büchi automaton. For $q \in Q$, let $\mathcal{A}_q$ the automaton $(Q, A, E, \{q\}, F)$. The automaton $\mathcal{A}$ is unambiguous iff the sets $\mathrm{L}(\mathcal{A}_q)$ are pairwise disjoint. The automaton $\mathcal{A}$ is complete iff $A^\omega \subset \bigcup_{q \in Q} \mathrm{L}(\mathcal{A}_q)$*

In particular, the automaton $\mathcal{A}$ is unambiguous and complete iff the family of sets $\mathrm{L}(\mathcal{A}_q)$ for $q \in Q$ is a partition of A^ω. It can be effectively verified that the two sets recognized by the automata $\mathcal{A}_q$ and $\mathcal{A}_{q'}$ are disjoint for $q \neq q'$. It can then be checked if the automaton is unambiguous. Furthermore, this test can be performed in polynomial time. The set $\bigcup_{q \in Q} \mathrm{L}(\mathcal{A}_q)$ is recognized by the automaton $\mathcal{A}_Q = (Q, A, E, Q, F)$ whose all states are initial. The inclusion $A^\omega \subset \bigcup_{q \in Q} \mathrm{L}(\mathcal{A}_q)$ holds iff this automaton recognizes A^ω. This can be checked but it does not seem it can be performed in polynomial time.

We now come to examples. We use Proposition 2 to verify that the following two automata are unambiguous and complete. In the figures, a transition $p \xrightarrow{a} q$ of an automaton is represented by an arrow labeled by a from p to q. Initial states have a small incoming arrow while final states are marked by a double circle. A Büchi automaton which is complete but ambiguous is given is Example 4.

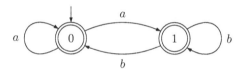

Fig. 1. CUBA of Example 2

Example 2. Let A be the alphabet $A = \{a, b\}$ and let $\mathcal{A}$ be the automaton pictured in Fig. 1. This automaton is unambiguous and complete since we have $\mathrm{L}(\mathcal{A}_0) = aA^\omega$ and $\mathrm{L}(\mathcal{A}_1) = bA^\omega$. It recognizes the set aA^ω of ω-words beginning with an a.

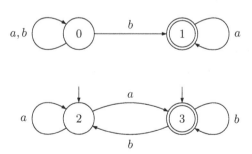

Fig. 2. CUBA of Example 3

The following example shows that a CUBA may have several connected components.

Example 3. Let A be the alphabet $A = \{a, b\}$ and let $\mathcal{A}$ be the automaton pictured in Fig. 2. It is unambiguous and complete since we have $\mathrm{L}(\mathcal{A}_0) = A^*ba^\omega$, $\mathrm{L}(\mathcal{A}_1) = a^\omega, \mathrm{L}(\mathcal{A}_2) = a(A^*b)^\omega$ and $\mathrm{L}(\mathcal{A}_3) = b(A^*b)^\omega$. It recognizes the set $(A^*b)^\omega$ of ω-words having an infinite number of b.

The automaton of the previous example has two connected components. Since it is unambiguous and complete any ω-word labels exactly one final path in this automaton. This final path is in the first component if the ω-word has finitely many b and it is the second component otherwise. This automaton shows that our definition of completeness for an unambiguous Büchi automaton is stronger than the usual one. Any connected component is complete in the usual sense if it is considered as a whole automaton. For any letter a and any state q in this component, there is exactly one incoming transition $p \xrightarrow{a} q$. However, each component is not complete according to our definition since not any ω-word labels a final path in this component.

In the realm of finite words, an automaton is usually made unambiguous by the usual subsets construction [4, p. 22]. This construction associates with an automaton $\mathcal{A}$ an equivalent deterministic automaton whose states are subsets of states of $\mathcal{A}$. Since left and right are symmetric for finite words, this construction can be reversed to get a codeterministic automaton which is also equivalent to $\mathcal{A}$. In the case of infinite words, the result of McNaughton [6] states that a Büchi automaton can be replaced by an equivalent Muller automaton which is deterministic. However, this construction cannot be reversed since ω-words are right-infinite. We have seen in Proposition 1 that a CUBA is codeterministic. The following theorem is the main result of the paper. It states that any rational set of ω-words is recognized by a CUBA. This theorem is thus the counterpart of McNaughton's result for codeterministic automata. Like Muller automata, CUBA make the complementation very easy to do. This will be shown in Sect. 5. The proof of Theorem 1 contains a new proof that the class of rational sets of ω-words is closed under complementation.

Theorem 1. *Any rational set of ω-words is recognized by a complete unambiguous Büchi automaton.*

There are two proofs of this result which are both rather long. Both proofs yield effective procedures which give a CUBA recognizing a given set of ω-words. The first proof is based on graphs and it directly constructs a CUBA from a Büchi automaton recognizing the set. The second proof is based on semigroups and it constructs a CUBA from a morphism from A^+ into a finite semigroup recognizing the set. An important ingredient of both proofs is the notion of a generalized Büchi automaton.

In a Büchi automaton, the set of final paths is the set of paths which go infinitely often through final states. In a generalized Büchi automaton, the set of final paths is given in a different way. A generalized Büchi automaton is equipped with an output function μ which maps any transition to a nonempty word over an alphabet B and with a fixed set K of ω-words over B. A path is final if the concatenation of the outputs of its transitions belongs to K. A generalized Büchi automaton can be seen as an automaton with an output function. We point out that usual Büchi automata are a particular case of generalized Büchi automata. Indeed, if the function μ maps any transition $p \xrightarrow{a} q$ to 1 if p or q is final and to 0 otherwise and if $K = (0^*1)^\omega$ is the set of ω-words over $\{0, 1\}$ having an infinite number of 1, a path in $\mathcal{A}$ is final if some final state occurs infinitely often in it.

The notions of unambiguity and completeness are then extended to generalized Büchi automata. A generalized Büchi automaton is said to be unambiguous (respectively complete) if any ω-word labels at most (respectively at least) one final path.

The generalized Büchi automata can be composed. If a set X is recognized by an automaton $\mathcal{A}$ whose fixed set K is recognized by automaton $\mathcal{B}$ which has a fixed set K', then X is also recognized by an automaton having the fixed set K' which can be easily constructed from $\mathcal{A}$ and $\mathcal{B}$. Furthermore, this composition is compatible with unambiguity and completeness. This means that if both automata $\mathcal{A}$ and $\mathcal{B}$ are unambiguous (respectively complete), then the automaton obtained by composition is also unambiguous (respectively complete).

4   Properties and Characterizations

In this section, we present some additional properties of CUBA. We first give another characterization of CUBA which involves loops going through final states. We present some consequences of this characterization. The characterization of CUBA given in Proposition 2 uses sets of ω-words. The family of sets of ω-words labeling a final path starting in the different states must be a partition of the set A^ω of all ω-words. The following proposition only uses sets of finite words to characterize UBA and CUBA.

Proposition 3. *Let $\mathcal{A} = (Q, A, E, I, F)$ be a Büchi automaton such that for any state q and any letter a, there exists exactly one incoming transition $p \xrightarrow{a} q$. Let S_q be the set of nonempty finite words w such that there is a path $q \xrightarrow{w} q$ going through a final state. The automaton $\mathcal{A}$ is unambiguous iff the sets S_q are pairwise disjoint. The automaton $\mathcal{A}$ is unambiguous and complete iff the family of sets S_q for $q \in Q$ is a partition of A^+. In this case, the final path labeled by the periodic ω-word w^ω is the path*

$$q \xrightarrow{w} q \xrightarrow{w} q \cdots$$

where q is the unique state such that $w \in S_q$.

The second statement of the proposition says that if that if the automaton $\mathcal{A}$ is supposed to be unambiguous, it is complete iff the inclusion $A^+ \subset \bigcup_{q \in Q} S_q$ holds. The assumption that the automaton is unambiguous is necessary. As the following example shows, it is not true in general that the automaton is complete iff the inclusion holds.

Example 4. The automaton of Fig. 3 is ambiguous since the ω-word b^ω labels two final paths. Since this automaton is deterministic and all states are final, it

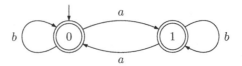

Fig. 3. CUBA of Example 4

is complete. However, it is not true that $A^+ \subset \bigcup_{q \in Q} S_q$. Indeed, no loop in this automaton is labeled by the finite word a.

Proposition 3 gives another method to check if a given Büchi automaton is unambiguous and complete. It must be first verified that for any state q and any letter a, there is exactly one incoming transition $p \xrightarrow{a} q$. Then, it must be checked if the family of sets S_q for $q \in Q$ forms a partition of A^+. The sets S_q are rational and a codeterministic automaton recognizing S_q can be easily deduced from the automaton $\mathcal{A}$. It is then straightforward to verify that the sets S_q form a partition of A^+.

The last statement of Proposition 3 says that the final path labeled by a periodic word is also periodic. It is worth mentioning that the same result does not hold for deterministic automata.

If follows from Proposition 3 that the trivial CUBA with one state (see Example 1) is the only CUBA which is deterministic.

5   Boolean Combinations

In this section, we show that CUBA have a fine behavior with the boolean operations. From CUBA recognizing two sets X and Y, CUBA recognizing the complement $A^\omega \setminus X$, the union $X \cup Y$ and the intersection $X \cap Y$ can be easily obtained. For usual Büchi automata or for Muller automata, automata recognizing the union and the intersection are easy to get. It is sufficient to consider the product of the two automata with some small additional memory. However, complementation is very difficult for general Büchi automata.

5.1   Complement

We begin with complementation which turns out to be a very easy operation for CUBA. Indeed, it suffices to change the initial states of the automaton to recognize the complement.

Proposition 4. Let $\mathcal{A} = (Q, A, E, I, F)$ be a CUBA recognizing a set X of ω-words. The automaton $\mathcal{A}' = (Q, A, E, Q \setminus I, F)$ where $Q \setminus I$ is the set of non initial states, is unambiguous and complete and it recognizes the complement $A^\omega \setminus X$ of X.

It must be pointed out that it is really necessary for the automaton $\mathcal{A}$ to be unambiguous and complete. Indeed, if $\mathcal{A}$ is ambiguous, it may happen that an ω-word x of X labels a final path starting at an initial state and another final path starting at a non initial state. In this case, the ω-word x is also recognized by the automaton $\mathcal{A}'$. If $\mathcal{A}$ is not complete, some ω-word x labels no final path. This ω-word which does not belong to X is not recognized by the automaton $\mathcal{A}'$.

By the previous result, the proof of Theorem 1 also provides a new proof of the fact that the family of rational sets of ω-words is closed under complementation.

5.2   Union and Intersection

In this section, we show how CUBA recognizing the union $X_1 \cup X_2$ and the intersection $X_1 \cap X_2$ can be obtained from CUBA recognizing X_1 and X_2.

We suppose that the sets X_1 and X_2 are respectively recognized by the CUBA $\mathcal{A}_1 = (Q_1, A, E_1, I_1, F_1)$ and $\mathcal{A}_2 = (Q_2, A, E_2, I_2, F_2)$. We will construct two CUBA $\mathcal{U} = (Q, A, E, I_{\mathcal{U}}, F)$ and $\mathcal{I} = (Q, A, E, I_{\mathcal{I}}, F)$ respectively recognizing the union $X_1 \cup X_2$ and the intersection $X_1 \cap X_2$. Both automata $\mathcal{U}$ and $\mathcal{I}$ share the same states set Q, the same transitions set E and the same set F of final states.

We first describe the states and the transitions of both automata $\mathcal{U}$ and $\mathcal{I}$. These automata are based on the product of the automata $\mathcal{A}_1$ and $\mathcal{A}_2$ but a third component is added. The final states may not appear at the same time in $\mathcal{A}_1$ and $\mathcal{A}_2$. The third component synchronizes the two automata by indicating in which of the two automata comes the first final state. The set Q of states is $Q = Q_1 \times Q_2 \times \{1, 2\}$. Each state is then a triple (q_1, q_2, ε) where q_1 is a state of $\mathcal{A}_1$, q_2 is a state of $\mathcal{A}_2$ and ε is 1 or 2. There is a transition $(q_1', q_2', \varepsilon') \xrightarrow{a} (q_1, q_2, \varepsilon)$ if $q_1' \xrightarrow{a} q_1$ and $q_2' \xrightarrow{a} q_2$ are transitions of $\mathcal{A}_1$ and $\mathcal{A}_2$ and if ε' is defined as follows.

$$\varepsilon' = \begin{cases} 1 & \text{if } q_1 \in F_1 \\ 2 & \text{if } q_1 \notin F_1 \text{ and } q_2 \in F_2 \\ \varepsilon & \text{otherwise} \end{cases}$$

This definition is not completely symmetric. When both q_1 and q_2 are final states, we choose to set $\varepsilon' = 1$. We now define the set F of final states as

$$F = \big\{ (q_1, q_2, \varepsilon) \big| q_2 \in F_2 \text{ and } \varepsilon = 1 \big\}.$$

This definition is also non symmetric.

It may be easily verified that any loop around a final state (q_1, q_2, ε) also contains a state $(q_1', q_2', \varepsilon')$ such that $q_2' \in F_2$. This implies that the function which maps a path γ to the pair (γ_1, γ_2) of paths in $\mathcal{A}_1$ and $\mathcal{A}_2$ is one to one from the set of final paths in $\mathcal{U}$ or $\mathcal{I}$ to the set of pairs of final paths in $\mathcal{A}_1$ and $\mathcal{A}_2$. Thus if both $\mathcal{A}_1$ and $\mathcal{A}_2$ are unambiguous and complete, then both automata $\mathcal{U}$ and $\mathcal{I}$ are also unambiguous and complete.

If q_1 and q_2 are the respective starting states of γ_1 and γ_2, the starting state of γ is then equal to (q_1, q_2, ε) with $\varepsilon \in \{1, 2\}$. We thus define the sets $I_{\mathcal{U}}$ and $I_{\mathcal{I}}$

of initial states of the automata $\mathcal{U}$ and $\mathcal{I}$ as follows.

$$I_{\mathcal{U}} = \left\{ (q_1, q_2, \varepsilon) \big| (q_1 \in I_1 \text{ or } q_2 \in I_2) \text{ and } \varepsilon \in \{1,2\} \right\}$$
$$I_{\mathcal{I}} = \left\{ (q_1, q_2, \varepsilon) \big| q_1 \in I_1 \text{ and } q_2 \in I_2 \text{ and } \varepsilon \in \{1,2\} \right\}$$

From these definitions, it is clear that both automata $\mathcal{U}$ and $\mathcal{I}$ are unambiguous and complete and that they respectively recognize $X_1 \cup X_2$ and $X_1 \cap X_2$.

Acknowledgment

The authors would like to thank Dominique Perrin, Jean-Éric Pin and Pascal Weil for helpful discussions and suggestions.

References

1. André Arnold. Rational omega-languages are non-ambiguous. *Theoretical Computer Science*, 26(1-2):221–223, 1983.
2. Danièle Beauquier and Dominique Perrin. Codeterministic automata on infinite words. *Information Processing Letters*, 20:95–98, 1985.
3. J. Richard Büchi. On a decision method in the restricted second-order arithmetic. In *Proc. Int. Congress Logic, Methodology and Philosophy of science, Berkeley 1960*, pages 1–11. Stanford University Press, 1962.
4. John E. Hopcroft and Jeffrey D. Ullman. *Introduction to Automata Theory, Languages and Computation*. Addison-Wesley, 1979.
5. Lawrence H. Landweber. Decision problems for ω-automata. *Math. Systems Theory*, 3:376–384, 1969.
6. Robert McNaughton. Testing and generating infinite sequences by a finite automaton. *Inform. Control*, 9:521–530, 1966.
7. A. W. Mostowski. Determinancy of sinking automata on infinite trees and inequalities between various Rabin's pair indices. *Inform. Proc. Letters*, 15(4):159–163, 1982.
8. A. W. Mostowski. Regular expressions for infinite trees and a standard form for automata. In A. Skowron, editor, *Computation theory*, volume 208 of *Lect. Notes in Comput. Sci.*, pages 157–168. Springer-Verlag, Berlin, 1984.
9. David Muller. Infinite sequences and finite machines. In Proc. of Fourth Annual IEEE Symp., editor, *Switching Theory and Logical Design*, pages 3–16, 1963.
10. Dominique Perrin. Finite automata. In J. van Leeuwen, editor, *Handbook of Theoretical Computer Science*, volume B, chapter 1. Elsevier, 1990.
11. Micheal Ozer Rabin. Decidability of second-order theories and automata on infinite trees. *Trans. Amer. Math. Soc.*, 141:1–35, 1969.
12. A. Prasad Sistla, Moshe Y. Vardi, and Pierre Wolper. The complementation problem for Büchi automata and applications to temporal logic. *Theoret. Comput. Sci.*, 49:217–237, 1987.
13. Wolfgang Thomas. Automata on infinite objects. In J. van Leeuwen, editor, *Handbook of Theoretical Computer Science*, volume B, chapter 4. Elsevier, 1990.
14. Moshe Y. Vardi. An automata-theoretic approach to linear temporal logic. In *Logics for Concurrency: Structure versus Automata*, volume 1043 of *Lect. Notes in Comput. Sci.*, pages 238–266. Springer-Verlag, 1996.

Linear Time Language Recognition on Cellular Automata with Restricted Communication

Thomas Worsch

Universität Karlsruhe, Fakultät für Informatik
worsch@ira.uka.de

Abstract. It is well-known that for classical one-dimensional one-way CA (OCA) it is possible to speed up language recognition times from $(1 + r)n$, $r \in \mathbf{R}_+$, to $(1 + r/2)n$. In this paper we show that this no longer holds for OCA in which a cell can comminucate only one bit (or more generally a fixed amount) of information to its neighbor in each step. For arbitrary real numbers $r_2 > r_1 > 1$ in time $r_2 n$ 1-bit OCA can recognize strictly more languages than those operating in time $r_1 n$. Thus recognition times may increase by an arbitrarily large constant factor when restricting the communication to 1 bit. For two-way CA there is also an infinite hierarchy but it is not known whether it is as dense as for OCA. Furthermore it is shown that for communication restricted CA two-way flow of information can be much more powerful than an arbitrary number of additional communication bits.

1 Introduction

The model of 1-bit CA results from the standard definition by restricting the amount of information which can be transmitted by a cell to its neighbors in one step to be only 1 bit. We call this the communication bandwidth.

Probably the first paper investigating 1-bit CA is the technical report by [2] where it is shown that even with this model solutions of the FSSP in optimal time are possible. More recently [4] has described 1-bit CA for several one- and two-dimensional problems (e.g. generation of Fibonacci sequences and determining whether two-dimensional patterns are connected) which again are running in the minimum time possible. Therefore immediately the questions arises about the consequences of the restriction to 1-bit information flow in the general case.

In Section 2 basic definitions are given and it is proved that each CA with s states can be simulated by a 1-bit CA with a slowdown by a factor of at most $\lceil \log s \rceil$. This seems to be some kind of folklore, but we include the proof for the sake of completeness and reference in later sections.

In Section 3 it is shown that for one-way CA (OCA) in general there *must* be a slowdown. More specifically there is a very fine hierarchy with an uncountable number of distinct levels (order isomorphic to the real numbers greater than 1) within the class of languages which can be recognized by 1-bit OCA in linear time.

In Section 4 we consider two-way CA with restricted communication.

G. Gonnet, D. Panario, and A. Viola (Eds.): LATIN 2000, LNCS 1776, pp. 417–426, 2000.

The results obtained are in contrast to those for cellular devices with unrestricted information flow. For example, general speedup theorems have been shown for iterative arrays [1] and cellular automata [3].

2   Simulation of k-bit CA by 1-bit CA

A deterministic CA is determined by a finite set of states Q, a neighborhood $N' = N \cup \{0\}$ and a local rule. (For a simpler notation below, we assume that $N = \{n_1, \ldots, n_{|N|}\}$ does not contain $\{0\}$). The local rule of C is of the form $\tau : Q \times Q^N \to Q$, i.e. each cell has the *full* information on the states of its neighbors.

In a k-bit CA B, each cell only gets k bits of information about the state of each neighbor. To this end there are functions $b_i : Q \to \mathbf{B}^k$ specified, where $\mathbf{B} = \{0, 1\}$. If a cell is in state q then $b_i(q)$ are the bits observed by neighbor n_i. We allow different bits to be seen by different neighbors. The local transformation of B is of the form $\tau : Q \times (\mathbf{B}^k)^N \to Q$.

Given a configuration $c : \mathbf{Z} \to Q$ and its successor configuration c' the new state of a cell i is $c_i' = \tau(c_i, b_1(c_{i+n_1}), \ldots, b_{|N|}(c_{i+n_{|N|}}))$.

As usual, for the recognition of formal languages over an input alphabet A one chooses $Q \supset A$ and a set of accepting final states $F \subseteq Q \smallsetminus A$. In the initial configuration for an input $x_1 \cdots x_n \in A^n$ cell i is in state x_i for $1 \leq i \leq n$ and all other cells are in a quiescent state q (satisfying $\tau(q, q^N) = q$). A configuration c is accepting iff $c_1 \in F$.

Given a k-bit CA C one can construct a 1-bit CA C' with the same neighborhood simulating C in the following sense: Each configuration c of C is also a legal configuration of C', and there is a constant l (independent of c) such that if c' is C's successor configuration of c then C' when starting in c reaches c' after l steps. The basic idea is to choose representations of states by binary words and to transmit them bit by bit to the neighbors before doing a "real" state transition.

Let $\mathbf{B}^{\leq i}$ denote $\mathbf{B}^0 \cup \cdots \cup \mathbf{B}^i$. Denote by $b_{i,j}(q)$ the j-th bit of $b_i(q)$, i.e. $b_i(q) = b_{i,k}(q) \cdots b_{i,1}(q)$.

Algorithm 1. As the set of states of C' choose $Q' = Q \times (\mathbf{B}^N)^{\leq k-1}$; i.e. each state q' consists of a $q \in Q$ and binary words $v_1, \ldots, v_{|N|}$ of identical length j for some $0 \leq j \leq k-1$. For each $q \in Q$ identify $(q, \varepsilon, \ldots, \varepsilon)$ with q so that Q can be considered a subset of Q'. (Here, ε is the empty word.) For $j \leq k-1$ and a $q' = (q, v_1, \ldots, v_{|N|}) \in Q \times (\mathbf{B}^N)^j$ define $b_i'(q') = b_{i,j+1}(q)$, where the b_i' are the functions describing the bit seen by neighbor n_i in C'.

The local transformation τ' of C' is defined as follows:

- If the length j if all v_e is $< k-1$ then $\tau'((q, v_1, \ldots, v_{|N|}), x_1, \ldots, x_{|N|}) = (q, x_1 v_1, \ldots, x_{|N|} v_{|N|})$.
- If the length j if all v_e is $= k-1$ then $\tau'((q, v_1, \ldots, v_{|N|}), x_1, \ldots, x_{|N|}) = (\tau(q, x_1 v_1, \ldots, x_{|N|} v_{|N|}), \varepsilon, \ldots, \varepsilon)$.

The above construction shows that the following lemma holds:

Lemma 2. *A k-bit CA can be simulated by a 1-bit CA with the same neighborhood and slowdown k.*

Since the states of a set Q can be unambiguously represented as binary words of length $\lceil \log_2 |Q| \rceil$, it is straightforward to see:

Corollary 3. *Each CA with s states can be simulated by a 1-bit CA with slowdown $\lceil \log_2 s \rceil$ having the same neighborhood and identical functions b_i for all neighbors.*

It should be observed that the above slowdown happens if the bit visible to other cells is the *same* for all neighbors. One could wonder whether the slowdown is always less if different bits are sent to different neighbors. However this is not the case. The proofs below for the lower bounds do not specifically make any use of the fact that all neighbors are observing the same bit; they work even if there were $|N|$ (possibly different) functions b_i for the neighboring cells.

On the other hand one should note that for certain CA there is a possibility for improvement, i.e. conversion to 1-bit CA with a smaller slowdown: Sometimes it is already known that neighbors do not need the full information about each state. In a typical case the set of states might be the Cartesian product of some sets and a neighbor only needs to know one component, as it is by definition the case in so-called partitioned CA. It is then possible to apply a similar construction as above, but only to that component. Since the latter can be described with less bits than the whole state, the construction results in a smaller slowdown.

We will make use of this and a related trick in Section 3.3.

3   A Linear-Time Hierarchy for 1-bit OCA

For a function $f : \mathbf{N}_+ \to \mathbf{N}_+$ denote by $\mathrm{OCA}_k(f(n))$ the family of languages which can be recognized by k-bit OCA in time $f(n)$. In this section we will prove:

Theorem 4. *For all real numbers $1 < r_1 < r_2$ holds:*

$$\mathrm{OCA}_1(r_1 n) \subsetneqq \mathrm{OCA}_1(r_2 n)$$

We will proceed in 3 major steps.

3.1   An Infinite Hierarchy

Let A_m be an input alphabet with exactly $m = 2^l - 1$ symbols. Hence l bits are needed to describe one symbol of $A_m \cup \{\square\}$, where $\square$ is the quiescent state. The case of alphabets with an arbitrary number of symbols will be considered later.

Denote by L_m the set $\{vv^R \mid v \in A_m^+\}$ of all palindromes of even length over A_m.

Lemma 5. *Each 1-bit OCA recognizing L_m needs at least time $(l-\varepsilon)n$ for every $\varepsilon > 0$.*

Proof. Consider a 1-bit OCA C recognizing L_m and an input length n. Denote by t the worst case computation time needed by C for inputs of length n.

Consider the boundary between cells k and $k+1$, $1 \le k < n$, which separates a left and a right part of an input. The computations in the left part are completely determined by the corresponding part of the input and the sequence B_r of bits received by cell k from the right during time steps $1, \ldots, t-k$. There are exactly 2^{t-k} such bit sequences. On the other hand there are $m^{(n-k)} = (2^l - 1)^{(n-k)}$ right parts of inputs of length n.

Assume that $2^{t-k} < (2^l - 1)^{(n-k)}$. Then there would exist two different words v_1 and v_2 of length $n - k$ resulting in the same bit string received by cell k during *any* computation for an input of one of the forms vv_1 or vv_2. Since we are considering OCA, the bit string is independent of any symbols to the left of cell $k + 1$. Therefore C would either accept or reject *both* inputs v_1v_1 or v_1v_2, although exactly one of them is in L_m. Contradiction.

Therefore $2^{t-k} \ge (2^l - 1)^{(n-k)}$. For sufficiently large n there is an arbitrarily small ε'' such that this implies $2^{t-k} \ge 2^{(l-\varepsilon'')(n-k)}$, i.e. $t - k \ge (l - \varepsilon'')(n - k)$, i.e. $t \ge ln + k - lk - \varepsilon''n + \varepsilon''k$. For an arbitrarily chosen $\varepsilon' > 0$ consider the case $k = \varepsilon'n$ (for sufficiently large n). One then gets $t \ge ln + \varepsilon'n - l\varepsilon'n - \varepsilon''n + \varepsilon''\varepsilon'n = ln - (\varepsilon'(l - 1 - \varepsilon'') + \varepsilon'')n$. If ε' and ε'' are chosen sufficiently small this is larger than $ln - \varepsilon n$ for a given ε.

Lemma 6. *L_m can be recognized by 1-bit OCA in time $(l + 1)n + O(1)$.*

Proof. Algorithm 7 below describes a $(l + 1)$-bit OCA recognizing the language in time $n + O(1)$. Hence the claim follows from Lemma 2.

Algorithm 7. We describe a $(l+1)$-bit OCA recognizing L_m. The set of states can be chosen to be of the form $Q_m = A_m \cup A_m \times (A_m \cup \{\Box\}) \times \mathbf{B}^2$. The local rule mapping the state q_c of a cell and the state q_r of its neighbor to $\tau(q_c, q_r)$ is chosen as follows. For the first step, with $a, b' \in A_m$:

$$\tau(a, b') = (a, b', 1, 1)$$

For later steps:

$$\tau((a, a', x, x'), (b, b', y, y')) = (a, b', x' \wedge [a = a'], y)$$

where $[a = a']$ is 1 if the symbols are equal and 0 otherwise. As can be seen immediately, the only information needed from the right neighbor is one symbol b' and one bit y. Hence an $(l + 1)$-bit OCA can do the job.

A closer look at the local rule reveals that the OCA above indeed recognizes palindromes in time $n + O(1)$ if one chooses as the set of final states $A_m \times \{\Box\} \times \{1\} \times \mathbf{B}$ (see [5] for details). Hence L_m can also be recognized by 1-bit OCA in time $(l + 1)n + O(1)$.

The upper bound of the previous lemma is not very close to the lower bound of Lemma 5, and it is not obvious how to improve at least one of them.

3.2   Reducing the Gap to $(1 \pm \varepsilon)n$

We will now define variants of the palindrome language for which gaps between upper and lower bound can be proved to be very small.

We will use vectors of length r of symbols from an alphabet A as symbols of a new alphabet A'. Although a vector of symbols is more or less the same as a word of symbols, we will use different notations for both concepts in order to make the construction a little bit clearer. Denote by $M^{\langle r \rangle}$ the set of vectors of length r of elements from a set M and by A^r the set of words of length r consisting of symbols from A. The obvious mapping $\langle x_1, \ldots, x_r \rangle \mapsto x_1 \cdots x_r$ induces a monoid homomorphism $h : (A^{\langle r \rangle})^* \to (A^r)^* \subseteq A^*$.

Definition 8. *For integers $m \geq 1$ and $r \geq 1$ let*

$$L_{m,r} = \{vh(v)^R \mid v \in (A_m^{\langle r \rangle})^+\}$$

$L_{m,r}$ is a language over the alphabet $A_m^{\langle r \rangle} \cup A_m$. The words in $L_{m,r}$ are still more or less palindromes where in the left part of a word groups of r elements from A_m are considered as *one* (vector) symbol. As a special case one has $L_{m,1} = L_m$ as defined earlier.

Lemma 9. *For each $\varepsilon > 0$ there is an $r \geq 1$ such that each 1-bit OCA recognizing $L_{m,r}$ needs at least time $(l - \varepsilon)n$.*

A proof can be given analogously to the proof of Lemma 5 above. One only has to observe that the border between cells k and $k+1$ must not lie within "the left part v" of an input. Therefore for small ε one must choose a sufficiently large r, e.g. $r > 1/\varepsilon$, to make sure that $|v| < \varepsilon|vh(v)^R|$.

Thus for sufficiently large r although $\lceil \log_2 |A_m| \rceil \cdot n$ is not a lower bound on the recognition time of $L_{m,r}$ by 1-bit OCA, it is "almost".

Lemma 10. *For each $\varepsilon > 0$ and $r = 1/\varepsilon$ the language $L_{m,r}$ can be recognized by a 1-bit OCA in time $(l + \varepsilon)n + O(1)$.*

Thus for sufficiently large r although $\lceil \log_2 |A_m| \rceil \cdot n$ is not an upper bound on the on the achievable recognition time on 1-bit OCA, it is "almost".

For the proof we use a construction similar to Algorithm 7.

Algorithm 11. The CA uses a few additional steps before and after the check for palindromes, where the check itself also has to be adapted to the different form of inputs.

- In the first step each cell sends one bit to its left neighbor indicating whether its input symbol is from A_m or $A_m^{\langle r \rangle}$. Thus, if the input is *not* in $(A_m^{\langle r \rangle})^* A_m^*$ this is detected by at least one cell and an error indicator is stored locally. It will be used later.
- One may therefore assume now that the input is of the indicated form, and we will call cells with an input symbol from A_m the "right" cells and those with a symbol from $A_m^{\langle r \rangle}$ the "left" cells.
  After the first step the rightmost of the left cells has indentified itself.

- With the second step an algorithm for palindrome checking is started. The modifications with respect to Algorithm 7 are as follows:
  - Each cell is counting modulo $lr + 1$ in each step. (This doesn't require any communication.)
  - During the first lr steps of a cycle the right cells are shifting r symbols to the left. In the $(lr + 1)$-st step they do not do anything.
  - During the first lr steps of a cycle the left cells are also shifting r symbols to the left. In addition they are accumulating what they receive in registers. In step lr step they are comparing whether the register contents "match" their own input symbol, and in step $lr + 1$ they are sending the result of the comparison, combined with the previously received comparison bit to their left neighbor.

  One should observe that the last point is the basic trick: the comparison bit has not to be transported one cell to the left each time a symbol has been received, but only every r symbols. Thus by increasing r the fraction of time needed for transmitting these bits can be made arbitrarily small.
- All the algorithms previously described have the following property: The part of the time space diagram containing all informations which are needed for the decision whether to accept or reject an input has the form of a triangle. Its longest line is a diagonal with some slope $n/t(n)$ (or $t(n)/n$ depending on how you look at it) leading from the rightmost input cell the leftmost one. Furthermore every cell can know when it has done its job because afterwards it only receives the encodings of the quiescent state.
- Therefore the following signal can be implemented easily: It starts at the rightmost input cell and collects the results of the checks done in the very first step. It is moved to the left immediately after a cell has transmitted at least one (encoding of the) quiescent state in a $(lr + 1)$-cycle. Thus this signal causes only one additional step to the overall recognition time.

Since the above algorithm needs $lr + 1$ steps per r input symbols from A_m and since the rightmost r symbols have to travel approximately n cells far, the total running time is $n \cdot (lr + 1)/r + O(1)$, i.e. $(l + 1/r)n + O(1)$ as required.

¿From the Lemmata 9 and 10 one can immediately deduce the following:

Corollary 12. *For each integer constant c the set of languages which can be recognized by 1-bit OCA in time cn is strictly included in the the set of languages which can be recognized by 1-bit OCA in time $(c + 2)n$.*

This has to be contrasted with unlimited OCA where there is no such infinite hierarchy within the family of languages which can recognized in linear time. One therefore gets the situation depicted in Figure 1.

In the top row one uses the fact that for each $i \geq 1$

$$OCA_1(2in) \subseteq OCA_1((2i + 1 - \varepsilon)n) \subsetneq OCA_1((2i + 1 + \varepsilon)n) \subseteq OCA_1((2i + 2)n)$$

and for each column one has to observe that

$$OCA_1(2in) \subsetneq OCA_1((2i + 2)n) \subseteq OCA((2i + 2)n) = OCA(2in) .$$

$$\mathrm{OCA}_1(2n) \subsetneqq \mathrm{OCA}_1(4n) \subsetneqq \mathrm{OCA}_1(6n) \subsetneqq \mathrm{OCA}_1(8n) \subsetneqq \cdots$$

$$\mathrm{OC\check{A}}(2n) = \mathrm{OC\check{A}}(4n) = \mathrm{OC\check{A}}(6n) = \mathrm{OC\check{A}}(8n) = \cdots$$

Fig. 1. A hierarchy for 1-bit OCA.

3.3 There Are Small Gaps Everywhere

Finally we will prove now that a small increase of the linear-time complexity already leads to an increased recognition power not only around $(r \pm \varepsilon)n$ for natural numbers r, but for all real numbers $r > 1$. Since the rational numbers are dense in $\mathbf{R}$ it suffices to prove the result for $r \in \mathbf{Q}$.

The basic idea is the following: The number l playing an important role in the previous sections is something like an "average number of bits needed per symbol". What we want to achieve below is an average number r of bits needed per symbol.

Assume that an arbirtrary rational number $r > 1$ has been fixed as well as the relatively prime natural numbers x and $y < x$ such that $r = x/y$. Then the above is more or less equivalent to saying that one needs x bits for every y symbols.

Therefore choose the smallest m such that $2^x < m^y$ and a set M of $2^x - 1$ different words from A_m^y. Then extend the alphabet and "mark" the first and last symbols of these words. These markings will only be used in Algorithm 15. For the sake of simplicity will ignore them in the following descriptions. In order to define the languages $L'_{m,r}$ to be used later we start with the languages $L_{m',r}$ considered in the previous section, where $m' = 2^x - 1$. Denote by $g_{x,y}$ a one-to-one mapping $g_{x,y} : A_{m'} \to M$ which is extended vectors of length r by considering it as a function mapping each r-tuple of symbols from $A_{m'}$ to word of length y of r-tuples of symbols from A_m and extending this further to a monoid homomorphism in the obvious way. Now choose

$$L_{x,y,m,r} = g_{x,y}(L_{m',r})$$

Lemma 13. *For each $\varepsilon > 0$ there is an $r \geq 1$ such that each 1-bit OCA recognizing $L_{x,y,m,r}$ needs at least time $(x/y - \varepsilon)n$.*

It is a routine exercise to adapt the proof of Lemma 9 to the new situation.

Lemma 14. *For each $\varepsilon > 0$ and $r = 1/\varepsilon$ the language $L_{x,y,m,r}$ can be recognized by a 1-bit OCA in time $(x/y + \varepsilon)n + O(1)$.*

Algorithm 15. Basically the same idea as in Algorithm 11 can be used. Two modifications are necessary.

The first one is a constant number of steps which have to be carried out in the very beginning. During these steps each cell collects the information about the $y - 1$ input symbols to its right, so that it knows which of the words $w \in M$ is to its right, assuming that it is the left end of one of it. From then on these marked cells play the role of all cells from Algorithm 11.

The second modification is an additional signal of appropriate speed which is sent from the right end of the input word. It checks that all the left end and right end of word markings are indeed distributed equidistantly over the whole input word. If this is not the case the input is rejected.

As a consequence there is an uncountable set of families of languages ordered by proper inclusion which is order isomorphic to the real numbers greater than 1 as already claimed at the beginning of this section:

Proof (of Theorem 4). Choose a rational number x/y and an $\varepsilon > 0$ such that $r_1 < x/y - \varepsilon < x/y + \varepsilon < r_2$. From Lemmata 13 and 14 follows that there is a language in $\mathrm{OCA}_1(x/y + \varepsilon) \setminus \mathrm{OCA}_1(x/y - \varepsilon)$ which is then also a witness for the properness of the above inclusion.

Therefore the hierarchy depicted in Figure 1 can be generalized to the following, where r_1 and r_2 are arbitrary real numbers satisfying $1 < r_1 < r_2$:

$$\cdots \subsetneq \mathrm{OCA}_1(r_1 n) \subsetneq \cdots \subsetneq \mathrm{OCA}_1(r_2 n) \subsetneq \cdots$$
$$\subsetneq \qquad\qquad\qquad \subsetneq$$
$$\cdots = \mathrm{OCA}(r_1 n) = \cdots = \mathrm{OCA}(r_2 n) = \cdots$$

Fig. 2. The very fine hierarchy for 1-bit OCA.

4   Two-Way CA

For two-way CA (CA for short) with 1-bit communications one has the following result:

Lemma 16. *Each 1-bit CA recognizing L_m needs at least time $\frac{l+2}{4} n = (1 + l/2)n/2$.*

Proof. Consider a 1-bit CA C recognizing L_m and an input length $n = 2k$. Denote by t the worst case computation time needed by C for inputs of length n.

Consider the boundary between cells $k = n/2$ and $k + 1$, which separates the two halves of an input. The computations in the left half are completely determined by the sequence B_r of bits received by cell k during the time steps $1, \ldots, t - n/2$ from the right, and the computations in the right half are completely determined by the sequence B_l of bits received by cell $k + 1$ during the time steps $1, \ldots, t - n/2$ from the left. There are exactly 2^{t-k} bit sequences B_l and 2^{t-k} bit sequences B_r. On the other hand there are $m^k = 2^{l \cdot k}$ left resp. right halves of inputs of length n.

Assume that $2^{2(t-k)} < 2^{l \cdot k}$. Then there would exist two different words v_1 and v_2 of length k resulting in the same bit strings B_l and B_r for the inputs $v_1 v_1$ and $v_2 v_2$. Therefore, in cells $1, \ldots, k$ the computation of C for the input $v_1 v_2$ would be the same as for $v_1 v_1$ and since C has to accept $v_1 v_1$, it would also accept $v_1 v_2$. Contradiction.

Therefore $2^{2(t-k)} \geq 2^{l \cdot k}$, i.e. $t - k \geq \frac{1}{2} l \cdot k$, i.e. $t \geq \frac{l+2}{4} n$.

On the other hand it is not difficult to construct a CA which shows:

Lemma 17. L_m *can be recognized by 1-bit CA in time* $(l+2)n/2$.

The straightforward construction of shifting the input symbols in both directions, accumulating comparison results everywhere and using the result of the middle cell suffices.

Lemmata 16 and 17 immediately give rise to an infinite hierarchy of complexity classes, but the gaps are large. For example one has

Corollary 18.

$$\mathrm{CA}_1(n) \subsetneqq \mathrm{CA}_1(3n) \subsetneqq \mathrm{CA}_1(3^2 n) \subsetneqq \mathrm{CA}_1(3^3 n) \subsetneqq \cdots$$

In fact the constants can be improved somewhat (using the lemmata above ultimately to c^j for any constant $c > 2$ if j is large enough). On the other hand it is unfortunately not clear at all how to results which are as sharp as for OCA.

Finally we point to the following relation between communication bounded OCA and CA. As mentioned above L_m can be recognized by 1-bit CA in time $(l+2)n/2$, but on the other hand it cannot be recognized by 1-bit OCA in time $(l - \varepsilon)n$. This is a gap of $(l - \varepsilon)n - (l+2)n/2 = (l - 2 - 2\varepsilon)n/2$ which can be made *arbitrarily large*! In other words:

Lemma 19. *For each constant $k > 1$ there are languages for which 1-bit CA can be faster than any 1-bit OCA recognizing it by a factor of k.*

Corollary 20. *For no constants $r > 1$ and $k > 1$ is $\mathrm{CA}_1(rn) \subseteq \mathrm{OCA}_1(krn)$. For no constants $r > 1$ and $k > 1$ is $\mathrm{CA}_1(rn) \subseteq \mathrm{OCA}_k(rn)$.*

Thus in a sense sometimes the ability to communicate in both directions is more powerful than any bandwidth for communication in only one direction.

5   Conclusion and Outlook

It has been shown that for all real numbers $r > 1$ and $\varepsilon > 0$ there are problems which can be solved on 1-bit OCA in time $(r + \varepsilon)n$, but not in time rn. As a consequence there are problems the solution of which on 1-bit OCA *must* be slower than on unlimited OCA by a factor of at least r.

It is therefore interesting, and in some way surprising, that certain problems which are considered to be nontrivial, e.g. the FSSP, can solved on 1-bit CA without any loss of time.

Two-way CA with the ability to communicate 1 bit of information in each direction are more powerful than one-way CA with the ability to communicate k bit in one direction. For certain formal languages the latter have to be slower by a constant factor which cannot be bounded.

Our current research on communication restricted CA is mainly concerned with two problem fields. One is the improvement of the results for two-way CA. In particular we suspect that the lower bound given in Lemma 16 can be improved. The other is an extension of the definitions to CA with an "average bandwith" of z bits, where $z > 1$ is allowed to be a rational number. We conjecture that for OCA there is also a dense hierarchy with respect to the bandwith (while keeping the time fixed). This is true if one restricts oneself to integers. For rational numbers there a some additional technical difficulties due to the not completely straightforward definition of z-bit CA.

Acknowledgements

The author would like to thank Hiroshi Umeo for valuable hints concerning preliminary versions of this paper and for asking the right questions. Interesting discussions with Thomas Buchholz and Martin Kutrib during IFIPCA 98 were also helpful.

References

1. S. N. Cole. Real-time computation by n-dimensional iterative arrays of finite-state machines. *IEEE Transactions on Computers*, C-18(4):349–365, 1969.
2. Jacques Mazoyer. A minimal time solution to the firing squad synchronization problem with only one bit of information exchanged. Technical Report TR 89-03, Ecole Normale Supérieure de Lyon, Lyon, 1989.
3. Alvy Ray Smith III. Cellular automata complexity trade-offs. *Information and Control*, 18:466–482, 1971.
4. Hiroshi Umeo. Cellular algorithms with 1-bit inter-cell communications. In Thomas Worsch and Roland Vollmar, editors, *MFCS'98 Satellite Workshop on Cellular Automata*, pages 93–104, 1998.
5. Roland Vollmar and Thomas Worsch. *Modelle der Parallelverarbeitung – eine Einführung*. Teubner, Stuttgart, 1995.

From Semantics to Spatial Distribution [*]

Luis R. Sierra Abbate[1][**], Pedro R. D'Argenio[2][***], and Juan V. Echagüe[1]

[1] Instituto de Computación. Universidad de la República. Uruguay
{echague,sierra}@fing.edu.uy
[2] Dept. of Computer Science. University of Twente. The Netherlands
dargenio@cs.utwente.nl

Abstract. This work studies the notion of locality in the context of
process specification. It relates naturally with other works where information about the localities of a program is obtained information from
its description written down in a programming language.
This paper presents a new approach for this problem. In our case, the
information about the system will be given in semantic terms using asynchronous transition systems. Given an asynchronous transition system we
build an algebra of localities whose models are possible implementations
of the known system. We present different results concerning the models
for the algebra of localities. In addition, our approach neatly considers
the relation of localities and non-determinism.

1 Introduction

In the framework of the so called true concurrency, the idea of causality has
been widely studied [13,12,8,7,15]. Localities, an idea somehow orthogonal to
causality, has become also interesting [1,4,5,10,11,9,3]. Causality states which
events are necessary for the execution of a new one, while localities observe in
which way the events are distributed. Both approaches have been shown not to
be equivalent or to coincide in a very discriminating point [6,17].

The idea of the work on localities is to state where an event occurs given
the already known structure of a process. Thus, the starting point is a process
written in a clearly defined syntax. For instance, consider the process

$$a.c.\textbf{stop} \parallel_c c.b.\textbf{stop} \tag{1}$$

where $\parallel_c$ is the CSP parallel composition: there are two processes running together, but they must synchronize in the action c. This process may execute
actions $a@ \bullet |\emptyset$, $b@\emptyset| \bullet$, and $c@ \bullet | \bullet$. The term in the right hand side of the @
indicates the places in which the action on the left side of @ occurs. In particular,
the $\bullet$ shows in which side of the parallel operation the action takes place. Notice
that a and b do not share any locality: a occurs at the left hand side of the

---

[*] This work is supported by the CONICYT/BID project 140/94 from Uruguay
[**] This author is supported by the PEDECIBA program
[***] This author is supported by the PROGRESS project TES-4999

G. Gonnet, D. Panario, and A. Viola (Eds.): LATIN 2000, LNCS 1776, pp. 427–436, 2000.

parallel composition while b occurs at the right hand side. On the other hand, the process

$$a.c.b.\textbf{stop} \tag{2}$$

presents the same sequence of actions a, c, and b, although in this case they occur exactly in the same place.

Besides, these works on localities have a nagging drawback: in some cases where non deterministic choice and parallel composition are involved, localities for actions do not seem to match our intuition. For instance, in the process

$$(a.\textbf{stop} \parallel b.\textbf{stop}) + (c.\textbf{stop} \parallel d.\textbf{stop}) \tag{3}$$

we have that $a@\bullet|\emptyset$ and $d@\emptyset|\bullet$. We could think that a and d do not share any resource, but in a causal-based model they are clearly in conflict: the occurrence of one of them forbids the occurrence of the other. From a causal point of view actions a and d must be sharing some locality.

The approach we chose is to deduce the distribution of events from the semantics of a given process. We use asynchronous transition systems [14,16] (ATS for short) to describe its behavior. Thus, in our case the architecture (i.e., the syntax) of the process is not known.

Our contribution consists of the statement and exploration of this original semantic-based approach. For each ATS we define an algebra of localities with a binary operation $\wedge$ that returns the common places of two events, and a constant 0 meaning "nowhere". The axioms of this algebra will give the minimal requirements needed for events to share or not to share some place. The axiomatization does not specify anything if such a statement cannot be deduced from the behavior. Thus, given the interpretation of the processes (1) and (2) we may deduce that a and c must have some common place, and we will write $a \wedge c \neq 0$.However, the axiomatization is not going to state whether $a \wedge b = 0$ or $a \wedge b \neq 0$. This will depend on the model chosen for the axiomatization, that gives the definitive criterion for the distribution of events: our models will be true implementations of ATS. We will show that our approach detects situations like the one described in process (3). In this case, we will have an explicit axiom saying that a and d share some common place, i.e, $a \wedge d \neq 0$.

In addition, we discuss different models for the algebra of localities of a given ATS. These models may be associated to a program whose specification was given in terms of the original ATS. First we introduce the *non-independence models* which consider whether two events are independent in the corresponding ATS. Then, we define models which take into account whether two events are adjacent.

Consider two events sharing a locality in a model $\mathcal{M}$ for a given ATS. If they share some locality in every possible model for this ATS, we call $\mathcal{M}$ a *minimal sharing model*. On the other hand, if two events share a locality in $\mathcal{M}$ only when they share a locality in any other model, then we call $\mathcal{M}$ a *maximal sharing model*. We show that the models concerning adjacency introduced in this work hold one of these properties.

The paper is organized as follows. Section 2 recalls the definition of ATS as well as some notions of graph theory. Section 3 introduces the algebra of localities. Six models for this algebra are presented in Section 4. Finally, conclusions and future works are given in Section 5.

2 Preliminaries

Asynchronous Transitions Systems Asynchronous transition systems [14,16] are a generalization of labeled transition systems. In ATSs, transitions are labeled with *events*, and each event represents a particular occurrence of an action. In addition, ATSs incorporate the idea of *independent* events. Two independent events can be executed in parallel, and so they cannot have resources in common. Formally, we define:

Definition 1. *Let* $\mathbf{A} = \{\alpha, \beta, \gamma, \ldots\}$ *be a set of* actions. *An* asynchronous transition system *is a structure* $T = (S, E, I, \longrightarrow, \ell)$ *where*

- $S = \{s, t, s', \ldots\}$ *is a set of* states *and* $E = \{a, b, c, \ldots\}$ *is a set of* events;
- $I \subseteq E \times E$ *is an irreflexive and symmetric relation of* independence. *We write* aIb *instead of* $(a, b) \in I$;
- $\longrightarrow \subseteq S \times E \times S$ *is the transition relation. We write* $s \xrightarrow{a} s'$ *instead of* $(s, a, s') \in \longrightarrow$;
- $\ell : E \to \mathbf{A}$ *is the* labeling function.

In addition, T *has to satisfy the following axioms,*

Determinism:	$s \xrightarrow{a} s' \wedge s \xrightarrow{a} s'' \implies s' = s''$
Forward stability:	$aIb \wedge s \xrightarrow{a} s' \wedge s \xrightarrow{b} s'' \implies \exists t \in E.\ s' \xrightarrow{b} t \wedge s'' \xrightarrow{a} t$
Commutativity:	$aIb \wedge s \xrightarrow{a} s' \wedge s' \xrightarrow{b} t \implies \exists s'' \in E.\ s \xrightarrow{b} s'' \wedge s'' \xrightarrow{a} t$

$\square$

Example 1. In the Introduction we have mentioned a couple of examples. We are going to use them as running examples. To simplify notation, we use the same name for events and actions.

We can represent both $a.c.b.\mathbf{stop}$ and $a.c.\mathbf{stop} \parallel_c c.b.\mathbf{stop}$ by the ATS in Figure 1. Notice that for the second process, we could have aIb although that is not actually relevant. However, it is important to notice that $\neg(aIc)$ and $\neg(bIc)$ in both cases.

The ATS for process $(a.\mathbf{stop} \parallel b.\mathbf{stop}) + (c.\mathbf{stop} \parallel d.\mathbf{stop})$ is depicted in Figure 2. Notice that aIb and cId while any other pair of events is not independent. Shadowing is used to show the independence relation between events. $\square$

Graphs A graph G consists of a finite set V of vertices together with a set X of unordered pairs of distinct vertices of V. The elements of X are the edges of G. We will note $\{v, w\} \in X$ as vw. We will write (V, X) for the graph G. Two vertices v and w are adjacent in G if $vw \in X$. Two edges e and f are adjacent if $e \cap f \neq \emptyset$.

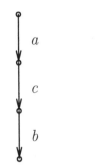

Fig. 1. The ATS for $a.c.b.\mathbf{stop}$ and $a.c.\mathbf{stop} \parallel_c c.b.\mathbf{stop}$

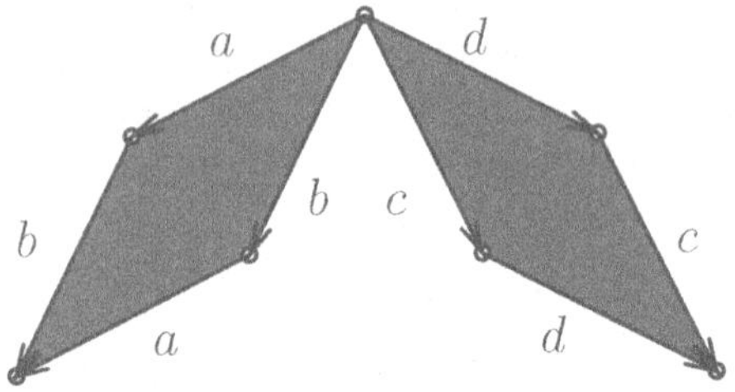

Fig. 2. The ATS for $(a.\mathbf{stop} \parallel b.\mathbf{stop}) + (c.\mathbf{stop} \parallel d.\mathbf{stop})$

Definition 2 (Subgraphs). *We call $H = (V', X')$ a subgraph of $G = (V, X)$, and note $H \subseteq G$, whenever $V' \subseteq V$ and $X' \subseteq X$. We write $\wp G$ for the set of all subgraphs of G. We write $\wp \wp G$ for the power set of $\wp G$.* □

A clique of a graph G is a maximal complete subgraph of G. As a complete graph is defined by its vertices, we will identify a clique with its corresponding set of vertices. We write $K(G)$ for the set of cliques of the graph G.

Lemma 1. *Let v and w be two vertices of $G = (V, X)$. Then, $vw \in X$ iff there exists a clique $K \in K(G)$ such that $vw \in X(K)$.*

3   The Algebra of Localities

In this section we explain how to obtain an *algebra of localities* from a given ATS. The algebra of localities is constructed over a semilattice by adding some particular axioms for each ATS.

Definition 3. *A semilattice is a structure $(\mathcal{L}, \wedge, 0)$ where $\wedge : \mathcal{L} \times \mathcal{L} \to \mathcal{L}$ and $0 \in \mathcal{L}$ satisfying the following axioms:*

$a \wedge b = b \wedge a$ *(commutativity)* $a \wedge (b \wedge c) = (a \wedge b) \wedge c$ *(associativity)*

$a \wedge a = a$ *(idempotence)* $a \wedge 0 = 0$ *(absorption)*
  □

Each element in the set $\mathcal{L}$ refers to a set of *"places"*. In particular, 0 means *"nowhere"*. The operation $\wedge$ gives the *"common places"* between the operands. The axioms make sense under this new nomenclature. Commutativity says that the common places of a and b are the same as the common places of b and a. Associativity says that the common places of a, b, and c are always the same regardless we consider first the common places of a and b, or the common places of b and c. According to idempotency, the common places of a and itself are again the places of a. Finally, absorption says that any element of $\mathcal{L}$ has no common place with nowhere.

Now we introduce the concept of adjacent events. Two events are adjacent if they label two consecutive transitions, or two outgoing transitions from the same state.

Definition 4. *Let* $T = (S, E, I, \longrightarrow, \ell)$ *be an ATS. Two events* $a, b \in E$ *are adjacent in* T, *notation* $adj(a, b)$, *if and only if there exist* $s, s', s'' \in S$ *such that*

$$s \xrightarrow{a} s' \xrightarrow{b} s'' \quad or \quad s \xrightarrow{b} s' \xrightarrow{a} s'' \quad or \quad s \xrightarrow{a} s' \ and \ s \xrightarrow{b} s''$$

$\square$

We are interested in independence relation between adjacent events. When two events are not adjacent an observer cannot differentiate whether they are independent. For instance, in the ATS of Figure 1 it is not relevant whether a and b are independent since that does not affect the overall behavior.

The carrier set of the algebra of localities associated to an ATS includes an appropriate interpretation of its events. Such an interpretation refers to "the places where an event happens".

Definition 5. *Let* $T = (S, E, I, \longrightarrow, \ell)$ *be an ATS. The* algebra of localities *associated to* T *is a structure* $\mathcal{A} = (\mathcal{L}, E, \wedge, 0)$ *satisfying:*

1. $E \subseteq \mathcal{L}$, *and* $(\mathcal{L}, \wedge, 0)$ *is a semilattice*
2. aIb *and* $adj(a, b) \implies a \wedge b = 0$
3. $\neg(aIb)$ *and* $adj(a, b) \implies a \wedge b \neq 0$

$\square$

Example 2. For the ATS of Figure 1 we obtain the following axioms:

$$a \wedge c \neq 0 \qquad\qquad c \wedge b \neq 0$$

Notice that the axiom system does not say whether $a \wedge b \neq 0$ or $a \wedge b = 0$. Thus, the algebra does not contradict the decision of implementing the ATS either with process $a.c.b.$**stop**, in which a and b occur in the same place, or with $a.c.$**stop** $\|_c c.b.$**stop**, in which a and b occur in different places.

For the ATS of Figure 2 we obtain the following axioms:

$$a \wedge b = 0 \qquad a \wedge c \neq 0 \qquad b \wedge c \neq 0$$
$$c \wedge d = 0 \qquad a \wedge d \neq 0 \qquad b \wedge d \neq 0$$

Notice that the axioms state that a and d must share some places. On the other hand, as we already said, other approaches to localities cannot identify such a conflict.

$\square$

4   Models for the Algebra of Localities

In this section we introduce several models for the algebra of localities associated to a given ATS, thus proving its soundness. Each of our models may be an implementation. The interpretation for the events will be based on the relations of independence and adjacency. The names of the models are taken from these basic relations.

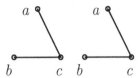

Fig. 3. I models for $a.c.b.$**stop** and $a.c.$**stop** $||_c$ $c.b.$**stop**

Fig. 4. I model for $(a.$**stop** $||$ $b.$**stop**$)$ + $(c.$**stop** $||$ $d.$**stop**$)$

The I Models The non-independence models (I models for short) for the algebra of localities associated to a given ATS assign common places to non-independent events. We define the non-independent models I and I2, based on cliques and edges respectively.

Let $T = (S, E, I, \longrightarrow, \ell)$ be an ATS. We define the graph $G^I = (E, \{\{a, b\} \subseteq E \mid \neg(aIb)\})$. We define the interpretation of an event a in the model I (I2) to be the set of cliques (edges) in G^I in which a appears.

$$[\![a]\!]^I \stackrel{\text{def}}{=} \{A \in K(G^I) \mid a \in A\} \qquad ([\![a]\!]^{I2} \stackrel{\text{def}}{=} \{A \in X(G^I) \mid a \in A\})$$

Each set $A \in [\![a]\!]^I$ is a different place where a may happen: each place is identified with the set of all events that can happen there. Moreover, an event can happen in several places simultaneously. The operation $\wedge$ of the algebra of localities is interpreted as the intersection $\cap$ between sets, and the constant 0 is interpreted as the empty set $\emptyset$.

Example 3. For the ATS in Figure 1 with aIb, we obtain the graph G^I on the left of Figure 3. This implementation uses two places or localities. One of them is shared by a and c, and the other by b and c. So, this model is well suited for the implementation $a.c.$**stop** $||_c$ $c.b.$**stop**. In this case, both I and I2 interpretations coincide. These could be written down as

$$[\![a]\!]^I = \{\{a, c\}\} \qquad [\![b]\!]^I = \{\{b, c\}\} \qquad [\![c]\!]^I = \{\{a, c\}, \{b, c\}\}$$

We have a new interpretation in case a and b are not independent. We can see it on the right of the Figure 3. Now, every event occurs in the same place. In other words, if $\neg(aIb)$, the I model implements $a.c.b.$**stop**.

$$[\![a]\!]^I = [\![b]\!]^I = [\![c]\!]^I = \{\{a, b, c\}\}$$

A different interpretation is established for model I2. In this case, we have

$$[\![a]\!]^{I2} = \{\{a, c\}, \{a, b\}\} \qquad [\![b]\!]^{I2} = \{\{b, c\}, \{a, b\}\} \qquad [\![c]\!]^{I2} = \{\{a, c\}, \{b, c\}\}$$

This model implements the program $a.b.$**stop** $||$ $a.c.$**stop** $||$ $c.b.$**stop** that uses three localities.

For the ATS of Figure 2 we have G^I depicted in Figure 4. The execution of a requires two places, one shared with c and the other with d. Thus, the event

a prevents the execution of d by occupying a place required by this event. This reflects the fact that selection between non independent events occurs actually in a place. For this implementation, we have

$$[\![a]\!]^I = \{\{a,c\},\{a,d\}\} \qquad\qquad [\![b]\!]^I = \{\{b,c\},\{b,d\}\}$$
$$[\![c]\!]^I = \{\{a,c\},\{b,c\}\} \qquad\qquad [\![d]\!]^I = \{\{a,d\},\{b,d\}\}$$

$\square$

Now we prove that non-independence models are indeed models for the algebra of localities.

Theorem 1 (Soundness). *Let $T = (S,E,I,\longrightarrow,\ell)$ be an ATS, $\mathcal{A}$ its algebra of localities, and $[\![E]\!]^{I(I2)} = \{[\![a]\!]^{I(I2)} \mid a \in E\}$. Then,*

$$\mathcal{M}^I \stackrel{\text{def}}{=} \left(\wp\wp\left(G^I\right), [\![E]\!]^I, \cap, \emptyset\right) \text{ and } \mathcal{M}^{I2} \stackrel{\text{def}}{=} \left(\wp\wp\left(G^I\right), [\![E]\!]^{I2}, \cap, \emptyset\right)$$

are models for $\mathcal{A}$.

Proof. By definition, $[\![E]\!]^{I2} \subseteq \wp\wp\left(G^I\right)$. Moreover, $\left(\wp\wp\left(G^I\right), \cap, \emptyset\right)$ is a well known semilattice.

Suppose that aIb and $adj(a,b)$. They are not adjacent in G^I, and so there is no edge between a and b in G^{I2}. Thus, $[\![a]\!]^{I2} \cap [\![b]\!]^{I2} = \emptyset$.

Finally, suppose that $\neg(aIb)$ and $adj(a,b)$. Then, $ab \in X(G^I)$, and hence $[\![a]\!]^{I2} \cap [\![b]\!]^{I2} \neq \emptyset$.

The proof for model I is similar, taking into account Lemma 1. $\square$

We can see that, although localities may change, the relation between these two models remain substantially unchanged. More explicitly, two events sharing resources in any of these models will share resources in the other.

Theorem 2. $\mathcal{M}^I \models a \wedge b \neq 0$ *if and only if* $\mathcal{M}^{I2} \models a \wedge b \neq 0$

Minimal Sharing Models: IJ and IJ2 In the models IJ and IJ2 we assign common places to events that are both adjacent and non-independent. We will show they are minimal sharing in the following sense : whenever two events share a place for this models, they will share a place in any other model.

Let $T = (S,E,I,\longrightarrow,\ell)$ be an ATS. Taking adjacent events into account we define the graph $G^{IJ} = (E, \{\{a,b\} \subseteq E \mid \neg(aIb) \text{ and } adj(a,b)\})$. As before, we define the interpretation of an event a to be the set of cliques or edges in G^{IJ} where a appears.

$$[\![a]\!]^{IJ} \stackrel{\text{def}}{=} \{A \in K(G^{IJ}) \mid a \in A\} \qquad\qquad [\![a]\!]^{IJ2} \stackrel{\text{def}}{=} \{A \in X(G^{IJ}) \mid a \in A\}$$

Theorem 3 (Soundness). *Let $T = (S,E,I,\longrightarrow,\ell)$ be an ATS and let $\mathcal{A}$ be its algebra of localities. Then,*

$$\mathcal{M}^{IJ} \stackrel{\text{def}}{=} \left(\wp\wp\left(G^{IJ}\right), [\![E]\!]^{IJ}, \cap, \emptyset\right) \text{ and } \mathcal{M}^{IJ2} \stackrel{\text{def}}{=} \left(\wp\wp\left(G^{IJ}\right), [\![E]\!]^{IJ2}, \cap, \emptyset\right)$$

are models for $\mathcal{A}$.

Theorem 4. $\mathcal{M}^{IJ} \models a \wedge b \neq 0$ *if and only if* $\mathcal{M}^{IJ2} \models a \wedge b \neq 0$

These models enjoy the following property: if two events are distributed (i.e., do not share a place) in some model for the algebra of localities of a given ATS, they are also distributed in these models. This justifies calling them minimal sharing models. The following theorem states the counter positive of that property.

Theorem 5. *Let* $T = (S, E, I, \longrightarrow, \ell)$ *be an ATS and let* $\mathcal{A}$ *be its algebra of localities. Let* $\mathcal{M}$ *be any model for* $\mathcal{A}$. *Then, for all events* $a, b \in E$,

$$\mathcal{M}^{IJ2} \models a \wedge b \neq 0 \implies \mathcal{M} \models a \wedge b \neq 0$$

Proof. Suppose $\mathcal{M}^{IJ2} \models a \wedge b \neq 0$, that is $[\![a]\!]^{IJ2} \cap [\![b]\!]^{IJ2} \neq \emptyset$. Thus $ab \in X(G^{IJ})$, which implies $\neg(aIb)$ and $adj(a, b)$. So, by Definition 5, $\mathcal{A} \vdash a \wedge b \neq 0$. Hence, for any model $\mathcal{M}$ of $\mathcal{A}$, $\mathcal{M} \models a \wedge b \neq 0$. $\qquad\square$

An easy application of Theorem 4 give us this corollary:

Corollary 1. $\mathcal{M}^{IJ}$ *is a minimal sharing model.* $\qquad\square$

Maximal Sharing Models: InJ and InJ2 In a similar way we construct a model of maximal sharing. In this case, two events share places unless they must execute independently. We call them InJ models because they may require non adjacency.

Let $T = (S, E, I, \longrightarrow, \ell)$ be an ATS. We define the graph $G^{InJ} = (E, \{\{a, b\} \subseteq E \mid \neg (aIb \text{ and } adj(a, b))\})$. We define the interpretation of an event a to be the set of cliques or edges in G^{InJ} where a appears.

$$[\![a]\!]^{InJ} \overset{\text{def}}{=} \{A \in K(G^{InJ}) \mid a \in A\} \qquad [\![a]\!]^{InJ2} \overset{\text{def}}{=} \{A \in X(G^{InJ}) \mid a \in A\}$$

Theorem 6 (Soundness). *Let* $T = (S, E, I, \longrightarrow, \ell)$ *be an ATS and let* $\mathcal{A}$ *be its algebra of localities. Then,*

$$\mathcal{M}^{InJ} \overset{\text{def}}{=} \left(\wp\wp\left(G^{InJ}\right), [\![E]\!]^{InJ}, \cap, \emptyset \right) \quad and$$
$$\mathcal{M}^{InJ2} \overset{\text{def}}{=} \left(\wp\wp\left(G^{InJ}\right), [\![E]\!]^{InJ2}, \cap, \emptyset \right)$$

are models for $\mathcal{A}$.

Theorem 7. $\mathcal{M}^{InJ2} \models a \wedge b \neq 0$ *if and only if* $\mathcal{M}^{InJ} \models a \wedge b \neq 0$

This model describes maximal sharing in the sense that if two events are distributed in it, they are distributed in any other model. The following theorems state this property for the InJ models.

Theorem 8. *Let* $T = (S, E, I, \longrightarrow, \ell)$ *be an ATS and let* $\mathcal{A}$ *be its algebra of localities. Let* $\mathcal{M}$ *be any model for* $\mathcal{A}$. *Then, for all events* $a, b \in E$,

$$\mathcal{M}^{InJ2} \models a \wedge b = 0 \implies \mathcal{M} \models a \wedge b = 0$$

Corollary 2. $\mathcal{M}^{InJ}$ *is a maximal sharing model.* □

Example 4. We can see on the right of Figure 3 the graph G^{InJ} for the ATS of Figure 1, no matter whether a and b are independent. Thus, we obtain the following interpretation in the maximal sharing model.

$$[\![a]\!]^{InJ} = [\![b]\!]^{InJ} = [\![c]\!]^{InJ} = \{\{a, b, c\}\}$$

Thus, $\mathcal{M}^{InJ} \models a \wedge b \neq 0$. However, from Example 3 we know that when aIb, $\mathcal{M}^{I} \models a \wedge b = 0$. So, we have that I models are not maximal sharing models.

We have that for the same ATS when a and b are not independent, $\mathcal{M}^{I} \models a \wedge b \neq 0$ and $\mathcal{M}^{IJ} \models a \wedge b = 0$. Thus, I models are not minimal sharing models either. □

5 Conclusions

In this work we have exploited the information about localities hidden in the ATS definition. Such information helps us to find implementations of systems with certain properties, like maximal or minimal sharing of localities.

The way to state how the locality of events are related is by means of the algebra of localities. We have introduced several models for this algebra and showed that this is not a trivial set of models. Figure 5 summarizes our result in Section 4. The up-going arrows in the picture mean that sharing on the lower models implies sharing on the upper models.

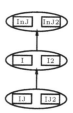

Fig. 5. Models of localities

We also have shown that our semantic approach exposes clearly difficulties arisen in syntactic language oriented approaches when dealing with non deterministic choices.

As a consequence of this work we can extract locality information from a specification written in terms of ATS. So, ATS formalism appears as a good candidate to become a theoretical assembler for distributed programming. At least, there are three interesting directions to continue this work. One of them is to go on a deeper comprehension of locality models. The nature of the hierarchy of models seems far away from being trivial, requiring more detailed studies on its structure. We believe that research in this direction will allow us to detect not only minimal sharing models, but also models with some constraints which require less localities to work.

We may develop the same strategy for other semantic formalisms, that is, to associate an algebra of localities and to obtain a model as before. Event structures [12], from where the notion of independence can be easily derived, would be a good candidate to study.

Another direction for future work would be to extend ATS with new characteristics. Time is a natural factor to consider in this extensions, as far as resources are used for events during certain time. A relation between a not yet defined timed ATS and timed graphs [2] would enable us to move into timed systems, where tools and methods for automatic verification have been developed.

Another way for continuing our work is the development of a toolkit for description of systems based in ATS. We believe that semantic studies in programming must come together with software development, and so implementation of good toolkits for both theoretical and practical developments will become more important in future.

References

1. L. Aceto. A static view of localities. *Formal Aspects of Computing*, 1994.
2. R. Alur and D. L.Dill. A theory of timed automata. *Theoretical Computer Science*, 126, 1994.
3. R. M. Amadio. An asynchronous model of locality, failure, and process mobility. Technical Report 3109, INRIA Sophia Antipolis, Feb. 1997.
4. G. Boudol, I. Castellani, M. Hennessy, and A. Kiehn. Observing localities. *Theoretical Computer Science*, 114:31–61, 1993.
5. G. Boudol, I. Castellani, M. Hennesy, and A. Kiehn. A theory of processes with localities. *Formal Aspects of Computing*, 6(2):165–200, 1994.
6. I. Castellani. Observing distribution in processes: static and dynamic localities. *Int. Journal of Foundations of Computer Science*, 1995.
7. P. Degano, R. D. Nicola, and U. Montanari. Partial orderings descriptions and observations of nondeterministic concurrent processes. In *REX School and Workshop on Linear Time, Branching Time and Partial Order in Logics and Models for Concurrency*, 1989.
8. R. v. Glabbeek. *Comparative Concurrency Semantics and Refinement of Actions.* PhD thesis, Free University, Amsterdam, 1990.
9. U. Montanari, M. Pistore, and D. Yankelevich. Efficient minimization up to location equivalence. In *Programming Languages and Systems – ESOP'96*, 1996.
10. U. Montanari and D. Yankelevich. A parametric approach to localities. In *Proceedings 19th ICALP,* Vienna, 1992.
11. U. Montanari and D. Yankelevich. Location equivalence in a parametric setting. *Theoretical Computer Science*, 149:299–332, 1995.
12. M. Nielsen, G. Plotkin, and G. Winskel. Petri nets, event structures and domains, part I. *Theoretical Computer Science*, 13(1):85–108, 1981.
13. W. Reisig. *Petri nets – an introduction.* EATCS Monographs on Theoretical Computer Science, Volume 4. Springer-Verlag, 1985.
14. M. Shields. Deterministic asynchronous automata. In *Formal Methods in Programming*. North-Holland, 1985.
15. W. Vogler. Bisimulation and action refinement. *Theoretical Computer Science*, 114:173–200, 1993.
16. G. Winskel and M. Nielsen. Models for concurrency. Technical Report DAIMI PB-492, Comp. Sci. Dept., Aarhus Univ., Nov. 1992.
17. D. Yankelevich. *Parametric views of process description languages.* PhD thesis, University of Pisa, 1993.

On the Expressivity and Complexity of Quantitative Branching-Time Temporal Logics

F. Laroussinie, Ph. Schnoebelen, and M. Turuani

Lab. Spécification & Vérification
ENS de Cachan & CNRS UMR 8643
61, av. Pdt. Wilson, 94235 Cachan Cedex France
email: {fl,phs,turuani}@lsv.ens-cachan.fr

Abstract. We investigate extensions of *CTL* allowing to express quantitative requirements about an abstract notion of time in a simple discrete-time framework,
and study the expressive power of several relevant logics.
When only subscripted modalities are used, polynomial-time model checking is
possible even for the largest logic we consider, while introducing freeze quantifiers
leads to a complexity blow-up.

1 Introduction

Temporal logic is widely used as a formal language for specifying the behaviour of
reactive systems (see [7]). This approach allows *model checking*, i.e. the automatic
verification that a finite state system satisfies its expected behavourial specifications.
The main limitation to model checking is the state-explosion problem but, in practice,
symbolic model checking techniques [5] have been impressively successful, and model
checking is now commonly used in the design of critical reactive systems.

Real-time. While temporal logics only deal with "before and after" properties, *real-
time temporal logics* and more generally *quantitative temporal logics* aim at expressing
quantitative properties of the time elapsed during computations. Popular real-time logics
are based on timed transition systems and appear in several tools (e.g., HyTech, Uppaal,
Kronos). The main drawback is that model checking is expensive [2,4].

Efficient model checking. By contrast, some real-time temporal logics retain usual discrete Kripke structures as models and allow to refer to quantitative information with
"bounded" modalities such as "$\mathsf{AF}_{\leq 10}\,A$" meaning that A will inevitably occur *in at
most 10 steps*. A specific aspect of this framework is that the underlying Kripke structures
have no inherent concept of time. It is the designer of the Kripke structure who decides
to encode the flow of elapsing time by this or that event, so that the temporal logics in use
are more properly called *quantitative temporal logics* than real-time logics. [8] showed
that $RTCTL$ (i.e. CTL plus bounded modalities "$\mathsf{A\_U}_{\leq k}\,\_$" and "$\mathsf{E\_U}_{\leq k}\,\_$" in the Kripke
structure framework) still enjoys the bilinear model checking time complexity of CTL.

G. Gonnet, D. Panario, and A. Viola (Eds.): LATIN 2000, LNCS 1776, pp. 437–446, 2000.

Our contribution. One important question is how far can one go along the lines of *RTCTL*-like logics while still allowing efficient model checking ? Here we study two quantitative extensions of *CTL*, investigate their expressive power and evaluate the complexity of model checking.

The first extension, called $TCTL_s$, s for "subscripts", is basically the most general logic along the lines of the *RTCTL* proposal : it allows combining "$\leq k$", "$\geq k$" and "$= k$" (so that modalities counting w.r.t. intervals are possible). We show this brings real improvements in expressive power, and model checking is still in polynomial time. This extends results for *RTCTL* beyond the increased expressivity: we use a finer measure for size of formula ($\mathsf{EF}_{=k}$ has size in $O(\log k)$ and not k) and do not require that one step uses one unit of time.

The second extension, called $TCTL_c$, c for "clocks", uses formula clocks, a.k.a. freeze quantifiers [3], and is a more general way of counting events. $TCTL_c$ can still be translated directly into *CTL* but model checking is expensive.

The results on expressive power formalize natural intuitions which (as far as we know) have never been proven formally, even in the dense time framework [1]. Furthermore, in our discrete time framework our results on expressive power must be stated in terms of how succinctly can one logic express this or that property. Such proofs are scarce in the literature (one example is [13]).

Related work. $TCTL_s$ and $TCTL_c$ are similar to (and inspired from) logics used in dense real-time frameworks (though, in the discrete framework we use here, their behaviour is quite different). Our results on complexity of model checking build on ideas from [6,11,4,10].

Other branching-time extensions of *RTCTL* have been considered. Counting with regular patterns makes model checking intractable [9]. Merging different time scales makes model checking NP-complete [10]. Allowing parameters makes model checking exponential in the number of parameters [10].

Another extension with freeze variables can be found in [14] where richer constraints on number of occurrences of events can be stated (rending satisfiability undecidable). On the other hand, the "until" modality is not included and the expressive power of different kinds of constraints is not investigated.

Plan of the paper. We introduce the basic notions and definitions in § 2. We discuss expressive power in § 3 and model checking in § 4. We assume the reader is familiar with standard notions of branching-time temporal logic (see [7]) and structural complexity (see [12]). Complete proofs appear in a full version of the paper, available from the authors.

2   *CTL* + Discrete Time

We write $\mathbb{N}$ for the set of natural numbers, and $AP = \{A, B, \ldots\}$ for a finite set of *atomic propositions*. Temporal formulae are interpreted over states in Kripke structures. Formally,

---

[1] See e.g. the conjecture at the end of [1] which becomes an unproved statement in [2].

Definition 2.1. *A Kripke structure (a "KS") is a tuple $S = \langle Q_S, R_S, l_S \rangle$ where $Q_S = \{q_1, \ldots\}$ is a non-empty set of states, $R_S \subseteq Q_S \times Q_S$ is a total transition relation, and $l_S : Q_S \to 2^{AP}$ labels every state with the propositions it satisfies.*

Below, we drop the "S" subscript in our notations whenever no ambiguity will arise. A *computation* in a KS is an infinite sequence π of the form $q_0 q_1 \ldots$ s.t. $(q_i, q_{i+1}) \in R$ for all $i \in \mathbb{N}$. For $i \in \mathbb{N}$, $\pi(i)$ (resp. $\pi_{|i}$) denotes the i-th state, q_i (resp. i-th prefix: $q_0 q_1, \ldots, q_i$). We write $\Pi(q)$ for the set of all computations starting from q. Since R is total, $\Pi(q)$ is never empty.

The flow of time. We assume a special atomic proposition $tick \in AP$ that describes the elapsing of time in the model. The intuition is that states labeled by $tick$ are states where we observe that time has just elapsed, that the *clock just ticked*. Equivalently, we can see all transitions as taking 1 time unit if they reach a state labeled by $tick$, and as being instantaneous otherwise [2]. In pictures, we use different grey levels to distinguish $tick$ states from non-$tick$ ones.

Given a computation $\pi = q_0 q_1 \ldots$ and $i \geq 0$, $\mathsf{Time}(\pi_{|i})$ denotes $|\{j \mid 0 < j \leq i \wedge tick \in l(q_j)\}|$, the time it took to reach q_i from q_0 along π.

2.1 $TCTL_s$

Syntax. $TCTL_s$ formulae are given by the following grammar:

$$\varphi, \psi ::= \neg\varphi \mid \varphi \wedge \psi \mid \mathsf{EX}\varphi \mid \mathsf{E}\varphi\mathsf{U}_I\,\psi \mid \mathsf{A}\varphi\mathsf{U}_I\,\psi \mid A \mid B \mid \ldots$$

where I can be any finite union $[a_1, b_1[\cup \cdots \cup [a_n, b_n[$ of disjoint integer intervals with $0 \leq a_1 < b_1 < a_2 < b_2 < \cdots a_n < b_n \leq \omega$.

Standard abbreviations include $\top, \bot, \varphi \vee \psi, \varphi \Rightarrow \psi, \ldots$ as well as $\mathsf{EF}_I\,\varphi$ (for $\mathsf{E}\top\mathsf{U}_I\,\varphi$), $\mathsf{AF}_I\,\varphi$ (for $\mathsf{A}\top\mathsf{U}_I\,\varphi$), $\mathsf{EG}_I\,\varphi$ (for $\neg\mathsf{AF}_I\,\neg\varphi$), and $\mathsf{AG}_I\,\varphi$ (for $\neg\mathsf{EF}_I\,\neg\varphi$).

Moreover we let $\mathsf{U}_{<k}$ stand for $\mathsf{U}_{[0,k[}$, $\mathsf{U}_{>k}$ for $\mathsf{U}_{[k+1,\omega[}$, and $\mathsf{U}_{=k}$ for $\mathsf{U}_{[k,k+1[}$. The usual CTL operators are included since the usual U corresponds to $\mathsf{U}_{<\omega}$.

Semantics. Figure 1 defines when a state q in some KS S, satisfies a $TCTL_s$ formula φ, written $q \models \varphi$, by induction over the structure of φ.

We let $TCTL_s[<]$, $TCTL_s[<,=]$, etc. denote the fragments of $TCTL_s$ where only simple constraints using only $<$ (resp. $<$ or $=$, etc.) are allowed. E.g., $RTCTL$ is $TCTL_s[<]$ (with the proviso that our KS's have $tick$'s).

2.2 $TCTL_c$

$TCTL_c$ uses freeze quantifiers [3]. Here "clocks" are introduced in the formula, set to zero when they are bound, and can be referenced "later" in arbitrary ways. This standard construct gives more flexibility than subscripts.

---

[2]   Thus KS's with $tick$'s can be seen as *discrete timed structures*, i.e. KS's where edges $(q, q') \in R$ are labeled by a natural number: the time it takes to follow the edge. While discrete timed structures are more natural, KS's with $tick$ are an essentially equivalent framework where technicalities are simpler since they do not need labels on the edges.

$$
\begin{aligned}
&q \models A && \text{iff } A \in l(q), \\
&q \models \neg\varphi && \text{iff } q \not\models \varphi, \\
&q \models \varphi \wedge \psi && \text{iff } q \models \varphi \text{ and } q \models \psi, \\
&q \models \mathsf{EX}\varphi && \text{iff there exists } \pi \in \Pi(q) \text{ s.t. } \pi \models \mathsf{X}\varphi, \\
&q \models \mathsf{E}\varphi\mathsf{U}_I \psi && \text{iff there exists } \pi \in \Pi(q) \text{ s.t. } \pi \models \varphi\mathsf{U}_I \psi \\
&q \models \mathsf{A}\varphi\mathsf{U}_I \psi && \text{iff for all } \pi \in \Pi(q), \text{ we have } \pi \models \varphi\mathsf{U}_I \psi \\
&\pi \models \mathsf{X}\varphi && \text{iff } \pi(1) \models \varphi, \\
&\pi \models \varphi\mathsf{U}_I \psi && \text{iff there exists } i \geq 0 \text{ s.t. } \mathsf{Time}(\pi_{|i}) \in I \\
&&& \text{and } \pi(i) \models \psi \text{ and } \pi(j) \models \varphi \text{ for all } 0 \leq j < i,
\end{aligned}
$$

Fig. 1. Semantics of $TCTL_s$

Syntax. For a set $Cl = \{x, y, \ldots\}$ of *clocks*, $TCTL_c$ formulae are given by the following grammar:

$$\varphi, \psi ::= \neg\varphi \mid \varphi \wedge \psi \mid \mathsf{EX}\varphi \mid \mathsf{E}\varphi\mathsf{U}\psi \mid \mathsf{A}\varphi\mathsf{U}\psi \mid x \underline{\text{ in }} \varphi \mid x \sim k \mid A \mid B \mid \cdots$$

where $\sim \in \{=, \leq, <, \geq, >\}$ and $k \in \mathbb{N}$. Constraints referring to clocks are restricted to the simple form $x \sim k$, in the spirit of $TCTL_s$.

An occurrence of a formula clock x in some $x \sim k$ is *bound* if it is in the scope of a "$x \underline{\text{ in }}$" freeze quantifier, otherwise it is *free*. A formula is *closed* if it has no free variables. Only closed formulae express properties of states in KS's.

Semantics. $TCTL_c$ formulae are interpreted over a state of a KS S together with a *valuation* $v : Cl \to \mathbb{N}$ of the clocks free in φ.

$$
\begin{aligned}
&q, v \models A && \text{iff } A \in l(q), \\
&q, v \models \neg\varphi && \text{iff } q, v \not\models \varphi, \\
&q, v \models \varphi \wedge \psi && \text{iff } q, v \models \varphi \text{ and } q, v \models \psi, \\
&q, v \models \mathsf{EX}\varphi && \text{iff there exists } \pi \in \Pi(q) \text{ s.t. } \pi, v \models \mathsf{X}\varphi \\
&q, v \models \mathsf{E}\varphi\mathsf{U}\psi && \text{iff there exists } \pi \in \Pi(q) \text{ s.t. } \pi, v \models \varphi\mathsf{U}\psi \\
&q, v \models \mathsf{A}\varphi\mathsf{U}\psi && \text{iff for all } \pi \in \Pi(q) \text{ we have } \pi, v \models \varphi\mathsf{U}\psi \\
&q, v \models x \underline{\text{ in }} \varphi && \text{iff } q, v[x \leftarrow 0] \models \varphi \\
&q, v \models x \sim k && \text{iff } v(x) \sim k \\
&\pi, v \models \mathsf{X}\varphi && \text{iff } \pi(1), v + d \models \varphi \text{ with } d = \mathsf{Time}(\pi_{|1}) \\
&\pi, v \models \varphi\mathsf{U}\psi && \text{iff there exists } i \geq 0 \text{ s.t. } \pi(i), v + d_i \models \psi \text{ and} \\
&&& \pi(j), v + d_j \models \varphi \text{ for all } 0 \leq j < i \text{ (where } d_l \stackrel{\text{def}}{=} \mathsf{Time}(\pi_{|l}))
\end{aligned}
$$

Fig. 2. Semantics of $TCTL_c$

Figure 2 defines when $q, v \models \varphi$ in some KS S by induction over the structure of φ. For $m \in \mathbb{N}$, $v + m$ denotes the valuation which maps each clock $x \in Cl$ to the value $v(x) + m$, and $v[x \leftarrow 0]$ is v where now x evaluates to 0.

Clearly the $TCTL_s$ operators can be defined with $TCTL_c$ operators:

$$\mathsf{E}\varphi\mathsf{U}_I \psi \stackrel{\text{def}}{=} x \underline{\text{ in }} \left(\mathsf{E}\varphi\mathsf{U}(I(x) \wedge \psi) \right) \qquad \mathsf{A}\varphi\mathsf{U}_I \psi \stackrel{\text{def}}{=} x \underline{\text{ in }} \left(\mathsf{A}\varphi\mathsf{U}(I(x) \wedge \psi) \right)$$

where, for I of the form $[a_1, b_1[\cup \cdots \cup [a_n, b_n[$, $I(x)$ denotes the *clocks constraint*
$$\bigvee_{i=1}^{n} \left((a_i \leq x) \wedge (x < b_i) \right).$$ Hence $TCTL_s$ can be seen as a fragment of $TCTL_c$ where
only one formula clock is allowed (and used in restricted ways).

A standard observation for logics such as $TCTL_c$ is that the actual values recorded
in v are only relevant up to a certain point depending on the formula at hand. Let M_φ
denote the largest constant appearing in φ (largest k in the "$x \sim k$"s) and, for $m \in \mathbb{N}$,
let $v \equiv_m v'$ when for any $x \in Cl$, either $v(x) = v'(x)$ or $v(x) > m < v'(x)$ (i.e. v and
v' agree, or are beyond m).

Lemma 2.2. *If $v \equiv_m v'$ and $m \geq M_\varphi$, then $q, v \models \varphi$ iff $q, v' \models \varphi$.*

Proof. Easy induction over the structure of φ, using the fact that $v \equiv_m v'$ entails
$v + k \equiv_m v' + k$ and $v[x \leftarrow 0] \equiv_m v'[x \leftarrow 0]$. $\square$

Remark 2.3. A related property is used by Emerson et al. in their study of $RTCTL$:
when checking whether $q \models \varphi$ inside some KS with $|Q| = m$ states, it is possible to
replace by m any constant k larger than m in the subscripts of φ. We emphasize that this
property does not hold for $TCTL_s[=]$ (it does hold for $TCTL_s[<, >]$). $\square$

The size of our formulae is the length of the string [3] used to write them down in
a sufficiently succinct way, e.g., $| A\alpha U_I \beta |$ is $1 + |\alpha| + |\beta| + |I|$. For I of
the form $[a_1, b_1[\cup \cdots \cup [a_n, b_n[$, we have $|I| \stackrel{\text{def}}{=} \lceil \log a_1 \rceil + \cdots + \lceil \log b_n \rceil$ (assuming
$\log(0) = \log(\omega) = 0$). $ht(\varphi)$ denotes the temporal height of formula φ. As usual, it is
the maximal number of nested modalities in φ. Obviously, $ht(\varphi)$ is smaller than the size
of φ (even when viewed as a dag).

3 Expressivity

Formally, $TCTL_s$ or $TCTL_c$ do not add expressive power to CTL:

Theorem 3.1. *Any closed $TCTL_c$ (or $TCTL_s$) formula is equivalent to a CTL formula.*

Proof. With any $TCTL_c$ formula φ, and valuation v, we associate a CTL formula $(\varphi)^v$
s.t. $q, v \models \varphi$ iff $q \models (\varphi)^v$ for any state q *of any Kripke structure*. Then, if φ has no free
clock variables, any $(\varphi)^v$ is a CTL equivalent to φ. The definition of $(\varphi)^v$ is given by
the following rewrite rules:

$$(\varphi \wedge \psi)^v \stackrel{\text{def}}{=} \varphi^v \wedge \psi^v \qquad\qquad (x \sim k)^v \stackrel{\text{def}}{=} \begin{cases} \top & \text{if } v(x) \sim k, \\ \bot & \text{otherwise} \end{cases}$$
$$(\neg\varphi)^v \stackrel{\text{def}}{=} \neg\varphi^v$$
$$(A)^v \stackrel{\text{def}}{=} A \qquad\qquad (x \underline{\text{ in }} \varphi)^v \stackrel{\text{def}}{=} \varphi^{v[x \leftarrow 0]}$$

---

[3] We sometimes see a formula as a dag, where identical subformulae are only counted once.
Such cases are stated explicitly.

$$(\mathsf{AF}\varphi)^v \stackrel{\text{def}}{\equiv} \begin{cases} \mathsf{AF}\varphi^v & \text{if } v+1 \equiv_{M_\varphi} v, \\ \varphi^v \vee \mathsf{AX}\Big[\mathsf{A}(\neg tick) \ \mathsf{U} \ \Big((\neg tick \wedge \varphi^v) \vee (tick \wedge (\mathsf{AF}\varphi)^{v+1})\Big)\Big] & \text{otherwise} \end{cases}$$

$$(\mathsf{E}\varphi \ \mathsf{U} \ \psi)^v \stackrel{\text{def}}{\equiv} \begin{cases} \mathsf{E}\varphi^v \ \mathsf{U} \ \psi^v & \text{if } v+1 \equiv_{M_{\varphi,\psi}} v, \\ \psi^v \vee \Big[\varphi^v \wedge \mathsf{EX}\Big(\mathsf{E}(\varphi^v \wedge \neg tick) \ \mathsf{U} \ \Big((\psi^v \wedge \neg tick) \vee (tick \wedge (\mathsf{E}\varphi\mathsf{U}\psi)^{v+1})\Big)\Big)\Big] \\ \qquad\qquad \text{otherwise} \end{cases}$$

This gives a well-founded definition for $(\_)^v$ since in the right-hand sides either $(\_)^v$ is recursively applied over subformulae, or $(\_)^{v+1}$ is applied on the same formula (or both). But moving from $(\_)^v$ to $(\_)^{v+1}$ is only done until $v \equiv_M v+1$, which is bound to eventually happen. Then it is a routine matter to check that the correctness invariant (i.e., "$q, v \models \varphi$ iff $q \models (\varphi)^v$") is preserved by these rules. $\qquad\square$

The translation we just gave is easy to describe but the resulting $(\varphi)^v$ formulae have enormous size. It turns out that this cannot be avoided. Even more, we can say that moving from CTL to $TCTL_s[<]$ to $TCTL_s$ to ... allows writing new formulae that have no succinct equivalent at the previous level.

Theorem 3.2. *1. $TCTL_s[<]$ can be exponentially more succinct than CTL,*
2. $TCTL_s[<,>]$ can be exponentially more succinct than $TCTL_s[<]$.

The proof is given by the following lemmas.

Lemma 3.3. *Any CTL formula equivalent to $\mathsf{EF}_{<n} A$ (a log n-sized formula) has temporal height at least n.*

Proof. Consider the KS described in Figure 3. One easily shows (by structural induction over φ) that for any CTL formula φ, $ht(\varphi) \leq i$ implies $\alpha_i \models \varphi$ iff $\alpha_{i+1} \models \varphi$. On the

$$\alpha_i \models tick \wedge \neg A$$
$$\gamma \models \neg tick \wedge A$$

Fig. 3. $\alpha_n \models \mathsf{EF}_{<n+1} A$ and $\alpha_{n+1} \not\models \mathsf{EF}_{<n+1} A$

other hand, $\alpha_j \models \mathsf{EF}_{<n} A$ iff $j < n$. Thus any CTL equivalent to $\mathsf{EF}_{<n} A$ must have temporal height larger than n. $\qquad\square$

Lemma 3.4. *Any $TCTL_s[<]$ formula equivalent to $\mathsf{EF}_{>n} A$ (a log n-sized formula) has temporal height at least n.*

Proof. Consider the KS described in Figure 4. One easily shows (by structural induction over φ) that for any formula φ in $TCTL_s[<]$, $ht(\varphi) \leq i$ implies $\alpha_i \models \varphi$ iff $\alpha_{i+1} \models \varphi$ and $\beta_i \models \varphi$ iff $\beta_{i+1} \models \varphi$. On the other hand, $\alpha_j \models \mathsf{EF}_{>n} A$ iff $j > n$. Thus any $TCTL_s[<]$ equivalent to $\mathsf{EF}_{>n} A$ must have temporal height larger than n. $\qquad\square$

Let us mention two (natural) conjectures that would allow separating further fragments:

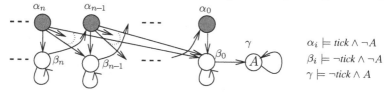

Fig. 4. $\alpha_n \not\models EF_{>n} A$ and $\alpha_{n+1} \models EF_{>n} A$

Conjecture 3.5.   1. $TCTL_s[<,>,=]$ can be exponentially more succinct than $TCTL_s[<,>]$,
2. $TCTL_c$ can be exponentially more succinct than $TCTL_s$.

We have not yet been able to find the required proofs, which are hard to build. The first point is based on the conjecture that any $TCTL_s[<,>]$ formula equivalent to $EF_{=k} A$ has temporal height at least k. For the second one, we conjecture that any $TCTL_s$ formula equivalent to $x \underline{in} EF(A \wedge EF(B \wedge x = k))$ has size at least k.

We have explained how $TCTL_s$ becomes more and more expressive when we allow subscripts with $<$, then also with $>$, then also with $=$. Subscripts of the form "$= k$" are the main difference between $RTCTL$ and our proposal. They enhance expressivity and make model checking more complex (see § 4).

Once we have $TCTL_s[<,>,=]$, subscripts with intervals are just a convenient shorthand:

Theorem 3.6.   *$TCTL_s$ is not more succinct than $TCTL_s[<,>,=]$.*

Proof. For I of the form $\bigcup_{i=1\ldots n}[a_i, b_i[$, we denote by $I-k$ the set $\bigcup_{i=1\ldots n}[a_i-k, b_i-k[$ (after the obvious normalization if $k > a_1$).

Let φ be a $TCTL_s$ formula. We build an equivalent $TCTL_s[<,>,=]$ formula $\tilde{\varphi}$ with the following equivalences:

$$E \alpha U_I \beta \equiv \bigvee_{i=1\ldots n} E \alpha U_{=a_i} (E \alpha U_{<b_i - a_i} \beta)$$

$$A \alpha U_I \beta \equiv \begin{cases} A \alpha U_{=a_1} (A \alpha U_{I-a_1} \beta) & \text{if } a_1 > 0, \\[2mm] \begin{aligned} &\neg E(\neg\beta)U_{<b_1} (\neg \alpha \wedge \neg\beta) \\ &\wedge \neg E(\neg\beta)U_{=b_1} \left(\neg A \alpha U_{=a_2-b_1} (A \alpha U_{I-a_2} \beta)\right) \end{aligned} & \text{otherwise} \end{cases}$$

Correctness is easy to check. The size of $\tilde{\varphi}$, seen as a dag, is linear in the size of φ seen as a dag [4]. $\square$

4   Model Checking

For the logics we investigate, the *model checking problem* is the problem of computing whether $q \models \varphi$ for q a state of a KS S and φ a temporal formula. In this section we analyse the complexity of model checking problems for $TCTL_s$ and $TCTL_c$.

---

[4] Viewing formulae as dags is convenient here, and agree with our later use of Theorem 3.6 when we investigate efficient model checking for $TCTL_s$.

Given a KS S and a formula φ, the complexity of model checking can be evaluated in term of $|S|$ and $|\varphi|$. But more discriminating information can be obtained by also looking at the *program complexity* of model checking (i.e., the complexity when φ is fixed and S, q is the only input) and the *formula complexity* (i.e., when S, q is fixed and φ is the only input).

While $TCTL_s$ model checking can be done efficiently, this is not true for $TCTL_c$ (even when considering a fixed KS).

Theorem 4.1. *Let $S = \langle Q, R, l \rangle$ be a KS and φ a $TCTL_s$ formula. There exists a model checking algorithm running in time $O\big((|Q|^3 + |R|) \times |\varphi|\big)$. Moreover if φ belongs to $TCTL_s[<, >]$, the algorithm runs in time $O\big((|Q| + |R|) \times |\varphi|\big)$.*

Proof (Idea). The algorithm extends the classical algorithms for CTL and $RTCTL$ (see [8]) with procedures dealing with $TCTL_s[<, =, >]$ operators (as seen in Theorem 3.6, formulae with interval subscripts can be decomposed). The most expensive procedure concerns the $\mathsf{EU}_=$ case where we compute the transitive closures of relations, hence the (quite naive) $O(|Q|^3 + |R|)$. The $TCTL_s[<, >]$ fragment uses only procedures in $O\big((|Q| + |R|) \times |\varphi|\big)$. $\qquad\square$

Theorem 4.2. *The model checking problem for $TCTL_c$ is PSPACE-complete. The formula complexity of $TCTL_c$ model checking is PSPACE-complete.*

Proof. To prove this result, it is sufficient to show that $TCTL_c$ model checking is in PSPACE[5] and that the formula complexity is PSPACE-hard. The proof of this last point relies on ideas from [4]: let P be an instance of QBF (Quantified Boolean Formula, a PSPACE-complete problem). W.l.o.g. P is some $Q_1 p_1 \ldots Q_n p_n . \varphi$ (with $Q_i \in \{\exists, \forall\}$ and φ a propositional formula over $p_1, \ldots, p_n$). We reduce P to a model checking problem $S, q \models \Phi$ where S is the simple KS $(\{q\}, \{q \to q\}, \{l(q) = tick\})$ and Φ is the following $TCTL_c$ formula:

$$t \underline{\text{ in }} \mathsf{EF}\Big[t=1 \wedge O_1\Big(x_1 \underline{\text{ in }} \mathsf{EF}\Big[t=2 \wedge \ldots \Big(t=i \wedge O_i(x_i \underline{\text{ in }} \mathsf{EF}(t=i+1 \wedge \ldots$$
$$\mathsf{EF}(t=n+1 \wedge \tilde{\varphi}) \ldots)\Big]\Big)\Big]$$

where O_i is $\mathsf{EF}_{\leq 1}$ (resp. $\mathsf{EG}_{\leq 1}$) if Q_i is $\exists$ (resp. $\forall$) and $\tilde{\varphi}$ is φ where occurrences of p_i have been replaced by $x_i = n + 1 - i$. Observe that any clock x_i is reset at time i or $i + 1$ and depending on this reset time the atomic propositions p_i will be interpreted as true or false after the $n+1$-*th* transition. The operator $\mathsf{EF}_{\leq 1}$ (resp. $\mathsf{EG}_{\leq 1}$) allows to quantify existentially (resp. universally) over these two reset times. Clearly Φ is valid iff $S, q \models \Phi$. $\qquad\square$

In practice, one can easily use any CTL model checker for model checking $TCTL_c$ formulae, and the resulting algorithm runs in time $O(|S| . M^{|Cl|} . |\varphi|)$. For example, with SMV, one just adds one variable for each formula clock and update them in the obvious way. This is much more practical than an approach based on Theorem 3.1 and the complexity is not too frightening for formulae with $|Cl| = 1$ (only one clock), a fragment already more expressive than $TCTL_s$.

---

[5] This uses standard arguments, see the long version for details.

A theoretical view. The following table gives a synthetic summary of complexity measures for model checking CTL, $TCTL_s$ and $TCTL_c$, showing that model checking the full $TCTL_s$ is as tractable as model checking CTL in both arguments. On the other hand, model checking $TCTL_c$ requires polynomial space even for a fixed Kripke structure.

	CTL	$TCTL_s$	$TCTL_c$
Complexity of model checking	P-complete		PSPACE-complete
Formula complexity	LOGSPACE		PSPACE-complete
Program complexity	NLOGSPACE-complete		

Filling the table. Model checking $TCTL_s$ is in P as we just saw. P-hardness results from the obvious reading of the circuit-value problem (with proper alternation) as a model checking problem for the **EX** fragment of CTL. The formula complexity of model checking CTL is LOGSPACE and this result can be easily extended to $TCTL_s$. The program complexity of model checking $TCTL_s$ and $TCTL_c$ is NLOGSPACE-complete since we proved (Theorem 3.1) that these logics can be translated into CTL, for which the NLOGSPACE-complete complexity is given in [11].

Symbolic model checking. When it comes to symbolic model checking (i.e., when S is given under the form of a synchronized product of k structures $S_1, \ldots, S_k$), CTL model checking becomes PSPACE-complete [11], this is also true for $TCTL_s$ and $TCTL_c$:

Theorem 4.3. *The symbolic model checking problem for $TCTL_s$ and $TCTL_c$ is PSPACE-complete.*

5   Conclusion

We investigated the expressive power and the complexity of model checking for $TCTL_s$ and $TCTL_c$, two quantitative extensions of CTL along the lines of $RTCTL$ [8,10].

The expressive power must be measured in a framework where, strictly speaking, everything can be translated into CTL.

We showed that $TCTL_s$, while more succinct than $RTCTL$, still allows an efficient model checking algorithm. By contrast $TCTL_c$, the extension of CTL with freeze quantifiers leads to a complexity blow-up.

References

1. R. Alur, C. Courcoubetis, and D. Dill. Model-checking for real-time systems. In *Proc. 5th IEEE Symp. Logic in Computer Science (LICS'90), Philadelphia, PA, USA, June 1990*, pages 414–425, 1990.
2. R. Alur, C. Courcoubetis, and D. Dill. Model-checking in dense real-time. *Information and Computation*, 104(1):2–34, 1993.
3. R. Alur and T. A. Henzinger. A really temporal logic. *Journal of the ACM*, 41(1):181–203, 1994.
4. L. Aceto and F. Laroussinie. Is your model checker on time ? In *Proc. 24th Int. Symp. Math. Found. Comp. Sci. (MFCS'99), Szklarska Poreba, Poland, Sep. 1999*, volume 1672 of *Lecture Notes in Computer Science*, pages 125–136. Springer, 1999.

5. J. R. Burch, E. M. Clarke, K. L. McMillan, D. L. Dill, and L. J. Hwang. Symbolic model checking: 10^{20} states and beyond. *Information and Computation*, 98(2):142–170, 1992.
6. S. Demri and Ph. Schnoebelen. The complexity of propositional linear temporal logics in simple cases (extended abstract). In *Proc. 15th Ann. Symp. Theoretical Aspects of Computer Science (STACS'98), Paris, France, Feb. 1998*, volume 1373 of *Lecture Notes in Computer Science*, pages 61–72. Springer, 1998.
7. E. A. Emerson. Temporal and modal logic. In J. van Leeuwen, editor, *Handbook of Theoretical Computer Science, vol. B*, chapter 16, pages 995–1072. Elsevier Science, 1990.
8. E. A. Emerson, A. K. Mok, A. P. Sistla, and J. Srinivasan. Quantitative temporal reasoning. In *Proc. 2nd Int. Workshop Computer-Aided Verification (CAV'90), New Brunswick, NJ, USA, June 1990*, volume 531 of *Lecture Notes in Computer Science*, pages 136–145. Springer, 1991.
9. E. A. Emerson and R. J. Trefler. Generalized quantitative temporal reasoning: An automata-theoretic approach. In *Proc. 7th Int. Joint Conf. Theory and Practice of Software Development (TAPSOFT'97), Lille, France, Apr. 1997*, volume 1214 of *Lecture Notes in Computer Science*, pages 189–200. Springer, 1997.
10. E. A. Emerson and R. J. Trefler. Parametric quantitative temporal reasoning. In *Proc. 14th IEEE Symp. Logic in Computer Science (LICS'99), Trento, Italy, July 1999*, pages 336–343, 1999.
11. O. Kupferman, M. Y. Vardi, and P. Wolper. An automata-theoretic approach to branching-time model checking, 1998. Full version of the CAV'94 paper, accepted for publication in J. ACM.
12. C. H. Papadimitriou. *Computational Complexity*. Addison-Wesley, 1994.
13. T. Wilke. CTL+ is exponentially more succint than CTL. In *Proc. 19th Conf. Found. of Software Technology and Theor. Comp. Sci. (FST&TCS'99), Chennai, India, Dec. 1999*, volume 1738 of *Lecture Notes in Computer Science*. Springer, 1999.
14. J. Yang, A. K. Mok, and F. Wang. Symbolic model checking for event-driven real-time systems. *ACM Transactions on Programming Languages and Systems*, 19(2):386–412, 1997.

A Theory of Operational Equivalence for Interaction Nets

Maribel Fernández[1] and Ian Mackie[2]

[1] LIENS (CNRS UMR 8548), École Normale Supérieure
45 Rue d'Ulm, 75005 Paris, France. maribel@dmi.ens.fr
[2] CNRS-LIX (UMR 7650), École Polytechnique
91128 Palaiseau Cedex, France. mackie@lix.polytechnique.fr

Abstract. The notion of contextual equivalence is fundamental in the theory of programming languages. By setting up a notion of bisimilarity, and showing that it coincides with contextual equivalence, one obtains a simple coinductive proof technique for showing that two programs are equivalent in all contexts. In this paper we apply these (now standard) techniques to interactions nets, a graphical programming language characterized by local reduction. This work generalizes previous studies of operational equivalence in interaction nets since it can be applied to untyped systems, thus all systems of interaction nets are captured.

1 Introduction

Interaction nets, introduced by Lafont [7], are graph rewriting systems that generalize the multiplicative proof nets of linear logic, and can be seen both as a high-level programming language or as a low-level implementation language. A program consists of a net (a graph built from a set of agents and wires) and a set of interaction rules that describe the way in which the net will be reduced. We are interested in the problem of defining an equivalence relation between programs that compute the same results, or in other words, that behave in the same way, in all contexts. In that case, one program can be replaced by the other, for example for efficiency reasons, without altering the operational semantics of the system. To define this equivalence relation we first need to develop an operational theory of interaction nets specifying in a precise way how programs are executed (i.e. a strategy of evaluation of nets and a notion of value).

In [2] we proposed a way of adapting the coinductive techniques, used successfully for the functional and object-oriented programming paradigms, to give a notion of operational equivalence for the interaction paradigm. The language of interaction nets that was studied focussed on the notion of type, which is natural if interaction nets are seen as a programming paradigm. In particular, types allow us to distinguish values from programs. However, some applications of interaction nets do not fit into the typed framework in a natural way. For instance, systems based on the interaction combinators [8], or the systems of interaction used for the encoding of the λ-calculus [9], are untyped. Although it is possible to develop a type system for them [6], a natural approach would be to develop an operational theory of equivalence of interaction nets that does not rely on the notion of types. The same remark can be made in the case of functional languages based on the

G. Gonnet, D. Panario, and A. Viola (Eds.): LATIN 2000, LNCS 1776, pp. 447–456, 2000.

λ-calculus, where two different approaches can be found in the literature, depending on whether the calculus is typed (see for instance [10]) or untyped (see for instance [1]).

In this paper we present an operational theory for *untyped* interaction nets, including a notion of contextual equivalence and an associated bisimilarity relation which permits the use of coinductive techniques in the proofs of operational equivalence. To express these notions we use the textual calculus of interaction nets presented in [3] instead of the graphical language, since it allows us to give a concise and formal presentation. We leave the use of diagrams for the examples and intuitive explanations.

A system of interaction nets is a user-defined language, in the same spirit as systems based on term-rewriting. Our results are applicable to *any* system of interaction nets; we are not restricted to one specific set of rules. If the system is typed, the information provided by types can be used to obtain a more refined equivalence relation between nets, recovering the results of [2]. We remark that interaction nets are also used as an object language for the coding of other rewriting systems. The λ-calculus is perhaps the most studied example of this (see e.g. [4,9]). Our results are also applicable here, so we have a proof technique for optimizations of such systems.

The paper is organized as follows. In the next section we set up the definition of interaction nets and define our evaluation strategy. Section 3 sets up to notion of bisimilarity. In Section 4 we give some examples of use of this relation. In Section 5 we formalize the notion of contextual equivalence, and Section 6 shows that this coincides with bisimilarity. Finally we conclude the paper in Section 7.

2   Background: Interaction Nets

We begin by presenting the textual calculus of interaction nets that we will use for the rest of the paper; we refer the reader to [3] for a more detailed description and examples.

Let Σ be a set of symbols, called *agents*, ranged over by $\alpha, \beta, \ldots$, each with a given *arity*, one *principal port* and a number of *auxiliary ports* equal to its arity. Σ can be partitioned into a set $\mathcal{C}$ of *constructors* and a set $\mathcal{D}$ of *destructors*, depending on the application. Let N be a disjoint set of names, ranged over by x, y, z, etc. *Terms* are defined by the grammar: $t ::= x \mid \alpha(t_1, \ldots, t_n)$, where $x \in N$, $\alpha \in \Sigma$, $arity(\alpha) = n$ and $t_1, \ldots, t_n$ are terms, with the restriction that each name may appear at most twice. $\mathcal{N}(t)$ denotes the set of names occurring in t. If a name occurs twice in a term, we say that it is *bound*, otherwise it is *free*. We write $\boldsymbol{t}$ for a list of terms $t_1, \ldots, t_n$. Graphically, a term of the form $\alpha(\boldsymbol{t})$ can be seen as a tree with connections between its leaves: the principal port of α (indicated by an arrow) is at the root, and the terms $t_1, \ldots, t_n$ are the subtrees connected to the auxiliary ports of α. A free variable represents a free port, and a bound variable represents a wire connecting two auxiliary ports.

If t and u are terms, then the (unordered) pair $t = u$ is an *equation*. $\Delta, \Theta, \ldots$ will be used to range over multisets of equations. The graphical representation of an equation is a pair of trees connected by their roots (principal ports).

Interaction rules are pairs of terms written as $\alpha(t) \bowtie \beta(u)$, where $(\alpha, \beta) \in \Sigma^2$ is the *active pair* of the rule. All names occur exactly twice in a rule, and there is one rule for each pair of agents.

Definition 1 (Configurations). *A* configuration *is a pair:* $c = (\mathcal{R}, \langle t \mid \Delta \rangle)$, *where* $\mathcal{R}$ *is a set of rules,* t *a list* $t_1, \ldots, t_n$ *of terms, and* Δ *a multiset of equations. Each variable occurs at most twice in* c. *If a name occurs once in* c *then it is* free, *otherwise it is* bound. *For simplicity we sometimes omit* $\mathcal{R}$ *when there is no ambiguity. We use* c, c' *to range over configurations. We call* t *the* head *and* Δ *the* body *of a configuration.*

Intuitively, $(\mathcal{R}, \langle t \mid \Delta \rangle)$ represents a net that we evaluate using $\mathcal{R}$. To draw the net we simply draw the trees for the terms in $\langle t \mid \Delta \rangle$, connect the common variables together, and connect the roots of the trees corresponding to the members of an equation together. The roots of the terms in the head of the configuration and the free names correspond to free ports in the *interface* of the net. Note that the head of the configuration may contain all or just some of the ports in the interface of the net, called *observable*. For this reason, the head is called the *observable interface* of the configuration.

We work modulo α-equivalence for bound names as usual, but also for free names. Configurations that differ only on the names of the free variables are equivalent, since they represent the same net.

Computation is performed by rewriting configurations using the following rewrite system, where if r is a rule, $\hat{r}$ denotes a fresh generic *instance* of r, that is, a copy of r where we introduce a new set of names:

Indirection: If $x \in \mathcal{N}(u)$, then $x = t, u = v \longrightarrow u[t/x] = v$.
Interaction: If $r \in \mathcal{R}$ and $\hat{r} = \alpha(t'_1, \ldots, t'_n) \bowtie \beta(u'_1, \ldots, u'_m)$, then

$$\alpha(t_1, \ldots, t_n) = \beta(u_1, \ldots, u_m) \longrightarrow$$
$$t_1 = t'_1, \ldots, t_n = t'_n, u_1 = u'_1, \ldots, u_m = u'_m$$

Context: If $\Delta \longrightarrow \Delta'$, then $\langle t \mid \Gamma, \Delta, \Gamma' \rangle \longrightarrow \langle t \mid \Gamma, \Delta', \Gamma' \rangle$.
Collect: If $x \in \mathcal{N}(t)$, then $\langle t \mid x = u, \Delta \rangle \longrightarrow \langle t[u/x] \mid \Delta \rangle$.

This rewrite system generates an *equational theory*, the corresponding equivalence relation is denoted by $c \longleftrightarrow^* c'$. The reduction relation $\longrightarrow$ is strongly confluent [3] since there is one rule for each pair of agents. Various strategies of evaluation are defined in [3]. The *values* that we use in this paper, called *interface normal forms*, have terms rooted by agents in the head whenever this is possible.

Definition 2 (Interface Normal Form). *A* configuration $(\mathcal{R}, \langle t \mid \Delta \rangle)$ *is in* interface normal form *(INF) if each* t_i *in* t *is of one of the following canonical forms:*

- $\alpha(s)$. *E.g.* $\langle S(x) \mid x = Z, \Delta \rangle$.
- x *where either* $x \in \mathcal{N}(t_j)$ *for some* $j \neq i$, *or* $x \in \mathcal{N}(u)$ *for some* $y = u \in \Delta$ *such that* $y \in N$ *is free* (x *is in an open path). E.g.* $\langle x, x \mid \Delta \rangle$
- x *where* $x \in \mathcal{N}(u)$ *for some* $y = u \in \Delta$ *such that* $y \in \mathcal{N}(u)$ (x *occurs in a cycle).*
  E.g. $\langle x \mid y = \alpha(\beta(y), x), \Delta \rangle$.

We denote by INF_i *the set of configurations where the* ith *port in the head is canonical.*

Computing interface normal forms suggests that we do the minimum work required to bring principal ports to the interface. This strategy is defined by the inference rules:

Axiom:

$$\frac{c \in \mathit{INF}}{c \Downarrow c}$$

Collect:

$$\frac{\langle t_1, \ldots, t, \ldots, t_n \mid \Delta \rangle \Downarrow c}{\langle t_1, \ldots, x, \ldots, t_n \mid x = t, \Delta \rangle \Downarrow c}$$

Indirection: if $x \in \mathcal{N}(u)$ and $y \in \mathcal{N}(t, u = v)$

$$\frac{\langle t_1, \ldots, y, \ldots, t_n \mid u[t/x] = v, \Delta \rangle \Downarrow c}{\langle t_1, \ldots, y, \ldots, t_n \mid x = t, u = v, \Delta \rangle \Downarrow c}$$

Interaction: if $x \in \mathcal{N}(\alpha(t) = \beta(u))$, $r \in \mathcal{R}$, $\hat{r} = \alpha(t') \bowtie \beta(u')$

$$\frac{\langle s_1, \ldots, x, \ldots, s_n \mid \overrightarrow{t = t'}, \overrightarrow{u = u'}, \Delta \rangle \Downarrow c}{\langle s_1, \ldots, x, \ldots, s_n \mid \alpha(t) = \beta(u), \Delta \rangle \Downarrow c}$$

This system is deterministic [3]. If $c \Downarrow v$ can be derived with these rules we say that v is the interface normal form of c. We write $c \Downarrow_i v$ (i.e. the position i in the head of v is canonical) if the rules Indirection and Interaction are only applied at position i in the head of the configuration and the axiom is replaced by

$$\frac{c \in \mathit{INF}_i}{c \Downarrow_i c}$$

Example 1 (Combinators). The interaction combinators [8] are a universal system of interaction built from the 0-ary agent ϵ and the binary agents δ and γ, with the rules:

$$\delta(x,y) \bowtie \delta(x,y) \qquad \delta(\epsilon,\epsilon) \quad \bowtie \quad \epsilon$$
$$\gamma(x,y) \bowtie \gamma(y,x) \qquad \gamma(\epsilon,\epsilon) \quad \bowtie \quad \epsilon$$
$$\epsilon \quad \bowtie \quad \epsilon \qquad \delta(\gamma(a,b),\gamma(c,d)) \bowtie \gamma(\delta(a,c),\delta(b,d))$$

The configuration $\langle x \mid \gamma(y,x) = \delta(\epsilon,y) \rangle$ gives a non-terminating sequence of reductions, but has an interface normal form: $\langle \delta(a,b) \mid \epsilon = \gamma(c,a), \delta(c,d) = \gamma(d,b) \rangle$.

3   Bisimilarity

In functional languages, we consider two functions equivalent when we can apply them to the same arguments and obtain the same results. In other words, we perform some form of experiment on the objects under test, and compare the results. For interaction nets, we take this general idea as inspiration. The way that we can make experiments with a net is to interact with it on a free principal port. Connecting nets on free principal ports is our analogue of applying a function to an argument. After evaluation, we can observe whether some principal ports are at the interface, which is analogous to observing whether a λ-term has evaluated to an abstraction. Only the *observable ports* (in the head of the configuration) are available for the experiments. Configurations with the same number of terms in the head will be said *comparable*.

Two configurations that cannot be distinguished by any experiment will be called *bisimilar*. We will show that the bisimilarity relation can be defined as the greatest postfixpoint of an operator (which allows us to use coinductive techniques to prove that two configurations are bisimilar), and more important, it coincides with the contextual equivalence, that is, two bisimilar configurations cannot be distinguished by any context and can therefore be exchanged without altering the behaviour of the system. We begin with some basic definitions to formalise these ideas.

Definition 3 (Visible Interface). *A configuration* $c = \langle t_1, \ldots, t_n \mid \Delta \rangle \in INF_i$ *has a visible interface at position* i *if either* t_i *is not a variable or there is an open path starting at* t_i *and finishing at some* $t_j = \alpha'(\boldsymbol{u}) \in \boldsymbol{t}$, *that is,* $t_i = \alpha(\boldsymbol{u})$ *or* $t_i = x$ *and there is an open path to* $t_j = \alpha'(\boldsymbol{u})$. *The* visible agent *at position* i *is* α *in the first case,* α' *in the second case. The rest of the net is called the kernel:* $\mathcal{K}_i(c) = \langle t_1, \ldots, t_{k-1}, \boldsymbol{u}, t_{k+1}, \ldots, t_n \mid \Delta \rangle$ *where* $k = i$ *or* $k = j$ *depending on whether we are in the first or second case. The set of* new observable positions *in* $\mathcal{K}_i(c)$, *denoted* $NP_\mathcal{K}(c, i)$, *is the set of positions of the terms* $\boldsymbol{u}$ *if* $k = i$, *otherwise it contains just the new position of* t_j. *We denote by* $\mathcal{V}_i$ *the set of all the configurations with a visible interface at position* i.

If v *and* v' *are comparable and have the same visible agent at position* i, *we write* $SVA_i(v, v')$. *If the visible agents are different, but they are not both constructors, we write* $\neg Constr_i(v, v')$.

Example 2. Let $c = \langle I(x), x \mid \rangle$. Since $t_1 = I(x)$ is not a variable, $c \in \mathcal{V}_1$. Since $t_2 = x$ and there is an open path to $t_1 = I(x)$, $c \in \mathcal{V}_2$. The visible agent is I for both positions, and $\mathcal{K}_1(c) = \mathcal{K}_2(c) = \langle x, x \mid \rangle$.

When we have different agents in the visible interfaces of the nets under test, and they are not constructors, we need to see if these agents behave in the same way for each possible agent interacting with them. For this we use closings.

Definition 4 (Closing). *A closing at position* i *of a configuration* $c = \langle \boldsymbol{t} \mid \Delta \rangle \in \mathcal{V}_i$, *denoted by* $cl_i(c)$, *is obtained from* c *by one of the following operations, where* $k = i$, *or* $k = j$ *if there is an open path starting at position* i *and finishing at position* j *in* c:

1. *replace* $t_k \equiv \alpha(\boldsymbol{s})$ *in* $\boldsymbol{t}$, *by a list of new variables* $z_1, \ldots, z_p \in \mathcal{N}(\boldsymbol{u})$, $p \geq 0$, *and add to* Δ *the equation* $t_k = \alpha'(\boldsymbol{u})$, *where* α' *is any agent, and the terms in* $\boldsymbol{u}$ *are either new variables (in which case they can appear twice in* $\alpha'(\boldsymbol{u})$ *or once in* $\alpha'(\boldsymbol{u})$ *and once in* $\boldsymbol{z}$*) or elements of* $\boldsymbol{t}$, *in which case they are erased from* $\boldsymbol{t}$.
   The set $NP_{cl}(c, i)$ *of* new observable positions *in* $cl_i(c)$ *contains the positions of the variables* $z_1, \ldots, z_p$ *in the new head if* $i = k$, *otherwise it contains just the new position of* t_i.
2. *erase* $t_k \equiv \alpha(\boldsymbol{s})$ *and another term* t_p *in* $\boldsymbol{t}$ *and add* $t_k = t_p$ *to* Δ. *In this case* $NP_{cl}(c, i) = \emptyset$ *if* $i = k$, *otherwise it contains just the new position of* t_i.

By abuse of notation, we will denote by $cl_i(c')$ *the result of applying to a configuration* c' *comparable with* c *the operations that define a closing at position* i *for* c.

Graphically, the first operation corresponds to connecting the principal port of an agent α' to the kth observable port in the interface of the net, and connecting some auxiliary ports of α' between them (if a variable appears twice in u), or to other observable ports in the net (if u contains terms in t). The second operation corresponds to simply adding a wire connecting the observable ports k and p.

We consider a complete lattice $(\mathsf{Rel}, \subseteq)$ where Rel is the set of binary relations between pairs $(c, i), (c', i)$ such that c, c' are comparable configurations whose heads have at least i elements (i.e. we can talk of the ith observable port). The operators $\langle \mathcal{R} \rangle$ and $[\mathcal{R}]$ for $\mathcal{R} \in \mathsf{Rel}$ will be used to define *similarity* and *bisimilarity* respectively.

Definition 5 (Operators). *Let c, c' be comparable configurations with at least i terms in the head.*

$$(c, i) \langle \mathcal{R} \rangle (c', i) \overset{\text{def}}{\Longleftrightarrow} c \Downarrow_i v \in \mathcal{V}_i \Rightarrow \exists v', (c' \Downarrow_i v' \text{ and}$$
$$\textit{either } SVA_i(v, v') \text{ and } \forall p \in NP_{\mathcal{K}}(v, i), (\mathcal{K}_i(v), p) \mathcal{R} (\mathcal{K}_i(v'), p)$$
$$\textit{or } \neg Constr_i(v, v'), \forall cl_i(v), \forall p \in NP_{cl}(v, i), (cl_i(v), p) \mathcal{R} (cl_i(v'), p))$$
$$(c, i)[\mathcal{R}](c', i) \overset{\text{def}}{\Longleftrightarrow} (c, i) \langle \mathcal{R} \rangle (c', i) \text{ and } (c', i) \langle \mathcal{R} \rangle (c, i)$$

Property 1. $\langle \cdot \rangle, [\cdot]$ are monotone operators.

Definition 6 (Similarity, Bisimilarity).

- *A relation $\mathcal{S} \in \mathsf{Rel}$ such that $\mathcal{S} \subseteq \langle \mathcal{S} \rangle$ (i.e. $\mathcal{S}$ is a post-fixpoint of $\langle \cdot \rangle$) is a* simulation. *The greatest such $\mathcal{S}$ is called a* similarity, *and written as $\precsim$. If c, c' are comparable configurations with n elements in the head, then $c \precsim c'$ if $(c, i) \precsim (c', i)$, $1 \leq i \leq n$.*
- *A relation $\mathcal{B} \in \mathsf{Rel}$ such that $\mathcal{B} \subseteq [\mathcal{B}]$ (i.e. $\mathcal{B}$ is a post-fixpoint of $[\cdot]$) is a* bisimulation. *The greatest such $\mathcal{B}$ is called a* bisimilarity, *and written as $\simeq$. If c, c' are comparable configurations with n elements in the head, then $c \simeq c'$ if $(c, i) \simeq (c', i)$, $1 \leq i \leq n$.*

Note that $\langle \cdot \rangle$ and $[\cdot]$ posses a greatest post-fixpoint by the Tarski-Knaster Fixed Point Theorem. Moreover, $\precsim$ and $\simeq$ are fixed points, i.e. $\precsim = \langle \precsim \rangle$ and $\simeq = [\simeq]$.

Remark 1. The main difference with the typed approach resides in the definition of closings and the way they are used in the definition of the operators $\langle \cdot \rangle$ and $[\cdot]$. Here closings are applied "on demand" whereas they are a static notion in the typed framework. More precisely, in a typed net a closing is built just by connecting agents to *all* the free *input* ports. The Subject Reduction property ensures that reduction will not create new free input ports. Instead, here we close one principal port at a time, and since reduction might create a new free principal port, closings are applied in a dynamic way.

The relations $\precsim, \simeq$ can be defined by levels, as done by Abramsky for the untyped λ-calculus [1].

Proposition 1 (Coinduction Principle). *Let c, c' be comparable configurations with n observable ports. To prove $c \simeq c'$ it suffices to find a bisimulation $\mathcal{B}$ such that $(c, i)\mathcal{B}(c', i)$ for $1 \leq i \leq n$.*

By coinduction we can show that the equational theory is included in the bisimilarity relation. In Section 4 we give more examples of application of coinduction to prove bisimilarity, in particular we will show that this inclusion is strict.

Theorem 1 (Bisimilarity includes the equational theory). $c \leftrightarrow^* c' \Rightarrow c \simeq c'$.

4 Examples

The Identity agent and a wire. Let I be the identity agent defined by rules
$$I(\alpha(x_1, \ldots, x_n)) \bowtie \alpha(I(x_1), \ldots, I(x_n))$$
for any $\alpha \in \Sigma$. We can prove $\langle I(x), x \mid \rangle \simeq \langle x, x \mid \rangle$ by coinduction. Take a symmetric R containing the pairs $((c, i), (c', i))$ such that c' is obtained from c by erasing the I agents at the root of a term in the head, or at the root of a member of an equation. We show that R is a bisimulation: if $c \Downarrow_i v \in \mathcal{V}_i$, then $c' \Downarrow_i v'$, and either they have the same visible agent α at position i, in which case the kernels are in the relation for all the new observable positions, or if they differ, then one is rooted by I and the other is just a variable. In that case the closings are in the relation, which is sufficient since I is not a constructor.

Copying before erasing or just erasing. In the system of the interaction combinators, replacing a net of the form

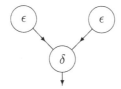

by the agent ϵ seems an intuitive optimization. We can prove that they are bisimilar by coinduction. The Main Theorem 2 tells us then that these configurations are contextually equivalent— the optimization is correct.

Agents γ and δ. The following nets are bisimilar:

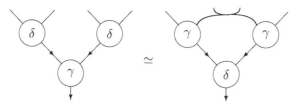

To show it using the coinduction principle we consider a relation containing $\simeq$ and these pairs, for any closing of the free principal port. The interesting closings are built by adding an agent ϵ, γ, or δ (the closings using just wires do not reduce to a value with a visible interface). The case of ϵ is trivial. For the other cases, by reducing to interface normal form we obtain configurations that have the same visible agents and whose kernels are easily shown to be bisimilar, hence contained in our relation.

η-rules for γ and δ. The following nets are bisimilar:

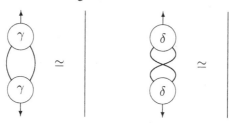

Note that these last two equivalences are neither included in the equational theory (since the nets are different normal forms) nor provable using the path semantics developed by Lafont [8].

5   Contextual Equivalence

We define a set of operations that build a context for a configuration, in the same way that closings were defined by operations. But there are more operations in the case of contexts, and we can have a sequence of operations instead of just one operation.

Definition 7 (Context). *A context at position i for a configuration $c = \langle t \mid \Delta \rangle$ is defined by a (possibly empty) sequence of operations, where $k = i$, or $k = j$ if there is an open path starting at position i and finishing at position j in c. Non-empty sequences (i.e. contexts) are defined inductively, there are three cases according to the first operation used.*

1. *Addition of agent by principal port: This operation replaces t_k in t by a list of new variables $z_1, \ldots, z_p \in \mathcal{N}(u)$, and adds to Δ the equation $t_k = \alpha(u)$, where α is any agent, and the terms in u are either new variables (in which case they can occur twice in $\alpha(u)$ or once in $\alpha(u)$ and once in z) or elements of t, in which case they are erased from t.*
   In this case the rest of the sequence is the concatenation of contexts at the positions of the variables $z_1, \ldots, z_p$ in the new head and at the new position of t_i if $i \neq k$.
2. *Addition of agent by auxiliary port: This operation replaces t_k in t by a list of new variables $z_1, \ldots, z_p$ occurring free in $y = \alpha(u)$ and adds this equation to Δ, where α is any agent, and the terms in u are either new variables (in which case they can occur twice in $\alpha(u)$ or once in $\alpha(u)$ and once in z) or elements of t, in which case they are erased from t. The term t_k must occur in u.*
   Also in this case the rest of the sequence is composed of contexts at the positions of the variables $z_1, \ldots, z_p$ in the new head and at the new position of t_i if $i \neq k$.
3. *Addition of a wire: erase t_k and another term t_p in t and add $t_k = t_p$ to Δ. In this case the rest of the sequence is empty if $i = k$, otherwise it is a context at the new position of t_i.*

We denote by $op_{i,j}(c)$ the result of applying an operation as above to the configuration c at position i, using the positions j in t. We denote by $C_i[c]$ the configuration resulting of applying the context C, defined by a sequence of operations as above, to the configuration c at position i, and by $C(c, i)$ a generic context for c at position i. We will also denote by $C_i[c']$ the result of applying to a configuration c' comparable with c the operations that define a context at position i for c.

The set $NP_C(c, i)$ of new observable positions of $C_i[c]$ is computed as follows: we start with the set $\{i\}$, and compute a new set each time we perform an operation. The first and second operations add the positions of the variables $z_1, \ldots, z_p$ in the new head, and if $i = k$ they erase i, otherwise they replace i by the new position of t_i in the head. The third operation simply erases the position i from the set if $i = k$, otherwise replaces i by the new position of t_i.

Graphically, the first two operations correspond to connecting an agent α to an observable port of the net (using the principal port of α in the first one, and an auxiliary port in the second one). The third operation corresponds to adding a wire connecting the observable ports k and p. Closings are particular cases of contexts defined by one operation of the first or third class.

Definition 8 (Contextual Preorder and Contextual Equivalence). *Let c, c' be comparable configurations with n elements in the head.*

$$c \leq c' \overset{\text{def}}{\iff} \forall i \in [1 \ldots n], (c, i) \leq (c', i)$$
$$(c, i) \leq (c', i) \overset{\text{def}}{\iff} \forall C(c, i), \forall p \in NP_C(c, i), C_i[c] \Downarrow_p v \in V_p \Rightarrow \exists v', (C_i[c'] \Downarrow_p v'$$
$$\text{and either } SVA_p(v, v') \text{ or } \neg Constr_p(v, v'))$$
$$(c, i) = (c', i) \overset{\text{def}}{\iff} (c, i) \leq (c', i) \text{ and } (c', i) \leq (c, i)$$
$$c = c' \overset{\text{def}}{\iff} \forall i \in [1 \ldots n], (c, i) = (c', i)$$

6 Main Result

We will show that the notions of contextual equivalence and bisimilarity coincide, if the interaction net system has "enough contexts" to extract the kernels of all values.

Definition 9. *A system of interaction is complete if for any $v \in V_i$ with visible agent α at position i, there exists a context C^α such that $\forall p \in NP_K(v, i), (K_i(v), p) \simeq (C_i^\alpha[v], p)$ and $p \in NP_{C^\alpha}(v, i)$.*

Theorem 2. *If the interaction net system is complete, $\leq$ (resp. $=$) coincides with $\precsim$ (resp. $\simeq$). Otherwise $\precsim$ (resp. $\simeq$) is included in $\leq$ (resp. $=$).*

Proof. To prove $\precsim \subseteq \leq$ it is sufficient to show that $\precsim$ is preserved by context. Following Howe [5] we prove that $\precsim$ is a precongruence (a preorder preserved by context) using an auxiliary relation $\precsim^*$, the *precongruence candidate*, defined as follows.

Let c, c' be comparable configurations with n elements in their heads.

$$c \precsim^* c' \overset{\text{def}}{\iff} \forall i \in [1 \ldots n], (c, i) \precsim^* (c', i)$$
$$(c, i) \precsim^* (c', i) \overset{\text{def}}{\iff} \text{either} \quad (c, i) \precsim (c', i),$$
$$\text{or} \quad c = op_{p,j}(d), i \in NP_{op}(d, p),$$
$$(d, q) \precsim^* (d', q), \forall q \in j \text{ and}$$
$$(op_{p,j}(d'), i) \precsim (c', i).$$

The precongruence candidate enjoys the following properties.

Property 2.   1. $\precsim \subseteq \precsim^*$.
  2. $\precsim^*$ is reflexive.
  3. $c \precsim^* c', c' \precsim c'' \Rightarrow c \precsim^* c''$
  4. $\precsim^*$ is preserved by context: $c \precsim^* c' \Rightarrow \forall i, \forall C(c, i), C_i[c] \precsim^* C_i[c']$.

To show that $\precsim$ is a precongruence it is sufficient to prove that it coincides with $\precsim^*$, for which it remains to prove $\precsim^* \subseteq \precsim$. This follows, by coinduction, from:

Proposition 2.   *1.* $v \in \mathcal{V}_i, (v, i) \precsim^* (c', i) \Rightarrow (v, i) \langle \precsim^* \rangle (c', i)$
   2. $c \precsim^* c', c \Downarrow_i v \in \mathcal{V}_i \Rightarrow v \precsim^* c'$

This concludes the proof of the first inclusion: $\precsim \subseteq \leq$. Now we prove $\leq \subseteq \precsim$ by coinduction, showing $\leq \subseteq \langle \leq \rangle$. Assume $(c, i) \leq (c', i)$. By definition of $\leq$, using an empty context, $c \Downarrow_i v \in \mathcal{V}_i \Rightarrow \exists v', \ c' \Downarrow_i v'$ and either $SVA_i(v, v')$ or $\neg Constr_i(v, v')$. In the latter case we are done, since closings are particular cases of contexts. In the first case, we know by completeness that $(\mathcal{K}_i(v), p) \simeq (C_i^\alpha[v], p), \forall p \in NP_\mathcal{K}(v, i)$. Moreover, since bisimilarity includes the equational theory (Theorem 1), and $(c, i) \leq (c', i)$: $(C_i^\alpha[v], p) \simeq (C_i^\alpha[c], p) \leq (C_i^\alpha[c'], p) \simeq (C_i^\alpha[v'], p)$. Again by completeness (since $SVA_i(v, v')$), $(C_i^\alpha[v'], p) \simeq (\mathcal{K}_i(v'), p)$. Since we have already proved $\simeq \subseteq =$, we get $(\mathcal{K}_i(v), p) \leq (\mathcal{K}_i(v'), p), \forall p \in NP_\mathcal{K}(v, i)$ as required. $\square$

7   Conclusion

In this paper we have presented a notion of bisimilarity for (untyped) interaction nets. This notion has been shown to coincide with the contextual equivalence, thus we have a simple proof technique for showing when two nets are equivalent in all contexts. One of the main applications that we see for this work are general correctness proofs for optimizations in interaction net implementations of various systems, such as the λ-calculus or term rewriting systems.

References

1. Samson Abramsky. The lazy λ-calculus. In David A. Turner, editor, *Research Topics in Functional Programming*, chapter 4, pages 65–117. Addison Wesley, 1990.
2. Maribel Fernández and Ian Mackie. Coinductive techniques for operational equivalence of interaction nets. In *Proceedings of the 13th Annual IEEE Symposium on Logic in Computer Science (LICS'98)*, pages 321–332. IEEE Computer Society Press, June 1998.
3. Maribel Fernández and Ian Mackie. A calculus for interaction nets. In *Proceedings of the first International Conference on Principles and Practice of Declarative Programming (PPDP'99)*, Lecture Notes in Computer Science, Springer-Verlag, September 1999.
4. Georges Gonthier, Martín Abadi, and Jean-Jacques Lévy. The geometry of optimal lambda reduction. In *Proceedings of the 19th ACM Symposium on Principles of Programming Languages (POPL'92)*, pages 15–26. ACM Press, January 1992.
5. Douglas J. Howe. Proving congruence of bisimulation in functional programming languages. *Information and Computation*, 124(2):103–112, 1996.
6. Lionel Khalil. Mémoire de DEA SPP, 1999. Available at http://www.dmi.ens.fr/~khalil.
7. Yves Lafont. Interaction nets. In *Proceedings of the 17th ACM Symposium on Principles of Programming Languages (POPL'90)*, pages 95–108. ACM Press, January 1990.
8. Yves Lafont. Interaction combinators. *Information and Computation*, 137(1):69–101, 1997.
9. Ian Mackie. YALE: Yet another lambda evaluator based on interaction nets. In *Proceedings of the 3rd ACM SIGPLAN International Conference on Functional Programming (ICFP'98)*, pages 117–128. ACM Press, September 1998.
10. Andrew M. Pitts. Operationally-based theories of program equivalence. In P. Dybjer and A. M. Pitts, editors, *Semantics and Logics of Computation*, Publications of the Newton Institute, pages 241–298. Cambridge University Press, 1997.

Run Statistics for Geometrically Distributed Random Variables (Extended Abstract)

Peter J. Grabner[1], Arnold Knopfmacher[2], and Helmut Prodinger[3] *

[1] Institut für Mathematik A
Technische Universität Graz
Steyrergasse 30
8010 Graz, Austria
grabner@weyl.math.tu-graz.ac.at

[2] The John Knopfmacher Centre for Applicable Analysis and Number Theory
Department of Computational and Applied Mathematics
University of the Witwatersrand, P. O. Wits
2050 Johannesburg, South Africa
arnoldk@gauss.cam.wits.ac.za,
WWW home page: http://www.wits.ac.za/science/number_theory/arnold.htm

[3] The John Knopfmacher Centre for Applicable Analysis and Number Theory
Department of Mathematics
University of the Witwatersrand, P. O. Wits
2050 Johannesburg, South Africa
helmut@gauss.cam.wits.ac.za,
WWW home page: http://www.wits.ac.za/helmut/index.htm

Abstract. For words of length n, generated by independent geometric random variables, we consider the mean and variance, and thereafter the distribution of the number of runs of equal letters in the words. In addition, we consider the mean length of a run as well as the length of the longest run over all words of length n.

1   Introduction

Let X denote a geometrically distributed random variable, i. e. $\mathbb{P}\{X = k\} = pq^{k-1}$ for $k \in \mathbb{N}$ and $q = 1 - p$. The combinatorics of n geometrically distributed independent random variables $X_1, \ldots, X_n$ has attracted recent interest, especially because of applications in computer science. We mention just two areas, the skip list [1,13,15,8] and probabilistic counting [3,6,7,9].

---

* The first named author is supported by the START-project Y96-MAT of the Austrian Science Foundation. Part of this work was done during his visit to the John Knopfmacher Centre for Applicable Analysis and Number Theory at the University of the Witwatersrand, Johannesburg, South Africa

G. Gonnet, D. Panario, and A. Viola (Eds.): LATIN 2000, LNCS 1776, pp. 457–462, 2000.

In [14] the number of left-to-right maxima was investigated for words $a_1 \ldots a_n$, where the letters a_i are independently generated according to the geometric distribution. In [10] the study of left-to-right maxima was continued, but now the parameters studied were the mean value and mean position of the r-th maximum.

In this article we study runs of consecutive equal letters in a string of n geometrically distributed independent random letters. For example in $w = 22211114431$ we have 5 runs of equal letters of respective lengths $3, 4, 2, 1, 1$. In the sequel we denote by $R_n(w)$ the number of runs in the word w, where w is of length n. Run statistics play a significant role in the behaviour of sorting algorithms, as explained at length in [12].

In section 2 we study the mean and variance of $R_n(w)$. Thereafter, in section 3 we study the distribution of the number of runs, which turns out to be Gaussian. Subsequently, in section 4 we study the average length of the runs per word. Finally, in section 5 we determine the mean and variance of the length of the longest run in a word of length n.

2   Moments of Number of Runs

In order to determine the mean and variance of the number of runs we will make use of the following decomposition of the set of all (non-empty) words. Here $\{\geq \mathbf{k}\}$ denotes the set $\{k, k+1, \ldots\}$; for a given set A we denote

$$A^+ = \bigcup_{k=1}^{\infty} A^k, \quad A^* = \varepsilon \cup A^+,$$

where ε stands for the empty word. We decompose the set of non-empty words according to runs of 1's, separated by words consisting of larger digits only

$$\{\geq \mathbf{1}\}^+ = (\varepsilon + \mathbf{1}^+)\left(\{\geq \mathbf{2}\}^+ \mathbf{1}^+\right)^* \{\geq \mathbf{2}\}^+(\varepsilon + \mathbf{1}^+) + \mathbf{1}^+ ; \tag{1}$$

here we find it more convenient to write $+$ instead of $\cup$.

We consider a probability generating function $F(z, u)$, where z labels the length of the word, and u counts the number of runs. We should always have $F(z, 1) = \frac{z}{1-z}$, and a replacement of z by qz, if we increase all letters by 1.

Then (1) translates into the functional equation

$$F(z, u) = \frac{F(qz, u)}{1 - F(qz, u)\dfrac{pzu}{1 - pz}}\left(\frac{pzu}{1 - pz} + 1\right)^2 + \frac{pzu}{1 - pz} . \tag{2}$$

Now we differentiate it w. r. t. u, plug in $u = 1$, set $G(z) = \frac{\partial}{\partial u}F(z, 1)$, and get

$$G(z) = G(qz)\frac{(1 - qz)^2}{(1 - z)^2} + \frac{pz(1 - pz)}{(1 - z)^2} .$$

Setting $H(z) = (1 - z)^2 G(z)$ yields

$$H(z) = H(qz) + pz(1 - pz) .$$

Comparing coefficients, we see that

$$[z]H(z) = 1 ,$$

$$[z^2]H(z) = -\frac{p^2}{1 - q^2} = -\frac{p}{1 + q} ,$$

and that the other coefficients are zero. Consequently,

$$H(z) = z - \frac{p}{1 + q} z^2 ,$$

and

$$G(z) = \frac{z - \frac{p}{1+q} z^2}{(1 - z)^2} .$$

This leads to

Proposition 1. *The mean value of the number of runs for $n \geq 1$ is given by*

$$\mu_n = \mathbb{E}R_n = [z^n]G(z) = \frac{2q}{1 + q} n + \frac{p}{1 + q}.$$

The computation of the variance is rather lengthier and requires that we differentiate (2) twice. This leads after some work to

Proposition 2. *The variance of the number of runs is given for $n \geq 2$ by*

$$\sigma_n^2 = \mathbb{V}R_n = \frac{2q(1 - q)^2(2 + q^2)}{(1 + q)^2(1 - q^3)} n - \frac{2q(1 - q)^2(3 - q + q^2)}{(1 + q)^2(1 - q^3)} .$$

3 Distribution of the Number of Runs

In this section we discuss a central limit theorem for the distribution of the number of runs. In order to derive this, we have to extract further information from the functional equation (2). We observe that the terms on the right-hand side are all simple rational functions, except for the terms containing $F(qz, u)$. By investigating the analytic properties of $F(z, u)$ it can be shown that $F(z, u)$ can be written as

$$F(z, u) = \frac{g(z, u)}{1 - f(u)z} + R(z, u), \tag{3}$$

where $g(z, u)$ and $R(z, u)$ are holomorphic in $|z| < 1 + \delta$, $|u - 1| < \delta$ for some $\delta > 0$. Now we are in the general framework of Hwang's quasi-power theorem (cf. [5]) and can deduce the following theorem.

Theorem 1. *The number of runs in words of length n produced by independent geometric random variables obeys a central limit law, more precisely*

$$\mathbb{P}\left(R_n(w) \leq \frac{2q}{1 + q} n + t\sqrt{\frac{2q(2 + q^2)}{1 - q^3} \frac{1 - q}{1 + q} \sqrt{n}} \right) = \Phi(t) + \mathcal{O}(n^{-\frac{1}{2}}). \tag{4}$$

4   Average Length of Runs

Given a string w of geometric random variables of length n with $R_n(w) = k$ runs we define the average length of a run to be $L_n(w) = \frac{n}{R_n(w)}$. It is of interest to determine the moments and the distribution of this parameter over all strings of length n. Intuitively, one expects that the mean length of a run should be close to n divided by the mean number of runs, which is

$$\frac{n}{\frac{2q}{1+q}n + \frac{p}{1+q}} = \frac{1+q}{2q} - \frac{1-q^2}{4q^2}\frac{1}{n} + \mathcal{O}\left(\frac{1}{n^2}\right).$$

In fact we obtain

Proposition 3. *For $n \geq 1$ the mean and variance of $L_n(w)$ are given respectively by*

$$\frac{1+q}{2q} + \mathcal{O}\left(\frac{1}{n}\right), \quad \frac{(1-q^2)^2(2+q^2)}{8q^3(1-q^3)}\frac{1}{n} + \mathcal{O}\left(\frac{1}{n^2}\right).$$

Moreover, $L_n(w)$ obeys a central limit theorem:

$$\mathbb{P}\left(L_n(w) - \frac{1+q}{2q} \leq \frac{(1-q^2)\sqrt{2+q^2}}{\sqrt{8q^3(1-q^3)}}\frac{t}{\sqrt{n}}\right) = \Phi(t) + \mathcal{O}(n^{-\frac{1}{2}}).$$

The proof makes use of the distribution obtained for the number of runs in Theorem 1.

5   Longest Runs

In this section we study the mean of the longest run $M_n(w)$ of equal digits in a string of length n. For this purpose we introduce the probability generating function $G_h(z)$ of all strings that have runs only of length less than h. Similar arguments as in the proof of (2) show that G_h satisfies

$$G_h(z) = \left(\frac{1-(pz)^h}{1-pz}\right)^2 \frac{G_h(qz)}{1 - G_h(qz)\frac{pz}{1-pz}(1-(pz)^{h-1})} + pz\frac{1-(pz)^{h-1}}{1-pz}. \quad (5)$$

In order to extract the asymptotic behaviour of the probability that a string of length n has runs of length at most h, we have to find the singularities of $G_h(z)$. Using bootstrapping we estimate ρ_h, the dominant singularity of the function G_h.

Combining this with estimates for G_h leads to

$$\mathbb{P}\left(M_n(w) < h\right) = (1 - pq^h)^n + \mathcal{O}(hq^h). \quad (6)$$

Using (6) and Abel summation we then find that the first and second moment of the longest run are given by

$$\mathbb{E}M_n(w) = \sum_{h \geq 1} \left(1 - \mathbb{P}\left(M_n(w) < h\right)\right) = \sum_{h \geq 1} \left(1 - (1 - pq^h)^n\right) + \mathcal{O}(1),$$

$$\mathbb{E}M_n(w)^2 = 2 \sum_{h \geq 1} h\left(1 - \mathbb{P}\left(M_n(w) < h\right)\right) - \mathbb{E}M_n(w) \tag{7}$$

$$= \sum_{h \geq 1} (2h - 1)\left(1 - (1 - pq^h)^n\right) + \mathcal{O}(1).$$

In order to compute the asymptotic behaviour of these two moments, we use the now classical exponential approximation technique (cf. [12]). Thereafter we make use of the Mellin transform and Mellin inversion formula to obtain finally

Proposition 4. *The mean value of the length of the longest run $M_n(w)$ in a string of n geometric random variables satisfies*

$$\mathbb{E}M_n(w) = \log_{\frac{1}{q}} n + \mathcal{O}(1).$$

Similarly, we could obtain an expression for the second moment

$$\mathbb{E}M_n(w)^2 + \mathcal{O}(1) = \log_{\frac{1}{q}}^2 n + \mathcal{O}(\log n). \tag{8}$$

References

1. L. Devroye. A limit theory for random skip lists. *Advances in Applied Probability,* 2:597–609, 1992.
2. P. Flajolet, X. Gourdon, and P. Dumas. Mellin transforms and asymptotics: Harmonic sums. *Theoretical Computer Science,* 144:3–58, 1995.
3. P. Flajolet and G. N. Martin. Probabilistic counting algorithms for data base applications. *Journal of Computer and System Sciences,* 31:182–209, 1985.
4. L. Guibas and A. Odlyzko. Long repetitive patterns in random sequences. *Zeitschrift für Wahrscheinlichkeitstheorie,* 53:241–262, 1980.
5. H.-K. Hwang. On convergence rates in the central limit theorems for combinatorial structures. *European Journal of Combinatorics,* 19:329–343, 1998.
6. P. Kirschenhofer and H. Prodinger. On the analysis of probabilistic counting. In E. Hlawka and R. F. Tichy, editors, *Number–theoretic Analysis,* volume 1452 of *Lecture Notes in Mathematics,* pages 117–120, 1990.
7. P. Kirschenhofer and H. Prodinger. A result in order statistics related to probabilistic counting. *Computing,* 51:15–27, 1993.
8. P. Kirschenhofer and H. Prodinger. The path length of random skip lists. *Acta Informatica,* 31:775–792, 1994.
9. P. Kirschenhofer, H. Prodinger, and W. Szpankowski. Analysis of a splitting process arising in probabilistic counting and other related algorithms. *Random Structures and Algorithms,* 9:379–401, 1996.
10. A. Knopfmacher and H. Prodinger. Combinatorics of geometrically distributed random variables: Value and position of the rth left-to-right maximum. *Discrete Mathematics, to appear.*

11. D. E. Knuth. The average time for carry propagation. *Indagationes Mathematicae*, 40:238–242, 1978.
12. D. E. Knuth. *The Art of Computer Programming*, volume 3: Sorting and Searching. Addison-Wesley, 1973. Second edition, 1998.
13. T. Papadakis, I. Munro, and P. Poblete. Average search and update costs in skip lists. *BIT*, 32:316–332, 1992.
14. H. Prodinger. Combinatorics of geometrically distributed random variables: Left-to-right maxima. *Discrete Mathematics*, 153:253–270, 1996.
15. W. Pugh. Skip lists: a probabilistic alternative to balanced trees. *Communications of the ACM*, 33:668–676, 1990.

Generalized Covariances of Multi-dimensional Brownian Excursion Local Times

Guy Louchard

Université Libre de Bruxelles, Département d'Informatique, CP 212, Boulevard du Triomphe, B-1050 Bruxelles, Belgium
Email: louchard@ulb.ac.be

Abstract. Expressions for the generalized covariances of multi-dimensional Brownian excursion local times are derived from corresponding densities transforms. Typical applications are moments of the cost of structures such as $M/G/1$ queue, Random trees, Markov stack or priority queue in Knuth's model. Brownian excursion area and a result of Biane and Yor are also revisited.

1 Introduction

Throughout this paper, the standard Brownian motion (BM) will be denoted by $x(t)$.

Fix $t > 0$ and denote the last zero of x before t and the first zero of x after t by

$$G(t) := \sup\{s : x \le t; x(s) = 0\}$$

and

$$D(t) := \inf\{s : s \ge t; x(s) = 0\}.$$

The processes restricted to $[G(t), t]$ and $[G(t), D(t)]$ are called the meandering process ending at t : $Z(u) := x^+(G(t) + u), 0 \le u \le L^-(t) := t - G(t)$ and the excursion process straddling t : $Y(u) := x^+(G(t) + u), 0 \le u \le L(t) := D(t) - G(t)$, respectively. The standard scale excursion (BE) is $X(u) := [Y(u)|L = 1]$; note that $Y(u) \stackrel{d}{=} \sqrt{\ell}X(u/\ell)$ when $L = \ell$. The distributions of G and L are well known: see Chung [2, Theorem 1].

The local time of $x(t)$ at a, denoted by

$$t^+(t, a) = \lim_{\epsilon \to 0} \frac{1}{\epsilon} \int_0^t I_{[a, a+\epsilon]}(x(t))dt,$$

and the local time of the standard scaled excursion X at a, denoted by $\tau^+(a)$, have been studied by several authors (note that for an excursion of length ℓ we have: $\tau^+(\ell, a) \stackrel{d}{=} \sqrt{\ell}\tau^+(a/\sqrt{\ell})$. See for instance Getoor and Sharpe [9], Knight [13], Cohen and Hooghiemstra [3], Hooghiemstra [12], Drmota and Gittenberger [4], Louchard [14], Gittenberger and Louchard [11]. Intuitively, the local time at a is the total time spent by the excursion in the neighbourhood of a.

G. Gonnet, D. Panario, and A. Viola (Eds.): LATIN 2000, LNCS 1776, pp. 463–472, 2000.

Applications of the BE are numerous: we will mention a few of them, emphasizing the meaning of the local time. For instance, consider a $M/G/1$ queuing system. There the customers arrive according to Poisson process $(\pi_t, t \geq 0)$ with rate α^{-1} where $\alpha > 0$. Denote the arriving time of the n-th customer by t_n and the service time by s_n which is assumed to be independent of the arrival process π_t. Then the actual waiting time process is defined by $w_1 := 0, w_{n+1} := \max\{0, w_n + s_n - (t_{n+1} - t_n)\}$ and the virtual waiting time process by

$$v_t := \max\{0, w_{\pi_t} + s_{\pi_t} - (t - t_{\pi_t})\}, t \geq 0.$$

Furthermore, denote the length of the first busy period by ℓ. Then Cohen and Hooghiemstra [3] have shown that for arbitrary $\delta > 0$ the following limit theorem holds:

$$\left((\frac{v_{su}}{\sqrt{2\alpha s}} \mid s < \ell \leq s + \delta), 0 \leq u \leq 1 \right) \xrightarrow{d} X(u), \quad s \to \infty.$$

In this context the BE local time process appears as the weak limit of the (suitably normalized) number of downcrossings of the virtual waiting time process, i.e.

$$d(v) = \#\{t : 0 \leq t \leq \ell, v_t = v\};$$

($\#A$ denotes the cardinality of A) conditioned on the number of customers served during the first busy period (see [3, Sec. 7]). Another BE application is the number of nodes at some level in a random tree. Consider a simply generated random tree (according to the notion of Meir and Moon [19] or, equivalently, the family tree of a Galton-Watson branching process conditioned on the total progeny. Then BE appears as the weak limit of the contour process of this tree, i.e. the process constructed of the distances of the nodes from the root when traversing the tree (for details see Gittenberger [10]). The local time corresponds here to the number of nodes at some level. The generation sizes of the branching processes converge weakly to BE local time. The external path length (EPL) of a random tree is given by the sum of distances from the root to the leafs.

Dynamical algorithms are also related the BE. The Stack structure of length $2n$ (see Flajolet [6] p. 126) is asymptotically equivalent to a BE (Louchard [15]). The priority queue in Knuth's model is combinatorially equivalent to a Markov Stack (see Louchard et al [17]). So the distribution of the size of this structure is asymptotically related, after suitable normalization, to the BE local time. The local time corresponds to the time spent by the structure at some level.

The cost G of structures such as $M/G/1$ queue busy period, Random tree, Markov stack or priority queue in Knuth's model is asymptotically given, for any cost function $g(\cdot)$, by $G = \int_0^1 g[X(u)]du$. For stacks and priority queue, the cost is related to the size. For the $M/G/1$ queue, the cost is related to the waiting time. For EPL, the cost is related to the distance to the leafs.

Moments of G are immediately related to the local time: we have

$$E[G^d] = d! \int_0^\infty dx_1 \int_{x_1}^\infty dx_2 \cdots \int_{x_{d-1}}^\infty dx_d g(x_1) g(x_2) \cdots g(x_d) \cdots K(x_1, x_2, \cdots x_d)$$

with $K(x_1, x_2, \cdots x_d) := E[\tau^+(x_1)\tau^+(x_2) \cdots \tau^+(x_d)]$ denoting the generalized co-variances. In this paper we obtain explicit expressions for $K(x_1, x_2 \cdots x_d)$. We revisit also two classical examples: the BE area ($g(x) = x$) related to the Airy distribution (which has a lot of applications in combinatorics and data structures) and a result of Biane and Yor [1] related to $g(x) = 1/x$.

The paper is organized as follows. Sec. 2 gives the basic formula's we need in the sequel, Sec. 3 provides an efficient algorithm for the generalized covariances computation. In Sec. 4, we consider two typical applications: the Brownian excursion area and the Biane and Yor formula.

2   Basic Formulas

In this section, we start from known results to derive expressions for the first generalized covariances $K(x_1, x_2, \cdots, x_d), d = 1 \cdots 4$.

In [11] we obtained the following result depending on some Laplace transforms.

$$E[e^{-\sum_1^d \beta_i \tau^+(x_i)}, 1 > m_{x_d}] = \frac{1}{\sqrt{2\pi i}} \int_S e^\alpha \Theta(d) d\alpha, x_d > x_{d-1} \cdots > x_1$$

where $S := [a - i\infty, a + i\infty], a > 0, m_x := \inf\{s : X(s) = x\}$,

$$\Theta(d) = \frac{\alpha^d}{2[F_1(d)]^2[\beta_d + C_1(d) + C_2(d)D_2(d)/F_1(d)]} \tag{1}$$

with some functions depending only on α and x.:

$$C_1(d) = \sqrt{\frac{\alpha}{2}} E(d, d-1)/Sh(d, d-1)$$

$$C_2(d) = -\frac{\alpha}{2Sh^2(d, d-1)}$$

$$C_3(d) = \sqrt{\frac{\alpha}{2}} \frac{Sh(d, d-2)}{Sh(d, d-1)Sh(d-1, d-2)}$$

$$C_4(d) = C_2(d-1)$$

$$C_5(d) = \sqrt{2}Sh(d, d-1)$$

and some functions depending also on β.:

$$F_1(d) = \beta_{d-1}D_2(d) + D_1(d)$$

$$D_2(d) = \beta_{d-2}D_4(d) + D_3(d)$$

$$\frac{D_1(d)}{D_2(d)} = C_3(d) + C_4(d)D_4(d)/D_2(d)$$

$$D_3(d) = C_5(d)D_1(d-1)$$

$$D_4(d) = C_5(d)D_2(d-1)$$

and

$$E(\ell,m) := e^{\sqrt{\cdot}[x_\ell - x_m]},$$
$$Sh(\ell,m) := \sinh[\sqrt{\cdot}(x_\ell - x_m)], Sh(\ell) := \sinh(\sqrt{\cdot}x_\ell)$$
$$\sqrt{\cdot} := \sqrt{2\alpha}.$$

Initialisations are given by

$$D_1(2) = \sqrt{\alpha}Sh(2)/\sqrt{2}$$
$$D_2(2) = Sh(1)Sh(2,1)$$

¿From (1), it is possible (with MAPLE) to derive explicit expressions for successive derivaties of $\Theta(d)$. For instance, with $E(i) := e^{\sqrt{\cdot}x_i}$,

$$\bar{K}(\alpha, x_1) = \left|\frac{\partial\Theta}{\partial\beta_1}\right|_{\beta_1=0} = 2E(1)^{-2}$$

$$\bar{K}(\alpha, x_1, x_2) = \left|\frac{\partial^2\Theta}{\partial\beta_2\partial\beta_1}\right|_{\beta_1,\beta_2=0} = \frac{-2\sqrt{2}}{\sqrt{\alpha}}E(2)^{-2}(E(1)^{-2} - 1) \qquad (2)$$

$$\bar{K}(\alpha, x_1, x_2, x_3) = \left|\frac{\partial^3\Theta}{\partial\beta_3\partial\beta_2\partial\beta_1}\right|_{\beta_1,\beta_2,\beta_3=0} =$$
$$2\frac{(3E(1)^{-2}E(2)^{-2} - E(2)^{-2} - 2E(1)^{-2})(E(1)^{-2} - 1)E(3)^{-2}}{E(1)^{-2}\alpha} \qquad (3)$$

$$\bar{K}(\alpha, x_1, x_2, x_3, x_4) = \left|\frac{\partial\Theta}{\partial\beta_4\partial\beta_3\partial\beta_2\partial\beta_1}\right|_{\beta_1,\beta_2,\beta_3,\beta_4=0} =$$
$$-\frac{4}{\sqrt{2}}E(4)^{-2}(E(1)^{-2} - 1) \cdot$$
$$\cdot(6E(1)^{-2}E(2)^{-4}E(3)^{-2} - 6E(1)^{-2}E(3)^{-2}E(2)^{-2} + 2E(1)^{-2}E(2)^{-2} + E(2)^{-4}$$
$$+E(1)^{-2}E(3)^{-2} - 3E(1)^{-2}E(2)^{-4} + 2E(3)^{-2}E(2)^{-2} - 3E(2)^{-4}E(3)^{-2})/$$
$$(E(2)^{-2}E(1)^{-2}\alpha^{3/2})$$

High order derivatives become difficult to compute, even with MAPLE. So another technique is obviously needed.

3   An Efficient Algorithm for Generalized Covariances Computation

In this section, we first derive a recurrence equation for some functions arising in the generalized covariances. This leads to some differential equations for related exponential generating functions. A simple matrix representation is finally obtained and it remains to invert the Laplace transforms.

3.1   A Recurrence Equation

Let us first differentiate Θ w.r.t. β_d (each time, after differentiation w.r.t. β_i, we set $\beta_i = 0$). This gives

$$\frac{-\alpha^d}{2[C_1 F_1 + C_2 D_2]^2}$$

We should write $C_1(d)$, etc. but we drop the $d-$ dependency to ease notations. Differentiating now w.r.t. β_{d-1}, this leads to

$$\frac{\alpha^d C_1 D_2}{[C_1 D_1 + C_2 D_2]^3} = \frac{\alpha^d C_1 [\beta_{d-2} D_4 + D_3]}{[C_7 \beta_{d-2} D_4 + C_7 D_3 + C_1 C_4 D_4]^3}$$

with $C_7(d) := C_1(d)C_3(d) + C_2(d) = C_1(d)C_1(d-1)$ after detailed computation. It is clear that the next differentiations will lead to some pattern. Indeed, set

$$H(d,i) := \frac{\partial^{d-2}}{\partial \beta_{d-2} \cdots \partial \beta_1} \frac{D_2(d)^i}{[C_1(d)D_1(d) + C_2(d)D_2(d)]^{i+2}} \Bigg|_{\beta_{d-2} \cdots \beta_1 = 0} \tag{4}$$

obviously

$$\left. \frac{\partial^d \Theta}{\partial \beta_d \cdots \partial \beta_1} \right|_{\beta_d \cdots \beta_1 = 0} = C_1(d)\alpha^d(-1)^d H(d,1) \tag{5}$$

Expanding (4), we derive (omitting the details)

$$H(d,i) = \frac{1}{C_1(d)^{i+2}C_1(d-1)^{i-1}C_5(d)^2} \cdot$$
$$\cdot \sum_{j=0}^{i-1} \binom{i-1}{j} (-1)^{i-1-j} C_2(d-1)^{i-1-j}[-2H(d-1,i-j)$$
$$+ (i+2)C_2(d-1)H(d-1,i-j+1)] \tag{6}$$

(6) is still too complicated. So we set first $H_1(d,i) := H(d,i)C_2(d)^{i-1}$. This leads to

$$H_1(d,i) = \frac{C_2(d)^{i-1}}{C_1(d)^{i+2}C_1(d-1)^{i-1}C_5(d)^2} \cdot$$
$$\cdot \sum_{j=0}^{i-1} \binom{i-1}{j} (-1)^{i-1-j}[-2H_1(d-1,i-j) + (i+2)H_1(d-1,i-j+1)]$$

But we remark that $\frac{C_2(d)}{C_1(d)C_1(d-1)} = -\frac{C_6(d-1)}{C_6(d)}$ with

$$C_6(d) := E(d)^2 - E(d-1)^2 \tag{7}$$

Then, we set $H_2(d,i) := H_1(d,i)C_6(d)^{i-1}$ and we obtain

$$H_2(d,i) = \frac{1}{C_1(d)^3 C_5(d)^2} \cdot$$

$$\cdot \sum_{j=0}^{i-1} \binom{i-1}{j} (-1)^j C_6(d-1)^j [-2H_2(d-1, i-j)$$

$$+ \frac{(i+2)}{C_6(d-1)} H_2(d-1, i-j+1)] \tag{8}$$

3.2 Some Generating Function

Eq. (8) is a perfect candidate for an exponential generating function (see Flajolet and Sedgewick [5]). We set

$$\varphi_2(d, v) := \sum_{1}^{\infty} \frac{H_2(d, i)v^{i-1}}{(i-1)!}$$

(8) leads to

$$\varphi_2(d, v) = \frac{1}{C_1(d)^3 C_5(d)^2} \cdot$$

$$\cdot \left\{ \left[-2\varphi_2(d-1, v) + \frac{1}{C_6(d-1)} \frac{\partial^2}{\partial v^2} [\varphi_2(d-1, v) \cdot v] \right] e^{-vC_6(d-1)} \right.$$

$$\left. + \frac{1}{C_6(d-1)} \partial_v \varphi_2(d-1, v) \cdot \partial_v [e^{-vC_6(d-1)} \cdot v] \right\}$$

With (7), we are led to set

$$\varphi_3(d, v) := \varphi_2(d, v)e^{vE(d-1)^2} \text{ and } H(d, 1) = \varphi_3(d, 0)$$

Before establishing the corresponding equation for φ_3, it is now time to find the effect of all our transforms on φ_2. Indeed, $H(2, i) = \gamma \delta_3^{i-1}$ (see (4) with

$$\delta_1 = D_2(2), \delta_2 = C_1(2)D_1(2) + C_2(2)D_2(2)$$

$$\gamma = \frac{\delta_1}{\delta_2^3} = \frac{-2E(2)^{-2}(E(1)^{-2} - E(2)^{-2})(E(1)^{-2} - 1)}{E(1)^{-2}\alpha^3}$$

$$\delta_3 = \frac{\delta_1}{\delta_2}$$

So

$$H_1(2, i) = \gamma \delta_4^{i-1}, \text{ with } \delta_4 = \delta_3 C_2(2)$$
$$H_2(2, i) = \gamma \delta_5^{i-1}, \text{ with } \delta_5 = \delta_4 C_6(2)$$
$$\varphi_2(2, v) = \gamma e^{v\delta_5}, \varphi_3(2, v) = \gamma e^{v\delta_6}, \text{ with } \delta_6 = \delta_5 + E(1)^2 = 1,$$

after all computations.

We see that it is convenient to finally set

$$\varphi_3(d, v) := \varphi_4(d, v)e^v, \varphi_4(2, v) = \gamma, H(d, 1) = \varphi_4(d, 0)$$

The differential equation for φ_4 is computed as follows (we omit the details)

$$\varphi_4(d,v) = \frac{2(E(d-1)^{-2} - E(d)^{-2})E(d)^{-2}}{E(d-1)^{-4}(E(d-2)^{-2} - E(d-1)^{-2})\sqrt{\cdot}^3}.$$

$$\left[\mu_1 v \frac{\partial^2}{\partial v^2}\varphi_4(d-1,v) + (\mu_2 + \mu_3 v)\frac{\partial}{\partial v}\varphi_4(d-1,v)\right.$$

$$\left. + (\mu_4 + \mu_5 v)\varphi_4(d-1,v)\right] \tag{9}$$

with

$$\mu_1 := E(d-2)^{-2}E(d-1)^{-2}$$
$$\mu_2 := 3E(d-2)^{-2}E(d-1)^{-2}$$
$$\mu_3 := 2E(d-2)^{-2}E(d-1)^{-2} - E(d-2)^{-2} - E(d-1)^{-2}$$
$$\mu_4 := -2E(d-2)^{-2} - E(d-1)^{-2} + 3E(d-2)^{-2}E(d-1)^{-2}$$
$$\mu_5 := E(d-2)^{-2}E(d-1)^{-2} - E(d-1)^{-2} - E(d-2)^{-2} + 1.$$

3.3   A Matrix Representation

It is now clear that $\varphi_4(d,v)$ is made of 2 parts: the first one is given by the product of γ with all coefficients in front of (9).

The other part is given, for each d, by a polynomial in v, the coefficients of which are given by the following algorithm.

Start with $vec_2[0] = 1, vec_2[i] = 0, i \geq 1$.

Construct a tri-diagonal band matrix A_d as follows: if we apply the differential operator of (9) i.e. $[\mu_1 v \frac{\partial^2}{\partial v^2} + (\mu_2 + \mu_3 v)\frac{\partial}{\partial v} + (\mu_4 + \mu_5)v]$ to a polynomial $\sum_0 a_i v^i$, we see that the new polynomial $\sum_0 \bar{a}_i v^i$ is given by

$$\bar{a}_0 = A_d[0,1]a_1 + A_d[0,0]a_0$$
$$\bar{a}_i = A_d[i,i+1]a_{i+1} + A_d[i,i]a_i + A_d[i,i-1]a_{i-1}, i \geq 1$$

with

$$A_d[i,i+1] := [i(i+1) + 3(i+1)]E(d-1)^{-2}E(d-2)^{-2}$$
$$A_d[i,i] := [(2i+3)E(d-1)^{-2}E(d-2)^{-2} - (i+2)E(d-2)^{-2}$$
$$-(i+1)E(d-1)^{-2}]$$
$$A_d[i,i-1] := [E(d-1)^{-2}E(d-2)^{-2} - E(d-1)^{-2} - E(d-2)^{-2} + 1] \tag{10}$$

All other elements of A_d are set to 0.

Successive applications of A_ℓ to vec_2 give the coefficients of the polynomial part of $\varphi_4(d,v)$:

$$vec_d := \prod_{\ell=3}^{d} A_\ell \, vec_2 \tag{11}$$

Now, by (5), $|\frac{\partial^d\Theta}{\partial\beta_d\cdots\partial\beta_1}|_{\beta_d\cdots\beta_1=0} = C_1(d)\alpha^d(-1)^d\varphi_4(d,0) = C_8(d)\ vec_d[0]$, where $C_8(d)$, after all simplification, is given, with (9) by

$$C_8(d) = \frac{-4(-1)^d}{\sqrt{\cdot}^{d-1}}\frac{E(d)^{-2}(E(1)^{-2}-1)}{E(1)^{-2}\cdots E(d-2)^{-2}}, d \geq 3 \tag{12}$$

Let us summarize our results in the following theorem

Theorem 1. $\bar{K}(\alpha, x_1, x_2 \cdots x_d) = |\frac{\partial^d\Theta}{\partial\beta_d\cdots\partial\beta_1}|_{\beta_d\cdots\beta_1=0} = C_8(d)\ vec_d[0]$ where $C_8(d)$ is given by (12), $vec_d = \prod_{\ell=3}^{d} A_\ell\ vec_\ell$, with $vec_2[0] = 1$, $vec_2[i] = 0, i \geq 1$ and the band matrix A_ℓ is given by (10)

The computation of our covariances is now trivial.

3.4   Inverting the Laplace Transforms

It is well known that $\mathcal{L}_\alpha[f(u)] = e^{-\sqrt{\cdot}a}$, with $f(u) = \frac{e^{-a^2/2u}a}{\sqrt{2\pi}u^{3/2}}$. Also $\mathcal{L}_\alpha[g(u)] = \frac{e^{-\sqrt{\cdot}a}\sqrt{2}}{2\sqrt{\alpha}}$, with $g(u) = \frac{e^{-a^2/2u}}{\sqrt{2\pi u}}$. Hence, from (2),

$$E[\tau^+(x_1)\tau^+(x_2)] = -4[e^{-2(x_1+x_2)^2} - e^{-2x_2^2}], x_2 \geq x_1 \tag{13}$$

We recover immediately Cohen and Hooghiemstra [3], (6.15)
  Similarly, from (3),

$$E[\tau^+(x_1)\tau^+(x_2)\tau^+(x_3)]$$
$$= 4\int_0^1 \left\{ 3[e^{-2[x_1+x_2+x_3]^2/t}(x_1+x_2+x_3) - e^{-2[x_2+x_3]^2/t}(x_2+x_3)] \right.$$
$$-[e^{-2[x_2+x_3]^2/t}(x_2+x_3) - e^{-2[x_3+x_2-x_1]^2}(x_2+x_3-x_1)]$$
$$\left. -2[e^{-2[x_1+x_3]^2/t}(x_1+x_3) - e^{-2[x_3]^2/t}(x_3)] \right\}\frac{dt}{t^{3/2}} \tag{14}$$

Next covariances lead to similar expressions, with multiple integrals on t.

4   Some Applications

In this section, we apply the generalized covariances to two classical problems: the Brownian Excursion Area and a result of Biane and Yor related to the cost function $g(u) = 1/x$. We can proceed either from the Laplace transforms or from the explicit covariances.

4.1   Brownian Excursion Area

In [14], [16] we proved that $W_d := E[\int_0^1 X(u)du]^d$ satisfies the following recurrence equation. Let $\gamma_k := (36\sqrt{2})^k W_k \Gamma(\frac{2k-1}{2})/2\sqrt{\pi}$.

Then

$$\gamma_n = \frac{12n}{6n-1}\varphi_n - \sum_{k=1}^{n-1} \varphi_k \binom{n}{k} \gamma_{n-k} \qquad (15)$$

with $\varphi_k := \Gamma(3k + \frac{1}{2})/\Gamma(k + \frac{1}{2})$.

The corresponding distribution, also called the Airy distribution has been the object of recent renewed interest (see Spencer [20], Flajolet et al [8], where many examples are given).

(15) leads to $W_1 = \sqrt{2\pi}/4, W_2 = 5/12, W_3 = \sqrt{2\pi}15/128...$ From (2) we compute the Laplace transforms

$$2\int_0^\infty x_1 dx_1 \int_{x_1}^\infty x_2 dx_2 \frac{-2\sqrt{2}}{\sqrt{\alpha}} E(2)^{-2}[E(1)^{-2} - 1] = \frac{5\sqrt{2}}{32\alpha^{5/2}}$$

Inverting, this leads to $5/12$ as expected.

Similarly with $G(\alpha)$ given by (3), $3!\int_0^\infty x_1 dx_1 \int_{x_1}^\infty x_2 dx_2 \int_{x_2}^\infty x_3 dx_3 G(\alpha) = \frac{6.15}{128\alpha^4}$. Inverting, this leads to $\sqrt{2\pi}15/128$ as expected.

An interesting question is how to derive the recurrence (15) from the matrix representation given by Theorem 1.

4.2   A Formula of Biane and Yor

In [1], Biane and Yor proved that

$$Y := \int_0^1 \frac{dt}{X(t)} \stackrel{\mathcal{D}}{\equiv} 2\xi, \text{ where } \xi := \sup_{[0,1]} X(t)$$

With our techniques, we prove in the full report [18] that all moments of both sides are equal.

5   Conclusion

We have constructed a simple and efficient algorithm to compute the generalized covariances $K(x_1, x_2, \cdots x_d)$. Another challenge would be to derive the cross-moments of any order:

$$K(x_1, i_1, x_2, i_2 \cdots x_d, i_d) = E[\tau^+(x_1)^{i_1}\tau^+(x_2)^{i_2}\cdots\tau^+(x_d)^{i_d}]$$

It appears that the first two derivatives

$$\partial_{\beta_d}^{i_d}\partial_{\beta_{d-1}}^{i_{d-1}}\Theta|_{\beta_d=\beta_{d-1}=0}$$

lead to a linear combination of terms of type

$$D_2(d)^\ell D_4(d)^r/(C_1(d)D_1(d) + C_2(d)D_2(d))^m$$

The next derivative $\partial_{\beta_{d-2}}^{i_{d-2}}\Theta$, after all simplifications (and setting $\beta_{d-2} = 0$) lead to terms of type

$$H(d-1, s, t) := \frac{D_2(k-1)^s}{[C_1(d-1)D_1(d-1) + C_2(d-1)D_2(d-1)]^t}$$

and this pattern appears in all successive derivatives.

So, we can, in principle, construct (complicated) recurrence equations for $H(\cdot, s, t)$ and recover our K by linear combinations. This is quide tedious and up to now, we couldn't obtain such a simple generating function as in Sec. 3.2 .

References

1. Biane, Ph.,Yor, M.: Valeurs principales associées aux temps locaux Browniens. Bull. Sc. math. 2e, **111** (1987) 23–101.
2. Chung, K.L.: Excursions in Brownian motion. Ark. Mat. **14** (1976) 155–177.
3. Cohen, J.W., Hooghiemstra, G.: Brownian excursion, the $M/M/1$ queue and their occupation times. Math. Operat. Res. **6** (1981) 608–629.
4. Drmota, M., Gittenberger, B.: On the profile of random trees. Rand. Str. Alg. **10** (1997) 421–451.
5. Flajolet, Ph., Sedgewick, R.: An Introduction to the Analysis of Algorithms. Addison-Wesley, U.S.A.(1996)
6. Flajolet, Ph. Analyse d'algorithmes de manipulation d'arbres et de fichiers. Université Pierre-et-Marie-Curie, Paris (1981)
7. Flajolet, Ph., Odlyzko, A.: Singularity analysis of generating functions. Siam J. Disc. Math. **3**, 2 (1990) 216–240.
8. FljoletT, Ph., Poblete, P., Viola, A.: On the analysis of linear probing hashing. Algorithmica **22** (1998) 490–515.
9. Getoor, R.K., Sharpe, M.J.: Excursions of Brownian motion and Bessel processes. Z. Wahrscheinlichkeitsth. **47** (1979) 83–106.
10. Gittenberger, B.: On the countour or random trees. SIAM J. Discr. Math., to appear.
11. Gittenberger, B., Louchard, G.: The Brownian excursion multi-dimensional local time density. To appear in JAP. (1998)
12. Hooghiemstra, G.: On the explicit form of the density of Brownian excursion local time. Proc. Amer. Math. Soc. **84** (1982) 127–130.
13. Knight, F.B.: On the excursion process of Brownian motion. Zbl. Math. **426**,abstract 60073, (1980).
14. Louchard, G.: Kac's formula, Levy's local time and Brownian Excursion. J. Appl. Prob. **21** (1984) 479–499.
15. Louchard, G.: Brownian motion and algorithm complexity. BIT **26** (1986) 17–34.
16. Louchard, G.: The Brownian excursion area: a numerical analysis. Comp & Maths. with Appls. **10**, 6, (1984) 413–417.
17. Louchard, G., Randrianarimanana, B., Schott, R.: Dynamic algorithms in D.E. Knuth's model: a probabilistic analysis. Theoret. Comp. Sci. **93** (1992) 201–225.
18. Louchard, G.: Generalized covariances of multi-dimensional Brownian excursion local times. **TR 396**, Département d'Informatique (1999)
19. Meir, A. and Moon, J.W.: On the altitude of nodes in random trees. Can. J. Math **30** (1978) 997–1015.
20. Spencer, J.: Enumerating graphs and Brownian motion. Comm. Pure and Appl. Math. **Vol. L** (1997) 0291–0294.

Combinatorics of Geometrically Distributed Random Variables: Length of Ascending Runs

Helmut Prodinger [*]

The John Knopfmacher Centre for Applicable Analysis and Number Theory
Department of Mathematics
University of the Witwatersrand, P. O. Wits
2050 Johannesburg, South Africa
helmut@gauss.cam.wits.ac.za,
WWW home page: http://www.wits.ac.za/helmut/index.htm

Abstract. For n independently distributed geometric random variables we consider the average length of the m–th run, for fixed m and $n \to \infty$. One particular result is that this parameter approaches $1 + q$. In the limiting case $q \to 1$ we thus rederive known results about runs in permutations.

1 Introduction

Knuth in [6] has considered the average length L_k of the kth ascending run in random permutations of n elements (for simplicity, mostly the instance $n \to \infty$ was discussed).

This parameter has an important impact on the behaviour of several sorting algorithms.

Let X denote a geometrically distributed random variable, i. e. $\mathbb{P}\{X = k\} = pq^{k-1}$ for $k \in \mathbb{N}$ and $q = 1 - p$.

In a series of papers we have dealt with the combinatorics of geometric random variables, and it turned out that in the limiting case $q \to 1$ the results (when they made sense) where the same as in the instance of permutations. Therefore we study the concept of ascending runs in this setting. We are considering infinite words, with letters $1, 2, \cdots$, and they appear with probabilities $p, pq, pq^2, \cdots$. If we decompose a word into ascending runs

$$a_1 < \cdots < a_r \geq b_1 < \cdots < b_s \geq c_1 < \cdots < c_t \geq \cdots,$$

then r is the length of the first, s of the second, t of the third run, and so on.

We are interested in the averages of these parameters.

---

[*] This research was partially conducted while the author was a guest of the projet Algo at INRIA, Rocquencourt. The funding came from the Austrian–French "Amadée" cooperation.

G. Gonnet, D. Panario, and A. Viola (Eds.): LATIN 2000, LNCS 1776, pp. 473–482, 2000.

2 Words with Exactly m Runs

As a preparation, we consider the probability that a random word of length n has m ascending runs. For $m = 1$, this is given by

$$[z^n] \prod_{i \geq 1} (1 + pq^{i-1}z) \quad \text{for } n \geq 1.$$

But the product involved here is well known in the theory of partitions; the usual notation is

$$(a)_n = (a; q)_n = (1-a)(1-aq)(1-aq^2) \cdots (1-aq^{n-1}) \quad \text{and}$$
$$(a)_\infty = (a; q)_\infty = (1-a)(1-aq)(1-aq^2) \cdots .$$

Therefore

$$\prod_{i \geq 1} (1 + pq^{i-1}z) = (-pz)_\infty = \sum_{n \geq 0} \frac{p^n q^{\binom{n}{2}} z^n}{(q)_n},$$

the last equality being the celebrated identity of Euler [2]. This was already noted in [7]. If we set

$$\Lambda_m(z) = \sum_{n \geq 0} [\text{Pr. that a word of length } n \text{ has (exactly) } m \text{ ascending runs}] z^n,$$

then $\Lambda_0(z) = 1$ and $\Lambda_1(z) = (-pz)_\infty - 1$.

Now for general m we should consider $(-pz)_\infty^m$. Indeed, words with exactly m ascending runs have a unique representation in this product. However, this product contains also words with *less* than m runs, and we have to subtract that.

A word with $m - 1$ ascending runs is $n + 1$ times as often contained as in $\Lambda_{m-1}(z)$. This is so because we can choose any gap between two letters (also on the border) in $n+1$ ways. Such a gap means that we deliberately cut a run into pieces. Then, however, everything is unique. In terms of generating functions, this is $D(z\Lambda_{m-1}(z))$. (We write $D = \frac{d}{dz}$.) For $m - 2$ ascending runs, we can select 2 gaps in $\binom{n+2}{2}$ ways, which amounts to $\frac{1}{2}D(z^2\Lambda_{m-2}(z))$, and so on.

Therefore we have the following recurrence:

$$\Lambda_m(z) = P^m - \sum_{k=1}^{m} \frac{1}{k!} D^k \big(z^k \Lambda_{m-k}(z)\big), \qquad \Lambda_0(z) = 1; \tag{1}$$

here are the first few values. We use the abbreviations $P = (-pz)_\infty$ and $P_k = z^k D^k P$.

$$\Lambda_0 = 1,$$
$$\Lambda_1 = P - 1$$
$$\Lambda_2 = (P-1)P - P_1,$$
$$\Lambda_3 = (P-1)P^2 + P_1 - 2PP_1 + \frac{1}{2}P_2,$$
$$\Lambda_4 = (P-1)P^3 + 2PP_1 - \frac{1}{2}P_2 - 3P^2P_1 + P_1^2 + PP_2 - \frac{1}{6}P_3.$$

In the limiting case $q \to 1$ we can specify these quantities explicitly. This was obtained by experiments after a few keystrokes with trusty Maple.—Instead of P we just have e^z, and that definitely makes life much easier, since all the derivatives are still P.

Theorem 1. *The sequence $\Lambda_m(z)$ is defined as follows,*

$$\Lambda_m(z) := e^{mz} - \sum_{k=1}^{m} \frac{1}{k!} D^k\big(z^k \Lambda_{m-k}(z)\big), \qquad \Lambda_0(z) := 1.$$

Then we have for $m \geq 1$

$$\Lambda_m(z) = \sum_{j=0}^{m} e^{jz} \frac{z^{m-j-1}(-1)^{m-j}j^{m-j-1}\big(j(z-1)+m\big)}{(m-j)!}.$$

Proof. First notice that if we write

$$\lambda_m(z) = \sum_{j=0}^{m} e^{jz} \frac{(-jz)^{m-j}}{(m-j)!},$$

for $m \geq 0$, then $\Lambda_m(z) = \lambda_m(z) - \lambda_{m-1}(z)$. And the equivalent formula is

$$\sum_{k=0}^{m} \frac{1}{k!} D^k\big(z^k \lambda_{m-k}(z)\big) = \sum_{j=1}^{m} e^{jz},$$

which we will prove by induction, the basis being trivial (as usual). Now

$$\sum_{k=0}^{m} \frac{1}{k!} D^k\big(z^k \lambda_{m-k}(z)\big) = \sum_{k=0}^{m} \frac{1}{k!} D^k \sum_{j=1}^{m-k} e^{jz} z^{m-j} \frac{(-j)^{m-k-j}}{(m-k-j)!}$$

$$= \sum_{k=0}^{m} \frac{1}{k!} \sum_{i=0}^{k} \binom{k}{i} \sum_{j=1}^{m-k} j^{k-i} e^{jz} z^{m-j-i} \frac{(m-j)!}{(m-j-i)!} \frac{(-j)^{m-k-j}}{(m-k-j)!},$$

and we have to prove that the coefficient of e^{jz} therein is 1. Writing $M := m - j$ it is

$$\sum_{k=0}^{M} \frac{1}{k!} \sum_{i=0}^{k} binomki j^{k-i} z^{M-i} \frac{M!}{(M-i)!} \frac{(-j)^{M-k}}{(M-k)!}$$

$$= \sum_{i=0}^{M} \frac{M!}{(M-i)!} (jz)^{M-i} \sum_{k=i}^{M} \frac{1}{k!} \binom{k}{i} \frac{(-1)^{M-k}}{(M-k)!}$$

$$= \sum_{i=0}^{M} \binom{M}{i} (jz)^{M-i} \frac{1}{(M-i)!} \sum_{k=i}^{M} \binom{M-i}{k-i} (-1)^{M-k}$$

$$= \sum_{i=0}^{M} \binom{M}{i} (jz)^{M-i} \frac{(-1)^{M-i}}{(M-i)!} \delta_{M,i} = 1.$$

Thus

$$\sum_{k=1}^{m} \frac{1}{k!} D^k \left(z^k \Lambda_{m-k}(z) \right) = \sum_{k=1}^{m} \frac{1}{k!} D^k \left(z^k \lambda_{m-k}(z) \right) - \sum_{k=1}^{m} \frac{1}{k!} D^k \left(z^k \Lambda_{m-1-k}(z) \right)$$

$$= \sum_{j=1}^{m} e^{jz} - \lambda_m(z) - \sum_{j=1}^{m-1} e^{jz} + \Lambda_{m-1}(z) = e^{mz} - \Lambda_m(z),$$

and the result follows. $\qquad\square$

Remark. Since every word has *some* number of ascending runs, we must have that

$$\sum_{m \geq 0} \Lambda_m(z) = \frac{1}{1-z}.$$

In the limiting case $q \to 1$ we will give an independent proof. For the general case, see the next sections. Consider

$$\Lambda_0(z) + \Lambda_1(z) + \cdots + \Lambda_m(z) = \lambda_m(z);$$

for $m = 6$ we get e. g.

$$\lambda_6(z) = 1 + z + z^2 + z^3 + z^4 + z^5 + z^6 + \frac{5039}{5040} z^7 + \frac{5009}{5040} z^8 + \frac{38641}{40320} z^9 + O\left(z^{10}\right).$$

Now this is no coincidence since we will prove that

$$[z^n]\lambda_m(z) = 1 \quad \text{for } n \leq m.$$

This amounts to proving that[1]

$$\frac{1}{n!} \sum_{k=0}^{n} \binom{n}{k} (m-k)^n (-1)^k = 1.$$

Notice that

$$\left\{ \begin{matrix} h \\ n \end{matrix} \right\} = \frac{1}{n!} \sum_{k=0}^{n} \binom{n}{k} (-1)^{n-k} k^h$$

are Stirling subset numbers, and they are zero for $h < n$. Therefore, upon expanding $(m-k)^n$ by the binomial theorem, almost all the terms are annihilated, only $(-k)^n$ survives. But then it is a Stirling number $\left\{ {n \atop n} \right\} = 1$. It is not too hard to turn this proof involving "discrete convergence" into one with "ordinary convergence."   □

3   The Average Length of the mth Run

We consider the parameter "combined lengths" of the first m runs in infinite strings and its probability generating function.

Note carefully hat now the elements of the probability space are infinite words, as opposed to the previous chapter where we had words of length n. This is not unlike the situation in the paper [4].

To say that this parameter is larger or equal to n is the same as to say that a word of length n (the first n letters of the infinite word) has $\leq m$ ascending runs. Therefore the sought probability generating function is

$$F_m(z) = \frac{z-1}{z} (\Lambda_0(z) + \cdots + \Lambda_m(z)) + \frac{1}{z}.$$

Now

$$F'_m(1) = \Lambda_0(1) + \cdots + \Lambda_m(1) - 1 = \Lambda_1(1) + \cdots + \Lambda_m(1)$$

is the expected value of the combined lengths of the first m runs. Thus $\Lambda_m(1)$ is the expected value of the length of the mth run ($m \geq 1$).

In the limiting case $q \to 1$ we can say more since we know $\Lambda_m(z)$ explicitly:

$$L_m = \Lambda_m(1) = m \sum_{j=0}^{m} \frac{(-1)^{m-j} j^{m-j-1}}{(m-j)!} e^j,$$

and this is exactly the formula that appears in [6] for the instance of permutations.

There, we also learn that $L_m \to 2$; in the general case, we will see in the next sections that $L_m \to 1+q$.

---

[1] A gifted former student of electrical engineering (Hermann A.) contacted me years after he was in my Discrete Mathematics course, telling me that he found this (or an equivalent) formula, but could not prove it. He was quite excited, but after I emailed him how to prove it, I never heared from him again.

4   Solving the Recursion

The effect of the operator $\frac{1}{k!}D^k(z^k f)$ can also be described by the *Hadamard product* (see [3] for definitions) of f with the series

$$T_k = \sum_n \binom{n+k}{n} z^n.$$

The reformulation of the recursion is then

$$P^m = \sum_{k=0}^{m} T_k \odot \Lambda_{m-k}.$$

We want to invert this relation to get the $\Lambda_m(z)$'s from the powers of P. We get

$$\Lambda_m(z) = \sum_{k=0}^{m} U_k \odot P^{m-k},$$

where

$$U_k = [w^k]\ 1 \Big/ \sum_{j \geq 0} T_j w^j.$$

Lemma 1.

$$U_k = (-1)^k \sum_n \binom{n+1}{k} z^n.$$

Proof. Since

$$U_n = -\sum_{k=0}^{n-1} U_k T_{n-k},$$

it is best to prove the formula by induction, the instance $n = 0$, i.e. $U_0 = 1/(1-z)$ being obvious.

The righthandside multiplies the coefficient of z^n by

$$-\sum_{k=0}^{n-1}(-1)^l \binom{n+1}{l}\binom{n+k-l}{n} = (-1)^{k+1}\sum_{l=0}^{k-1}\binom{n+1}{l}\binom{-n-1}{k-l}$$

$$= (-1)^k \binom{n+1}{k},$$

which finishes the proof.

Therefore we get the formula

Proposition 1.

$$[z^n]\Lambda_m(z) = [z^n]\sum_{k=0}^{m}(-1)^k \binom{n+1}{k} P^{m-k};$$

for $n < m$ this is zero by the combinatorial interpretation or directly.

Now this form definitely looks like an outcome of the inclusion–exclusion principle. Indeed, there are $n + 1$ gaps between the n letters, and one can pick k of them where the condition that a new run should start is *violated*. Since we don't need that, we confine ourselves to this brief remark.

Let us prove that

$$\sum_{m \geq 0} \Lambda_m(z) = \frac{1}{1-z} :$$

$$[z^n] \sum_{m \geq 0} \Lambda_m(z) = [z^n] \sum_{0 \leq k \leq m \leq n} (-1)^k \binom{n+1}{k} P^{m-k}$$

$$= [z^n] \sum_{k=1}^{n} (-1)^{n-k} \binom{n+1}{k+1} \sum_{i=0}^{k} P^i$$

$$= [z^n] \sum_{k=1}^{n} (-1)^{n-k} \binom{n+1}{k+1} \sum_{i=0}^{k} \sum_{j=0}^{i} \binom{i}{j} (P-1)^i$$

$$= [z^n](P-1)^n = 1.$$

(Note that $P - 1 = z + \ldots$.)

Let us also rederive the formula for $\Lambda_m(1)$ in the limiting case $q \to 1$; since

$$[z^n] \Lambda_m(z) = \sum_{k=0}^{m} (-1)^{m-k} \binom{n+1}{m-k} \frac{k^n}{n!}$$

we have

$$\sum_{n \geq 0} [z^n] \Lambda_m(z) = \sum_{k=0}^{m} \frac{(-1)^{m-k}}{(m-k)!} \sum_{n \geq 0} \left(\frac{1}{(n-m+k)!} + \frac{m-k}{(n+1-m+k)!} \right) k^n$$

$$= \sum_{k=0}^{m} \frac{(-1)^{m-k}}{(m-k)!} \left(k^{m-k} + (m-k)k^{m-k-1} \right) e^k$$

$$= m \sum_{k=0}^{m} \frac{(-1)^{m-k}}{(m-k)!} k^{m-k-1} e^k.$$

5   A Double Generating Function

From Proposition 1 we infer that

$$
\Lambda_m(z) = \sum_{k=0}^{m} \frac{(-1)^k}{k!} D^k(zP^{m-k})
$$

$$
= \sum_{k=0}^{m} \frac{(-1)^k}{k!} \left(zD^k P^{m-k} + kD^{k-1}P^{m-k} \right)
$$

$$
= z \sum_{k=0}^{m} \frac{(-1)^k}{k!} D^k P^{m-k} - \sum_{k=0}^{m-1} \frac{(-1)^k}{k!} D^k P^{m-1-k}.
$$

Now these forms look like convolutions, and thus we introduce the double generating function

$$
R(w, z) = \sum_{m \geq 0} \Lambda_m(z) w^m.
$$

Upon summing we find

$$
R(w, z) = ze^{-wD}\frac{1}{1-wP} - we^{-wD}\frac{1}{1-wP} = \frac{z-w}{1-w\big(-p(z-w)\big)_\infty};
$$

the last step was by noticing that e^{aD} is the shift operator E^a (see e. g. [1]).

It is tempting to plug in $w = 1$, because of the summation of the $\Lambda_m(z)$'s, but this is forbidden, because of divergence.

The instance $(q \to 1)$

$$
R(w, 1) = \frac{1-w}{1-we^{1-w}}
$$

differs from Knuth's

$$
\frac{w(1-w)}{e^{w-1}-w} + w
$$

just by 1, because in [6] $L_0 = 0$, whereas here it is $L_0 = 1$.

Theorem 2. *The generating function of the numbers L_m is given by*

$$
\frac{1-w}{1-w\prod_{i \geq 0}\big(1-(w-1)pq^i\big)}.
$$

The dominant singularity is at $w = 1$, and the local expansion starts as

$$\frac{1+q}{1-w} + \cdots.$$

From this, singularity analysis [5] entails that

$$L_m = 1 + q + O(\rho^{-m})$$

for some $\rho > 1$ that depends on q.

Experiments indicate the expansion $(m \to \infty)$

$$L_m = 1 + q - q^{m+1} + 2mq^{m+2} - (1+2m^2)q^{m+3} - \frac{2m^4+10m^2-15m+9}{3}q^{m+4} + \cdots$$

so that it seems that one could take $\rho = \frac{1}{q} - \epsilon$. However, that would require a proof.

6   Weakly Ascending Runs

Relaxing the conditions, we might now say that $\cdots > a_1 \leq \cdots \leq a_r > \cdots$ is a run (of length r).

Many of the previous considerations carry over, so we only give a few remarks. The recursion (1) stays the same, but with $P = 1/(pz)_\infty$. With this choice of P the formula

$$[z^n]\Lambda_m(z) = [z^n]\sum_{k=0}^{m}(-1)^k\binom{n+1}{k}P^{m-k}$$

still holds.

The bivariate generating function is

$$\frac{z-w}{1-w/\big(p(z-w)\big)_\infty}.$$

The poles of interest are the solutions of

$$\prod_{i\geq 0}\big(1 + (w-1)pq^i\big) = w;$$

the dominant one is $w = 1$ with a local expansion

$$\left(1+\frac{1}{q}\right)\frac{1}{1-w} + \cdots,$$

from which we can conclude that $L_m \to 1 + \frac{1}{q}$.

And the experiments indicate that

$$L_m = 1 + \tfrac{1}{q} - (-1)^m q^{2m-1}\left(1 - 2(m-1)q + (2m^2 - 5m + 4)q^2 + \cdots\right).$$

References

1. M. Aigner. *Combinatorial Theory.* Springer, 1997. Reprint of the 1979 edition.
2. G. E. Andrews. *The Theory of Partitions*, volume 2 of *Encyclopedia of Mathematics and its Applications.* Addison–Wesley, 1976.
3. L. Comtet. *Advanced Combinatorics.* Reidel, Dordrecht, 1974.
4. P. Flajolet, D. Gardy, and L. Thimonier. Birthday paradox, coupon collectors, caching algorithms, and self–organizing search. *Discrete Applied Mathematics*, 39:207–229, 1992.
5. P. Flajolet and A. Odlyzko. Singularity analysis of generating functions. *SIAM Journal on Discrete Mathematics*, 3:216–240, 1990.
6. D. E. Knuth. *The Art of Computer Programming*, volume 3: Sorting and Searching. Addison-Wesley, 1973. Second edition, 1998.
7. H. Prodinger. Combinatorial problems of geometrically distributed random variables and applications in computer science. In V. Strehl and R. König, editors, *Publications de l'IRMA (Straßbourg)*, volume 30, pages 87–95, 1993.

Author Index